Applied Mathematical Sciences

Volume 90

More information about this series at http://www.springer.com/series/34

Kenneth R. Meyer • Daniel C. Offin

Introduction to Hamiltonian Dynamical Systems and the N-Body Problem

Third Edition

 Springer

Kenneth R. Meyer
Department of Mathematical Sciences
University of Cincinnati
Cincinnati, OH, USA

Daniel C. Offin
Department of Mathematic and Statistics
Queen's University
Kingston, ON, Canada

ISSN 0066-5452
Applied Mathematical Sciences
ISBN 978-3-319-85218-8
DOI 10.1007/978-3-319-53691-0

ISSN 2196-968X (electronic)

ISBN 978-3-319-53691-0 (eBook)

Mathematics Subject Classification (2010): 34C15, 34K12, 37E40, 70H08

1st edition: © Springer Science+Business Media New York 1992
2nd edition: © Springer-Verlag New York 2009
3rd edition: © Springer International Publishing AG 2017
Softcover reprint of the hardcover 3rd edition 2017

Printed on acid-free paper

This Springer imprint is published by Springer Nature
The registered company is Springer International Publishing AG
The registered company address is: Gewerbestrasse 11, 6330 Cham, Switzerland

Preface

The theory of Hamiltonian systems is a vast subject which can be studied from many different viewpoints. This book develops the basic theory of Hamiltonian differential equations from a dynamical systems point of view.

In this third edition, the first four chapters give a sound grounding in Hamiltonian theory and celestial mechanics making it suitable for an advanced undergraduate or beginning graduate course. It contains an expanded treatment of the restricted three-body problem including a new derivation, a treatment of Hill's regions, discrete symmetries, and more. It also has a detailed presentation of the symmetries, the moment map, Noether's theorem, and the Meyer-Marsden-Weinstein reduction theorem with applications to the three-body problem. Also included is an introduction to singular reduction and orbifolds with application to bifurcation of periodic solutions along with an introduction of the lemniscate functions used for bounded and stability results.

This edition retains and advances the treatment of such advanced topics as parametric stability, conical forms for Hamiltonian matrices, symplectic geometry, Maslov index, bifurcation theory, variational methods, and stability results including KAM theory.

It assumes a basic knowledge of linear algebra, advanced calculus, and differential equations but does not assume the advanced topics such as Lebesgue integration, Banach spaces, or Lie algebras. Some theorems which require long technical proofs are stated without proof but only on rare occasions.

Cincinnati, OH, USA
Kingston, ON, Canada

Kenneth R. Meyer
Daniel C. Offin

Acknowledgments

This new edition was read by many individuals who made valuable suggestions and many corrections. Our thanks go to Nancy Diemler, Scott Dumas, Dick Hall, Herb Halpern, Phil Korman, Pat McSwiggen, Carolyn Meyer, David Minda, Jesús Palacián, Dieter Schmidt, Nages Shanmugalingam, and Patricia Yanguas.

We extend our very special thanks to Jesús Palacián, Dieter Schmidt, and Patricia Yanguas for the new and corrected figures.

The first author acknowledges the continuing support of the Charles Phelps Taft Foundation at the University of Cincinnati.

Invitation

Please send your corrections, suggestions, or comments to:
KEN.MEYER@UC.EDU,
Ken Meyer
Department of Mathematical Sciences
University of Cincinnati
Cincinnati, Ohio 45221, USA.

Contents

1. **Beginnings**   **1**
   1.1   Hamiltonian Equations 1
   1.2   The Poisson Bracket 3
   1.3   Harmonic Oscillator 4
   1.4   Flows on a Sphere 6
   1.5   Dense Flow on the Torus 14
   1.6   Lemniscate Functions 15
   1.7   Forced Nonlinear Oscillator 17
   1.8   Newtonian System 18
   1.9   Euler–Lagrange Equations 18
   1.10   Spherical Pendulum 24
   1.11   Problems 26

2. **Hamiltonian Systems**   **29**
   2.1   Linear Equations 29
   2.2   Symplectic Linear Spaces 36
   2.3   Canonical Forms 41
   2.4 $Sp(2, \mathbb{R})$ 47
   2.5   Floquet-Lyapunov Theory 50
   2.6   Symplectic Transformations 52
        2.6.1   The Variational Equations 55
        2.6.2   Poisson Brackets 56
   2.7   Symplectic Scaling 57
        2.7.1   Equations Near an Equilibrium Point 58
   2.8   Problems 58

3. **Celestial Mechanics**   **61**
   3.1   The N-Body Problem 61
   3.2   The 2-Body Problem 62
   3.3   The Kepler Problem 64
   3.4   Symmetries and Integrals 66
        3.4.1   Symmetries 67

	3.4.2	The Classical Integrals	67
	3.4.3	Noether's Theorem	69
	3.4.4	Reduction	70
	3.4.5	Foundations	73
3.5	Equilibrium and Central Configurations		73
	3.5.1	Equilibrium Solutions	73
	3.5.2	Central Configurations	74
	3.5.3	Rotating Coordinates	78
3.6	Total Collapse		79
3.7	Problems		80

4. The Restricted Problem   **83**
4.1	Defining	83
4.2	Discrete Symmetry	86
4.3	Equilibria of the Restricted Problem	87
4.4	Hill's Regions	89
4.5	Spectrum at the Equilibrium Points	90
4.6	Mass Ratios	93
4.7	Canonical Forms for the Matrix at $\mathcal{L}_4$	94
4.8	Other Restricted Problems	99
	4.8.1	Hill's Lunar Equations	99
	4.8.2	Elliptic Restricted Problem	101
4.9	Problems	102

5. Topics in Linear Theory   **103**
5.1	Parametric Stability	103
5.2	Logarithm of a Symplectic Matrix	108
	5.2.1	Functions of a Matrix	109
	5.2.2	Logarithm of a Matrix	110
	5.2.3	Symplectic Logarithm	112
5.3	Spectral Decomposition	113
5.4	Normal Forms for Hamiltonian Matrices	117
	5.4.1	Zero Eigenvalue	117
	5.4.2	Pure Imaginary Eigenvalues	122
5.5	Topology of $Sp(2n, \mathbb{R})$	129
	5.5.1	Angle Function	129
	5.5.2	Fundamental Group	131
5.6	Maslov Index	133
5.7	Problems	141

6. Local Geometric Theory   **143**
6.1	The Dynamical System Point of View	143
6.2	Discrete Dynamical Systems	147
	6.2.1	Diffeomorphisms and Symplectomorphisms	147
	6.2.2	The Time τ–Map	149
	6.2.3	The Period Map	149

6.3   Flow Box Theorems 151
6.4   Periodic Solutions and Cross Sections 155
    6.4.1   Equilibrium Points 155
    6.4.2   Periodic Solutions 156
    6.4.3   A Simple Example 158
    6.4.4   Systems with Integrals 159
6.5   The Stable Manifold Theorem 161
6.6   Problems 167

7.   Symplectic Geometry   **169**
7.1   Exterior Algebra 169
    7.1.1   Linear Symplectic Form 174
7.2   Tangent and Cotangent Spaces 174
7.3   Vector Fields and Differential Forms 177
    7.3.1   Poincaré's Lemma 181
    7.3.2   Changing Variables 182
7.4   Symplectic Manifold 183
    7.4.1   Darboux's Theorem 185
7.5   Lie Groups 185
7.6   Group Actions 187
7.7   Moment Maps and Reduction 188
7.8   Integrable Systems 190
7.9   Problems 192

8.   Special Coordinates   **195**
8.1   Differential Forms and Generating Functions 195
    8.1.1   The Symplectic Form 195
    8.1.2   Generating Functions 196
    8.1.3   Mathieu Transformations 197
8.2   Jacobi Coordinates 197
8.3   Action–Angle Variables 200
    8.3.1   d'Alembert Character 201
8.4   General Action–Angle Coordinates 202
8.5   Lemniscate Coordinates 205
8.6   Polar Coordinates 206
    8.6.1   Kepler's Problem in Polar Coordinates 206
    8.6.2   The 3-Body Problem in Jacobi–Polar
             Coordinates 207
8.7   Spherical Coordinates 208
8.8   Complex Coordinates 211
    8.8.1   Levi–Civita Regularization 212
8.9   Delaunay and Poincaré Elements 214
    8.9.1   Planar Delaunay Elements 214
    8.9.2   Planar Poincaré Elements 216
    8.9.3   Spatial Delaunay Elements 217

8.10   Pulsating Coordinates 219
       8.10.1   The Elliptic Restricted 3-Body Problem 221
8.11   Problems 223

9.   Poincaré's Continuation Method   225
9.1   Continuation of Solutions............................. 225
9.2   Lyapunov Center Theorem 227
       9.2.1   Lyapunov Families at the Euler and Lagrange
               points 228
9.3   Poincaré's Orbits 229
9.4   Hill's Orbits 230
9.5   Comets 232
9.6   From the Restricted to the Full Problem 233
9.7   Some Elliptic Orbits 236
9.8   Problems 237

10.   Normal Forms   239
10.1   Normal Form Theorems 239
       10.1.1   Normal Form at an Equilibrium Point 239
       10.1.2   Normal Form at a Fixed Point 242
10.2   Forward Transformations 245
       10.2.1   Near-Identity Symplectic Change of Variables 245
       10.2.2   The Forward Algorithm 246
       10.2.3   The Remainder Function 248
10.3   The Lie Transform Perturbation Algorithm 251
       10.3.1   Example: Duffing's Equation.................... 251
       10.3.2   The General Algorithm 253
       10.3.3   The General Perturbation Theorem............... 253
10.4   Normal Form at an Equilibrium 258
10.5   Normal Form at $\mathcal{L}_4$ 266
10.6   Normal Forms for Periodic Systems 267
10.7   Problems 276

11.   Bifurcations of Periodic Orbits   279
11.1   Bifurcations of Periodic Solutions....................... 279
       11.1.1   Elementary Fixed Points 280
       11.1.2   Extremal Fixed Points. 281
       11.1.3   Period Doubling 283
       11.1.4 k-Bifurcation Points 287
11.2   Schmidt's Bridges 290
11.3   Bifurcations in the Restricted Problem 292
11.4   Hamiltonian–Hopf Bifurcation 295
11.5   Problems 301

12. Stability and KAM Theory   **305**
  12.1 Lyapunov and Chetaev's Theorems 307
  12.2 Local Geometry 311
  12.3 Moser's Invariant Curve Theorem 312
  12.4 Morris' Boundedness Theorem 315
  12.5 Arnold's Stability Theorem 316
  12.6 Singular Reduction 320
  12.7 2:-1 Resonance 330
  12.8 3:-1 Resonance 332
  12.9 1:-1 Resonance 334
  12.10 Stability of Fixed Points 339
  12.11 Applications to the Restricted Problem 341
        12.11.1 Invariant Curves for Small Mass................. 341
        12.11.2 The Stability of Comet Orbits 341
  12.12 Problems 343

13. Variational Techniques   **345**
  13.1 The N-Body and the Kepler Problem Revisited 347
  13.2 Symmetry Reduction for Planar 3-Body Problem 349
  13.3 Reduced Lagrangian Systems 352
  13.4 Discrete Symmetry with Equal Masses 356
  13.5 The Variational Principle 357
  13.6 Isosceles 3-Body Problem.............................. 360
  13.7 A Variational Problem for Symmetric Orbits............. 361
  13.8 Instability of the Orbits and the Maslov Index 365
  13.9 Remarks 371

References   **373**

Index   **381**

1. Beginnings

Newton's second law of motion gives rise to a system of second-order differential equations in $\mathbb{R}^n$ and so to a system of first-order equations in $\mathbb{R}^{2n}$, an even-dimensional space. If the forces are derived from a potential function, the equations of motion of the mechanical system have many special properties, most of which follow from the fact that the equations of motion can be written as a Hamiltonian system. The Hamiltonian formalism is the natural mathematical structure to develop the theory of conservative mechanical systems such as the equations of celestial mechanics.

This chapter defines a Hamiltonian system of ordinary differential equations, gives some basic results about such systems, and presents several classical examples. This discussion is informal. Some of the concepts introduced in the setting of these examples are fully developed later.

1.1 Hamiltonian Equations

A Hamiltonian system is a system of $2n$ ordinary differential equations of the form

$$\dot{q} = H_p, \qquad \dot{p} = -H_q,$$

$$\frac{dq_i}{dt} = \frac{\partial H}{\partial p_i}(t, q, p), \quad \frac{dp_i}{dt} = -\frac{\partial H}{\partial q_i}(t, q, p), \quad i = 1, \ldots, n, \tag{1.1}$$

where $H = H(t, q, p)$, the Hamiltonian, is a smooth real-valued function defined for $(t, q, p) \in \mathcal{O}$, an open set in $\mathbb{R}^1 \times \mathbb{R}^n \times \mathbb{R}^n$. The vectors[1] $q = (q_1, \ldots, q_n)$ and $p = (p_1, \ldots, p_n)$ are traditionally called the position and momentum vectors, respectively, because that is what these variables represent in many classical examples[2].

---

[1] Vectors are to be understood to be column vectors, but will be written as row vectors in text.

[2] In Lagrangian mechanics the position variable is commonly denoted by q. Later p was defined as momentum in Hamiltonian mechanics, so the common usage today is (q, p).

© Springer International Publishing AG 2017
K.R. Meyer, D.C. Offin, *Introduction to Hamiltonian Dynamical Systems and the N-Body Problem*, Applied Mathematical Sciences 90, DOI 10.1007/978-3-319-53691-0_1

The variable t is called time, and the dot represents differentiation with respect to t, i.e., $\dot{} = d/dt$. This notation goes back to Newton who created a fluxion from a fluent by placing a dot over it.

The variables q and p are said to be conjugate variables: p is conjugate to q. The concept of conjugate variable grows in importance as the theory develops. The integer n is the number of *degrees of freedom* of the system.

In general, introduce the $2n$ vector z, the $2n \times 2n$ skew-symmetric matrix J, and the gradient by

$$z = \begin{bmatrix} q \\ p \end{bmatrix}, \quad J = J_n = \begin{bmatrix} 0 & I \\ -I & 0 \end{bmatrix}, \quad \nabla H = \begin{bmatrix} \dfrac{\partial H}{\partial z_1} \\ \vdots \\ \dfrac{\partial H}{\partial z_{2n}} \end{bmatrix},$$

where 0 is the $n \times n$ zero matrix and I is the $n \times n$ identity matrix. In this notation (1.1) becomes

$$\dot{z} = J \nabla H(t, z). \tag{1.2}$$

One of the basic results from the general theory of ordinary differential equations is the existence and uniqueness theorem. This theorem states that for each $(t_0, z_0) \in \mathcal{O}$, there is a unique solution $z = \phi(t, t_0, z_0)$ of (1.2) defined for t near t_0 that satisfies the initial condition $\phi(t_0, t_0, z_0) = z_0$. ϕ is defined on an open neighborhood $\mathcal{Q}$ of $(t_0, t_0, z_0) \in \mathbb{R}^{2n+2}$ into $\mathbb{R}^{2n}$. The function $\phi(t, t_0, z_0)$ is smooth in all its displayed arguments, and so ϕ is C^∞ if the equations are C^∞, and it is analytic if the equations are analytic. $\phi(t, t_0, z_0)$ is called the general solution. See Chicone (1999), Hubbard and West (1990), or Hale (1972) for details of the theory of ordinary differential equations.

In the special case when H is independent of t so that $H : \mathcal{O} \to \mathbb{R}^1$ where $\mathcal{O}$ is some open set in $\mathbb{R}^{2n}$, the differential equations (1.2) are autonomous, and the Hamiltonian system is called *conservative*. It follows that $\phi(t - t_0, 0, z_0) = \phi(t, t_0, z_0)$ holds, because both sides satisfy Equation (1.2) and the same initial conditions. Usually the t_0 dependence is dropped and only $\phi(t, z_0)$ is considered, where $\phi(t, z_0)$ is the solution of (1.2) satisfying $\phi(0, z_0) = z_0$. The solutions are pictured as parameterized curves called *orbits* in $\mathcal{O} \subset \mathbb{R}^{2n}$, and the set $\mathcal{O}$ is called the phase space. By the existence and uniqueness theorem, there is a unique curve through each point in $\mathcal{O}$; and by the uniqueness theorem, two such solution curves cannot cross in $\mathcal{O}$. The space obtained by identifying an orbit to a point is called the *orbit space*. As we shall see in this chapter, orbit spaces can be very nice or very bad.

An integral for (1.2) is a smooth function $F : \mathcal{O} \to \mathbb{R}^1$ which is constant along the solutions of (1.2); i.e., $F(\phi(t, z_0)) = F(z_0)$ is constant. The classical conserved quantities of energy, momentum, etc. are integrals. The level surfaces $F^{-1}(c) \subset \mathbb{R}^{2n}$, where c is a constant, are invariant sets, i.e., they are sets such that if a solution starts in the set, it remains in the set.

In general, the level sets are manifolds of dimension $2n - 1$, and so with an integral F, the solutions lie on the set $F^{-1}(c)$, which is of dimension $2n - 1$. If you were so lucky as to find $2n - 1$ independent integrals, $F_1, \ldots, F_{2n-1}$, then holding all these integrals fixed would define a curve in $\mathbb{R}^{2n}$, the solution curve. In the classical sense, the problem has been integrated. This happens in some special cases, for example when looking at two harmonic oscillators with rationally related frequencies in Section 1.4 or in the Kepler problem in Section 3.3. We say such a system in *completely integrable*.

A classical example is angular momentum for the central force problem. The equation $\ddot{q} = k(|q|)q$ where $k : \mathbb{R}^1 \to \mathbb{R}^1$ can be written as a Hamiltonian system

$$\dot{q} - p = \frac{\partial H}{\partial p}, \quad \dot{p} - k(|q|)q = -\frac{\partial H}{\partial q},$$

where

$$H = \frac{1}{2}|p|^2 - K(|q|), \quad K(\beta) = \int^{\beta} k(\tau)d\tau.$$

The angular momentum $A = q \times p$ is a vector of integrals since

$$\dot{A} = \dot{q} \times p + q \times \dot{p} = p \times p + q \times k(|q|)q = 0.$$

1.2 The Poisson Bracket

Many of the special properties of Hamiltonian systems are formulated in terms of the Poisson bracket operator, so this operator plays a central role in the theory developed here. Let H, F, and G be smooth functions from $\mathcal{O} \subset \mathbb{R}^1 \times \mathbb{R}^n \times \mathbb{R}^n$ into $\mathbb{R}^1$, and define the Poisson bracket of F and G by

$$\begin{aligned}
\{F, G\} = \nabla F^T J \nabla G &= \frac{\partial F^T}{\partial q} \frac{\partial G}{\partial p} - \frac{\partial F^T}{\partial p} \frac{\partial G}{\partial q} \\
&= \sum_{i=1}^{n} \left(\frac{\partial F}{\partial q_i}(t, p, q) \frac{\partial G}{\partial p_i}(t, q, p) - \frac{\partial F}{\partial p_i}(t, q, p) \frac{\partial G}{\partial q_i}(t, q, p) \right).
\end{aligned} \tag{1.3}$$

Clearly $\{F, G\}$ is a smooth map from $\mathcal{O}$ to $\mathbb{R}^1$ as well, and one can easily verify that $\{\cdot, \cdot\}$ is skew-symmetric and bilinear. A little tedious calculation verifies Jacobi's identity:

$$\{F, \{G, H\}\} + \{G, \{H, F\}\} + \{H, \{F, G\}\} = 0. \tag{1.4}$$

By a common abuse of notation, let $H(t) = H(t, \phi(t, t_0, z_0))$, where ϕ is the solution of (1.2). By the chain rule,

$$\frac{d}{dt} H(t) = \frac{\partial H}{\partial t}(t, \phi(t, t_0, z_0)) + \{H, H\}(t, \phi(t, t_0, z_0)). \tag{1.5}$$

Hence $dH/dt = \partial H/\partial t$.

Theorem 1.2.1. *Let F, G, and H be as above and independent of time t. Then*

1. *F is an integral for (1.2) if and only if $\{F, H\} = 0$.*
2. *H is an integral for (1.2).*
3. *If F and G are integrals for (1.2), then so is $\{F, G\}$.*

Proof. (1) follows directly from the definition of an integral and from (1.5). (2) follows from (1) and from the fact that the Poisson bracket is skew-symmetric, so $\{H, H\} = 0$. (3) follows from the Jacobi identity (1.4).

If $\{F, G\} = 0$ then F and G are in *involution*. If a Hamiltonian H of n degree of freedom has n independent integrals in involution then the Hamiltonian system is *integrable*. For example the planar central force problem already discussed has 2 degrees of freedom and the two integrals of energy, H, and angular momentum, A, with $\{H, A\} = 0$.

In many of the examples given below, the Hamiltonian H is the total energy of a physical system; when it is, the theorem says that energy is a conserved quantity.

In the conservative case when H is independent of t, a critical point of H as a function (i.e., a point where the gradient of H is zero) is an equilibrium (or critical, rest, stationary) point of the system of differential equations (1.1) or (1.2), i.e., a constant solution.

For the rest of this section, let H be independent of t. An equilibrium point ζ of system (1.2) is stable if for every $\epsilon > 0$, there is a $\delta > 0$ such that $\|\zeta - \phi(t, z_0)\| < \epsilon$ for all t whenever $\|\zeta - z_0\| < \delta$. Note that "all t" means both positive and negative t, and that stability is for both the future and the past.

Theorem 1.2.2 (Dirichlet). *If ζ is a strict local minimum or maximum of H, then ζ is stable.*

Proof. Without loss of generality, assume that $\zeta = 0$ and $H(0) = 0$. Because $H(0) = 0$ and 0 is a strict minimum for H, there is an $\eta > 0$ such that $H(z)$ is positive for $0 < \|z\| \leq \eta$. (In the classical literature, one says that H is positive definite.) Let $\kappa = \min(\epsilon, \eta)$ and $M = \min\{H(z) : \|z\| = \kappa\}$, so $M > 0$. Because $H(0) = 0$ and H is continuous, there is a $\delta > 0$ such that $H(z) < M$ for $\|z\| < \delta$. If $\|z_0\| < \delta$, then $H(z_0) = H(\phi(t, z_0)) < M$ for all t. $\|\phi(t, z_0)\| < \kappa \leq \epsilon$ for all t, because if not, there is a time t' when $\|\phi(t', z_0)\| = \kappa$, and $H(\phi(t', z_0)) \geq M$, a contradiction. See Dirichlet (1846).

1.3 Harmonic Oscillator

The harmonic oscillator is the second-order, linear, autonomous, ordinary differential equation

$$\ddot{x} + \omega^2 x = 0, \tag{1.6}$$

where ω is a positive constant. It can be written as a system of two first order equations by introducing the conjugate variable $u = \dot{x}/\omega$ and as a Hamiltonian system by letting $H = (\omega/2)(x^2 + u^2)$ (energy in physical problems). The equations become

$$\dot{x} = \omega u = \frac{\partial H}{\partial u},$$
$$\dot{u} = -\omega x = -\frac{\partial H}{\partial x}. \tag{1.7}$$

The variable u is a scaled velocity, and thus the x, u plane is essentially the position-velocity plane, or the phase space of physics. The basic existence and uniqueness theorem of differential equations asserts that through each point (x_0, u_0) in the plane, there is a unique solution passing through this point at any particular epoch t_0. The general solution is given by

$$\begin{bmatrix} x(t, t_0, x_0, u_0) \\ u(t, t_0, x_0, u_0) \end{bmatrix} = \begin{bmatrix} \cos\omega(t - t_0) & -\sin\omega(t - t_0) \\ \sin\omega(t - t_0) & \cos\omega(t - t_0) \end{bmatrix} \begin{bmatrix} x_0 \\ u_0 \end{bmatrix}. \tag{1.8}$$

The solution curves are parameterized circles. The reason that one introduces the scaled velocity instead of using the velocity itself, as is usually done, is so that the solution curves become circles instead of ellipses. In dynamical systems the geometry of the family of curves is of prime importance. Because the system is independent of time, it admits H as an integral by Theorem 1.2.1 (or note $\dot{H} = \omega x \dot{x} + \omega u \dot{u} = 0$). Because a solution lies in the set where $H = $ constant, which is a circle in the x, u plane, the integral alone gives the geometry of the solution curves in the plane. See Figure 1.1. The origin is a local minimum for H and so the origin is stable.

Introduce polar coordinates, $r^2 = x^2 + u^2$, $\theta = \tan^{-1} u/x$, so that equations (1.7) become

$$\dot{r} = 0, \qquad \dot{\theta} = -\omega. \tag{1.9}$$

This shows again that the solutions lie on circles about the origin because, $\dot{r} = 0$. The circles are swept out clockwise with constant angular velocity ω.

But the equations in polar coordinates do not look like Hamiltonian equations since $H(r, \theta) = \frac{1}{2} r^2$, so a better choice would be

$$I = \frac{1}{2}(x^2 + u^2), \quad \theta = \tan^{-1} u/x,$$

so that $H(I, \theta) = \frac{1}{2}\omega I$ and the equations become

$$\dot{I} = \frac{\partial H}{\partial \theta} = 0, \quad \dot{\theta} = -\frac{\partial H}{\partial I} = -\omega.$$

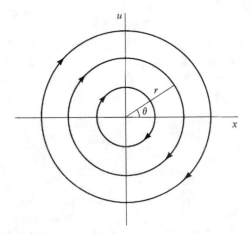

Figure 1.1. Phase portrait of the harmonic oscillator.

Thus (I, θ) are the natural polar coordinates in the study of Hamiltonian systems and are called *action–angle variables*. Coordinates that preserve the Hamiltonian character of the problem are called *symplectic* coordinates or *canonical* coordinates. This is the first of many such special coordinate systems we shall see.

1.4 Flows on a Sphere

This section introduces some interesting geometry of Hamiltonian flows in an understandable setting. Consider a pair of harmonic oscillators

$$\ddot{x}_1 + \omega_1^2 x_1 = 0, \qquad \ddot{x}_2 + \omega_2^2 x_2 = 0,$$

which as a system is

$$\dot{x}_1 = \omega_1 u_1 = \frac{\partial H}{\partial u_1}, \qquad \dot{x}_2 = \omega_2 u_2 = \frac{\partial H}{\partial u_2},$$

$$\dot{u}_1 = -\omega_1 x_1 = -\frac{\partial H}{\partial x_1}, \qquad \dot{u}_2 = -\omega_2 x_2 = -\frac{\partial H}{\partial x_2},$$

(1.10)

with Hamiltonian $H = \dfrac{\omega_1}{2}(x_1^2 + u_1^2) + \dfrac{\omega_2}{2}(x_2^2 + u_2^2)$. In action angle variables

$$I_1 = \frac{1}{2}(x_1^2 + u_1^2), \qquad \theta_1 = \tan^{-1} u_1/x_1,$$

$$I_2 = \frac{1}{2}(x_2^2 + u_2^2), \qquad \theta_2 = \tan^{-1} u_2/x_2,$$

the Hamiltonian becomes

$$H = \omega_1 I_1 + \omega_2 I_2$$

and the equations are

$$\dot{I}_1 = 0, \qquad \dot{\theta}_2 = -\omega_1,$$

$$\dot{I}_2 = 0, \qquad \dot{\theta}_2 = -\omega_2. \tag{1.11}$$

This is an integrable system since it has two degrees of freedom and I_1, I_2 are two independent integrals in involution, $\{I_1, I_2\} = 0$.

Let[3] $\omega_1 > 0$, $\omega_2 > 0$ and consider the set where $H = 1$ which is an ellipsoid in $\mathbb{R}^4$ or topologically a three sphere S^3. We can use action-angle variables to introduce coordinates on the sphere S^3, provided we are careful to observe the conventions of polar coordinates: (i) $I_1 \geq 0$, $I_2 \geq 0$; (ii) θ_1 and θ_2 are defined modulo 2π; and (iii) $I_1 = 0$ or $I_2 = 0$ corresponds to a point.

Starting with the symplectic coordinates $I_1, \theta_1, I_2, \theta_2$ for $\mathbb{R}^4$, we note that since $H = \omega_1 I_1 + \omega_2 I_2 = 1$ we can discard I_2 and make the restriction $0 \leq I_1 \leq 1/\omega_1$ (note that I_2 will return). We use I_1, θ_1, θ_2 as coordinates on S^3. Now I_1, θ_1 with $0 \leq I_1 \leq 1/\omega_1$ are just coordinates for the closed unit disk in $\mathbb{R}^2$ which is drawn in green in Figure 1.2(a). For each point of the open disk, there is a circle with coordinate θ_2 (defined mod 2π), but $I_2 = 0$ when $I_1 = 1/\omega_1$, so the circle collapses to a point over the boundary of the disk.

The geometric model of S^3 is given by two solid cones with points on the boundary cones identified as shown in Figure 1.2(a). Through each point in the open unit disk with coordinates I_1, θ_1 there is a line segment (the red dashed line) perpendicular to the disk. The angular coordinate θ_2 is measured downward on this segment: $\theta_2 = 0$ is the disk, $\theta_2 = -\pi$ is the upper boundary cone, and $\theta_2 = +\pi$ is the lower boundary cone. Each point on the upper boundary cone with coordinates $I_1, \theta_1, \theta_2 = -\pi$ is identified with the point on the lower boundary cone with coordinates $I_1, \theta_1, \theta_2 = +\pi$. This is our model for the sphere S^3.

There are two special orbits in this model. The first one is right up the center where $I_1 = 0$ which is periodic with period $2\pi/\omega_2$ and the second one is around theedge where $I_1 = 1/\omega_1$ or $I_2 = 0$ which is periodic with period $2\pi/\omega_1$. Both of them are drawn in purple in Figure 1.2(b). These periodic solutions are special and are usually called the normal modes. All the others wrap around a torus T^2, where $I_1 = \tilde{I}$, and $\tilde{I}$ is a constant such that $0 < \tilde{I} < 1/\omega_1$, as illustrated in Figure 1.2(b).

Coordinates on the torus T^2 are θ_1 and θ_2, and the equations are

$$\dot{\theta}_1 = -\omega_1, \qquad \dot{\theta}_2 = -\omega_2 \tag{1.12}$$

---

[3]Here we only consider the case where the Hamiltonian is positive definite. The case where the ω's are of different sign leads to a different geometry. See some of the examples in Chapter 12.

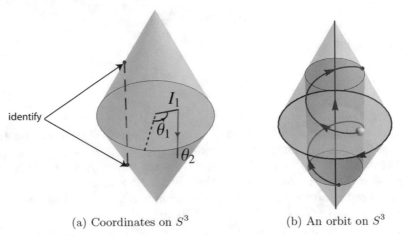

(a) Coordinates on S^3 (b) An orbit on S^3

Figure 1.2. A model of S^3

the standard linear equations on a torus.

If ω_1/ω_2 is rational, then $\omega_1 = r\delta$ and $\omega_2 = s\delta$, where r and s are relatively prime integers and δ is a nonzero real number. In this case the solution of (1.12) through α_1, α_2 at epoch $t = 0$ is $\theta_1(t) = \alpha_1 - \omega_1 t$, $\theta_2(t) = \alpha_2 - \omega_2 t$, and so if $\tau = 2\pi/\delta$, then $\theta_1(\tau) = \alpha_1 - r2\pi$ and $\theta_1(\tau) = \alpha_2 - s2\pi$. That is, the solution is periodic with period τ on the torus, and this corresponds to a periodic solution of (1.10).

If ω_1/ω_2 is irrational, then none of the solutions are periodic. In fact, the solutions of (1.12) are dense lines on the torus (see Section 1.5), and this corresponds to the fact that the solutions of (1.10) are quasiperiodic but not periodic.

The two normal mode periodic solutions are linked circles in S^3, i.e., one of the circles intersects a disk bounded by the other circle in an algebraically nontrivial way. The circle where $I_1 = 1/\omega_1$ is the boundary of the green disk in Figure 1.2(b) and the circle $I_1 = 0$ intersects this disk once. It turns out that the number of intersections is independent of the bounding disk provided one counts the intersections algebraically.

Consider the special case when $\omega_1 = \omega_2 = 1$. In this case every solution is periodic, and so its orbit is a circle in the 3-sphere. Other than the two normal modes, on each orbit as θ_1 increases by 2π, so does θ_2. Thus each such orbit hits the open disk where $\theta_2 = 0$ (the green disk) in one yellow point. We can identify each such orbit with the unique yellow point where it intersects the disk. One normal mode meets the disk at the center, so we can identify it with the center. The other normal mode is the outer boundary circle of the disk which is a single orbit. When we identify this circle with a point, the closed disk whose outer circle is identified with a point becomes a 2-sphere.

Theorem 1.4.1. *The 3-sphere, S^3, is the union of circles. Any two of these circles are linked. The quotient space obtained by identifying a circle with a point, the orbit space, is a 2-sphere.*

This is known as the Hopf fibration of the three-sphere and is our first example where the orbit space is nice.

When $\omega_1 = r$ and $\omega_2 = s$ with r and s relatively prime integers, all solutions are periodic, and the 3-sphere is again a union of circles, but it is not locally a product near the normal modes. The non-normal nodes solutions are r, s-torus knots. They link r times with one normal node and s times with the other.

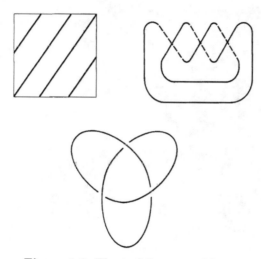

Figure 1.3. The trefoil as a toral knot.

These links follow by a slight extension of the ideas of the previous proposition. An r, s-torus knot is a closed curve that wraps around the standard torus in $\mathbb{R}^3$ in the longitudinal direction r times and in the meridional direction s times.

Figure 1.3 shows that the 2,3-torus knot is the classical trefoil or cloverleaf knot. The first diagram in Figure 1.3 is the standard model of a torus: a square with opposite sides identified. The line with slope $3/2$ is shown wrapping three times around one way and twice around the other. Think of folding the top half of the square back and around and then gluing the top edge to the bottom to form a cylinder. Add two extra segments of curves to connect the right and left ends of the curve to get the second diagram in Figure 1.3. Smoothly deform this to get the last diagram in Figure 1.3, the standard presentation of the trefoil. See Rolfsen (1976) for more information on knots.

To see what the orbit space looks like when ω_1/ω_2 is rational, consider the case when $\omega_1 = 2$, $\omega_2 = 3$. Select two cross sections to the flow defined by equations (1.10), i.e., two disks such that the solutions cross the disks transversally and each orbit intersects at least one of the disks (maybe more than once). Refer to Figure 1.4.

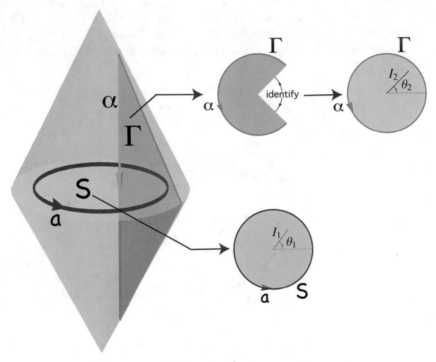

Figure 1.4. Sections in S^3 when $\omega_1 = 2$, $\omega_2 = 3$

The first section, S, is the disk where $\theta_2 = 0$ and $0 \leq I_1 \leq 3/(4\,q)$ and all θ_1. Denote the boundary of S by the oriented circle a. The section S is a 2-disk with symplectic action-angle coordinates (I_1, θ_1). Orbits cross S upwardly since $\dot{\theta}_2 = -p$ and θ_2 increases downwardly. (Remember how we measure θ_2.)

The second section, Γ, is the disk where $\theta_1 = 0$ and $1/(4\,q) \leq I_1 \leq 1/q$. At first sight Γ, the region bounded by the orange curve on the left picture of Figure 1.4, looks like a triangle but when the identification is taken into account as illustrated in the figure it is indeed a disk. Specifically, Γ is a 2-disk with symplectic action-angle coordinates (I_2, θ_2). Orbits cross Γ since $\dot{\theta}_1 = -q$. Denote the boundary of Γ by the oriented circle α. Here one sees the reason why we measure θ_2 downward, because with that convention I_2, θ_2 are correctly oriented in Γ.

Due to the overlap of these two sections every orbit crosses one or the other or both.

The section S is defined by $\theta_2 = 0$ and has coordinates I_1, θ_1. The first return time is $T = 2\pi/3$ and the section map is

$$I_1 \to I_1, \qquad \theta_1 \to \theta_1 - 4\pi/3.$$

A point not at the origin hits S three times and repeats. That is

$$\theta_1 \to \theta_1 - 4\pi/3 \to \theta_1 - 8\pi/3 \to \theta_1 - 12\pi/3 \equiv \theta_1.$$

Thus an orbit other than $I_1 = 0$ hits S three times equally spaced in θ_1, so we should take the sector $0 \le \theta_1 \le 2\pi/3$ and identify the lines $\theta_1 = 0$ and $\theta_1 - 2\pi/3$ as shown on the left-hand side of Figure 1.5 to get a sharp cone.

Similarly the section Γ is defined by $\theta_1 = 0$ and has coordinates I_2, θ_2. The first return time is $T = \pi$ and the section map is

$$I_2 \to I_2, \qquad \theta_2 \to \theta_2 - \pi.$$

A point not at the origin hits S two times and repeats. That is

$$\theta_2 \to \theta_2 - \pi \to \theta_2 - 2\pi \equiv \theta_2.$$

Thus an orbit other than $I_2 = 0$ hits Γ two times equally spaced in θ_2, so we should take the sector $0 \le \theta_2 \le \pi$ and identify the lines $\theta_2 = 0$ and $\theta_2 = \pi$ as shown on the right-hand side of Figure 1.5 to get a cone.

The boundary curves a, α, a', α' help in identifying the two sections, see Figure 1.5. Thus we have a model for the orbit space when the ratio of the ω's is rational. It is clearly a topological three-sphere, S^3, but it is not smooth. It has peaks, but the peaks are understandable. Later we shall show that it has a differentiable structure which makes it into what is called a *symplectic orbifold*. The orbit space is still nice.

There is another more algebraic way to understand the orbit space when $\omega_1 = r$, $\omega_2 = s$, where r and s are relatively prime positive integers. In this case the system is completely integrable since it has the three independent integrals

$$I_1, \ I_2, \ r\theta_2 - s\theta_1.$$

Holding these three integrals fixed defines a curve in $\mathbb{R}^4$, i.e., an orbit. To obtain a polynomial system of integrals set

$$
\begin{aligned}
a_1 &= I_1 = x_1^2 + u_1^2, \\
a_2 &= I_2 = x_2^2 + u_2^2, \\
a_3 &= I_1^{s/2} I_2^{r/2} \cos(r\theta_2 - s\theta_1) = \Re[(x_1 - u_1 i)^s (x_2 + u_2 i)^r], \\
a_4 &= I_1^{s/2} I_2^{r/2} \sin(r\theta_2 - s\theta_1) = \Im[(x_1 - u_1 i)^s (x_2 + u_2 i)^r].
\end{aligned}
\tag{1.13}
$$

Of course these polynomials are not independent but subject to the constraints

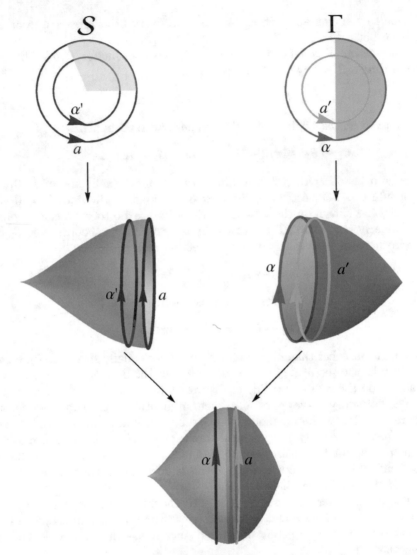

Figure 1.5. Gluing sections when $\omega_1 = 2$, $\omega_2 = 3$

$$a_3^2 + a_4^2 = a_1^s a_2^r, \ a_1 \geq 0, \ a_2 \geq 0. \tag{1.14}$$

This constraint comes from $\sin^2 \phi + \cos^2 \phi = 1$.

An orbit of the system is uniquely specified by the four invariants subject to the constraints (1.14) and so the orbit space is determined by (1.13), (1.14), and

$$H = r\, a_1 + s\, a_2 = h. \tag{1.15}$$

Solve (1.15) for a_2 and substitute into the constraint equation to get

$$a_3^2 + a_4^2 = a_1^s \left(\frac{h - r\,a_1}{s} \right)^r , \qquad (1.16)$$

which defines a surface in (a_1, a_3, a_4)-space. This surface is a representation of the orbit space. Note that (1.16) defines a surface of revolution, so let ρ, ψ be polar coordinates in the (a_3, a_4)-plane so that equation (1.16) becomes

$$\rho^2 = a_1^s \left(\frac{h - r\,a_1}{s} \right)^r . \qquad (1.17)$$

The surface in (a_1, a_3, a_4)-space is compact. See Figure 1.6 for the views of the (ρ, a_1)-sections of the orbit spaces.

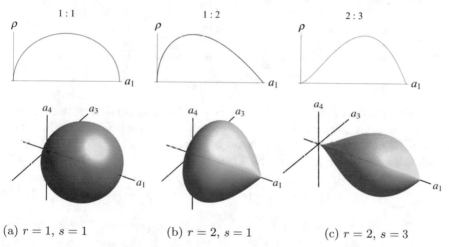

(a) $r = 1$, $s = 1$ (b) $r = 2$, $s = 1$ (c) $r = 2$, $s = 3$

Figure 1.6. Above: ρ versus a_1. Below: Orbit spaces

When $r = s = 1$, then by completing the square we have $\rho^2 + (a_1 - h/2)^2 = h^2/4$ which is a circle of radius $h/2$ centered at $(\rho, a_1) = (0, h/2)$, and so (1.17) defines a sphere in a-space of radius $h/2$ and center at $(a_1, a_3, a_4) = (h/2, 0, 0)$, see Figure 1.6(a). There are no peaks.

Now consider the case when $r \neq s$, then the right-hand side of equation (1.17) is positive for $0 < a_1 < h/s$ and so the surface of revolution defined by (1.16) is smooth for a_1 in that range. But the right-hand side is zero for $a_1 = 0$ and $a_1 = h/r$ and these are the candidates for peaks. Near $a_1 = 0$ from (1.17) we see that $\rho \sim c\,a_1^{s/2}$ where $c > 0$ is a constant. Thus the surface of revolution is only smooth at $a_1 = 0$ if $s = 1$. For $p > 1$ the surface at $a_1 = 0$ has a peak which is cone-like when $s = 2$ and is cusp-like for $s > 2$. The peak gets sharper for larger s.

Similarly near $a_1 = h/r$ we see that $\rho \sim c(h - rq\,a_1)^{r/2}$ where $c > 0$ is a constant. Thus the surface of revolution is only smooth at $a_1 = h/q$ if $r = 1$.

When $r > 1$ the surface has a peak at $a_1 = h/r$ and the peak gets sharper for larger r. The peak is cone-like when $r = 2$ and cusp-like for $r > 2$. See Figure 1.6(b) and 1.6(c).

1.5 Dense Flow on the Torus

In order to show that the solutions of (1.12) on the torus are dense when ω_1/ω_2 is irrational, the following simple lemmas from number theory are needed.

Lemma 1.5.1. *Let δ be any irrational number. Then for every $\epsilon > 0$, there exist integers q and p such that*

$$| q\delta - p | < \epsilon. \tag{1.18}$$

Proof. Case 1: $0 < \delta < 1$. Let $N \geq 2$ be an integer and

$$S_N = \{s\delta - r : 1 \leq s, r \leq N\}.$$

For each element of this set we have $| s\delta - r | < N$. Because δ is irrational, there are N^2 distinct members in the set S_N; so at least one pair is less than $4/N$ apart. (If not, the total length would be greater than $(N^2 - 1)4/N > 2N$.) Call this pair $s\delta - r$ and $s'\delta - r'$. Thus

$$0 < | (s - s')\delta - (r - r') | < \frac{4}{N} < \frac{4}{| s - s' |}. \tag{1.19}$$

Take $N > 4/\epsilon, q = s - s'$ and $p = r' - r$ to finish this case. The other cases follow from the above. If $-1 < \delta < 0$, then apply the above to $-\delta$; and if $| \delta | > 1$, apply the above to $1/\delta$.

Lemma 1.5.2. *Let δ be any irrational number and ξ any real number. Then for every $\epsilon > 0$ there exist integers p and q such that*

$$| q\delta - p - \xi | < \epsilon. \tag{1.20}$$

Proof. Let p' and q' be as given in Lemma 1.5.1, so $\eta = q'\delta - p'$ satisfies $0 < | \eta | < \epsilon$. There is a integer m such that $| m\eta - \xi | < \epsilon$. The lemma follows by taking $q = mq'$ and $p = mp'$.

Theorem 1.5.1. *Let ω_1/ω_2 be irrational. Then the solution curves defined by Equations (1.12) are dense on the torus.*

Proof. Measure the angles in revolutions instead of radians so that the angles θ_1 and θ_2 are defined modulo 1 instead of 2π. The solution of equations (1.12) through $\theta_1 = \theta_2 = 0$ at $t = 0$ is $\theta_1(t) = \omega_1 t, \theta_2(t) = \omega_2 t$. Let $\epsilon > 0$ and ξ be given. Then $\theta_1 \equiv \xi$ and $\theta_2 \equiv 0 \mod 1$ is an arbitrary point on the circle $\theta_2 \equiv 0$

mod 1 on the torus. Let $\delta = \omega_1/\omega_2$ and p, q be as given in Lemma 1.5.1. Let $\tau = q/\omega_2$, so $\theta_1(\tau) = \delta q, \theta_2(\tau) = q$. Thus, $| \theta_1(\tau) - p - \xi |< \epsilon$, but because p is an integer, this means that $\theta_1(\tau)$ is within ϵ of ξ; so the solution through the origin is dense on the circle $\theta_2 \equiv 0$ mod 1. The remainder of the proof follows by translation.

In this example orbits are dense on each other and so the orbit space in not even Hausdorff — an example of a bad orbit space.

1.6 Lemniscate Functions

Many classical functions can be defined as the solutions of a differential equation. Here we introduce a classical function which will be used in some stability studies in subsequent chapters. The sine lemniscate function, denoted by $sl\,(t)$, can be defined as the solution of the differential equation

$$\ddot{x} + 2x^3 = 0, \quad x(0) = 0, \quad \dot{x}(0) = 1. \tag{1.21}$$

It is clearly a Hamiltonian system with Hamiltonian

$$L = \frac{1}{2}(y^2 + x^4).$$

Since $x(t) = sl\,(t)$ lies in the oval $L = 1/2$ the function $sl\,(t)$ is periodic. The cosine lemniscate function, denoted by $cl\,(t)$, is defined by $cl\,(t) = \dot{sl}\,(t)$. Thus

$$cl^{\,2}(t) + sl^{\,4}(t) = 1 \tag{1.22}$$

and

$$sl\,(0) = 0, \quad \dot{sl}\,(0) = 1, \quad cl\,(0) = 1, \quad \dot{cl}\,(0) = 0.$$

Let τ be the least positive value such that $x(\tau) = 1$. Solving $y^2 + x^4 = 1$ for $y = \dot{x}$ and then separating variables one finds

$$\tau = \int_0^1 \frac{du}{\sqrt{1 - u^4}} = \frac{1}{4}\mathbf{B}\left(\frac{1}{4}, \frac{1}{2}\right) \approx 1.311028777\ldots,$$

where $\mathbf{B}$ is the classical Beta function.

Both $sl\,(t)$ and $-sl\,(-t)$ satisfy the equation (1.21) and initial conditions so $sl\,(t) = -sl\,(-t)$ or $sl\,(t)$ is odd and $cl\,(t)$ is even. Now $sl\,(\tau) = 1$ and $\dot{sl}\,(\tau) = 0$ so both $sl\,(\tau + t)$ and $sl\,(\tau - t)$ satisfy the equation and the same initial conditions so $sl\,(\tau + t) = sl\,(\tau - t)$ or $sl\,(t)$ is even about τ. Likewise $cl\,(t)$ is an odd about τ. In summary

$$sl\,(t) = -sl\,(-t), \quad cl\,(t) = cl\,(-t),$$

$$sl\,(\tau + t) = sl\,(\tau - t), \quad cl\,(\tau + t) = -cl\,(\tau - t),$$

$$sl\,(t) = sl\,(t + 4\tau), \quad cl\,(t) = cl\,(t + 4\tau).$$

Thus, sl is an odd 4τ-periodic function. It is also odd harmonic function, i.e., its Fourier series only contains terms like $\sin(j\pi t/4\tau)$ where j is an odd integer.

From the symmetry sl is increasing for $-\tau < t < \tau$. Also from the equation $\ddot{sl} > 0$ (so sl is convex) for $-\tau < t < 0$ and also $\ddot{sl} < 0$ (so sl is concave) for $0 < t < \tau$. The graph of $sl\,(t)$ has the same general form as $\sin(t)$ with τ playing the role of $\pi/2$. See the graph in Figure 1.7.

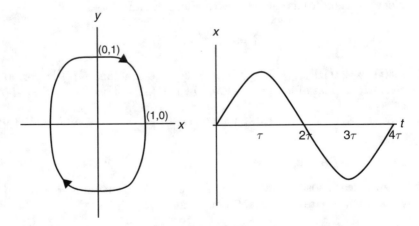

Figure 1.7. Graph of sl.

The traditional approach defines the inverse of sine lemniscate function as an indefinite integral and therefore the $sl\,(t)$ is the inverse of an elliptic integral. On page 524ff of Whittaker and Watson (1927) it is stated there that the idea of investigating this function occurred to C. F. Gauss on January 8, 1797 and it represents the first appearance of functions defined as the inverse of an integral.

1.7 Forced Nonlinear Oscillator

Consider the system

$$\ddot{x} + f(x) = g(t), \qquad (1.23)$$

where x is a scalar and f and g are smooth real-valued functions of a scalar variable. A mechanical system that gives rise to this equation is a spring-mass system with an external force. Here, x is the displacement of a particle of mass 1. The particle is connected to a nonlinear spring with restoring force $-f(x)$ and is subject to an external force $g(t)$. One assumes that these are the only forces acting on the particle and, in particular, that there are no velocity-dependent forces acting such as a frictional force. The inclusion of the t dependent force g makes the system nonautonomous.

An electrical system that gives rise to this equation is an LC circuit with an external voltage source. In this case, x represents the charge on a nonlinear capacitor in a series circuit that contains a linear inductor and an external electromotive force $g(t)$. In this problem, assume that there is no resistance in the circuit, and so there are no terms in $\dot{x}$.

This equation is equivalent to the system

$$\dot{x} = y = \frac{\partial H}{\partial y}, \qquad \dot{y} = -f(x) + g(t) = -\frac{\partial H}{\partial x}, \qquad (1.24)$$

where

$$H = \frac{1}{2}y^2 + F(x) - xg(t), \qquad F(x) = \int_0^x f(s)ds. \qquad (1.25)$$

Many named equations are of this form, for example: (i) the harmonic oscillator: $\ddot{x} + \omega^2 x = 0$; (ii) the pendulum equation: $\ddot{\theta} + \sin\theta = 0$; (iii) the forced Duffing's equation: $\ddot{x} + x + \alpha x^3 = \cos\omega t$.

In the case when the forcing term g is absent, $g \equiv 0$, H is an integral, and the solutions lie in the level curves of H. Therefore, the phase portrait is easily obtained by plotting the level curves.[4] In fact, these equations are integrable in the classical sense that they can be solved "up to a quadrature"; i.e., they are completely solved after one integration or quadrature. Let $h = H(x_0, y_0)$. Solve $H = h$ for y and separate the variables to obtain

$$y = \frac{dx}{dt} = \pm\sqrt{2h - 2F(x)},$$

$$t - t_0 = \pm \int_{x_0}^x \frac{d\tau}{\sqrt{2h - 2F(\tau)}}. \qquad (1.26)$$

---

[4]The level curves are easy to obtain by exploiting the fact that H has y^2 sitting by itself. To plot the curves $H = \ell/2$ for fixed ℓ first graph $z = -F(x) + \ell$ for various values of ℓ on the same plot and then take the square root of the graphs. In taking the square root of a graph remember that the square root of a crossing of the axis gives a smooth curve (the square root of $y = x$ is $y = \pm\sqrt{x}$ or $y^2 = x$) and the square root of a tangency is an $\times$ (the root of $y = x^2$ is $y = \pm x$).

Thus, the solution is obtained by performing the integration in (1.26) and then taking the inverse of the function so obtained. In general this is quite difficult, but when f is linear, the integral in (1.26) is elementary, and when f is quadratic or cubic, then the integral in (1.26) is elliptic.

1.8 Newtonian System

The n-dimensional analog of (1.23) is

$$M\ddot{x} + \nabla F(x) = g(t), \tag{1.27}$$

where x is an n-vector, M is a nonsingular, symmetric $n \times n$ matrix, F is a smooth function defined on an open domain $\mathcal{O}$ in $\mathbb{R}^n$, ∇F is the gradient of F, and g is a smooth n-vector valued function of t, for t in some open set in $\mathbb{R}^1$. Let $y = M\dot{x}$. Then (1.27) is equivalent to the Hamiltonian system

$$\dot{x} = \frac{\partial H}{\partial y} = M^{-1}y, \qquad \dot{y} = -\frac{\partial H}{\partial x} = -\nabla F(x) + g(t), \tag{1.28}$$

where the Hamiltonian is

$$H = \frac{1}{2}y^T M^{-1} y + F(x) - x^T g(t). \tag{1.29}$$

If x represents the displacement of a particle of mass m, then $M = mI$ where I is the identity matrix, y is the linear momentum of the particle, $\frac{1}{2}y^T M^{-1}y$ is the kinetic energy, $g(t)$ is an external force, and F is the potential energy. If $g(t) \equiv 0$, then H is an integral and is total energy. This terminology is used in reference to nonmechanical systems of the form (1.27) also. In order to write (1.28) as a Hamiltonian system, the correct choice of the variable conjugate to x is $y = M\dot{x}$, the linear momentum, and not $\dot{x}$, the velocity.

In the special case when $g \equiv 0$, a critical point of the potential is a critical point of H and hence an equilibrium point of the Hamiltonian system of equations (1.28). In many physical examples, M is positive definite. In this case, if x' is a local minimum for the potential F, then $(x', 0)$ is a local minimum for H and therefore a stable equilibrium point by Theorem 1.2.2.

It is tempting to think that if x' is a critical point of F and not a minimum of the potential, then the point $(x', 0)$ is an unstable equilibrium point. This is not true. See Laloy (1976) and Chapter 12 for a discussion of stability questions for Hamiltonian systems.

1.9 Euler–Lagrange Equations

Many of the laws of physics can be given as minimizing principles and this led the theologian-mathematician Leibniz to say that we live in the best of all possible worlds. In more modern times and circumstances, the physicist

Richard Feynman once said that of all mathematical-physical principles, the principle of least action is one that he has pondered most frequently.

Under mild smoothness conditions, one shows in the calculus of variations that minimizing the curve functional with fixed boundary constraints

$$F(q) = \int_\alpha^\beta L(q(t), \dot{q}(t))\, dt, \quad q(\alpha) = q_\alpha, \quad q(\beta) = q_\beta$$

leads to a function $q : [\alpha, \beta] \to \mathbb{R}^n$ satisfying the Euler–Lagrange equations

$$\frac{d}{dt} \frac{\partial L}{\partial \dot{q}} - \frac{\partial L}{\partial q} = 0. \tag{1.30}$$

These equations are also known as Euler's equations.

Here we use the symbol $\dot{q}$ with two meanings. The function L is a function of two variables and these two variables are denoted by $q, \dot{q}$, so $\partial L / \partial \dot{q}$ denotes the partial derivative of L with respect to its second variable. A solution of (1.30) is a smooth function of t, denoted by $q(t)$, whose derivative with respect to t is $\dot{q}(t)$.

In particular, if $q, \dot{q}$ are the position-velocity of a mechanical system subject to a system of holonomic[5] constraints and $K(\dot{q})$ is its kinetic energy, $P(q)$ its potential energy, and $L = K - P$ the Lagrangian, then (1.30) is the equation of motion of the system — see Arnold (1978), Siegel and Moser (1971), or almost any advanced texts on mechanics.

More generally, any critical point of the action functional $F(\cdot)$ leads to the same conclusion concerning the critical function $q(\cdot)$. Moreover, the boundary conditions for the variational problem may be much more general, including the case of periodic boundary conditions, which would replace the fixed endpoint condition with the restriction on the class of functions

$$q(\alpha) = q(\beta).$$

This is an important generalization, in as much as all the periodic solutions of the N-body problem can be realized as critical points of the action, subject to the periodic boundary condition. In fact, this observation leads one to look for such periodic solutions directly by finding appropriate critical points of the action functional, rather than by solving the boundary value problem connected with the Euler equations. This is called the direct method of the calculus of variations, which is a global method in that it does not require nearby known solutions for its application. This method has recently helped the discovery of some spectacular new periodic solutions of the N-body problem that are far from any integrable cases and which

---

[5] A holonomic mechanical system is such that the constrains define a submanifold in position space.

are discussed in subsequent chapters. We give a very simple example of this method below, together with some extensions of this method to the question of global stability of periodic solutions.

Here are the ingredients of the argument that relates the critical points of F to the Euler–Lagrange equations. Suppose that q_ϵ is a one parameter curve of functions through the critical function q that satisfies the boundary constraints. That is, $q_0(t) = q(t)$, $\alpha \le t \le \beta$, and $q_\epsilon(\alpha) = q_\alpha$, $q_\epsilon(\beta) = q_\beta$ in the case of fixed boundary conditions, or $q_\epsilon(\alpha) = q_\epsilon(\beta)$ in the case of periodic conditions. In either of these cases, one would naturally infer that the composite function $g(\epsilon) = F(q_\epsilon)$ has a critical point at $\epsilon = 0$. Assuming that we are able to differentiate under the integral sign, and that the variation vector field

$$\xi(t) = \left. \frac{\partial}{\partial \epsilon} q_\epsilon(t) \right|_{\epsilon=0}$$

is smooth, we find that

$$dF(q) \cdot \xi = \left. \frac{\partial}{\partial \epsilon} F(q_\epsilon) \right|_{\epsilon=0} = \int_\alpha^\beta \left(\frac{\partial L}{\partial x} \cdot \xi + \frac{\partial L}{\partial \dot{x}} \cdot \dot{\xi} \right) dt$$

$$= \left. \frac{\partial L}{\partial \dot{x}} \cdot \xi \right|_\alpha^\beta + \int_\alpha^\beta \left(-\frac{d}{dt} \frac{\partial L}{\partial \dot{x}} + \frac{\partial L}{\partial x} \right) \cdot \xi \, dt.$$

(1.31)

The last line of (1.31) is done using an integration by parts. It is not difficult to see that if the function q is critical for the functional F with either set of boundary conditions, then the boundary terms and the integral expression must vanish independently for an arbitrary choice of the variation vector field $\xi(t)$. This leads to two conclusions: first, that the Euler–Lagrange equations (1.30) must vanish identically on the interval $\alpha \le t \le \beta$ and second, that the transversality conditions

$$\left. \frac{\partial L}{\partial \dot{x}} \cdot \xi \right|_\alpha^\beta = 0$$

(1.32)

should also hold for the critical function q at the endpoints α, β. In the case of fixed boundary conditions, these transversality conditions don't give any additional information because $\xi(\alpha) = \xi(\beta) = 0$. In the case of periodic boundary conditions, they imply that

$$\frac{\partial L}{\partial \dot{q}}(\alpha) = \frac{\partial L}{\partial \dot{q}}(\beta),$$

(1.33)

because $\xi(\alpha) = \xi(\beta)$. As we show below, this guarantees that a critical point of the action functional with periodic boundary conditions is just the configuration component of a periodic solution of Hamilton's equations.

We have shown in (1.31) that we can identify critical points of the functional $F(\cdot)$ with solutions of the Euler equations (1.30) subject to various boundary constraints. One powerful and important application of this is that the Euler–Lagrange equations are invariant under general coordinate transformations.

Theorem 1.9.1. *If the transformation* $(x, \dot{x}) \rightarrow (q, \dot{q})$ *is a local diffeomorphism with*

$$q = q(x), \quad \dot{q} = \frac{\partial q}{\partial x}(x) \cdot \dot{x},$$

then the Euler–Lagrange equations (1.30) *transform into an equivalent set of Euler–Lagrange equations*

$$\frac{d}{ds}\frac{\partial \tilde{L}}{\partial \dot{x}} - \frac{\partial \tilde{L}}{\partial x} = 0,$$

where the new Lagrangian is defined by the coordinate transformation

$$\tilde{L}(x, \dot{x}) = L(q(x), \frac{\partial q}{\partial x}(x) \cdot \dot{x}).$$

Proof. The argument rests on two simple observations. First, the condition that $F(q)$ takes a critical value is independent of coordinates; and second, the functional $F(q)$ transforms in a straightforward manner

$$F(q(t)) = \int_\alpha^\beta L(q(t), \dot{q}(t))dt$$

$$= \int_\alpha^\beta L(q(x(t)), \frac{\partial q}{\partial x}(x(t)) \cdot \dot{x}(t))dt$$

$$= \int_\alpha^\beta \tilde{L}(x(t), \dot{x}(t))dt = \tilde{F}(x(t)).$$

From this, we conclude that the critical points of $F(\cdot)$ correspond to critical points of $\tilde{F}(\cdot)$ under the coordinate transformation. The conclusion of the theorem follows, because we have shown in (1.31) that critical points of $F(\cdot)$ are solutions of the Euler equations for the Lagrangian L, and critical points of $\tilde{F}(\cdot)$ are solutions of the Euler equations for the Lagrangian $\tilde{L}$.

Sometimes L depends on t and we wish to change the time variable also. By the same reasoning, if the transformation $(x, x', s) \rightarrow (q, \dot{q}, t)$ is

$$q = q(x, s), \quad t = t(x, s), \quad \dot{q} = \dot{q}(x, x', s) = \frac{q_x(x, s)x' + q_s(x, s)}{t_x(x, s)x' + t_s(x, s)},$$

then the Euler–Lagrange equations (1.30) become

$$\frac{d}{ds}\frac{\partial \tilde{L}}{\partial x'} - \frac{\partial \tilde{L}}{\partial x} = 0,$$

where $' = d/ds$ and

$$\tilde{L}(x, x', s) = L(q(x, s), \dot{q}(x, x', s), t(x, s)).$$

We consider one interesting example here, whereby the variational structure of certain solutions is directly tied to the stability type of these solutions. We follow this thread of an idea in later examples, especially when we apply the variational method to finding symmetric periodic solutions of the N-body problem. The mathematical pendulum is given by specifying a constrained mechanical system in the plane with Cartesian coordinates (x, y). The gravitational potential energy is $U(x, y) = mgy$ and the kinetic energy $K(\dot{x}, \dot{y}) = \frac{1}{2}(\dot{x}^2 + \dot{y}^2)$. The constraint requires the mass m to lie at a fixed distance l from the point $(0, l)$ so that $x^2 + (y - l)^2 = l^2$. Introducing a local angular coordinate $\theta \mod 2\pi$ on the circle $x^2 + (y - l)^2 = l^2$ and expressing the Lagrangian in these coordinates we find the Lagrangian and the resulting Euler–Lagrange equations,

$$L(\theta, \dot{\theta}) = \frac{1}{2}ml^2\dot{\theta}^2 + mgl(1 + \cos(\theta)), \quad ml^2\ddot{\theta} = -mgl\sin(\theta).$$

The equations in θ follow by the invariance of Euler–Lagrange equations (1.30), see Theorem (1.9.1). The factor mgl is subtracted from the potential to make the action positive, and doesn't affect the resulting differential equations. The action of the variational problem is the integral of the Lagrangian, so we study the periodic problem

$$F(q) = \int_0^T \left(\frac{1}{2}ml^2\dot{q}^2 + mgl(1 + \cos(q)) \right) dt, \quad q(0) = q(T).$$

We make the simple observation that the absolute minimizer of the action corresponds to a global maximum of the potential, and the global minimum of the potential corresponds to a mountain pass critical point of the action functional

$$F(\pm\pi) \leq F(q), \quad F(0) = \min \max_{\deg q = 1} F(q).$$

The first inequality may be easily verified, because the kinetic energy is positive and the potential takes a maximum value at $\pm\pi$. In the second case, the maximum is taken with respect to loops in the configuration variable, which make one circuit of the point 0 before closing. This is described by the topological degree $= 1$. The minimum is then taken over all such loops, including the limit case when the loop is stationary at the origin.

It is interesting to observe here that the global minimum of the action functional corresponds to a hyperbolic critical point, and the stable critical point (see Dirichlet's theorem (1.2.2)) corresponds to a mountain pass type critical point. This fact is not isolated, and we discuss a theory to make this kind of prediction concerning stability and instability in much more general settings when we discuss the Maslov index in Section 5.6.

One could consider the forced pendulum equations, as was done in Section 1.7. Here the analysis and the results become essentially more interesting, because there are no longer any equilibrium solutions; however, the direct method of the calculus of variations leads to some very interesting global results for this simple problem, which we describe briefly. The Euler equation and the action functional become

$$ml^2\ddot{\theta} = -mgl\sin(\theta) + f(t), \quad f(t+T) = f(t),$$

$$F(q) = \int_0^T \left(\frac{1}{2}ml^2\dot{q}^2 + mgl(1 + \cos(q)) + qf(t)\right) dt, \quad q(0) = q(T).$$

In this problem, the question of stable and unstable periodic solutions becomes an interesting nonelementary research topic. The first question one encounters here is the question of existence of T-periodic solutions. There is a simple analytical condition on the forcing term f that guarantees the existence of both minimizing and mountain pass critical points for the action functional. This condition is that the mean value of f is zero, $\int_0^T f\,dt = 0$. By our earlier discussion on the equality of periodic solutions and critical functions, this guarantees nontrivial harmonic oscillations of the forced pendulum equation (see Mawhin and Willem (1984)). The related question concerning which forcing terms f can admit such T-periodic solutions of the Euler equations is an open problem.

The next question one might well ask is whether our earlier observation on stability and other dynamical properties is again related to the variational structure of minimizing and mountain pass critical points, when they can be shown to exist. This question is pursued when we discuss the Maslov index, but it suffices to say that the minimizing critical periodic curves continue to represent unstable periodic motion, and mountain pass critical points can be used to show the existence of an infinite family of subharmonic oscillations of period kT when the minimizing solution is nondegenerate, see Offin (1990). There is at this time no global method of determining whether such a harmonic oscillation is stable. The next section elaborates on this pendulum example in a more geometrical setting.

The Euler–Lagrange equations often have an equivalent Hamiltonian formulation.

Theorem 1.9.2. *If the transformation* $(q, \dot{q}, t) \to (q, p, t)$ *is a diffeomorphism, with p defined by*

$$p = \frac{\partial L}{\partial \dot{q}},$$

then the Euler–Lagrange equation (1.30) *is equivalent to the Hamiltonian system*

$$\dot{q} = \frac{\partial H}{\partial p}, \qquad \dot{p} = -\frac{\partial H}{\partial q},$$

with $H(q, p, t) = p^T\dot{q} - L(q, \dot{q}, t)$.

Proof. First,

$$\dot{p} = \frac{d}{dt}\frac{\partial L}{\partial \dot{q}} = \frac{\partial L}{\partial q} = -\frac{\partial H}{\partial \dot{q}},$$

and second,

$$\frac{\partial H}{\partial p} = \dot{q} + p\,\frac{\partial \dot{q}}{\partial p} - \frac{\partial L}{\partial \dot{q}}\frac{\partial \dot{q}}{\partial p} = \dot{q}.$$

In the Hamiltonian formulation, the transversality conditions (1.32) become $p \cdot \xi|_\alpha^\beta = 0$, which for periodic boundary conditions becomes $p(\alpha) = p(\beta)$.

1.10 Spherical Pendulum

In the spherical pendulum, a particle, or bob, of mass m is constrained to move on a sphere of radius l by a massless rigid rod. The only forces acting on the bob are the frictionless constraint force and the constant gravitational force. This defines a holonomic system. Let the origin of the coordinate system be at the center of the sphere, so the constraints are $\|q\| = l$, $q \cdot \dot{q} = 0$. Thus position space is the sphere, and phase space (position-velocity space) is the set of all tangent vectors to the sphere, the tangent bundle of the sphere. The Newtonian formulation of the problem is

$$m\ddot{q} = -mgle_3 + R(q, \dot{q}),$$

where g is the acceleration due to gravity, $e_3 = (0,0,1)$, and $R(q, \dot{q})$ is the force of constraint. Because the constraint must be normal to the sphere, we have $R(q, \dot{q}) = r(q, \dot{q})q$, with r a scalar. The constraint implies that $q \cdot \dot{q} = 0$ and $\dot{q} \cdot \dot{q} + q \cdot \ddot{q} = 0$. The latter restriction can be used to find R explicitly yielding a very ugly system. Fortunately, we can ignore this by introducing appropriate coordinates and using the coordinate invariance of the Euler–Lagrange equations.

The problem is symmetric about the q_3 axis, and so admits one component of angular momentum, $A_3 = m(q \times \dot{q}) \cdot e_3$, as an integral, because

$$\frac{dA_3}{dt} = m(\dot{q} \times \dot{q}) \cdot e_3 + (q \times -mge_3 + r(q, \dot{q})q) \cdot e_3 = 0.$$

Lemma 1.10.1. *If $A_3 = 0$, then the motion is planar.*

Proof. $A_3 = q_1\dot{q}_2 - q_2\dot{q}_1 = 0$, or $dq_1/q_1 = dq_2/q_2$ which implies $q_1 = c\,q_2$.

The planar pendulum is of the form discussed in Section 1.7 and is treated in the problems also, so we assume that $A_3 \neq 0$ and thus the bob does not go through the north or south poles.

Because the bob does not go through the poles, there is no problem in using spherical coordinates. For simplicity, let $m = l = g = 1$. The kinetic and potential energies are

$$K = \frac{1}{2}\|\dot{q}\|^2 = \frac{1}{2}\{\dot{\phi}^2 + \sin^2\phi\,\dot{\theta}^2\}, \quad P = \{q \cdot e_3 + 1\} = \{1 - \cos\phi\},$$

and the Lagrangian is $L = K - P$. Instead of writing the equations in Lagrangian form, proceed to the Hamiltonian form by introducing the conjugate variables Θ and Φ with

$$\Theta = \frac{\partial L}{\partial \dot{\theta}} = \sin^2\phi\,\dot{\theta}, \quad \Phi = \frac{\partial L}{\partial \dot{\phi}} = \dot{\phi},$$

so that

$$H = \frac{1}{2}\left\{\Phi^2 + \frac{\Theta^2}{\sin^2\phi}\right\} + \{1 - \cos\phi\},$$

and the equations of motion are

$$\dot{\theta} = \frac{\partial H}{\partial \Theta} = \frac{\Theta}{\sin^2\phi}, \quad \dot{\Theta} = -\frac{\partial H}{\partial \theta} = 0,$$

$$\dot{\phi} = \frac{\partial H}{\partial \Phi} = \Phi, \quad \dot{\Phi} = -\frac{\partial H}{\partial \phi} = \Theta^2 \csc^2\phi \cot\phi - \sin\phi.$$

Thus we have defined symplectic coordinates $(\theta, \phi, \Theta, \Phi)$ for this problem. A general discussion of the special coordinates used in Hamiltonian mechanics appears in Chapter 8, and Section 8.7 deals with spherical coordinates in particular.

H is independent of θ, so $\dot{\Theta} = -\partial H/\partial\theta = 0$ and Θ is an integral of motion. This is an example of the classic maxim "the variable conjugate to an ignorable coordinate is an integral"; i.e., θ is ignorable, so Θ is an integral. Θ is the component of angular momentum about the e_3 axis.

The analysis starts by ignoring θ and setting $\Theta = c \neq 0$, so the Hamiltonian becomes

$$H = \frac{1}{2}\Phi^2 + A_c(\phi), \quad A_c(\phi) = \frac{c^2}{\sin^2\phi} + \{1 - \cos\phi\}, \qquad (1.34)$$

which is the Hamiltonian of one degree of freedom with a parameter c. H is of the form kinetic energy plus potential energy, where A_c is the potential energy, so it can be analyzed by the method of Section 1.7. The function A_c is called the amended potential.

This is an example of what is called reduction. The system admits a symmetry, i.e., rotation about the e_3 axis, which implies the existence of an integral, namely Θ. Holding the integral fixed and identifying symmetric configurations by ignoring θ leads to a Hamiltonian system of fewer degrees of freedom. This is the system on the reduced space.

It is easy to see that $A_c(\phi) \to +\infty$ as $\phi \to 0$ or $\phi \to \pi$ and that A_c has a unique critical point, a minimum, at $0 < \phi_c < \pi$ (a relative equilibrium). Thus $\phi = \phi_c$, $\Phi = 0$ is an equilibrium solution for the reduced system, but

$$\theta = \frac{c}{\sin^2 \phi_c} t + \theta_0, \quad \Theta = c, \quad \phi = \phi_c, \quad \Phi = 0$$

is a periodic solution of the full system.

The level curves of (1.34) are closed curves encircling ϕ_c, and so all the other solutions of the reduced system are periodic. For the full system, these solutions lie on the torus which is the product of one of these curves in the ϕ, Φ-space and the circle $\Theta = c$ with any θ. The flows on these tori can be shown to be equivalent to the linear flow discussed in Section (1.5).

1.11 Problems

1. Let x, y, z be the usual coordinates in $\mathbb{R}^3$, $r = xi + yj + zk$, $X = \dot{x}$, $Y = \dot{y}$, $Z = \dot{z}$, $R = \dot{r} = Xi + Yj + Zk$.
   a) Compute the three components of angular momentum $mr \times R$.
   b) Compute the Poisson bracket of any two of the components of angular momentum and show that it is $\pm m$ times the third component of angular momentum.
   c) Show that if a system admits two components of angular momentum as integrals, then the system admits all three components of angular momentum as integrals.
2. A Lie algebra $\mathcal{A}$ is a vector space with a product $* : \mathcal{A} \times \mathcal{A} \to \mathcal{A}$ that satisfies
   a) $a * b = -b * a$ (anticommutative),
   b) $a * (b + c) = a * b + a * c$ (distributive),
   c) $(\alpha a) * b = \alpha(a * b)$ (scalar associative),
   d) $a * (b * c) + b * (c * a) + c * (a * b) = 0$ (Jacobi's identity),
   where $a, b, c \in \mathcal{A}$ and $\alpha \in \mathbb{R}$ or $\mathbb{C}$.
   a) Show that vectors in $\mathbb{R}^3$ form a Lie algebra where the product $*$ is the cross product.
   b) Show that smooth functions on an open set in $\mathbb{R}^{2n}$ form a Lie algebra, where $f * g = \{f, g\}$, the Poisson bracket.
   c) Show that the set of all $n \times n$ matrices, $gl(n, \mathbb{R})$, is a Lie algebra, where $A * B = AB - BA$, the Lie product.
3. The pendulum equation is $\ddot{\theta} + \sin\theta = 0$.
   a) Show that $2I = \frac{1}{2}\dot{\theta}^2 + (1 - \cos\theta) = \frac{1}{2}\dot{\theta}^2 + 2\sin^2(\theta/2)$ is an integral.
   b) Sketch the phase portrait.
4. Let $H : \mathbb{R}^{2n} \to \mathbb{R}$ be a globally defined conservative Hamiltonian, and assume that $H(z) \to \infty$ as $\|z\| \to \infty$. Show that all solutions of $\dot{z} = J\nabla H(z)$ are bounded. (Hint: Think like Dirichlet.)
5. If a particle is constrained to move on a surface in $\mathbb{R}^3$ without friction, then the force of constraint acts normal to the surface. If there are no external forces, then the particle is said to be a free particle on the surface, and the only force acting on the particle is the force of constraint. In the

free case, acceleration is normal to the surface. In differential geometry, a curve on a surface that minimizes distance (at least locally) is called a geodesic, and it can be shown that geodesics are characterized by the fact that their acceleration is normal to the surface. Thus, a free particle moves on a geodesic.

a) Consider a free particle on the 2-sphere $S^2 = \{x \in \mathbb{R}^3 : \|x\| = 1\}$. It moves to satisfy an equation of the form $\ddot{x} = \lambda x$, where λ is the scalar of proportionality. Show that $\lambda = -\|\dot{x}\|^2$, $x^T \dot{x} = 0$, and λ is constant along a solution (i.e., $d\lambda/dt = 0$).

b) Show that if the initial velocity is nonzero, then the solutions are great circles.

c) Show that the set of unit tangent vectors to S^2, called the unit tangent bundle of S^2 and denoted by $T_1 S^2$, is an invariant set and is given by $\{(x, y) \in \mathbb{R}^3 \times \mathbb{R}^3 : \|x\| = 1, \|y\| = 1, x^T y = 0\}$.

d) Show that the unit tangent bundle of the two sphere is the same as $SO(3, \mathbb{R})$, the special orthogonal group. $SO(3, \mathbb{R})$ is the set of all 3×3 orthogonal matrices with determinant equal $+1$, or the set of all 3×3 rotation matrices. (Hint: Think of the orthonormal frame consisting of the unit tangent, normal, and binormal vectors.)

2. Hamiltonian Systems

The study of Hamiltonian systems starts with the study of linear systems and the associated linear algebra. This will lead to basic results on periodic systems and variational equations of nonlinear systems.

The basic linear algebra introduced in this chapter will be the cornerstone of many of the later results on nonlinear systems. Some of the more advanced results which require a knowledge of multilinear algebra or the theory of analytic functions of a matrix are relegated to Chapter 5 or to the literature.

2.1 Linear Equations

A familiarity with the basic theory of linear algebra and linear differential equations is assumed. Let $gl(m, \mathbb{F})$ denote the set of all $m \times m$ matrices with entries in the field $\mathbb{F}$ ($\mathbb{R}$ or $\mathbb{C}$) and $Gl(m, \mathbb{F})$ the set of all nonsingular $m \times m$ matrices with entries in $\mathbb{F}$. $Gl(m, \mathbb{F})$ is a group under matrix multiplication and is called the general linear group. Let $I = I_m$ and $0 = 0_m$ denote the $m \times m$ identity and zero matrices, respectively. In general, the subscript is clear from the context and so will be omitted.

In this theory a special role is played by the $2n \times 2n$ matrix

$$J = \begin{bmatrix} 0 & I \\ -I & 0 \end{bmatrix}. \tag{2.1}$$

Note that J is orthogonal and skew-symmetric; i.e.,

$$J^{-1} = J^T = -J. \tag{2.2}$$

Here as elsewhere the superscript T denotes the transpose.

Let z be a coordinate vector in $\mathbb{R}^{2n}$, $\mathbb{I}$ an interval in $\mathbb{R}$, and $S : \mathbb{I} \to gl(2n, \mathbb{R})$ be continuous and symmetric. A linear Hamiltonian system is the system of $2n$ ordinary differential equations

$$\dot{z} = J\frac{\partial H}{\partial z} = JS(t)z = A(t)z, \tag{2.3}$$

© Springer International Publishing AG 2017
K.R. Meyer, D.C. Offin, *Introduction to Hamiltonian Dynamical Systems and the N-Body Problem*, Applied Mathematical Sciences 90, DOI 10.1007/978-3-319-53691-0_2

where

$$H = H(t,z) = \frac{1}{2}z^T S(t)z, \tag{2.4}$$

$A(t) = JS(t)$. The Hamiltonian H is a quadratic form in z with coefficients that are continuous in $t \in \mathbb{I} \subset \mathbb{R}$. If S, and hence H, is independent of t, then H is an integral for (2.3) by Theorem 1.2.1.

Let $t_0 \in \mathbb{I}$ be fixed. From the theory of differential equations, for each $z_0 \in \mathbb{R}^{2n}$, there exists a unique solution $\phi(t, t_0, z_0)$ of (2.3) for all $t \in \mathbb{I}$ that satisfies the initial condition $\phi(t_0, t_0, z_0) = z_0$. Let $Z(t, t_0)$ be the $2n \times 2n$ fundamental matrix solution of (2.3) that satisfies $Z(t_0, t_0) = $ I. Then $\phi(t, t_0, z_0) = Z(t, t_0)z_0$.

In the case where S and A are constant, we take $t_0 = 0$ and

$$Z(t) = e^{At} = \exp At = \sum_{i=1}^{\infty} \frac{A^n t^n}{n!}. \tag{2.5}$$

A matrix $A \in gl(2n, \mathbb{F})$ is called *Hamiltonian* (or sometimes infinitesimally symplectic), if

$$A^T J + JA = 0. \tag{2.6}$$

The set of all $2n \times 2n$ Hamiltonian matrices is denoted by $sp(2n, \mathbb{R})$.

Theorem 2.1.1. *The following are equivalent:*

(i) A is Hamiltonian,
(ii) $A = JR$ where R is symmetric,
(iii) JA is symmetric.

Moreover, if A and B are Hamiltonian, then so are A^T, αA ($\alpha \in \mathbb{F}$), $A \pm B$, and $[A, B] \equiv AB - BA$.

Proof. $A = J(-JA)$ and (2.6) is equivalent to $(-JA)^T = (-JA)$; thus (i) and (ii) are equivalent. Because $J^2 = -I$, (ii) and (iii) are equivalent. Thus the coefficient matrix $A(t)$ of the linear Hamiltonian system (2.1) is a Hamiltonian matrix.

The first three parts of the next statement are easy. Let $A = JR$ and $B = JS$, where R and S are symmetric. Then $[A, B] = J(RJS - SJR)$ and $(RJS - SJR)^T = S^T J^T R^T - R^T J^T S^T = -SJR + RJS$ so $[A, B]$ is Hamiltonian.

In the 2×2 case,

$$A = \begin{bmatrix} \alpha & \beta \\ \gamma & \delta \end{bmatrix}$$

so,

$$A^T J + JA = \begin{bmatrix} 0 & \alpha + \delta \\ -\alpha - \delta & 0 \end{bmatrix}.$$

Thus, a 2×2 matrix is Hamiltonian if and only if its trace, $\alpha + \delta$, is zero. If you write a second-order equation $\ddot{x} + p(t)\dot{x} + q(t)x = 0$ as a system in the usual way with $\dot{x} = y$, $\dot{y} = -q(t)x - p(t)y$, then it is a linear Hamiltonian system when and only when $p(t) \equiv 0$. The $p(t)\dot{x}$ is usually considered the friction term.

Now let A be a $2n \times 2n$ matrix and write it in block form

$$A = \begin{bmatrix} a & b \\ c & d \end{bmatrix}$$

so,

$$A^T J + JA = \begin{bmatrix} c - c^T & a^T + d \\ -a - d^T & -b + b^T \end{bmatrix}.$$

Therefore, A is Hamiltonian if and only if $a^T + d = 0$ and b and c are symmetric.

The function $[\cdot, \cdot] : gl(m, \mathbb{F}) \times gl(m, \mathbb{F}) \to gl(m, \mathbb{F})$ of Theorem 2.1.1 is called the *Lie product*. The second part of this theorem implies that the set of all $2n \times 2n$ Hamiltonian matrices, $sp(2n, \mathbb{R})$, is a Lie algebra. We develop some interesting facts about Lie algebras of matrices in the Problem section.

A $2n \times 2n$ matrix T is called symplectic with multiplier μ if

$$T^T J T = \mu J, \tag{2.7}$$

where μ is a nonzero constant. If $\mu = +1$, then T is simply said to be symplectic. The set of all $2n \times 2n$ symplectic matrices is denoted by $Sp(2n, \mathbb{R})$.

Theorem 2.1.2. *If T is symplectic with multiplier μ, then T is nonsingular and*

$$T^{-1} = -\mu^{-1} J T^T J. \tag{2.8}$$

If T and R are symplectic with multiplier μ and ν, respectively, then T^T, T^{-1}, and TR are symplectic with multipliers μ, μ^{-1}, and $\mu\nu$, respectively.

Proof. Because the right-hand side, μJ, of (2.7) is nonsingular, T must be also. Formula (2.8) follows at once from (2.7). If T is symplectic, then (2.8) implies $T^T = -\mu J T^{-1} J$; so, $T J T^T = TJ(-\mu J T^{-1} J) = \mu J$. Thus T^T is symplectic with multiplier μ. The remaining facts are proved in a similar manner.

This theorem shows that $Sp(2n, \mathbb{R})$ is a group, a closed subgroup of $Gl(2n, \mathbb{R})$. Weyl says that originally he advocated the name "complex group" for $Sp(2n, \mathbb{R})$, but it became an embarrassment due to the collisions with the word "complex" as in complex numbers. "I therefore proposed to replace it by the corresponding Greek adjective 'symplectic.'" See page 165 in Weyl (1948).

In the 2×2 case

$$T = \begin{bmatrix} \alpha & \beta \\ \gamma & \delta \end{bmatrix}$$

and so

$$T^T J T = \begin{bmatrix} 0 & \alpha\delta - \beta\gamma \\ -\alpha\delta + \beta\gamma & 0 \end{bmatrix}.$$

So a 2×2 matrix is symplectic (with multiplier μ) if and only if it has determinant $+1$ (respectively μ). Thus a 2×2 symplectic matrix defines a linear transformation which is orientation-preserving and area-preserving.

Now let T be a $2n \times 2n$ matrix and write it in block form

$$T = \begin{bmatrix} a & b \\ c & d \end{bmatrix} \tag{2.9}$$

and so

$$T^T J T = \begin{bmatrix} a^T c - c^T a & a^T d - c^T b \\ b^T c - d^T a & b^T d - d^T b \end{bmatrix}.$$

Thus T is symplectic with multiplier μ if and only if $a^T d - c^T b = \mu I$ and $a^T c$ and $b^T d$ are symmetric. Formula (2.8) gives

$$T^{-1} = \mu^{-1} \begin{bmatrix} d^T & -b^T \\ -c^T & a^T \end{bmatrix}. \tag{2.10}$$

This reminds one of the formula for the inverse of a 2×2 matrix!

Theorem 2.1.3. *The fundamental matrix solution $Z(t, t_0)$ of a linear Hamiltonian system (2.3) is symplectic for all $t, t_0 \in \mathbb{I}$. Conversely, if $Z(t, t_0)$ is a continuously differential function of symplectic matrices, then Z is a matrix solution of a linear Hamiltonian system.*

Proof. Let $U(t) = Z(t, t_0)^T J Z(t, t_0)$. Because $Z(t_0, t_0) = I$, it follows that $U(t_0) = J$. Now

$$\dot{U}t) = \dot{Z}^T J Z + Z^T J \dot{Z} = Z^T (A^T J + J A) Z = 0;$$

so, $U(t) \equiv J$.

If $Z^T J Z = J$ for $t \in \mathbb{I}$, then $\dot{Z}^T J Z + Z^T J \dot{Z} = 0$; so, $(\dot{Z} Z^{-1})^T J + J(\dot{Z} Z^{-1}) = 0$. This shows that $A = \dot{Z} Z^{-1}$ is Hamiltonian and $\dot{Z} = AZ$.

Corollary 2.1.1. *The constant matrix A is Hamiltonian if and only if e^{At} is symplectic for all t.*

Change variables by $z = T(t)u$ in system (2.3). Equation (2.3) becomes

$$\dot{u} = (T^{-1}AT - T^{-1}\dot{T})u. \qquad (2.11)$$

In general this equation will not be Hamiltonian, however:

Theorem 2.1.4. *If T is symplectic with multiplier μ^{-1}, then* (2.11) *is a Hamiltonian system with Hamiltonian*

$$H(t, u) = \frac{1}{2}u^T (\mu T^T S(t)T + R(t))u,$$

where

$$R(t) = JT^{-1}\dot{T}.$$

Conversely, if (2.11) *is Hamiltonian for every Hamiltonian system* (2.3), *then U is symplectic with constant multiplier μ.*

The function $\frac{1}{2}u^T R(t)u$ is called the remainder.

Proof. Because $TJT^T = \mu^{-1}J$ for all t, $\dot{T}JT^T + TJ\dot{T}^T = 0$ or $(T^{-1}\dot{T})J + J(T^{-1}\dot{T})^T = 0$; so, $T^{-1}\dot{T}$ is Hamiltonian. Also $T^{-1}J = \mu JT^T$; so, $T^{-1}AT = T^{-1}JST = \mu JT^T ST$, and so, $T^{-1}AT = J(\mu T^T ST)$ is Hamiltonian also.

For the converse let (2.11) always be Hamiltonian. By taking $A \equiv 0$ we have that $T^{-1}\dot{T} = B(t)$ is Hamiltonian or T is a matrix solution of the Hamiltonian system

$$\dot{v} = vB(t). \qquad (2.12)$$

So, $T(t) = KV(t, t_0)$, where $V(t, t_0)$ is the fundamental matrix solution of (2.12), and $K = T(t_0)$ is a constant matrix. By Theorem 2.1.3, V is symplectic.

Consider the change of variables $z = T(t)u = KV(t, t_0)u$ as a two-stage change of variables: first $w = V(t, t_0)u$ and second $z = Kw$. The first transformation from u to w is symplectic, and so, by the first part of this theorem, preserves the Hamiltonian character of the equations. Because the first transformation is reversible, it would transform the set of all linear Hamiltonian systems onto the set of all linear Hamiltonian systems. Thus the second transformation from w to z must always take a Hamiltonian system to a Hamiltonian system.

If $z = Kw$ transforms all Hamiltonian systems $\dot{z} = JCz$, C constant and symmetric, to a Hamiltonian system $\dot{w} = JDw$, then $JD = K^{-1}JCK$ is Hamiltonian, and $JK^{-1}JCK$ is symmetric for all symmetric C. Thus

$$JK^{-1}JCK = (JK^{-1}JCK)^T = K^T CJK^{-T}J,$$
$$C(KJK^T J) = (JKJK^T)C,$$
$$CR = R^T C,$$

where $R = KJK^TJ$. Fix i, $1 \le i \le 2n$ and take C to be the symmetric matrix that has $+1$ at the i,i position and zero elsewhere. Then the only nonzero row of CR is the ith, which is the ith row of R and the only nonzero column of R^TC is the ith, which is the ith column of R^T. Because these must be equal, the only nonzero entry in R or R^T must be on the diagonal. So R and R^T are diagonal matrices. Thus $R = R^T = \text{diag}(r_1, \ldots, r_{2n})$, and $RC - CR = 0$ for all symmetric matrices C. But $RC - CR = ((r_j - r_i)c_{ij}) = (0)$. Because c_{ij}, $i < j$, is arbitrary, $r_i = r_j$, or $R = -\mu I$ for some constant μ. $R = KJK^TJ = -\mu I$ implies $KJK^T = \mu J$.

This change of variables preserves the Hamiltonian character of a linear system of equations. The general problem of which changes of variables preserve the Hamiltonian character is discussed in detail in Chapter 2.6.

The fact that the fundamental matrix of (2.3) is symplectic means that the fundamental matrix must satisfy the identity (2.7). There are many functional relations in (2.7); so, there are functional relations between the solutions. Theorem 2.1.5 given below is just one example of how these relations can be used. See Meyer and Schmidt (1982) for some other examples.

Let $z_1, z_2 : \mathbb{I} \to \mathbb{R}^{2n}$ be two smooth functions where $\mathbb{I}$ is an interval in $\mathbb{R}^1$; we define the Poisson bracket of z_1 and z_2 to be

$$\{z_1, z_2\}(t) = z_1^T(t)Jz_2(t); \tag{2.13}$$

so, $\{z_1, z_2\} : \mathbb{I} \to \mathbb{R}$ is smooth. The Poisson bracket is bilinear and skew-symmetric. Two functions z_1 and z_2 are said to be in *involution* if $\{z_1, z_2\} \equiv 0$. A set of n linearly independent functions that are pairwise in involution are said to be a *Lagrangian set*. In general, the complete solution of a $2n$-dimensional system requires $2n$ linearly independent solutions, but for a Hamiltonian system a Lagrangian set of solutions suffices.

Theorem 2.1.5. *If a Lagrangian set of solutions of* (2.3) *is known, then a complete set of $2n$ linearly independent solutions can be found by quadrature. (See* (2.14).)

Proof. Let $C = C(t)$ be the $2n \times n$ matrix whose columns are the n linearly independent solutions. Because the columns are solutions, $\dot{C} = AC$; because they are in involution, $C^TJC = 0$; and because they are independent, C^TC is an $n \times n$ nonsingular matrix. Define the $2n \times n$ matrix by $D = JC(C^TC)^{-1}$. Then $D^TJD = 0$ and $C^TJD = -I$, and so $P = [D, C]$ is a symplectic matrix. Therefore,

$$P^{-1} = \begin{bmatrix} -C^TJ \\ D^TJ \end{bmatrix};$$

change coordinates by $z = P\zeta$ so that

$$\dot{\zeta} = P^{-1}(AP - \dot{P})\zeta = \begin{bmatrix} C^TAD + C^TJ\dot{D} & 0 \\ -D^TAD - D^TJ\dot{D} & 0 \end{bmatrix}\zeta.$$

All the submatrices above are $n \times n$. The one in the upper left-hand corner is also zero, which can be seen by differentiating $C^T J D = -I$ to get $\dot{C}^T J D + C^T J \dot{D} = (AC)^T J D + C^T J \dot{D} = C^T S D + C^T J \dot{D} = 0$. Therefore,

$$\dot{u} = 0, \qquad \text{where } \zeta = \begin{bmatrix} u \\ v \end{bmatrix},$$
$$\dot{v} = -D^T (AD + J\dot{D})u,$$

which has a general solution $u = u_0, v = v_0 - V(t)u_0$, where

$$V(t) = \int_{t_0}^{t} D^T (AD + J\dot{D}) dt. \tag{2.14}$$

A symplectic fundamental matrix solution of (2.3) is

$$Z = [D - CV, C].$$

Thus the complete set of solutions is obtained by performing the integration or quadrature in the formula above.

This result is closely related to the general result given in a later chapter which says that when k integrals in involution for a Hamiltonian system of n degrees of freedom are known then the integrals can be used to reduce the number of degrees of freedom by k, and hence to a system of dimension $2n - 2k$.

Recall that a nonsingular matrix T has two polar decompositions, $T = PO = O'P'$, where P and P' are positive definite matrices and O and O' are orthogonal matrices. These representations are unique. P is the unique positive definite square root of TT^T; P' is the unique positive definite square root of $T^T T$, $O = (TT^T)^{-1/2}T$; and $O' = T(T^T T)^{-1/2}$.

Theorem 2.1.6. *If T is symplectic, then the P, O, P', O' of the polar decomposition given above are symplectic also.*

Proof. The formula for T^{-1} in (2.8) is an equivalent condition for T to be symplectic. Let $T = PO$. Because $T^{-1} = -JT^T J$, $O^{-1}P^{-1} = -JO^T P^T J = (J^T O^T J)(J^T P^T J)$. In this last equation, the left-hand side is the product of an orthogonal matrix O^{-1} and a positive definite matrix P^{-1}, similarly the right-hand side a product of an orthogonal matrix $J^{-1}OJ$ and a positive definite matrix $J^T PJ$. By the uniqueness of the polar representation, $O^{-1} = J^{-1}O^T J = -JO^T J$ and $P^{-1} = J^T PJ = -JP^T J$. By (2.8) these last relations imply that P and O are symplectic. A similar argument shows that P' and O' are symplectic. $\qquad\blacksquare$

Theorem 2.1.7. *The determinant of a symplectic matrix is $+1$.*

Proof. Depending on how much linear algebra you know, this theorem is either easy or difficult. In Section 5.3 and Chapter 7 we give alternate proofs.

Let T be symplectic. Formula (2.7) gives $\det(T^T J T) = \det T^T \det J \det T = (\det T)^2 = \det J = 1$ so $\det T = \pm 1$. The problem is to show that $\det T = +1$.

The determinant of a positive definite matrix is positive; so, by the polar decomposition theorem it is enough to show that an orthogonal symplectic matrix has a positive determinant. So let T be orthogonal also.

Using the block representation in (2.9) for T, formula (2.10) for T^{-1}, and the fact that T is orthogonal, $T^{-1} = T^T$, one has that T is of the form

$$T = \begin{bmatrix} a & b \\ -b & a \end{bmatrix}.$$

Define P by

$$P = \frac{1}{\sqrt{2}} \begin{bmatrix} I & iI \\ I & -iI \end{bmatrix}, \quad P^{-1} = \frac{1}{\sqrt{2}} \begin{bmatrix} I & I \\ -iI & iI \end{bmatrix}.$$

Compute $PTP^{-1} = \mathrm{diag}((a - bi), (a + bi))$, so

$$\det T = \det PTP^{-1} = \det(a - bi) \det(a + bi) > 0.$$

2.2 Symplectic Linear Spaces

What is the matrix J? There are many different answers to this question depending on the context in which the question is asked. In this section we answer this question from the point of view of abstract linear algebra. We present other answers later on, but certainly not all.

Let $\mathbb{V}$ be an m-dimensional vector space over the field $\mathbb{F}$ where $\mathbb{F} = \mathbb{R}$ or $\mathbb{C}$. A bilinear form is a mapping $B : \mathbb{V} \times \mathbb{V} \to \mathbb{F}$ that is linear in both variables. A bilinear form is skew-symmetric or alternating if $B(u, v) = -B(v, u)$ for all $u, v \in \mathbb{V}$. A bilinear form B is nondegenerate if $B(u, v) = 0$ for all $v \in \mathbb{V}$ implies $u = 0$. An example of an alternating bilinear form on $\mathbb{F}^m$ is $B(u, v) = u^T S v$, where S is any skew-symmetric matrix, e.g., J.

Let B be a bilinear form and $e_1, \ldots, e_m$ a basis for $\mathbb{V}$. Given any vector $v \in \mathbb{V}$, we write $v = \Sigma \alpha_i e_i$ and define an isomorphism $\Phi : \mathbb{V} \to \mathbb{F}^m : v \to a = (\alpha_1, \ldots, \alpha_m)$.[1] Define $s_{ij} = B(e_i, e_j)$ and S to be the $m \times m$ matrix $S = [s_{ij}]$, the matrix of B in the basis (e). Let $\Phi(u) = b = (\beta_1, \ldots, \beta_m)$; then $B(u, v) = \Sigma\Sigma \alpha_i \beta_j B(e_i, e_j) = b^T S a$. So in the coordinates defined by the basis (e_i), the bilinear form is just $b^T S a$ where S is the matrix $[B(e_i, e_j)]$. If B is alternating, then S is skew-symmetric, and if B is nondegenerate, then S is nonsingular and conversely.

If you change the basis by $e_i = \Sigma q_{ij} f_j$ and Q is the matrix $Q = (q_{ij})$, then the bilinear form B has the matrix R in the basis (f), where $S = QRQ^T$.

---

[1] Remember vectors are column vectors, but sometimes written as row vectors in the text.

One says that R and S are congruent (by Q). If Q is any elementary matrix so that premultiplication of R by Q is an elementary row operation, then postmultiplication of R by Q^T is the corresponding column operation. Thus S is obtained from R by performing a sequence of row operations and the same sequence of column operations and conversely.

Theorem 2.2.1. *Let S be any skew-symmetric matrix; then there exists a nonsingular matrix Q such that*

$$R = QSQ^T = \mathrm{diag}(\mathrm{K}, \mathrm{K}, \dots, \mathrm{K}, 0, 0, \dots, 0),$$

where

$$K = \begin{bmatrix} 0 & 1 \\ -1 & 0 \end{bmatrix}.$$

Or given an alternating form B there is a basis for $\mathbb{V}$ such that the matrix of B in this basis is R.

Proof. If $S = 0$, we are finished. Otherwise, there is a nonzero entry that can be transferred to the first row by interchanging rows. Perform the corresponding column operations. Now bring the nonzero entry in the first row to the second column (the (1,2) position) by column operations and perform the corresponding row operations.

Scale the first row and the first column so that $+1$ is in the (1,2) and so that -1 is in the (2,1) position. Thus the matrix has the 2×2 matrix K in the upper left-hand corner. Using row operations we can eliminate all the nonzero elements in the first two columns below the first two rows. Performing the corresponding column operation yields a matrix of the form $\mathrm{diag}(K, S')$, where S' is an $(m - 2) \times (m - 2)$ skew-symmetric matrix. Repeat the above argument on S'.

Note that the rank of a skew-symmetric matrix is always even; thus, a nondegenerate, alternating bilinear form is defined on an even dimensional space.

A symplectic linear space, or just a symplectic space, is a pair, $(\mathbb{V}, \omega)$ where $\mathbb{V}$ is a $2n$-dimensional vector space over the field $\mathbb{F}$, $\mathbb{R}$ or $\mathbb{C}$, and ω is a nondegenerate alternating bilinear form on $\mathbb{V}$. The form ω is called the *symplectic form* or the symplectic inner product. Throughout the rest of this section we shall assume that $\mathbb{V}$ is a symplectic space with symplectic form ω. The standard example is $\mathbb{F}^{2n}$ and $\omega(x, y) = x^T J y$. In this example we shall write $\{x, y\} = x^T J y$ and call the space $(\mathbb{F}^{2n}, J)$ or simply $\mathbb{F}^{2n}$, if no confusion can arise.

A symplectic basis for $\mathbb{V}$ is a basis $v_1, \dots, v_{2n}$ for $\mathbb{V}$ such that $\omega(v_i, v_j) = J_{ij}$, the i, j^{th} entry of J. A symplectic basis is a basis so that the matrix of ω is just J. The standard basis $e_1, \dots, e_{2n}$, where e_i is 1 in the ith position and zero elsewhere, is a symplectic basis for $(\mathbb{F}^{2n}, J)$. Given two symplectic spaces $(\mathbb{V}_i, \omega_i), i = 1, 2$, a symplectic isomorphism or an isomorphism is a linear isomorphism $L : \mathbb{V}_1 \to \mathbb{V}_2$ such that $\omega_2(L(x), L(y)) = \omega_1(x, y)$ for all $x, y \in \mathbb{V}_1$; that is, L preserves the symplectic form. In this case we say that the two spaces are symplectically isomorphic or symplectomorphic.

Corollary 2.2.1. *Let $(\mathbb{V}, \omega)$ be a symplectic space of dimension $2n$. Then $\mathbb{V}$ has a symplectic basis. $(\mathbb{V}, \omega)$ is symplectically isomorphic to $(\mathbb{F}^{2n}, J)$, or all symplectic spaces of dimension $2n$ are isomorphic.*

Proof. By Theorem 2.2.1 there is a basis for $\mathbb{V}$ such that the matrix of ω is $\mathrm{diag}(K, \ldots, K)$. Interchanging rows $2i$ and $n + 2i - 1$ and the corresponding columns brings the matrix to J. The basis for which the matrix of ω is J is a symplectic basis; so, a symplectic basis exists.

Let $v_1, \ldots, v_{2n}$ be a symplectic basis for $\mathbb{V}$ and $u \in \mathbb{V}$. There exist constants α_i such that $u = \sum \alpha_i v_i$. The linear map $L : \mathbb{V} \to \mathbb{F}^{2n} : u \to (\alpha_1, \ldots, \alpha_{2n})$ is the desired symplectic isomorphism.

The study of symplectic linear spaces is really the study of one canonical example, e.g., $(\mathbb{F}^{2n}, J)$. Or put another way, J is just the coefficient matrix of the symplectic form in a symplectic basis. This is one answer to the question "What is J?".

If $\mathbb{V}$ is a vector space over $\mathbb{F}$, then f is a linear functional if $f : \mathbb{V} \to \mathbb{F}$ is linear, $f(\alpha u + \beta v) = \alpha f(u) + \beta f(v)$ for all $u, v \in \mathbb{V}$, and $\alpha, \beta \in \mathbb{F}$. Linear functionals are sometimes called 1-forms or covectors. If $\mathbb{E}$ is the vector space of displacements of a particle in Euclidean space, then the work done by a force on a particle is a linear functional on $\mathbb{E}$. The usual geometric representation for a vector in $\mathbb{E}$ is a directed line segment. Represent a linear functional by showing its level planes. The value of the linear functional f on a vector v is represented by the number of level planes the vector crosses. The more level planes the vector crosses, the larger is the value of f on v.

The set of all linear functionals on a space $\mathbb{V}$ is itself a vector space when addition and scalar multiplication are just the usual addition and scalar multiplication of functions. That is, if f and f' are linear functionals on $\mathbb{V}$ and $\alpha \in \mathbb{F}$, then define the linear functionals $f + f'$ and αf by the formulas $(f + f')(v) = f(v) + f'(v)$ and $(\alpha f)(v) = \alpha f(v)$. The space of all linear functionals is called the dual space (to $\mathbb{V}$) and is denoted by $\mathbb{V}^*$.

When $\mathbb{V}$ is finite dimensional so is $\mathbb{V}^*$ with the same dimension. Let $u_1, \ldots, u_m$ be a basis for $\mathbb{V}$; then for any $v \in \mathbb{V}$, there are scalars $f^1, \ldots, f^m$ such that $v = f^1 u_1 + \cdots + f^m u_m$. The f^i are functions of v so we write $f^i(v)$, and they are linear. It is not too hard to show that $f^1, \ldots, f^m$ forms a basis for $\mathbb{V}^*$; this basis is called the dual basis (dual to $u_1, \ldots, u_m$). The defining property of this basis is $f^i(u_j) = \delta^i_j$ (the Kronecker delta function, defined by $\delta^i_j = 1$ if $i = j$ and zero otherwise).

If $\mathbb{W}$ is a subspace of $\mathbb{V}$ of dimension r, then define $\mathbb{W}^0 = \{f \in \mathbb{V}^* : f(e) = 0 \text{ for all } e \in \mathbb{W}\}$. $\mathbb{W}^0$ is called the annihilator of $\mathbb{W}$ and is easily shown to be a subspace of $\mathbb{V}^*$ of dimension $m - r$. Likewise, if $\mathbb{W}$ is a subspace of $\mathbb{V}^*$ of dimension r then $\mathbb{W}^0 = \{e \in \mathbb{V} : f(e) = 0 \text{ for all } f \in \mathbb{W}^*\}$ is a subspace of $\mathbb{V}$

of dimension $m - r$. Also $\mathbb{W}^{00} = \mathbb{W}$. See any book on vector space theory for a complete discussion of dual spaces with proofs.

Because ω is a bilinear form, for each fixed $v \in \mathbb{V}$ the function $\omega(v, \cdot)$: $\mathbb{V} \to \mathbb{R}$ is a linear functional and so is in the dual space $\mathbb{V}^*$. Because ω is nondegenerate, the map $\flat : \mathbb{V} \to \mathbb{V}^* : v \to \omega(v, \cdot) = v^\flat$ is an isomorphism. Let $\# : \mathbb{V}^* \to \mathbb{V} : v \to v^\#$ be the inverse of $\flat$. Sharp, $\#$, and flat, $\flat$, are musical symbols for raising and lowering notes and are used here because these isomorphisms are index raising and lowering operations in the classical tensor notation.

Let $\mathbb{U}$ be a subspace of $\mathbb{V}$. Define $\mathbb{U}^\perp = \{ v \in \mathbb{V} : \omega(v, \mathbb{U}) = 0 \}$. Clearly $\mathbb{U}^\perp$ is a subspace, $\{ \mathbb{U}, \mathbb{U}^\perp \} = 0$ and $\mathbb{U} = \mathbb{U}^{\perp\perp}$.

Lemma 2.2.1. $\mathbb{U}^\perp = \mathbb{U}^{0\#}$. $\dim \mathbb{U} + \dim \mathbb{U}^\perp = \dim \mathbb{V} = 2n$.

Proof.

$$\begin{aligned}
\mathbb{U}^\perp &= \{ x \in \mathbb{V} : \omega(x, y) = 0 \text{ for all } y \in \mathbb{U} \} \\
&= \{ x \in \mathbb{V} : x^\flat(y) = 0 \text{ for all } y \in \mathbb{U} \} \\
&= \{ x \in \mathbb{V} : x^\flat \in \mathbb{U}^0 \} \\
&= \mathbb{U}^{0\#}.
\end{aligned}$$

The second statement follows from $\dim \mathbb{U} + \dim \mathbb{U}^0 = \dim \mathbb{V}$ and the fact that $\#$ is an isomorphism.

A symplectic subspace $\mathbb{U}$ of $\mathbb{V}$ is a subspace such that ω restricted to this subspace is nondegenerate. By necessity $\mathbb{U}$ must be of even dimension, and so, $(\mathbb{U}, \omega)$ is a symplectic space.

Proposition 2.2.1. *If $\mathbb{U}$ is symplectic, then so is $\mathbb{U}^\perp$, and $\mathbb{V} = \mathbb{U} \oplus \mathbb{U}^\perp$. Conversely, if $\mathbb{V} = \mathbb{U} \oplus \mathbb{W}$ and $\omega(\mathbb{U}, \mathbb{W}) = 0$, then $\mathbb{U}$ and $\mathbb{W}$ are symplectic.*

Proof. Let $x \in \mathbb{U} \cap \mathbb{U}^\perp$; so, $\omega(x, y) = 0$ for all $y \in \mathbb{U}$, but $\mathbb{U}$ is symplectic so $x = 0$. Thus $\mathbb{U} \cap \mathbb{U}^\perp = 0$. This, with Lemma 2.2.1, implies $\mathbb{V} = \mathbb{U} \oplus \mathbb{U}^\perp$.

Now let $\mathbb{V} = \mathbb{U} \oplus \mathbb{W}$ and $\omega(\mathbb{U}, \mathbb{W}) = 0$. If ω is degenerate on $\mathbb{U}$, then there is an $x \in \mathbb{U}$, $x \neq 0$, with $\omega(x, \mathbb{U}) = 0$. Because $\mathbb{V} = \mathbb{U} \oplus \mathbb{W}$ and $\omega(\mathbb{U}, \mathbb{W}) = 0$, this implies $\omega(x, \mathbb{V}) = 0$ or that ω is degenerate on all of $\mathbb{V}$. This contradiction yields the second statement.

The next lemma is a technical result used much later.

Lemma 2.2.2. *Let $\mathbb{U}$ be a subspace of $\mathbb{V}^*$ and $\mathbb{Z} = \mathbb{U}^0 / (\mathbb{U}^\# \cap \mathbb{U}^0)$. For $[x], [y] \in \mathbb{Z}$ the bilinear form given by $\{ [x], [y] \} = \{ x, y \}$ is a well-defined symplectic inner product on $\mathbb{Z}$.*

Proof. If $v \in \mathbb{U}^{\#}$ and $u \in \mathbb{U}^0$, then $\{v, u\} = 0$ by definition. Thus if $x, y \in \mathbb{U}^0$ and $\xi, \psi \in \mathbb{U}^{\#} \cap \mathbb{U}^0$ one has

$$\{[x + \xi], [y + \psi]\} = \{x + \xi, y + \psi\} = \{x, y\} = \{[x], [y]\},$$

so the form is well defined.

Now assume that $\{[x], [y]\} = 0$ for all $[y] \in \mathbb{Z}$. Then $\{x, y\} = 0$ for all $y \in \mathbb{U}^0$, or $\{x, \cdot\} \in \mathbb{U}$ and thus $x \in \mathbb{U}^{\#}$ or $[x] = 0$. Thus the form is nondegenerate on $\mathbb{Z}$.

A Lagrangian space $\mathbb{U}$ is a subspace of $\mathbb{V}$ of dimension n such that ω is zero on $\mathbb{U}$, i.e., $\omega(u, w) = 0$ for all $u, w \in \mathbb{U}$. A direct sum decomposition $\mathbb{V} = \mathbb{U} \oplus \mathbb{W}$ where $\mathbb{U}$, and $\mathbb{W}$ are Lagrangian spaces, is called a Lagrangian splitting, and $\mathbb{W}$ is called the Lagrangian complement of $\mathbb{U}$. In $\mathbb{R}^2$ any line through the origin is Lagrangian, and any other line through the origin is a Lagrangian complement.

Lemma 2.2.3. *Let $\mathbb{U}$ be a Lagrangian subspace of $\mathbb{V}$, then there exists a Lagrangian complement of $\mathbb{U}$.*

Proof. The example above shows the complement is not unique. Let $\mathbb{V} = \mathbb{F}^{2n}$ and $\mathbb{U} \subset \mathbb{F}^{2n}$. Then $\mathbb{W} = J\mathbb{U}$ is a Lagrangian complement to $\mathbb{U}$. If $x, y \in \mathbb{W}$ then $x = Ju, y = Jv$ where $u, v \in \mathbb{U}$, or $\{u, v\} = 0$. But $\{x, y\} = \{Ju, Jv\} = \{u, v\} = 0$, so $\mathbb{W}$ is Lagrangian. If $x \in \mathbb{U} \cap J\mathbb{U}$ then $x = Jy$ with $y \in \mathbb{U}$. So $x, Jx \in \mathbb{U}$ and so $\{x, Jx\} = -\|x\|^2 = 0$ or $x = 0$. Thus $\mathbb{U} \cap \mathbb{W} = \phi$.

Lemma 2.2.4. *Let $\mathbb{V} = \mathbb{U} \oplus \mathbb{W}$ be a Lagrange splitting and $x_1, \ldots, x_n$ any basis for $\mathbb{U}$. Then there exists a unique basis $y_1, \ldots, y_n$ of $\mathbb{W}$ such that $x_1, \ldots, x_n, y_1, \ldots, y_n$ is a symplectic basis for $\mathbb{V}$.*

Proof. Define $\phi_i \in \mathbb{W}^0$ by $\phi_i(w) = \omega(x_i, w)$ for $w \in \mathbb{W}$. If $\sum \alpha_i \phi_i = 0$, then $\omega(\sum \alpha_i x_i, w) = 0$ for all $w \in \mathbb{W}$ or $\omega(\sum \alpha_i x_i, \mathbb{W}) = 0$. But because $\mathbb{V} = \mathbb{U} \oplus \mathbb{W}$ and $\omega(\mathbb{U}, \mathbb{U}) = 0$, it follows that $\omega(\sum \alpha_i x_i, \mathbb{V}) = 0$. This implies $\sum \alpha_i x_i = 0$, because ω is nondegenerate, and this implies $\alpha_i = 0$, because the $x_i s$ are independent. Thus $\phi_1, \ldots, \phi_n$ are independent, and so, they form a basis for $\mathbb{W}^0$. Let $y_1, \ldots, y_n$ be the dual basis in $\mathbb{W}$; so, $\omega(x_i, y_j) = \phi_i(y_j) = \delta_{ij}$.

A linear operator $L : \mathbb{V} \to \mathbb{V}$ is called Hamiltonian, if

$$\omega(Lx, y) + \omega(x, Ly) = 0 \tag{2.15}$$

for all $x, y \in \mathbb{V}$. A linear operator $L : \mathbb{V} \to \mathbb{V}$ is called symplectic, if

$$\omega(Lx, Ly) = \omega(x, y) \tag{2.16}$$

for all $x, y \in \mathbb{V}$. If $\mathbb{V}$ is the standard symplectic space $(\mathbb{F}^{2n}, J)$ and L is a matrix, then (2.15) means $x^T (L^T J + JL)y = 0$ for all x and y. But this implies that L is a Hamiltonian matrix. On the other hand, if L satisfies (2.16) then

$x^T L^T J L y = x^T J y$ for all x and y. But this implies L is a symplectic matrix. The matrix representation of a Hamiltonian (respectively, symplectic) linear operator in a symplectic coordinate system is a Hamiltonian (respectively, symplectic) matrix.

Lemma 2.2.5. *Let $\mathbb{V} = \mathbb{U} \oplus \mathbb{W}$ be a Lagrangian splitting and $A : \mathbb{V} \to \mathbb{V}$ a Hamiltonian (respectively, symplectic) linear operator that respects the splitting; i.e., $A : \mathbb{U} \to \mathbb{U}$ and $A : \mathbb{W} \to \mathbb{W}$. Choose any basis of the form given in Lemma 2.2.4; the matrix representation of A in these symplectic coordinates is of the form*

$$\begin{bmatrix} B^T & 0 \\ 0 & -B \end{bmatrix} \quad \left(\text{respectively,} \quad \begin{bmatrix} B^T & 0 \\ 0 & B^{-1} \end{bmatrix} \right). \tag{2.17}$$

Proof. A respects the splitting and the basis for $\mathbb{V}$ is the union of the bases for $\mathbb{U}$ and $\mathbb{W}$, therefore the matrix representation for A must be in block-diagonal form. A Hamiltonian or symplectic matrix which is in block-diagonal form must be of the form given in (2.17).

2.3 Canonical Forms

In this section we obtain some canonical forms for Hamiltonian and symplectic matrices in some simple cases. The complete picture is very detailed and special, so see Sections 5.3 and 5.4 if more extensive material is needed. We start with only real matrices, but sometimes we need to go into the complex domain to finish the arguments. We simply assume that all our real spaces are embedded in a complex space of the same dimension.

If A is Hamiltonian and T is symplectic, then $T^{-1}AT$ is Hamiltonian also. Thus if we start with a linear constant coefficient Hamiltonian system $\dot{z} = Az$ and make the change of variables $z = Tu$, then in the new coordinates the equations become $\dot{u} = (T^{-1}AT)u$, which is again Hamiltonian. If $B = T^{-1}AT$, where T is symplectic, then we say that A and B are symplectically similar. This is an equivalence relation. We seek canonical forms for Hamiltonian and symplectic matrices under symplectic similarity. In as much as it is a form of similarity transformation, the eigenvalue structure plays an important role in the following discussion.

Because symplectic similarity is more restrictive than ordinary similarity, one should expect more canonical forms than the usual Jordan canonical forms. Consider, for example, the two Hamiltonian matrices

$$A_1 = \begin{bmatrix} 0 & 1 \\ -1 & 0 \end{bmatrix} \quad \text{and} \quad A_2 = \begin{bmatrix} 0 & -1 \\ 1 & 0 \end{bmatrix} \tag{2.18}$$

both of which could be the coefficient matrix of a harmonic oscillator. In fact, they are both the real Jordan forms for the harmonic oscillator.

The reflection $T = \text{diag}(1, -1)$ defines a similarity between these two; i.e., $T^{-1}A_1T = A_2$. The determinant of T is not $+1$, therefore T is not symplectic. In fact, A_1 and A_2 are not symplectically equivalent. If $T^{-1}A_1T = A_2$, then $T^{-1}\exp(A_1t)T = \exp(A_2t)$, and T would take the clockwise rotation $\exp(A_1t)$ to the counterclockwise rotation $\exp(A_2t)$. But, if T were symplectic, its determinant would be $+1$ and thus would be orientation preserving. Therefore, T cannot be symplectic.

Another way to see that the two Hamiltonian matrices in (2.18) are not symplectically equivalent is to note that $A_1 = JI$ and $A_2 = J(-I)$. So the symmetric matrix corresponding to A_1 is I, the identity, and to A_2 is $-I$. I is positive definite, whereas $-I$ is negative definite. If A_1 and A_2 where symplectically equivalent, then I and $-I$ would be congruent, which is clearly false.

A polynomial $p(\lambda) = a_m\lambda^m + a_{m-1}\lambda^{m-1} + \cdots + a_0$ is even if $p(-\lambda) = p(\lambda)$, which is the same as $a_k = 0$ for all odd k. If λ_0 is a zero of an even polynomial, then so is $-\lambda_0$; therefore, the zeros of a real even polynomial are symmetric about the real and imaginary axes. The polynomial $p(\lambda)$ is a reciprocal polynomial if $p(\lambda) = \lambda^m p(\lambda^{-1})$, which is the same as $a_k = a_{m-k}$ for all k. If λ_0 is a zero of a reciprocal polynomial, then so is λ_0^{-1}; therefore, the zeros of a real reciprocal polynomial are symmetric about the real axis and the unit circle (in the sense of inversion).

Proposition 2.3.1. *The characteristic polynomial of a real Hamiltonian matrix is an even polynomial. Thus if λ is an eigenvalue of a Hamiltonian matrix, then so are $-\lambda$, $\overline{\lambda}$, $-\overline{\lambda}$.*

The characteristic polynomial of a real symplectic matrix is a reciprocal polynomial. Thus if λ is an eigenvalue of a real symplectic matrix, then so are λ^{-1}, $\overline{\lambda}$, $\overline{\lambda}^{-1}$

Proof. Recall that $\det J = 1$. Let A be a Hamiltonian matrix; then $p(\lambda) = \det(A - \lambda I) = \det(JA^TJ - \lambda I) = \det(JA^TJ + \lambda JJ) = \det J \det(A + \lambda I) \det J = \det(A + \lambda I) = p(-\lambda)$.

Let T be a symplectic matrix; by Theorem 2.1.7 $\det T = +1$. $p(\lambda) = \det(T - \lambda I) = \det(T^T - \lambda I) = \det(-JT^{-1}J - \lambda I) = \det(-JT^{-1}J + \lambda JJ) = \det(-T^{-1} + \lambda I) = \det T^{-1} \det(-I + \lambda T) = \lambda^{2n}\det(-\lambda^{-1}I + T) = \lambda^{2n}p(\lambda^{-1})$.

Actually we can prove much more. By (2.6), Hamiltonian matrix A satisfies $A = J^{-1}(-A^T)J$; so, A and $-A^T$ are similar, and the multiplicity of the eigenvalues λ_0 and $-\lambda_0$ are the same. In fact, the whole Jordan block structure will be the same for λ_0 and $-\lambda_0$.

By (2.8), symplectic matrix T satisfies $T^{-1} = J^{-1}T^TJ$; so, T^{-1} and T^T are similar, and the multiplicity of the eigenvalues λ_0 and λ_0^{-1} are the same. The whole Jordan block structure will be the same for λ_0 and λ_0^{-1}.

Consider the linear constant coefficient Hamiltonian system of differential equations

$$\dot{x} = Ax, \tag{2.19}$$

where A is a Hamiltonian matrix and $Z(t) = e^{At}$ is the fundamental matrix solution. By the above it is impossible for all the eigenvalues of A to be in the left half-plane, and, therefore, it is impossible for all the solutions to be exponentially decaying. Thus the origin cannot be asymptotically stable.

Henceforth, let A be a real Hamiltonian matrix and T a real symplectic matrix. First we develop the theory for Hamiltonian matrices and then the theory of symplectic matrices. Because eigenvalues are sometimes complex, it is necessary to consider complex matrices at times, but we are always be concerned with the real answers in the end.

First consider the Hamiltonian case. Let λ be an eigenvalue of A, and define subspaces of $\mathbb{C}^{2n}$ by $\eta_k(\lambda) = \text{kernel}\,(A - \lambda I)^k$, $\eta^\dagger(\lambda) - \cup_1^{2n}\eta_k(\lambda)$. The eigenspace of A corresponding to the eigenvalue λ is $\eta(\lambda) = \eta_1(\lambda)$, and the generalized eigenspace is $\eta^\dagger(\lambda)$. If $\{x, y\} = x^T Jy = 0$, then x and y are J-orthogonal.

Lemma 2.3.1. *Let λ and μ be eigenvalues of A with $\lambda + \mu \neq 0$, then $\{\eta(\lambda), \eta(\mu)\} = 0$. That is, the eigenvectors corresponding to λ and μ are J-orthogonal.*

Proof. Let $Ax = \lambda x$, and $Ay = \mu y$, where $x, y \neq 0$. $\lambda\{x, y\} = \{Ax, y\} = x^T A^T Jy = -x^T JAy = -\{x, Ay\} = -\mu\{x, y\}$; and so, $(\lambda + \mu)\{x, y\} = 0$.

Corollary 2.3.1. *Let A be a $2n \times 2n$ Hamiltonian matrix with distinct eigenvalues $\lambda_1, \ldots, \lambda_n, -\lambda_1, \ldots, -\lambda_n$; then there exists a symplectic matrix S (possibly complex) such that $S^{-1}AS = \text{diag}(\lambda_1, \ldots, \lambda_n, -\lambda_1, \ldots, -\lambda_n)$.*

Proof. $\mathbb{V} = \mathbb{U} \oplus \mathbb{W}$ is a Lagrange splitting where

$$\mathbb{U} = \eta_1(\lambda_1) \oplus \cdots \oplus \eta_1(\lambda_n), \quad \mathbb{W} = \eta_1(-\lambda_1) \oplus \cdots \oplus \eta_1(-\lambda_n),$$

and A respects this splitting. Choose a symplectic basis for $\mathbb{V}$ by Lemma 2.2.4. Changing to that basis is effected by a symplectic matrix G; i.e., $G^{-1}AG = \text{diag}(B^T, -B)$, where B has eigenvalues $\lambda_1, \ldots, \lambda_n$. Let C be such that $C^{-T}B^T C^T = \text{diag}(\lambda_1, \ldots, \lambda_n)$ and define a symplectic matrix by $Q = \text{diag}(C^T, C^{-1})$. The required symplectic matrix is $S = GQ$.

If complex transformations are allowed, then the two matrices in (2.18) can both be brought to $\text{diag}(i, -i)$ by a symplectic similarity, and thus one is symplectically similar to the other. However, as we have seen they are not similar by a real symplectic similarity. Let us investigate the real case in detail.

A subspace $\mathbb{U}$ of $\mathbb{C}^n$ is called a *complexification* (of a real subspace) if $\mathbb{U}$ has a real basis. If $\mathbb{U}$ is a complexification, then there is a real basis $x_1, \ldots, x_k$ for $\mathbb{U}$, and for any $u \in \mathbb{U}$, there are complex numbers $\alpha_1, \ldots, \alpha_k$ such that $u = \alpha_1 x_1 + \cdots + \alpha_n x_n$. But then $\overline{u} = \overline{\alpha}_1 x_1 + \cdots + \overline{\alpha}_n x_n \in \mathbb{U}$ also.

Conversely, if $\mathbb{U}$ is a subspace such that $u \in \mathbb{U}$ implies $\overline{u} \in \mathbb{U}$, then $\mathbb{U}$ is a complexification. Because if $x_1, \ldots, x_k$ is a complex basis with $x_j = u_j + v_j i$, then $u_j = (x_j + \overline{x}_j)/2$ and $v_j = (x_j - \overline{x}_j)/2i$ are in $\mathbb{U}$, and the totality of

$u_1, \ldots, u_k, v_1, \ldots, v_k$ span $\mathbb{U}$. From this real spanning set, one can extract a real basis. Thus $\mathbb{U}$ is a complexification if and only if $\mathbb{U} = \overline{\mathbb{U}}$ (i.e., $u \in \mathbb{U}$ implies $\overline{u} \in \mathbb{U}$).

Until otherwise said let A be a real Hamiltonian matrix with distinct eigenvalues $\lambda_1, \ldots, \lambda_n, -\lambda_1, \ldots, -\lambda_n$ so 0 is not an eigenvalue. The eigenvalues of A fall into three groups: (1) the real eigenvalues $\pm\alpha_1, \ldots, \pm\alpha_s$, (2) the pure imaginary $\pm\beta_1 i, \ldots, \pm\beta_r i$, and (3) the truly complex $\pm\gamma_1 \pm \delta_1 i, \ldots, \pm\gamma_t \pm \delta_t i$. This defines a direct sum decomposition

$$\mathbb{V} = (\oplus_j \mathbb{U}_j) \oplus (\oplus_j \mathbb{W}_j) \oplus (\oplus_j \mathbb{Z}_j), \tag{2.20}$$

where

$$\mathbb{U}_j = \eta(\alpha_j) \oplus \eta(-\alpha_j)$$

$$\mathbb{W}_j = \eta(\beta_j i) \oplus \eta(-\beta_j i)$$

$$\mathbb{Z}_j = \{\eta(\gamma_j + \delta_j i) \oplus \eta(\gamma_j - \delta_j i)\} \oplus \{\eta(-\gamma_j - \delta_j i) \oplus \eta(-\gamma_j + \delta_j i)\}.$$

Each of the summands in the above is an invariant subspace for A. By Lemma 2.3.1, each space is J-orthogonal to every other, and so by Proposition 2.2.1 each space must be a symplectic subspace. Because each subspace is invariant under complex conjugation, each is the complexification of a real space. Thus we can choose symplectic coordinates for each of the spaces, and A in these coordinates would be block diagonal. Therefore, the next task is to consider each space separately.

Lemma 2.3.2. *Let A be a 2×2 Hamiltonian matrix with eigenvalues $\pm\alpha$, α real and positive. Then there exists a real 2×2 symplectic matrix S such that*

$$S^{-1}AS = \begin{bmatrix} \alpha & 0 \\ 0 & -\alpha \end{bmatrix}. \tag{2.21}$$

The Hamiltonian is $H = \alpha\xi\eta$.

Proof. Let $Ax = \alpha x$, and $Ay = -\alpha y$, where x and y are nonzero. Because x and y are eigenvectors corresponding to different eigenvalues, they are independent. Thus $\{x, y\} \neq 0$. Let $u = \{x, y\}^{-1} y$: so, x, u is a real symplectic basis, $S = [x, u]$ is a real symplectic matrix, and S is the matrix of the lemma.

Lemma 2.3.3. *Let A be a real 2×2 Hamiltonian matrix with eigenvalues $\pm\beta i$, $\beta > 0$. Then there exists a real 2×2 symplectic matrix S such that*

$$S^{-1}AS = \begin{bmatrix} 0 & \beta \\ -\beta & 0 \end{bmatrix}, \quad or \quad S^{-1}AS = \begin{bmatrix} 0 & -\beta \\ \beta & 0 \end{bmatrix}. \tag{2.22}$$

The Hamiltonian is $H = \frac{\beta}{2}(\xi^2 + \eta^2)$ or $H = -\frac{\beta}{2}(\xi^2 + \eta^2)$.

Proof. Let $Ax = i\beta x$, and $x = u + vi \neq 0$. So $Au = -\beta v$ and $Av = \beta u$. Because $u + iv$ and $u - iv$ are independent, u and v are independent. Thus $\{u, v\} = \delta \neq 0$. If $\delta = \gamma^2 > 0$, then define $S = [\gamma^{-1}u, \gamma^{-1}v]$ to get the first option in (2.22), or if $\delta = -\gamma^2 < 0$, then define $S = [\gamma^{-1}v, \gamma^{-1}u]$ to get the second option.

Sometimes it is more advantageous to have a diagonal matrix than to have a real one; yet you want to keep track of the real origin of the problem. This is usually accomplished by reality conditions as defined in the next lemma.

Lemma 2.3.4. *Let A be a real 2×2 Hamiltonian matrix with eigenvalues $\pm\beta i$, $\beta \neq 0$. Then there exist a 2×2 matrix S and a matrix R such that*

$$S^{-1}AS = \begin{bmatrix} i\beta & 0 \\ 0 & -i\beta \end{bmatrix}, \quad R = \begin{bmatrix} 0 & 1 \\ 1 & 0 \end{bmatrix}, \quad S^T J S = \pm 2iJ, \quad \overline{S} = SR. \quad (2.23)$$

Proof. Let $Ax = i\beta x$, where $x \neq 0$. Let $x = u + iv$ as in the above lemma. Compute $\{x, \overline{x}\} = 2i\{v, u\} \neq 0$. Let $\gamma = 1/\sqrt{|\{u, v\}|}$ and $S = [\gamma x, \gamma \overline{x}]$.

If S satisfies (2.23), then S is said to satisfy reality conditions with respect to R. The matrix S is no longer a symplectic matrix but is what is called a symplectic matrix with multiplier $\pm 2i$. We discuss these types of matrices later. The matrix R is used to keep track of the fact that the columns of S are complex conjugates. We could require $S^T J S = +2iJ$ by allowing an interchange of the signs in (2.23).

Lemma 2.3.5. *Let A be a 2×2 Hamiltonian matrix with eigenvalue 0. Then there exists a real 2×2 symplectic matrix S such that*

$$S^{-1}AS = \begin{bmatrix} 0 & \delta \\ 0 & 0 \end{bmatrix} \quad (2.24)$$

where $\delta = +1$, -1, or 0.
   The Hamiltonian is $H = \delta\eta^2$.

Lemma 2.3.6. *Let A be a 4×4 Hamiltonian matrix with eigenvalue $\pm\gamma \pm \delta i$, $\gamma \neq 0$, $\delta \neq 0$. Then there exists a real 4×4 symplectic matrix S such that*

$$S^{-1}AS = \begin{bmatrix} B^T & 0 \\ 0 & -B \end{bmatrix},$$

where B is a real 2×2 matrix with eigenvalues $+\gamma \pm \delta i$.

Proof. $\mathbb{U} = \eta(\gamma_j + \delta_j i) \oplus \eta(\gamma_j - \delta_j i)$ is the complexification of a real subspace and by Lemma 2.3.1 is Lagrangian. A restricted to this subspace has eigenvalues $+\gamma \pm \delta i$. A complement to $\mathbb{U}$ is $\mathbb{W} = \eta(-\gamma_j + \delta_j i) \oplus \eta(-\gamma_j - \delta_j i)$. Choose any real basis for $\mathbb{U}$ and complete it by Lemma 2.2.5. The result follows from Lemma 2.2.5.

In particular you can choose coordinates so that B is in real Jordan form; so,

$$B = \begin{bmatrix} \gamma & \delta \\ -\delta & \gamma \end{bmatrix}.$$

There are many cases when $n > 1$ and A has eigenvalues with zero real part; i.e., zero or pure imaginary. These cases are discussed in detail in Section 5.4. In the case where the eigenvalue zero is of multiplicity 4 the canonical forms are

$$\begin{bmatrix} 0 & 1 & 0 & 0 \\ 0 & 0 & 0 & 0 \\ 0 & 0 & 0 & 0 \\ 0 & 0 & -1 & 0 \end{bmatrix}, \quad \begin{bmatrix} 0 & 1 & 0 & 0 \\ 0 & 0 & 0 & \pm 1 \\ 0 & 0 & 0 & 0 \\ 0 & 0 & -1 & 0 \end{bmatrix}. \tag{2.25}$$

The corresponding Hamiltonians are

$$\xi_2 \eta_1, \qquad \xi_2 \eta_1 \pm \eta_2^2/2.$$

In the case of a double eigenvalue $\pm \alpha i$, $\alpha \neq 0$, the canonical forms in the 4×4 case are

$$\begin{bmatrix} 0 & 0 & \alpha & 0 \\ 0 & 0 & 0 & \pm \alpha \\ -\alpha & 0 & 0 & 0 \\ 0 & \mp \alpha & 0 & 0 \end{bmatrix}, \quad \begin{bmatrix} 0 & \alpha & 0 & 0 \\ -\alpha & 0 & 0 & 0 \\ \pm 1 & 0 & 0 & \alpha \\ 0 & \pm 1 & -\alpha & 0 \end{bmatrix}. \tag{2.26}$$

The corresponding Hamiltonians are

$$(\alpha/2)(\xi_1^2 + \eta_1^2) \pm (\alpha/2)(\xi_2^2 + \eta_2^2), \qquad \alpha(\xi_2 \eta_1 - \xi_1 \eta_2) \mp (\xi_1^2 + \xi_2^2)/2.$$

Next consider the symplectic case. Let λ be an eigenvalue of T, and define subspaces of $\mathbb{C}^{2n}$ by $\eta_k(\lambda) = \text{kernel }(T - \lambda I)^k$, $\eta^\dagger(\lambda) = \cup_1^{2n} \eta_k(\lambda)$. The eigenspace of T corresponding to the eigenvalue λ is $\eta(\lambda) = \eta_1(\lambda)$, and the generalized eigenspace is $\eta^\dagger(\lambda)$. Because the proof of the next set of lemmas is similar to those given just before, the proofs are left as problems.

Lemma 2.3.7. *If λ and μ are eigenvalues of the symplectic matrix T such that $\lambda \mu \neq 1$; then $\{\eta(\lambda), \eta(\mu)\} = 0$. That is, the eigenvectors corresponding to λ and μ are J-orthogonal.*

Corollary 2.3.2. *Let T be a $2n \times 2n$ symplectic matrix with distinct eigenvalues $\lambda_1, \ldots, \lambda_n, \lambda_1^{-1}, \ldots, \lambda_n^{-1}$; then there exists a symplectic matrix S (possibly complex) such that*

$$S^{-1}TS = \text{diag}(\lambda_1, \ldots, \lambda_n, \lambda_1^{-1}, \ldots, \lambda_n^{-1}).$$

If complex transformations are allowed, then the two matrices

$$\begin{bmatrix} \alpha & \beta \\ -\beta & \alpha \end{bmatrix}, \quad \text{and} \quad \begin{bmatrix} \alpha & -\beta \\ \beta & \alpha \end{bmatrix}, \quad \alpha^2 + \beta^2 = 1,$$

can both be brought to $\mathrm{diag}(\alpha + \beta\mathrm{i}, \alpha - \beta\mathrm{i})$ by a symplectic similarity, and thus, one is symplectically similar to the other. However, they are not similar by a real symplectic similarity. Let us investigate the real case in detail.

Until otherwise said, let T be a real symplectic matrix with distinct eigenvalues $\lambda_1, \ldots, \lambda_n, \lambda_1^{-1}, \ldots, \lambda_n^{-1}$, so 1 is not an eigenvalue. The eigenvalues of T fall into three groups: (1) the real eigenvalues, $\mu_1^{\pm 1}, \ldots, \mu_s^{\pm 1}$, (2) the eigenvalues of unit modulus, $\alpha \pm \beta_1 i, \ldots, \alpha_r \pm \beta_r i$, and (3) the complex eigenvalues of modulus different from one, $(\gamma_1 \pm \delta_1 i)^{\pm 1}, \ldots, (\gamma_t \pm \delta_t i)^{\pm 1}$. This defines a direct sum decomposition

$$\mathbb{V} = (\oplus_j \mathbb{U}_j) \oplus (\oplus_j \mathbb{W}_j) \oplus (\oplus_j \mathbb{Z}_j), \qquad (2.27)$$

where

$$\mathbb{U}_j = \eta(\mu_j) \oplus \eta(\mu_j^{-1})$$

$$\mathbb{W}_j = \eta(\alpha_j + \beta_j i) \oplus \eta(\alpha_j - \beta_j i)$$

$$\mathbb{Z}_j = \{\eta(\gamma_j + \delta_j i) \oplus \eta(\gamma_j - \delta_j i)\} \oplus \{\eta(\gamma_j + \delta_j i)^{-1} \oplus \eta(\gamma_j - \delta_j i)^{-1}\}.$$

Each of the summands in (2.27) is invariant for T. By Lemma 2.3.7 each space is J-orthogonal to every other, and so each space must be a symplectic subspace. Because each subspace is invariant under complex conjugation, each is the complexification of a real space. Thus we can choose symplectic coordinates for each of the spaces, and T in these coordinates would be block diagonal. The 2×2 case will be considered in detail in the next section and the general case is postponed to Chapter 5.

2.4 $Sp(2, \mathbb{R})$

It is time to stop and look back at the simplest case in order to foreshadow complexity that is still to come. A 2×2 matrix T is symplectic when $\det T = +1$ which also means that T is in the special linear group $Sl(2, \mathbb{R})$. So in this lowest dimension $Sp(2, \mathbb{R}) = Sl(2, \mathbb{R})$ and these groups have some interesting properties, so much so that in 1985 the famed algebraist Serge Lang wrote a book whose sole title was $\mathbf{SL}_2(\mathbf{R})$.[2]

One of the topics of this chapter is devoted to the question of when two Hamiltonian or two symplectic matrices T and S are symplectic similar, i.e., when does there exist a symplectic matrix P such that $S = P^{-1}TP$. The Hamiltonian case has been discussed, so let us look at this question in more detail for 2×2 symplectic matrices.

Consider the action Φ of $Sp(2, \mathbb{R})$ on itself defined by

$$\Phi : Sp(2, \mathbb{R}) \times Sp(2, \mathbb{R}) \to Sp(2, \mathbb{R}) : (P, T) \mapsto P^{-1}TP,$$

---

[2]Lang's $\mathbf{SL}_2(\mathbf{R})$ is the same as our $Sl(2, \mathbb{R})$.

and the orbit of T is

$$\mathbb{O}(T) = \{S \in Sp(2, \mathbb{R}) : S = P^{-1}TP \text{ for some } P \in Sp(2, \mathbb{R})\}.$$

Our task is to assign to each orbit a unique angle $\Theta(T)$ defined modulo 2π.

Let us look at the canonical forms for each case as we did for Hamiltonian matrices in the last section. In the 2×2 case the characteristic polynomial is of the form $\lambda^2 - \text{tr}(T)\lambda + 1$ so the trace determines the eigenvalues. The hyperbolic case is when $|\text{tr}(T)| > 2$ and the eigenvalues are real and not equal to ± 1, the elliptic case is when $|\text{tr}(T)| < 2$ and the eigenvalues are complex conjugates, and the parabolic case is when $|\text{tr}(T)| = 2$.

Lemma 2.4.1 (Hyperbolic). *Let T be a 2×2 symplectic matrix with eigenvalues $\mu^{\pm 1}$, μ real and $\mu \neq 1$. Then there exists a real 2×2 symplectic matrix S such that*

$$S^{-1}TS = \begin{bmatrix} \mu & 0 \\ 0 & \mu^{-1} \end{bmatrix}.$$

Proof. Let $Tx = \mu x$ and $Ty = \mu^{-1}y$. Since x and y are eigenvectors corresponding to different eigenvalues they are independent and thus $\{x, y\} \neq 0$. Let $u = \{x, y\}^{-1}y$ so $S = [x, u]$ is the symplectic matrix of the lemma.

In this case define $\Theta(T) = 0$ if $\mu > 0$ and $\Theta(T) = \pi$ if $\mu < 0$.

Lemma 2.4.2 (Elliptic). *Let T be a real 2×2 symplectic matrix with eigenvalues $\alpha \pm \beta i$, with $\alpha^2 + \beta^2 = 1$, and $\beta > 0$. Then there exists a real 2×2 symplectic matrix S such that*

$$S^{-1}TS = \begin{bmatrix} \alpha & \beta \\ -\beta & \alpha \end{bmatrix} \quad or \quad S^{-1}TS = \begin{bmatrix} \alpha & -\beta \\ \beta & \alpha \end{bmatrix}. \qquad (2.28)$$

Proof. Let $x = u + iv$ be an eigenvector corresponding to $\lambda = \alpha + i\beta$, so $Tu = \alpha u - \beta v$ and $Tv = \alpha v + \beta u$. Since $u + iv$ and $u - iv$ are independent so are u and v so $\{u, v\} = \delta \neq 0$. If $\delta = \gamma^2 > 0$ then define $S = [\gamma^{-1}u, \gamma^{-1}v]$ to get the first option. If $\delta = -\gamma^2 < 0$ then define $S = [\gamma^{-1}v, \gamma^{-1}v)]$ to get the second option.

Note that these two matrices in (2.28) are orthogonal/rotation matrices. In this case the eigenvalues are of the form $e^{\pm i\theta} = \alpha \pm \beta i$ and for a general matrix one cannot choose the plus or the minus sign, but here one can. The image of $(1, 0)^T$ by the second matrix in (2.28) is $(\alpha, \beta)^T$ which is in the upper half-plane so one should take $0 < \theta < \pi$ modulo 2π for this second matrix. Likewise for the first matrix one should take $-\pi < \theta < 0$ modulo 2π. Thus we have define $\Theta(T) = \theta$ in this case.

Since a symplectic change of variables is orientation preserving the two matrices is (2.28) are not symplectically similar.

Lemma 2.4.3 (Parabolic). *Let T be a real 2×2 symplectic matrix with eigenvalue $+1$ (respectively -1). Then there exists a real 2×2 symplectic matrix S such that*

$$S^{-1}TS = \begin{bmatrix} 1 & \delta \\ 0 & 1 \end{bmatrix}, \quad \left(\text{respectively,} \quad S^{-1}TS = \begin{bmatrix} -1 & -\delta \\ 0 & -1 \end{bmatrix} \right) \qquad (2.29)$$

where $\delta = +1, -1,$ or 0.

Proof. First let the eigenvalue be $+1$. If T is the identity matrix then clearly $\delta = 0$. Assume otherwise so T has the repeated eigenvalue 1 and its standard Jordan canonical form is $\begin{bmatrix} 1 & 1 \\ 0 & 1 \end{bmatrix}$. In other words there are independent vectors x, y such that $Ax = x$, $Ay = x + y$. Since they are independent $\{x, y\} = \delta\gamma^2$ where $\delta = 1$, or -1 and $\gamma \neq 0$.

Define $u = \delta\gamma^{-1}x$, $v = \gamma^{-1}y$, so $\{u, v\} = 1$ and T in these coordinates is of the form in the Lemma.

The eigenvalue -1 is similar.

In this case define $\Theta(T) = 0$ if T is the identity matrix, $\Theta(T) = \pi$ if $\delta = +1$ and $\Theta(T) = -\pi$ if $\delta = -1$

Each $T \in Sp(2, \mathbb{R})$ lies in one of the orbits defined in the above lemmas, so Θ is defined and continuous on $Sp(2, \mathbb{R})$.

Except in the parabolic case when the eigenvalues are ± 1 the orbits are closed. When $\delta \neq 0$ the orbit of $\begin{bmatrix} \pm 1 & \delta \\ 0 & \pm 1 \end{bmatrix}$ is not closed but has $\pm I$ (I the identity matrix) as a limit.

Let $\Gamma : [0, 1] \to Sp(2, \mathbb{R})$ be a smooth path of symplectic matrices with $\Gamma(0) = I$, the identity matrix. Let

$$\gamma : [0, 1] \to S^1 : t \mapsto \exp(i\Theta(\Gamma(t)))$$

where S^1 the unit circle in $\mathbb{C}$. Now lift this map to the universal cover $\mathbb{R}$ of $S^1 = \mathbb{R}/\mathbb{Z}$ to get the map $\bar{\gamma} : [0, 1] \to \mathbb{R}$ with the property that $\bar{\gamma}(0) = 0$. Now define the rotation of the path Γ by

$$\text{rot}(\Gamma) = \bar{\gamma}(1).$$

Note that if Γ^* is another such path with the same end points which is homotopic to Γ then $\text{rot}(\Gamma) = \text{rot}(\Gamma^*)$.

Let $\lfloor r \rfloor = \max\{n \in \mathbb{Z} : n \leq r\}$ be the floor function. The path Γ is nondegenerate if $\Gamma(1)$ does not have the eigenvalue 1. For a nondegenerate path Γ we define its Conley-Zehnder index $\chi = \chi(\Gamma)$ as follows
  (i) If $|\text{tr}(\Gamma(1))| \leq 2$, define $\chi = 2\lfloor \text{rot}(\Gamma(1)) \rfloor + 1$,
  (ii) If $|\text{tr}(\Gamma(1))| > 2$, define $\chi = 2\lfloor \text{rot}(\Gamma(1)) \rfloor$.

2.5 Floquet-Lyapunov Theory

In this section we introduce some of the vast theory of periodic Hamiltonian systems. A classical detailed discussion of periodic systems can be found in the two-volume set by Yakubovich and Starzhinskii (1975).

Consider a periodic, linear Hamiltonian system

$$\dot{z} = J\frac{\partial H}{\partial z} = JS(t)z = A(t)z, \tag{2.30}$$

where

$$H = H(t, z) = \frac{1}{2}z^T S(t)z, \tag{2.31}$$

and $A(t) = JS(t)$. Assume that A and S are continuous and T-periodic; i.e.

$$A(t + T) = A(t), \quad S(t + T) = S(t) \quad \text{for all } t \in \mathbb{R}$$

for some fixed $T > 0$. The Hamiltonian, H, is a quadratic form in z with coefficients which are continuous and T-periodic in $t \in \mathbb{R}$. Let $Z(t)$ be the fundamental matrix solution of (2.30) that satisfies $Z(0) = I$.

Lemma 2.5.1. $Z(t + T) = Z(t)Z(T)$ for all $t \in \mathbb{R}$.

Proof. Let $X(t) = Z(t + T)$ and $Y(t) = Z(t)Z(T)$. $\dot{X}(t) = \dot{Z}(t + T) = A(t + T)Z(t + T) = A(t)X(t)$; so, $X(t)$ satisfies (2.30) and $X(0) = Z(T)$. $Y(t)$ also satisfies (2.30) and $Y(0) = Z(T)$. By the uniqueness theorem for differential equations, $X(t) \equiv Y(t)$.

The above lemma only requires (2.30) to be periodic, not necessarily Hamiltonian. Even though the equations are periodic the fundamental matrix need not be so, and the matrix $Z(T)$ is the measure of the nonperiodicity of the solutions. $Z(T)$ is called the *monodromy matrix* of (2.30), and the eigenvalues of $Z(T)$ are called the (characteristic) multipliers of (2.30). The multipliers measure how much solutions are expanded, contracted, or rotated after a period. The monodromy matrix is symplectic by Theorem 2.1.3, and so the multipliers are symmetric with respect to the real axis and the unit circle by Proposition 2.3.1. Thus the origin cannot be asymptotically stable.

In order to understand periodic systems we need some information on logarithms of matrices. The complete proof is long, therefore the proof has been relegated to Section 5.2. Here we shall prove the result in the case when the matrices are diagonalizable.

A matrix R has a logarithm if there is a matrix Q such that $R = \exp Q$, and we write $Q = \log R$. The logarithm is not unique in general, even in the real case, because $I = \exp O = \exp 2\pi J$. Thus the identity matrix I has logarithm O the zero matrix and also $2\pi J$.

If R has a logarithm, $R = \exp Q$, then R is nonsingular and has a square root $R^{1/2} = \exp(Q/2)$. The matrix

$$R = \begin{bmatrix} -1 & 1 \\ 0 & -1 \end{bmatrix}$$

has no real square root and hence no real logarithm.

Theorem 2.5.1. *Let R be a nonsingular matrix; then there exists a matrix Q such that $R = \exp Q$. If R is real and has a square root, then Q may be taken as real. If R is symplectic, then Q may be taken as Hamiltonian.*

Proof. We only prove this result in the case when R is symplectic and has distinct eigenvalues because in this case we only need to consider the canonical forms of Section 2.3. See Section 5.2 for a complete discussion of logarithms of symplectic matrices.

Consider the cases. First

$$\log \begin{bmatrix} \mu & 0 \\ 0 & \mu^{-1} \end{bmatrix} = \begin{bmatrix} \log \mu & 0 \\ 0 & -\log \mu \end{bmatrix}$$

is a real logarithm when $\mu > 0$ and complex when $\mu < 0$. A direct computation shows that $\text{diag}(\mu, \mu^{-1})$ has no real square root when $\mu < 0$.

If α and β satisfy $\alpha^2 + \beta^2 = 1$, then let θ be the solution of $\alpha = \cos\theta$ and $\beta = \sin\theta$ so that

$$\log \begin{bmatrix} \alpha & \beta \\ -\beta & \alpha \end{bmatrix} = \begin{bmatrix} 0 & \theta \\ -\theta & 0 \end{bmatrix}.$$

Lastly, $\log\text{diag}(B^T, B^{-1}) = \text{diag}(\log B^T, -\log B)$ where

$$B = \begin{bmatrix} \gamma & \delta \\ -\delta & \gamma \end{bmatrix},$$

and

$$\log B = \log \rho \begin{bmatrix} 1 & 0 \\ 0 & 1 \end{bmatrix} + \begin{bmatrix} 0 & \theta \\ -\theta & 0 \end{bmatrix},$$

is real where $\rho = \sqrt{(\gamma^2 + \delta^2)}$, and $\gamma = \rho\cos\theta$ and $\delta = \rho\sin\theta$.

The monodromy matrix $Z(T)$ is nonsingular and symplectic so there exists a Hamiltonian matrix K such that $Z(T) = \exp(KT)$. Define $X(t)$ by $X(t) = Z(t)\exp(-tK)$ and compute

$$\begin{aligned} X(t+T) &= Z(t+T)\exp K(-t-T) \\ &= Z(t)Z(T)\exp(-KT)\exp(-Kt) \\ &= Z(t)\exp(-Kt) \\ &= X(t). \end{aligned}$$

Therefore, $X(t)$ is T-periodic. Because $X(t)$ is the product of two symplectic matrices, it is symplectic. In general, X and K are complex even if A and Z are real. To ensure a real decomposition, note that by Lemma 2.5.1, $Z(2T) = Z(T)Z(T)$; so, $Z(2T)$ has a real square root. Define K as the real solution of $Z(2T) = \exp(2KT)$ and $X(t) = Z(t)\exp(-Kt)$. Then X is $2T$ periodic.

Theorem 2.5.2. *(The Floquet–Lyapunov theorem) The fundamental matrix solution $Z(t)$ of the Hamiltonian (2.30) that satisfies $Z(0) = I$ is of the form $Z(t) = X(t)\exp(Kt)$, where $X(t)$ is symplectic and T-periodic and K is Hamiltonian. Real $X(t)$ and K can be found by taking $X(t)$ to be $2T$-periodic if necessary.*

Let Z, X, and K be as above. In Equation (2.30) make the symplectic periodic change of variables $z = X(t)w$; so,

$$\dot{z} = \dot{X}w + X\dot{w} = (\dot{Z}e^{-Kt} - Ze^{-Kt}K)w + Ze^{-Kt}\dot{w}$$

$$= AZe^{-Kt}w - Ze^{-Kt}Kw + Ze^{-Kt}\dot{w}$$

$$= Az = AXw = AZe^{-Kt}w$$

and hence

$$-Ze^{-Kt}Kw + Ze^{-Kt}\dot{w} = 0$$

or

$$\dot{w} = Kw. \tag{2.32}$$

Corollary 2.5.1. *The symplectic periodic change of variables $z = X(t)w$ transforms the periodic Hamiltonian system (2.30) to the constant Hamiltonian system (2.32). Real X and K can be found by taking $X(t)$ to be $2T$-periodic if necessary.*

The eigenvalues of K are called the (characteristic) exponents of (2.30) where K is taken as $\log(Z(T)/T)$ even in the real case. The exponents are the logarithms of the multipliers and so are defined modulo $2\pi i/T$.

2.6 Symplectic Transformations

It is now time to turn to the nonlinear systems and generalize the notion of symplectic transformations. Some examples and applications are given in this chapter and many more specialized examples are given in Chapter 8.

Let $\Xi : O \to \mathbb{R}^{2n} : (t, z) \to \zeta = \Xi(t, z)$ be a smooth function where O is some open set in $\mathbb{R}^{2n+1}$; Ξ is called a symplectic function (or transformation or map etc.) if the Jacobian of Ξ with respect to z, $D_2\Xi(t, z) = \partial\Xi/\partial z$, is a symplectic matrix at every point of $(t, z) \in O$. Sometimes we use the notation $D_2\Xi$ for the Jacobian of Ξ, and sometimes the notation $\partial\Xi/\partial z$ is used. In the first case we think of the Jacobian $D_2\Xi$ as a map from O into the space $\mathcal{L}(\mathbb{R}^{2n}, \mathbb{R}^{2n})$ of linear operators from $\mathbb{R}^{2n}$ to $\mathbb{R}^{2n}$, and in the second case, we think of $\partial\Xi/\partial z$ as the matrix

$$\frac{\partial\Xi}{\partial z} = \begin{bmatrix} \dfrac{\partial\Xi_1}{\partial z_1} & \cdots & \dfrac{\partial\Xi_1}{\partial z_{2n}} \\ \vdots & & \vdots \\ \dfrac{\partial\Xi_{2n}}{\partial z_1} & \cdots & \dfrac{\partial\Xi_{2n}}{\partial z_{2n}} \end{bmatrix}.$$

Thus Ξ is symplectic if and only if

$$\frac{\partial\Xi}{\partial z} J \frac{\partial\Xi}{\partial z}^T = J. \tag{2.33}$$

Recall that if a matrix is symplectic then so is its transpose, therefore we could just as easily transpose the first factor in (2.33). Because the product of two symplectic matrices is symplectic, the composition of two symplectic maps is symplectic by the chain rule of differentiation. Because a symplectic matrix is invertible, and its inverse is symplectic, the inverse function theorem implies that a symplectic map is locally invertible and its inverse, $Z(t, \zeta)$, is symplectic where defined. Because the determinant of a symplectic matrix is $+1$, the transformation is orientation and volume-preserving.

If the transformation $z \to \zeta = \Xi(t, z)$ is considered a change of variables, then one calls ζ symplectic or canonical coordinates. Consider a nonlinear Hamiltonian system

$$\dot{z} = J\nabla_z H(t, z), \tag{2.34}$$

where H is defined and smooth in some open set $O \subset \mathbb{R}^{2n+1}$. Make a symplectic change of variables from z to ζ by

$$\zeta = \Xi(t, z) \quad \text{with inverse } z = Z(t, \zeta) \tag{2.35}$$

(so $\zeta \equiv \Xi(t, Z(t, \zeta))$, $z \equiv Z(t, \Xi(t, z))$). Let $\mathbb{O} \in \mathbb{R}^{2n+1}$ be the image of O under this transformation. Then the Hamiltonian $H(t, z)$ transforms to the function $\hat{H}(t, \zeta) = H(t, Z(t, \zeta))$. Later we abuse notation and write $H(t, \zeta)$ instead of introducing a new symbol, but now we are careful to distinguish H and $\hat{H}$. The equation (2.34) transforms to

$$\dot{\zeta} = \frac{\partial \Xi}{\partial t}(t, z) + \frac{\partial \Xi}{\partial z}(t, z)\dot{z}$$

$$= \frac{\partial \Xi}{\partial t}(t, z) + \frac{\partial \Xi}{\partial z}(t, z)J\left(\frac{\partial H}{\partial z}(t, z)\right)^T$$

$$= \frac{\partial \Xi}{\partial t}(t, z) + \frac{\partial \Xi}{\partial z}(t, z)J\left(\frac{\partial \hat{H}}{\partial \zeta}(t, \zeta)\frac{\partial \Xi}{\partial z}(t, z)\right)^T \qquad (2.36)$$

$$= \frac{\partial \Xi}{\partial t}(t, z) + J\left(\frac{\partial \hat{H}}{\partial \zeta}\right)^T$$

$$= \left.\frac{\partial \Xi}{\partial t}(t, z)\right|_{z=Z(t,\zeta)} + J\nabla_\zeta \hat{H}(t, \zeta).$$

The notation in the second to last term in (2.36) means that you are to take the partial derivative with respect to t first and then substitute in $z = Z(t, \zeta)$. If the change of coordinates, Ξ, is independent of t, then the term $\partial \Xi/\partial t$ is missing in (2.36); so, the equation in the new coordinates is simply $\dot{\zeta} = J\nabla_\zeta \hat{H}$, a Hamiltonian system with Hamiltonian $\hat{H}$. In this case one simply substitutes the change of variables into the Hamiltonian H to get the new Hamiltonian $\hat{H}$. The Hamiltonian character of the equations is preserved. Actually the system (2.36) is still Hamiltonian even if Ξ depends on t, provided $\mathbb{O}$ is a nice set, as we show.

For each fixed t, let the set $\mathbb{O}_t = \{\zeta : (t, \zeta) \in \mathbb{O}\}$ be a ball in $\mathbb{R}^{2n}$. We show that there is a smooth function $R : \mathbb{O} \to \mathbb{R}^1$ such that

$$\left.\frac{\partial \Xi}{\partial t}(t, z)\right|_{z=Z(t,\zeta)} = J\nabla_\zeta R(t, \zeta). \qquad (2.37)$$

R is called the remainder function. Therefore, in the new coordinates, the equation (2.36) is Hamiltonian with Hamiltonian $R(t, \zeta) + H(t, \zeta)$. (In the case where $\mathbb{O}_t$ is not a ball, the above holds locally; i.e., at each point of $p \in \mathbb{O}$ there is a function R defined in a neighborhood of p such that (2.37) holds in the neighborhood, but R may not be globally defined as a single-valued function on all of $\mathbb{O}$.) By Corollary 7.3.1, we must show that J times the Jacobian of the left-hand side of (2.37) is symmetric. That is, we must show

$$\Gamma = \Gamma^T,$$

where

$$\Gamma(t, \zeta) = \left.J\frac{\partial^2 \Xi}{\partial t \partial z}(t, z)\right|_{z=Z(t,\zeta)} \frac{\partial Z}{\partial \zeta}(t, \zeta).$$

Differentiating (2.33) with respect to t gives

$$\frac{\partial^2 \Xi^T}{\partial t \partial z}(t,z) J \frac{\partial \Xi}{\partial z}(t,z) + \frac{\partial \Xi^T}{\partial z}(t,z) J \frac{\partial^2 \Xi}{\partial t \partial z}(t,z) = 0$$

$$\frac{\partial \Xi^{-T}}{\partial z}(t,z) \frac{\partial^2 \Xi^T}{\partial t \partial z}(t,z) J + J \frac{\partial^2 \Xi}{\partial t \partial z}(t,z) \frac{\partial \Xi^{-1}}{\partial z}(t,z) = 0.$$

(2.38)

Substituting $z = Z(t,\zeta)$ into (2.38) and noting that $(\partial \Xi^{-1}/\partial z)(t, Z(t,\zeta)) = \partial Z(t,\zeta)$ yields $-\Gamma^T + \Gamma = 0$. Thus we have shown the following.

Theorem 2.6.1. *A symplectic change of variables on $\mathbb{O}$ takes a Hamiltonian system of equations into a Hamiltonian system.*

A partial converse is also true. If a change of variables preserves the Hamiltonian form of all Hamiltonian equations, then it is symplectic. We do not need this result and leave it as an exercise.

2.6.1 The Variational Equations

Let $\phi(t,\tau,\zeta)$ be the general solution of (2.34); so, $\phi(\tau,\tau,\zeta) = \zeta$, and let $X(t,\tau,\zeta)$ be the Jacobian of ϕ with respect to ζ; i.e.,

$$X(t,\tau,\zeta) = \frac{\partial \phi}{\partial \zeta}(t,\tau,\zeta).$$

$X(t,\tau,\zeta)$ is called the monodromy matrix. Substituting ϕ into (2.34) and differentiating with respect to ζ gives

$$\dot{X} = JS(t,\tau,\zeta)X, \qquad S(t,\tau,\zeta) = \frac{\partial^2 H}{\partial x^2}(t, \phi(t,\tau,\zeta)). \qquad (2.39)$$

Equation (2.39) is called the variational equation and is a linear Hamiltonian system. Differentiating the identity $\phi(\tau,\tau,\zeta) = \zeta$ with respect to ζ gives $X(\tau,\tau,\zeta) = I$, the $2n \times 2n$ identity matrix; so, X is a fundamental matrix solution of the variational equation. By Theorem 2.1.3, X is symplectic.

Theorem 2.6.2. *Let $\phi(t,\tau,\zeta)$ be the general solution of the Hamiltonian system (2.34). Then for fixed t and τ, the map $\zeta \to \phi(t,\tau,\zeta)$ is symplectic. Conversely, if $\phi(t,\tau,\zeta)$ is the general solution of a differential equation $\dot{z} = f(t,z)$, where f is defined and smooth on $I \times O$, I an interval in $\mathbb{R}$ and O a ball in $\mathbb{R}^{2n}$, and the map $\zeta \to \phi(t,\tau,\zeta)$ is always symplectic, then the differential equation $\dot{z} = f(t,z)$ is Hamiltonian.*

Proof. The direct statement was proved above; now consider the converse. Let $\phi(t,\tau,\zeta)$ be the general solution of $\dot{z} = f(t,z)$, and let $X(t,\tau,\zeta)$ be the Jacobian of ϕ. X satisfies

$$X^T(t, \tau, \zeta) J X(t, \tau, \zeta) = J$$

Differentiate this with respect to t (first argument) and set $t = \tau$ so $\dot{X}^T(\tau, \tau, \zeta) J + J \dot{X}(\tau, \tau, \zeta) = 0$ so $\dot{X}(\tau, \tau, \zeta)$ is Hamiltonian.

X also satisfies $\dot{X}(t, \tau, \zeta) = f_z(t, \phi(t, \tau, \zeta)) X(t, \tau, \zeta)$ Set $t = \tau$ to get $\dot{X}(\tau, \tau, \zeta) = f_z(\tau, \zeta)$, so $f_z(\tau, \zeta)$ is Hamiltonian, and thus $-J\dot{X}$ is symmetric. But $X(t, \tau, \zeta) = \partial f / \partial z(t, \phi(t, \tau, \zeta))$; so, $-J\partial f / \partial z$ is symmetric. Because O is a ball, $-Jf$ is a gradient of a function H by Corollary 7.3.1. Thus $f(t, z) = J\nabla H(t, z)$.

This theorem says that the flow defined by an autonomous Hamiltonian system is volume-preserving. So, in particular, there cannot be an asymptotically stable equilibrium point, periodic solution, etc. This makes the stability theory of Hamiltonian systems difficult and interesting. In general, it is difficult to construct a symplectic transformation with nice properties using definition (2.33). The theorem above gives one method of assuring that a transformation is symplectic, and this is the basis of the method of Lie transforms explored in Chapter 9.

2.6.2 Poisson Brackets

Let $F(t, z)$ and $G(t, z)$ be smooth, and recall the definition of the Poisson bracket $\{F, G\}_z(t, z) = \nabla_z F(t, z)^T J \nabla_z G(t, z)$. Here we subscript the bracket to remind us it is a coordinate dependent definition. Let $\hat{F}(t, \zeta) = F(t, Z(t, \zeta))$ and $\hat{G}(t, \zeta) = G(t, Z(t, \zeta))$ where $Z(t, \zeta)$ is symplectic for fixed t; so,

$$\{\hat{F}, \hat{G}\}_\zeta(t, \zeta) = \nabla_\zeta \hat{F}(t, \zeta)^T J \nabla_\zeta \hat{G}(t, \zeta)$$

$$= \left(\frac{\partial Z^T}{\partial \zeta}(t, \zeta) \nabla_z F(t, Z(t, \zeta)) \right)^T J \frac{\partial Z^T}{\partial \zeta} \nabla_z G(t, Z(t, \zeta))$$

$$= \nabla_z F(t, Z(t, \zeta))^T \frac{\partial Z}{\partial \zeta} J \frac{\partial Z^T}{\partial \zeta}(t, \zeta) \nabla_z G(t, Z(t, \zeta))$$

$$= \nabla_z F(t, Z(t, \zeta))^T J \nabla_z G(t, Z(t, \zeta))$$

$$= \{F, G\}_z(t, Z(t, \zeta)).$$

This shows that the Poisson bracket operation is invariant under symplectic changes of variables. That is, you can commute the operations of computing Poisson brackets and making a symplectic change of variables.

Theorem 2.6.3. *Poisson brackets are preserved by a symplectic change of coordinates.*

Let $\zeta_i = \Xi_i(t, z)$ be the ith component of the transformation. In components, Equation (2.33) says

$$\{\Xi_i, \Xi_j\} = J_{ij}, \tag{2.40}$$

where $J = (J_{ij})$.

If the transformation (2.35) is given in the classical notation

$$Q_i = Q_i(q, p), \quad P_i = P_i(q, p), \tag{2.41}$$

then (2.40) becomes

$$\{Q_i, Q_j\} = 0, \qquad \{P_i, P_j\} = 0, \qquad \{Q_i, P_j\} = \delta_{ij}, \tag{2.42}$$

where δ_{ij} is the Kronecker delta.

Theorem 2.6.4. *The transformation (2.35) is symplectic if and only if (2.40) holds, or the transformation (2.41) is symplectic if and only if (2.42) holds.*

2.7 Symplectic Scaling

If instead of satisfying (2.33) a transformation $\zeta = \Xi(t, z)$ satisfies

$$J = \mu \frac{\partial \Xi}{\partial z} J \frac{\partial \Xi}{\partial z}^T,$$

where μ is some nonzero constant, then $\zeta = \Xi(t, z)$ is called a symplectic transformation (map, change of variables, etc.) with multiplier μ. Equation (2.34) becomes

$$\dot{\zeta} = \mu J \nabla_\zeta H(t, \zeta) + J \nabla_\zeta R(t, \zeta),$$

where all the symbols have the same meaning as in Section 2.6. In the time-independent case, you simply multiply the Hamiltonian by μ.

As an example consider scaling the universal gravitational constant G. When the N-body problem was introduced in Section 3.1, the equations contained the universal gravitational constant G. Later we set $G = 1$. This can be accomplished by a symplectic change of variables with multiplier. The change of variables $q = \alpha q', p = \alpha p'$ is symplectic with multiplier α^{-2}, and so the Hamiltonian of the N-body problem, (3.5), becomes

$$H = \sum_{i=1}^{N} \frac{\|p_i'\|^2}{2m_i} - \sum_{1 \le i < j \le N} \frac{G}{\alpha^3} \frac{m_i m_j}{\|q_i' - q_j'\|}.$$

If we take $\alpha^3 = G$, then in the prime coordinates the gravitational constant will be 1. q has the dimensions of distance, and p has the dimensions of distance-mass/time; and so the change of variables can be done by changing the units of distance only. A better way to make the universal gravitational constant unity is to change the unit of mass. The scaling given here is simply an example.

2.7.1 Equations Near an Equilibrium Point

Consider a Hamiltonian that has a critical point at the origin; so,

$$H(z) = \frac{1}{2}z^T S z + K(z),$$

where S is the Hessian of H at $z = 0$, and K vanishes along with its first and second partial derivatives at the origin. The change of variables $z = \epsilon w$ is a symplectic change of variables with multiplier ϵ^{-2}; so, the Hamiltonian becomes

$$H(w) = \frac{1}{2}w^T S w + \epsilon^{-2}K(\epsilon w) = \frac{1}{2}w^T S w + O(\epsilon).$$

In the above, the classical notation, $O(\epsilon)$, of perturbation theory is used. Because K is at least third order at the origin, there is a constant C such that $|\epsilon^{-2}K(\epsilon w)| \le C\epsilon$ for w in a neighborhood of the origin and ϵ small, which is written $\epsilon^{-2}K(\epsilon w) = O(\epsilon)$. The equations of motion become

$$\dot{w} = Aw + O(\epsilon), \qquad A = JS. \tag{2.43}$$

If $\|w\|$ is about 1 and ϵ is small, then z is small. Thus the above transformation is useful in studying the equations near the critical point. To the lowest order in ϵ the equations are linear; so, close to the critical point the linear terms are the most important terms. This is an example of what is called scaling variables, and ϵ is called the scale parameter. To avoid the growth of symbols, one often says: scale by $z \to \epsilon z$ which means replace z by ϵz everywhere. This would have the effect of changing w back to z in (2.43). It must be remembered that scaling is really changing variables.

2.8 Problems

1. Consider a quadratic form $H = (1/2)x^T S x$, where $S = S^T$ is a real symmetric matrix. The index of the quadratic form H is the dimension of the largest linear space where H is negative. Show that the index of H is the same as the number of negative eigenvalues of S. Show that if S is nonsingular and H has odd index, then the linear Hamiltonian system $\dot{x} = JSx$ is unstable. (Hint: Show that the determinant of JS is negative.)

2. Consider the linear fractional (or Möbius) transformation

$$\Phi : z \to w = \frac{1+z}{1-z}, \quad \Phi^{-1} : w \to z = \frac{w-1}{w+1}.$$

a) Show that Φ maps the left half-plane into the interior of the unit circle. What are $\Phi(0), \Phi(1), \Phi(i), \Phi(\infty)$?

b) Show that Φ maps the set of $m \times m$ matrices with no eigenvalue $+1$ bijectively onto the set of $m \times m$ matrices with no eigenvalue -1.

c) Let $B = \Phi(A)$ where A and B are $2n \times 2n$. Show that B is symplectic if and only if A is Hamiltonian.

d) Apply Φ to each of the canonical forms for Hamiltonian matrices to obtain canonical forms for symplectic matrices.

3. Consider the system (*) $M\ddot{q} + Vq = 0$, where M and V are $n \times n$ symmetric matrices and M is positive definite. From matrix theory there is a nonsingular matrix P such that $P^T M P = I$ and an orthogonal matrix R such that $R^T(P^T V P)R = \Lambda = \text{diag}(\lambda_1, \ldots, \lambda_n)$. Show that the above equation can be reduced to $\ddot{p} + \Lambda p = 0$. Discuss the stability and asymptotic behavior of these systems. Write (*) as a Hamiltonian system with Hamiltonian matrix $A = J\text{diag}(V, M^{-1})$. Use the above results to obtain a symplectic matrix T such that

$$T^{-1}AT = \begin{bmatrix} 0 & I \\ -\Lambda & 0 \end{bmatrix}.$$

(Hint: Try $T = \text{diag}(PR, P^{-T}R)$.)

4. Let M and V be as in Problem 5.

a) Show that if V has one negative eigenvalue, then some solutions of (*) in Problem 5 tend to infinity as $t \to \pm\infty$.

b) Consider the system (**) $M\ddot{q} + \nabla U(q) = 0$, where M is positive definite and $U : \mathbb{R}^n \to \mathbb{R}$ is smooth. Let q_0 be a critical point of U such that the Hessian of U at q_0 has one negative eigenvalue (so q_0 is not a local minimum of U). Show that q_0 is an unstable critical point for the system (**).

5. Let $H(t, z) = \frac{1}{2}z^T S(t)z$ and $\zeta(t)$ be a solution of the linear system with Hamiltonian H. Show that

$$\frac{d}{dt}H = \frac{\partial}{\partial t}H;$$

i.e.,

$$\frac{d}{dt}H(t, \zeta(t)) = \frac{\partial}{\partial t}H(t, \zeta(t)).$$

6. The general linear group, $Gl(m, \mathbb{F})$, is the set of all $m \times m$ nonsingular matrices. A matrix Lie group is a closed subgroup of $Gl(m, \mathbb{F})$. Show that the following are matrix Lie groups.

    a) $Sl(m, \mathbb{F})$ = special linear group = set of all $A \in Gl(m, \mathbb{F})$ with det $A = 1$.

    b) $O(m, \mathbb{F})$ = orthogonal group = set of all $m \times m$ orthogonal matrices.

    c) $So(m, \mathbb{F})$ = special orthogonal group = $O(m, \mathbb{F}) \cap Sl(m, \mathbb{F})$.

    d) $Sp(2n, \mathbb{F})$ = symplectic group = set of all $2n \times 2n$ symplectic matrices.

7. Show that the following are Lie subalgebras of $gl(m, \mathbb{F})$, see Problem 2 in Chapter 1.

    a) $sl(m, \mathbb{F})$ = set of $m \times m$ matrices with trace = 0. (sl = special linear.)

    b) $o(m, \mathbb{F})$ = set of $m \times m$ skew-symmetric matrices. (o = orthogonal.)

    c) $sp(2n, \mathbb{F})$ = set of all $2n \times 2n$ Hamiltonian matrices.

8. Let $Q(n, \mathbb{F})$ be the set of all quadratic forms in $2n$ variables with coefficients in $\mathbb{F}$, so $q \in Q(n, \mathbb{R})$, if $q(x) = \frac{1}{2}x^T S x$, where S is a $2n \times 2n$ symmetric matrix and $x \in \mathbb{F}^{2n}$.

    a) Prove that $Q(n, \mathbb{F})$ is a Lie algebra, where the product is the Poisson bracket.

    b) Prove that $\Psi : Q(n, \mathbb{F}) \to sp(2n, \mathbb{F}) : q(x) = \frac{1}{2}x^T S x \to JS$ is a Lie algebra isomorphism.

9. Show that the matrices

$$\begin{bmatrix} -1 & 1 \\ 0 & -1 \end{bmatrix} \quad \text{and} \quad \begin{bmatrix} -2 & 0 \\ 0 & -1/2 \end{bmatrix}$$

have no real logarithm.

10. Prove the theorem: $e^{At} \in \mathcal{G}$ for all t if and only if $A \in \mathcal{A}$ in the following cases:

    a) When $\mathcal{G} = Gl(m, \mathbb{R})$ and $\mathcal{A} = gl(m, \mathbb{R})$

    b) When $\mathcal{G} = Sl(m, \mathbb{R})$ and $\mathcal{A} = sl(m, \mathbb{R})$

    c) When $\mathcal{G} = O(m, \mathbb{R})$ and $\mathcal{A} = so(m, \mathbb{R})$

    d) When $\mathcal{G} = Sp(2n, \mathbb{R})$ and $\mathcal{A} = sp(2n, \mathbb{R})$

11. Consider the map $\Phi : \mathcal{A} \to \mathcal{G} : A \mapsto e^A = \sum_0^\infty A^n/n!$. Show that Φ is a diffeomorphism of a neighborhood of $0 \in \mathcal{A}$ onto a neighborhood of $I \in \mathcal{G}$ in the following cases:

    a) When $\mathcal{G} = Gl(m, \mathbb{R})$ and $\mathcal{A} = gl(m, \mathbb{R})$

    b) When $\mathcal{G} = Sl(m, \mathbb{R})$ and $\mathcal{A} = sl(m, \mathbb{R})$

    c) When $\mathcal{G} = O(m, \mathbb{R})$ and $\mathcal{A} = so(m, \mathbb{R})$

    d) When $\mathcal{G} = Sp(2n, \mathbb{R})$ and $\mathcal{A} = sp(2n, \mathbb{R})$

(Hint: The linearization of Φ is $A \mapsto I + A$. Think implicit function theorem.)

12. Show that $Gl(m, \mathbb{R})$ (respectively $Sl(m, \mathbb{R})$, $O(m, \mathbb{R})$, $Sp(2n, \mathbb{R})$) is a differential manifold of dimension m^2 (respectively, m^2, $m(m-1)/2$, $(2n^2 + n)$). (Hint: Use the problem above and group multiplication to move neighborhoods around.)

13. Show that if you scale time by $t \to \mu t$, then you should scale the Hamiltonian by $H \to \mu^{-1}H$.

3. Celestial Mechanics

Science as we know it today started with Newton's formulations of the three laws of motion, the universal law of gravity, and his solution of the 2-body problem. With a few simple principles and some mathematics, he could explain the three empirical laws of Kepler on the motion of Mars and the other planets. The sun and one planet can be considered as a 2-body problem in the first approximation; this was relatively easy for him to solve.

Newton then turned his attention to the motion of the moon, which requires three bodies: the sun, Earth, and moon, in the first approximation. His inability to solve the 3-body problem lead him to remark to the astronomer John Machin that "his head never ached but with his studies on the moon."[1] The 3-body problem thus became the most celebrated problem in mathematics.

3.1 The N-Body Problem

Consider N point masses moving in a Newtonian reference system, $\mathbb{R}^3$, with the only force acting on them being their mutual gravitational attraction. Let the ith particle have position vector q_i and mass $m_i > 0$.

Newton's second law says that the mass times the acceleration of the ith particle, $m_i \ddot{q}_i$, is equal to the sum of the forces acting on the particle. Newton's law of gravity says that the magnitude of force on particle i due to particle j is proportional to the product of the masses and inversely proportional to the square of the distance between them, $\mathcal{G}m_i m_j / \|q_j - q_i\|^2$ ($\mathcal{G}$ is the proportionality constant). The direction of this force is along a unit vector from particle i to particle j, $(q_j - q_i)/\|q_i - q_j\|$. Putting it all together yields the equations of motion

$$m_i \ddot{q}_i = \sum_{j=1, i \neq j}^{N} \frac{\mathcal{G}m_i m_j (q_j - q_i)}{\|q_i - q_j\|^3} = \frac{\partial U}{\partial q_i}, \qquad (3.1)$$

---

[1] Newton manuscript in Keynes collection, King's College, Cambridge, UK. MSS 130.6, Book 3; 130.5, Sheet 3.

© Springer International Publishing AG 2017   61
K.R. Meyer, D.C. Offin, *Introduction to Hamiltonian Dynamical Systems and the N-Body Problem*, Applied Mathematical Sciences 90, DOI 10.1007/978-3-319-53691-0_3

where

$$U = \sum_{1 \le i < j \le N} \frac{\mathcal{G}m_i m_j}{\|q_i - q_j\|}. \tag{3.2}$$

In the above, $\mathcal{G} = 6.67408 \times 10^{-11} m^3/\sec^2 kg$ is the gravitational constant, U is the self-potential or the negative of the potential.

Let $q = (q_1, \ldots, q_N) \in \mathbb{R}^{3N}$ and let M the $3n \times 3n$ diagonal matrix $M = \mathrm{diag}(m_1, m_1, m_1, \ldots, m_N, m_N, m_N)$; thus Equation (3.1) is of the form

$$M\ddot{q} - \frac{\partial U}{\partial q} = 0. \tag{3.3}$$

Let Δ be the subset of $\mathbb{R}^{3N}$ where $q_i = q_j$ for $i \ne j$, the collusion set. Equations (3.1) and (3.3) are systems of second order equations in $\mathbb{R}^{3N} \setminus \Delta$, the position space.

Define $p = (p_1, \ldots, p_N) \in \mathbb{R}^{3N}$ by $p = M\dot{q}$, so $p_i = m_i \dot{q}_i$ is the momentum of the ith particle. The equations of motion become

$$\dot{q}_i = p_i/m_i = \frac{\partial H}{\partial p_i}, \quad \dot{p}_i = \sum_{j=1, j \ne i}^{N} \frac{\mathcal{G}m_i m_j (q_j - q_i)}{\|q_i - q_j\|^3} = -\frac{\partial H}{\partial q_i}, \tag{3.4}$$

where the Hamiltonian is

$$H = T - U, \tag{3.5}$$

and T is kinetic energy

$$T = \sum_{i=1}^{N} \frac{\|p_i\|^2}{2m_i} = \frac{1}{2} p^T M^{-1} p \left(= \frac{1}{2} \sum_{i=1}^{N} m_i \|\dot{q}_i\|^2 \right). \tag{3.6}$$

Here again the correct conjugate of position q is momentum p.

The Hamiltonian (3.5) and the Hamiltonian equations (3.4) are defined on phase space $\mathbb{R}^{6N} \setminus \Delta$. All the above is written for the spacial problem in $\mathbb{R}^3$, but with just minor changes applies equally will for the planar problem in $\mathbb{R}^2$ and the collinear problem in $\mathbb{R}^1$.

3.2 The 2-Body Problem

Before tacking the general problem we will look at the two-body problem in detail since as Newton discovered it is solvable (well almost) whereas the three-body problem can lead to headaches.

First just look at the equations

$$m_1 \ddot{q}_1 = \mathcal{G} m_1 m_2 \frac{(q_2 - q_1)}{\|q_2 - q_1\|^3}, \quad m_2 \ddot{q}_2 = \mathcal{G} m_1 m_2 \frac{(q_1 - q_2)}{\|q_2 - q_1\|^3}.$$

Add and subtract these equations to get

$$m_1 \ddot{q}_1 + m_2 \ddot{q}_2 = 0, \quad \ddot{q}_1 - \ddot{q}_2 = -\mathcal{G}(m_1 + m_2) \frac{(q_1 - q_2)}{\|q_1 - q_2\|^3}$$

or

$$\ddot{C} = 0, \quad \ddot{u} = -\mu \frac{u}{\|u\|^3},$$

where $C = m_1 q_1 + m_2 q_2$, $u = q_1 - q_2$, and $\mu = \mathcal{G}(m_1 + m_2)$. The first equation integrates to $C = L_0 t + C_0$ where L_0 and C_0 are integration constants and the second equation which is called the Kepler problem contains all the interest.

Knowing the answer we can see the first use Jacobi coordinates (g, u, G, v) where

$$g = \nu_1 q_1 + \nu_2 q_2, \qquad G = p_1 + p_2,$$

$$u = q_2 - q_1, \qquad v = -\nu_2 p_1 + \nu_1 p_2,$$

and

$$\nu_1 = \frac{m_1}{m_1 + m_2}, \quad \nu_2 = \frac{m_2}{m_1 + m_2}, \quad \nu = m_1 + m_2, \quad M = \frac{m_1 m_2}{m_1 + m_2}.$$

So g is the center of mass, G is total linear momentum, u is the position of particle 2 as viewed from particle 1, and v is a scaled momentum.

By a direct computation one sees that this is a linear symplectic change of variables, so preserves the Hamiltonian character. A complete discussion of Jacob coordinates is given in Section 8.2.

The Hamiltonian of the 2-body problem in these Jacobi coordinates is

$$H = \frac{\|G\|^2}{2\nu} + \frac{\|v\|^2}{2M} - \frac{m_1 m_2}{\|u\|},$$

and the equations of motion are

$$\dot{g} = \frac{\partial H}{\partial G} = \frac{G}{\nu}, \qquad \dot{G} = -\frac{\partial H}{\partial g} = 0,$$

$$\dot{u} = \frac{\partial H}{\partial v} = \frac{v}{M}, \qquad \dot{v} = -\frac{\partial H}{\partial u} = -\frac{m_1 m_2 u}{\|u\|^3}.$$

This says that total linear momentum G is an integral and the center of mass g moves with constant linear velocity. By taking $g = G = 0$ as initial conditions we are reduced to a problem in the u, v variables alone. The equations reduce to

$$\ddot{u} = \frac{\mathcal{G}(m_1 + m_2)u}{\|u\|^3}.$$

This is just the central forced problem or the Kepler problem discussed in the next section. This says that the motion of one body, say the sun, when viewed from another, say the earth, is as if the earth were a fixed body with mass $m_1 + m_2$ and the sun were attracted to the earth by a central force. Maybe Ptolemy was right.

3.3 The Kepler Problem

Consider a two body problem where the first body is so massive (like the sun) that its position is fixed to the first approximation and the second body has mass 1. In this case, the equations describe the motion of the second body are

$$\ddot{q} = -\frac{\mu q}{\|q\|^3}, \tag{3.7}$$

where $q \in \mathbb{R}^3$ is the position vector of the second body and μ is the constant $\mathcal{G}m$ where $\mathcal{G}$ is the universal gravitational constant and m is the mass of the first body fixed at the origin. In this case by defining $p = \dot{q}$, this equation becomes Hamiltonian with

$$H = \frac{\|p\|^2}{2} - \frac{\mu}{\|q\|}, \tag{3.8}$$

and equations

$$\dot{q} = p, \quad \dot{p} = -\frac{\mu q}{\|q\|^3}. \tag{3.9}$$

Equation (3.7) or Hamiltonian (3.8) defines the Kepler problem. As we have just seen, the 2-body problem can be reduced to this problem.

Let $A = q \times p$, the angular momentum, and note

$$\dot{A} = \dot{q} \times p + q \times \dot{p} = p \times p + q \times (-\mu q/\|q\|^3) = 0$$

so A is constant along the solutions, and the three components of A are integrals.

If $A = 0$, then

$$\frac{d}{dt}\left(\frac{q}{\|q\|}\right) = \frac{(q \times \dot{q}) \times q}{\|q\|^3} = \frac{A \times q}{\|q\|^3} = 0. \tag{3.10}$$

The first equality above is a vector identity, so, if the angular momentum is zero, the motion is collinear. Letting the line of motion be one of the coordinate axes makes the problem a one degree of freedom problem and so solvable by formula (1.9). In this case the integrals are elementary, and one obtains simple formulas (see the Problem section).

If $A \neq 0$, then both q and $p = \dot{q}$ are orthogonal to A; and so, the motion takes place in the plane orthogonal to A known as the *invariant plane*. In this case, take one coordinate axis, say the last, to point along A, so, the motion is in a coordinate plane. The equations of motion in this coordinate plane have the same form as (3.7), but $q \in \mathbb{R}^2$. In the planar problem only the component of angular momentum perpendicular to the plane is nontrivial; so the problem is reduced to two degrees of freedom with one integral. Such a problem is solvable "up to quadrature." It turns out that the problem is solvable in terms of elementary functions, as we now show.

If $q = (r \cos \theta, r \sin \theta, 0)$ then $A = q \times p = q \times \dot{q} = (0, 0, r^2 \dot{\theta}) = (0, 0, c) \neq 0$. A standard formula that gives that the rate at which area is swept out by a radius vector is just $\frac{1}{2} r^2 \dot{\theta}$; thus, the particle sweeps out area at a constant rate of $c/2$. This is Kepler's second law.

There are many ways to solve the Kepler problem. One way is given here and other ways are given in Section 8.6.1 and Section 8.8.1.

Multiply the first equation in (3.10) by $-\mu$ to get

$$-\mu \frac{d}{dt} \left(\frac{q}{\|q\|} \right) = A \times \frac{-\mu q}{\|q\|^3} = A \times \dot{p}.$$

Integrating this identity gives

$$\mu \left(e + \frac{q}{\|q\|} \right) = p \times A, \tag{3.11}$$

where e is a vector integration constant. Dot the above by A and use $q \cdot A = 0$, to see that $e \cdot A = 0$. Thus if $A \neq 0$, then e lies in the invariant plane. If $A = 0$, then $e = -q/\|q\|$ and then e lies on the line of motion and e has length 1.

Let $A \neq 0$ for the rest of this section. Dot both sides of (3.11) with q to obtain

$$\mu(e \cdot q + \|q\|) = q \cdot p \times A = q \times p \cdot A = A \cdot A,$$

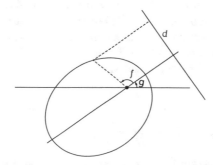

Figure 3.1. The elements of a Kepler motion.

and then,

$$e \cdot q + \|q\| = \frac{c^2}{\mu} \tag{3.12}$$

with $c = \|A\|$.

If $e = 0$, then $\|q\| = c^2/\mu$, a constant. As before $r^2\dot{\theta} = c$ where $q = (r\cos\theta, r\sin\theta, 0)$ so $\dot{\theta} = \mu^2/c^3$. Thus when $e = 0$, the particle moves on a circle with uniform angular velocity.

Now suppose that $e \neq 0$ and $\epsilon = \|e\|$. Let the plane of motion be illustrated in Figure 3.1. Let r, θ be the polar coordinates of the particle with angle θ measured from the positive q_1 axis. The angle from the positive q_1 axis to e is denoted by g, and the difference of these two angles by $f = \theta - g$. Thus, $e \cdot q = \epsilon r \cos f$ and Equation (3.12) becomes

$$r = \frac{c^2/\mu}{1 + \epsilon \cos f}. \tag{3.13}$$

Consider the line d illustrated in Figure 3.1 that is at a distance of $c^2/\mu\epsilon$ from the origin and perpendicular to e as illustrated. Equation (3.13) can be rewritten

$$r = \epsilon \left(\frac{c^2}{\mu\epsilon} - r\cos f \right),$$

which says that the distance of the particle from the origin is equal to ϵ times its distance from the line d. This gives Kepler's first law: the particle moves on a conic section of eccentricity ϵ with one focus at the origin. Recall that $0 < \epsilon < 1$ is an ellipse, $\epsilon = 1$ is a parabola, and $\epsilon > 1$ is a hyperbola.

Equation (3.13) shows that r is at its closest approach when $f = 0$ and so e points to the point of closest approach. This point is called the *perihelion* if the sun is at the origin or the *perigee* if the earth is at the origin. We use perigee. The angle g is called the *argument of the perigee* and the angle f is called the *true anomaly*.

Has the Kepler problem been solved? We know the motion takes place either on a line or on a conic section, but we have not given precise time information. Would you be satisfied if you asked the station master "Where is the train?" and he answered "On this track." It would be nice if we could give the position of the particle as a simple function of t, but that is not so easy. This is a practical problem for astronomers, but the details will not be needed here. For a concise discussion of the solution of this problem see Pollard (1966).

3.4 Symmetries and Integrals

In the quest to find a complete general solution of the N-body problem many special features have been noted. Much of this chapter gives is devoted to these features and sometimes to their generalizations.

3.4.1 Symmetries

We begin by observing that in equation (3.1) the q's only appear as differences on the right-hand side and so is unchanged if the same quantity is added to each q_i. The left-hand side is unchanged if $at + b$ is added to each q_i where $a, b \in \mathbb{R}^3$. Putting these two observations together leads to

If $q_i = \phi_i(t)$, $i = 1, \ldots, N$ is a solution of the N-body problem (3.1) then so is $q_i = \phi_i(t) + at + b$, $i = 1, \ldots, N$, where $a, b \in \mathbb{R}^3$ are constants.

In other words the problem is unchanged in a uniformly translating system and this is the basis of Newton's first law of motion.

Now note that equation (3.1) is unchanged if each q_i is replaced by Rq_i where $R \in O(3, \mathbb{R})$, the orthogonal group. This observation leads to

If $q_i = \phi_i(t)$, $i = 1, \ldots, N$ is a solution of the N-body problem (3.1) then so is $q_i = R\phi_i(t)$, $i = 1, \ldots, N$, where $R \in O(3, \mathbb{R})$.

Poincaré (1899) studied the planar 3-body problem with two small masses and found what he called relative periodic solutions, where after a fixed time T there is a fixed $R \in So(2, \mathbb{R})$ such that all the initial positions and velocities $(u_1, u_2, u_3, v_1, v_2, v_3)$ become $(Ru_1, Ru_2, Ru_3, Rv_1, Rv_2, Rv_3)$. By the above symmetry this pattern repeats every time t increases by T, and this is called a *relative periodic solution.*

The two observations given above are purely geometric and do not involve time, t, but there is a time related symmetry also. Note that the equation (3.1) is unchanged if q is replaced by $a^2 q$ and t by $a^3 t$ where a is a positive number. Thus,

If $q_i = \phi_i(t)$, $i = 1, \ldots, N$ is a solution of the N-body problem (3.1) then so is $q_i = a^2 \phi_i(a^3 t)$, $i = 1, \ldots, N$, where a is a positive number.

This is the general principle which gives Kepler's third law of planetary motion: *The square of the period of any planet is proportional to the cube of the semimajor axis of its orbit.*

Notice that a reflection $R \in O(3, \mathbb{R})$, $R^2 = I$, gives rise to a discrete symmetry of the N-body problem. The ramifications of this type of symmetry will be treated in Section 4.2 as it applies to the restricted 3-body problem. For the general case see Meyer (1981b).

3.4.2 The Classical Integrals

The N-body problem is a system of $6N$ first-order equations, so, a complete solution would require $6N - 1$ time-independent integrals plus one time-dependent integral. It is now fairly clear that for $N > 2$, it is too optimistic to expect so many global integrals. However, for all N there are ten integrals for the system.

Let

$$L = p_1 + \cdots + p_N$$

be the total linear momentum. From (3.4) it follows that $\dot{L} = 0$, because each term in the sum appears twice with opposite sign. This gives $\ddot{C} = 0$, where

$$C = m_1 q_1 + \cdots + m_N q_N$$

is the center of mass of the system because $\dot{C} = L$. Thus the total linear momentum is constant, and the center of mass of the system moves with uniform rectilinear motion. Integrating the center of mass equation gives $C = L_0 t + C_0$, where L_0 and C_0 are constants of integration. L_0 and C_0 are functions of the initial conditions, and thus are integrals of the motion. Thus we have six constants of motion or integrals, namely, the three components of L_0 and the three components of C_0.

Let

$$A = q_1 \times p_1 + \cdots q_N \times p_N$$

be the total angular momentum of the system. Then

$$\frac{dA}{dt} = \sum_1^N (\dot{q}_i \times p_i + q_i \times \dot{p}_i)$$

$$= \sum_1^N q_i \times m_i q_i + \sum_1^N \sum_1^N \frac{G m_i m_j q_i \times (q_j - q_i)}{\|q_i - q_j\|^3} = 0.$$

The first sum above is zero because $q_i \times q_i = 0$. In the second sum, use $q_i \times (q_j - q_i) = q_i \times q_j$ and then observe that each term in the remaining sum appears twice with opposite sign. Thus the three components of angular momentum are constants of the motion or integrals also. Remember that energy, H, is also an integral, so we have found the ten classical integrals of the N-body problem. The count has been for the spatial problem in $\mathbb{R}^3$, for the planar problem in $\mathbb{R}^2$ the count is 6. Thus,

The spatial N-body problem has the 10 integrals: 3 for linear momentum, 3 for center of mass, 3 for angular momentum and 1 for energy. The planar N-body problem has the 6 integrals: 2 for linear momentum, 2 for center of mass, 1 for angular momentum, and 1 for energy.

For the spatial 2-body problem the phase space is $\mathbb{R}^{12} \setminus \Delta$ and for the planar 2-body the phase space is $\mathbb{R}^8 \setminus \Delta$ where Δ is the lower dimensional collision set. Thus, in both cases the integrals reduce the motion to a two-dimensional variety. In general such a system can be solved by a single integration, i.e., by a quadrature. But as we have seen the 2-body problem has an extra integral making it solvable in the classical sense.

The count does not turn out well even for the planar 3-body problem where the phase space is $\mathbb{R}^{12} \setminus \Delta$ and so the integrals reduce the problem to a 6-dimensional variety.

It turns out that not all the symmetries of the N-body problem are accounted for by the 10 integrals. The problem is still unchanged when rotated about the angular momentum vector. This leads to one further reduction by identifying configurations that are obtained from one another by rotation about the angular momentum vector. Locally this can be accomplished by using Jacobi coordinates found in Section 3.4.4. Globally the description of the reduced space where the integrals are fixed and the rotational symmetry is eliminated leads to some interesting topology. See McCord et al. (1998) and the references therein.

3.4.3 Noether's Theorem

Noether (1918) showed that a Lagrangian system that admits a Lie group of symmetries must have integrals. This result has been gentrified many times, so here we present a version that applies to the N-body problem. A more general abstract version will be given later in Section 7.7.

Let ψ_t be a Hamiltonian flow on $\mathbb{R}^{2n}$, so,

i) For fixed t, the map $\psi_t : \mathbb{R}^{2n} \to \mathbb{R}^{2n}$ is symplectic.
ii) $\psi_0 = id$, the identity of $\mathbb{R}^{2n}$.
iii) $\psi_t \circ \psi_s = \psi_{t+s}$ for all $t, s \in \mathbb{R}$.

That is $\psi(t, \xi) = \psi_t(\xi)$ is the general solution of a Hamiltonian system $\dot{x} = J\nabla F(x)$, where $F : \mathbb{R}^{2n} \to \mathbb{R}$ is smooth. We say ψ_t is a *symplectic symmetry* for the Hamiltonian H if

$$H(x) = H(\psi(t, x)) = H(\psi_t(x)) \tag{3.14}$$

for all $x \in \mathbb{R}^{2n}$ and all $t \in \mathbb{R}$.

Theorem 3.4.1. *Let ψ_t be a symplectic symmetry for the Hamiltonian H. Then F is an integral for the Hamiltonian system with Hamiltonian H.*

Proof. Differentiate (3.14) with respect to t to get

$$0 = \frac{\partial H(\psi(t, x))}{\partial x} \frac{\partial \psi(t, x)}{\partial t}$$

$$= \frac{\partial H(\psi(t, x))}{\partial x} J \frac{\partial F(\psi(t, x))}{\partial x}$$

$$= \{H, F\}(\psi(t, x)).$$

Setting $t = 0$ yields $\{H, F\}(x) = 0$ or simply $\{H, F\} = 0$.

Consider the N-body problem with coordinates $z = (q, p) \in \mathbb{R}^{6N}$. The Hamiltonian 3.5 is invariant under translations. That is, if $b \in \mathbb{R}^3$, then $\psi_t : (q_1, \ldots, q_N, p_1, \ldots, p_N) \to (q_1 + tb, \ldots, q_N + tb, p_1, \ldots, p_N)$ is a symplectic symmetry for the N-body problem. The Hamiltonian that generates ψ_t is $F = b^T(p_1 + \cdots + p_N)$. So by the theorem, $F = b^T(p_1 + \cdots + p_N)$ is an integral for all b, and so linear momentum, $L = p_1 + \cdots + p_N$, is an integral for the N-body problem. In general translational invariance implies the conservation of linear momentum.

Another variation of Noether's theorem is:

Theorem 3.4.2. *Let $\mathcal{G}$ be a Lie group of symplectic matrices of $\mathbb{R}^{2n}$ and $\mathcal{A}$ its Lie algebra of Hamiltonian matrices. If $H : \mathbb{R}^{2n} \to \mathbb{R}$ is a smooth Hamiltonian which admits $\mathcal{G}$ as a group of symmetries, i.e.*

$$H(Gz) = H(z), \quad \text{for all} \ z \in \mathbb{R}^{2n} \ \text{and all} \ G \in \mathcal{G},$$

then for each $S \in \mathcal{A}$ the function $I_S = \frac{1}{2} z^T J S z$ is an integral.

Proof. If $S \in \mathcal{A}$ then $e^{St} \in \mathcal{G}$ and $H(e^{St} z) = H(z)$ for all $t \in \mathbb{R}$. Differentiate this with respect to t and then set $t = 0$ to get

$$\frac{\partial H}{\partial z}(z) S z = \nabla H^T(z) S z = -\nabla H^T(z) J J S z = -\{H, I_S\} = 0$$

where $I_S = \frac{1}{2} z^T J S z$. Thus, I_S is an integral for H for each $S \in \mathcal{A}$.

Let $\mathcal{G}$ be the subgroup of $Sp(6N, \mathbb{R})$ consisting of all matrices of the form $T = (A, \dots, A)$, where $A \in So(3, \mathbb{R})$ (the special orthogonal group or group of three-dimensional rotations). Then the Hamiltonian H of the N-body problem, (3.5), has $\mathcal{G}$ as a symmetry group. That is, $H(Tx) = H(x)$ for all $T \in \mathcal{G}$. This simply means that the equations are invariant under a rotation of coordinates. The algebra of Hamiltonian matrices, $\mathcal{A}$, for $\mathcal{G}$ is the set of all Hamiltonian matrices of the form $B = (C, \dots, C)$, where C is a 3×3 skew-symmetric matrix. So $e^{Bt} \in \mathcal{G}$, and $H(x) = H(e^{Bt} x)$ for all $x \in \mathbb{R}^{2n}$ and $t \in \mathbb{R}$. Thus $\psi_t(x) = e^{Bt} x$ is a symplectic symmetry for the Hamiltonian of the N-body problem. The Hamiltonian that generates $\psi_t(x) = e^{Bt} x$ is $F = \sum_{i=1}^{n} q_i^T C p_i$, and so by Noether's theorem it is an integral for the N-body problem. If we take the three choices for C as follows,

$$\begin{bmatrix} 0 & 0 & 0 \\ 0 & 0 & 1 \\ 0 & -1 & 0 \end{bmatrix}, \quad \begin{bmatrix} 0 & 0 & 1 \\ 0 & 0 & 0 \\ -1 & 0 & 0 \end{bmatrix}, \quad \begin{bmatrix} 0 & 1 & 0 \\ -1 & 0 & 0 \\ 0 & 0 & 0 \end{bmatrix}, \tag{3.15}$$

then the corresponding integrals are the three components of angular momentum. So the fact that the Hamiltonian is invariant under all rotations implies the law of conservation of angular momentum.

It turns out that considering symmetries and integrals separately misses some of the information that is known about the system. This interplay will be discussed after the next section.

3.4.4 Reduction

Symmetries give rise to integrals but in general not all the symmetry has been used and this gives rise to further reductions. Fixing a set of integrals defines an invariant subspace, I, for the dynamics and a subgroup $\mathcal{G}_I$ of the group of symmetries keeps this invariant space fixed. In general the dynamics make sense on the quotient space $I/\mathcal{G}_I$ and this space is symplectic. The general

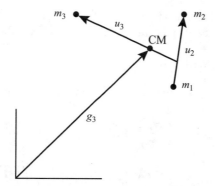

Figure 3.2. Jacobi coordinates for the 3-body problem.

case requires some additional theory which is given in Section 7.7, but for now we will illustrate reduction for the planar 3-body problem.

A concrete way to do the reduction is to use the Jacobi coordinates $(g_3, u_2, u_3, G_3, v_2, v_3)$ which are discussed in detail in Section 8.2. The variable g_2 points to the center of mass of the q_1, q_2 and g_3 points to the center of mass, CM, of the whole system, i.e.,

$$g_2 = \frac{m_2 q_2 + m_1 q_1}{m_1 + m_2}, \quad g_3 = \frac{m_2 q_2 + m_1 q_1 + m_3 q_3}{m_1 + m_2 + m_3}.$$

The other position coordinates u_2, u_3 are relative coordinates namely u_2 is the vector from q_2 to q_3 and u_3 is the vector from g_2 to q_3 namely

$$u_2 = q_3 - q_2, \quad u_3 = q_3 - g_2,$$

see Figure 3.2.

Conjugate variables (G_3, v_2, v_3) can be found by the methods discussed already in Lemma 2.2.4 but the details are not important for now except for the fact that $G_3 = p_1 + p_2 + p_3$ in total linear momentum.

The 3-body problem in Jacobi coordinates is

$$H = \frac{\|G_3\|^2}{2\nu} + \frac{\|v_2\|^2}{2M_2} + \frac{\|v_3\|^2}{2M_3} - \frac{m_1 m_2}{\|u_2\|} - \frac{m_1 m_3}{\|u_3 + \alpha_0 u_2\|} - \frac{m_2 m_3}{\|u_3 - \alpha_1 u_2\|},$$

where

$$M_2 = \frac{m_1 m_2}{m_1 + m_2}, \qquad M_3 = \frac{m_3(m_1 + m_2)}{m_1 + m_2 + m_3},$$

$$\alpha_0 = \frac{m_2}{m_1 + m_2}, \qquad \alpha_1 = \frac{m_1}{m_1 + m_2},$$

and $\nu = m_1 + m_2 + m_3$ is the total mass.

H is independent of g_3, the center of mass, so its conjugate momentum total linear momentum G_3 is an integral. We effect the first reduction by setting $g_3 = G_3 = 0$.

Introduce polar coordinates by

$$u_2 = (r_1 \cos\theta_1, r_1 \sin\theta_1), \quad u_3 = (r_2 \cos\theta_2, r_2 \sin\theta_2),$$

$$v_2 = \left(R_1 \cos\theta_1 - \left(\frac{\Theta_1}{r_1}\right)\sin\theta_1, R_1 \sin\theta_1 + \left(\frac{\Theta_1}{r_1}\right)\cos\theta_1\right),$$

$$v_3 = \left(R_2 \cos\theta_2 - \left(\frac{\Theta_2}{r_2}\right)\sin\theta_2, R_2 \sin\theta_2 + \left(\frac{\Theta_2}{r_2}\right)\cos\theta_2\right),$$

Here of course r, θ are the usual polar coordinates in the plane and R, Θ are radial and angular momenta respectively. (r, θ, R, Θ) are symplectic coordinates. A complete discussion of various special coordinates is given in Chapter 8, see in particular Section 8.6. The Hamiltonian becomes

$$H = \frac{1}{2M_2}\left\{r_1^2 + \left(\frac{\Theta_1^2}{r_1^2}\right)\right\} + \frac{1}{2M_3}\left\{r_2^2 + \left(\frac{\Theta_2^2}{r_2^2}\right)\right\} - \frac{m_0 m_1}{r_1}$$

$$- \frac{m_0 m_2}{\sqrt{r_2^2 + \alpha_0^2 r_1^2 - 2\alpha_0 r_1 r_2 \cos(\theta_2 - \theta_1)}}$$

$$- \frac{m_1 m_2}{\sqrt{r_2^2 + \alpha_1^2 r_1^2 - 2\alpha_1 r_1 r_2 \cos(\theta_2 - \theta_1)}}.$$

Because the Hamiltonian depends only on the difference of the two polar angles, make the symplectic change of coordinates

$$\phi_1 = \theta_2 - \theta_1, \qquad \theta_2 = -\theta_2 + 2\theta_1,$$

$$\Phi_1 = 2\Theta_2 + \Theta_1, \qquad \Phi_2 = \Theta_2 + \Theta_1,$$

so that the Hamiltonian becomes

$$H = \frac{1}{2M_1}\left\{R_1^2 + (2\Phi_1 - \Phi_2)^2/r_1^2\right\} \tag{3.16}$$

$$+ \frac{1}{2M_2}\left\{R_2^2 + (\Phi_1 - \Phi_2)^2/r_2^2\right\} + \frac{m_0 m_1}{r_1}$$

$$+ \frac{m_0 m_2}{\sqrt{r_2^2 + \alpha_0^2 r_1^2 - 2\alpha_0 r_1 r_2 \cos(\phi_1)}}$$

$$+ \frac{m_1 m_2}{\sqrt{r_2^2 + \alpha_1^2 r_1^2 + 2\alpha_1 r_1 r_2 \cos(\phi_1)}}.$$

Note that the Hamiltonian is independent of ϕ_2, so ϕ_2 is an ignorable coordinate, therefore, its conjugate momentum $\Phi_2 = \Theta_2 + \Theta_1$, total angular is an integral. The reduction to the reduced space is done by holding Φ_2 fixed and ignoring ϕ_2. The Hamiltonian 3.16 has three degrees of freedom, (r_1, r_2, ϕ_1), and one parameter Φ_2.

Two further reductions can be accomplished by holding the Hamiltonian (energy) fixed and eliminating time to get a nonautonomous system of differential equations of order 4. The reduction of the 3-body problem is classical, with the elimination of the node due to Jacobi. The general results about the symplectic nature of the reduced space are in Meyer (1973) and Marsden and Weinstein (1974) and will be discussed later in Section 7.7.

3.4.5 Foundations

For a mathematician the N-body problem requires the choice of a frame of reference, i.e., an $\mathbb{R}^3$. By the previous discussion two choices are equally good if they differ by a Newtonian transformation, i.e., by a uniform translation and a fixed rotation.

But the astronomer must find an inertial frame in space. That is he must find three mutually perpendicular vectors in space upon which to build a coordinate system, his $\mathbb{R}^3$, such that Newton's laws hold. An inertial frame is such that a body that is not acted upon by a force remains at rest or moves with uniform motion, i.e., where the Newton's first law of motion holds. So an inertial frame cannot accelerate or rotate. The traditional and practical definition is that an inertial frame is one based on the 'fixed stars'. Of course astronomers now believe there is no such thing as a fixed star, the stars are all moving relative to one another.

Indeed today we know that Newton's laws fail in some extreme situations, for example they fail to account for the advance of the perihelion of Mercury. These fundamental questions are addressed by appealing to Einstein's general theory of relativity.

However, if you want to keep track of a spy satellite or send a man to Mars you better rely on Newton and forget Einstein.

3.5 Equilibrium and Central Configurations

The N-body problem for $N > 2$ has resisted all attempts to be solved, but over the years many special types of solutions have been found using various mathematical techniques. The simplest type of solution one might look for is equilibrium or rest solutions.

3.5.1 Equilibrium Solutions

From (3.1) or (3.3), an equilibrium solution would have to satisfy

$$\frac{\partial U}{\partial q_i} = 0 \quad \text{for} \quad i = 1, \ldots, N. \tag{3.17}$$

However, U is homogeneous of degree -1, so by Euler's theorem on homogeneous functions,

$$\sum q_i \frac{\partial U}{\partial q_i} = -U. \tag{3.18}$$

Because U is the sum of positive terms, it is positive. If (3.17) were true, then the left side of (3.18) would be zero, which gives a contradiction. Thus there are no equilibrium solutions of the N-body problem.

3.5.2 Central Configurations

For a second type of simple solution to (3.1), try $q_i(t) = \phi(t)a_i$, where the a_is are constant vectors and $\phi(t)$ is a scalar-valued function. Substituting into (3.1) and rearranging yields

$$\mid \phi \mid^3 \phi^{-1} \ddot{\phi} m_i a_i = \sum_{j=1, j \neq i}^{N} \frac{\mathcal{G} m_i m_j (a_j - a_i)}{\|a_j - a_i\|^3}. \tag{3.19}$$

Because the right side is constant, the left side must be also; let the constant be λ. Therefore, (3.19) has a solution if there exist a scalar function $\phi(t)$, a constant λ, and constant vectors a_i such that

$$\ddot{\phi} = -\frac{\lambda \phi}{\mid \phi \mid^3}, \tag{3.20}$$

$$-\lambda m_i a_i = \sum_{j=1, j \neq i}^{N} \frac{\mathcal{G} m_i m_j (a_j - a_i)}{\|a_j - a_i\|^3}, \quad i = 1, \ldots, N. \tag{3.21}$$

Equation (3.20) is a simple ordinary differential equation (the one-dimensional Kepler problem!); and so has many solutions. For example, one solution is $\alpha t^{2/3}$, where $\alpha^3 = 9\lambda/2$. This solution goes from zero to infinity as t goes from zero to infinity. Equation (3.21) is a nontrivial system of nonlinear algebraic equations. The complete solution is known only for $N = 2, 3$, but there are many special solutions known for $N > 3$.

Now consider the planar N-body problem, where all the vectors lie in $\mathbb{R}^2$. Identify $\mathbb{R}^2$ with the complex plane $\mathbb{C}$ by considering the q_i, p_i, etc., as complex numbers. Seek a homographic solution of (3.1) by letting $q_i(t) = \phi(t)a_i$, where the a_is are constant complex numbers and $\phi(t)$ is a time-dependent complex-valued function. Geometrically, multiplication by a complex number is a rotation followed by a dilation or expansion, i.e., a homography. Thus we seek a solution such that the configuration of the particles is always homographically equivalent to a fixed configuration. Substituting this guess into (3.1) and rearranging gives the same equation (3.19), and the same argument gives Equations (3.20) and (3.21). Equation (3.20) is now the two-dimensional Kepler problem. That is, if you have a solution of (3.21) where the a_is are planar, then there is a solution of the N-body problem of the form $q_i = \phi(t)a_i$, where $\phi(t)$ is any solution of the planar Kepler problem, e.g., circular, elliptic, etc.

A configuration of the N particles given by constant vectors $a_1, \ldots, a_N$ satisfying (3.21) for some λ is called a central configuration (or c.c. for short). In the special case when the a_is are coplanar, a central configuration is also called a relative equilibrium because, as we show, they become equilibrium solutions in a rotating coordinate system. Central configurations are important in the study of the total collapse of the system because it can be shown that the limiting configuration of a system as it tends to a total collapse is a central configuration. See Saari (1971, 2005).

Note that any uniform scaling of a c.c. is also a c.c. In order to measure the size of the system, we define the *moment of inertia* of the system as

$$I = \frac{1}{2} \sum_{i=1}^{N} m_i \|q_i\|^2. \tag{3.22}$$

Then (3.21) can be rewritten as

$$\frac{\partial U}{\partial q}(a) + \lambda \frac{\partial I}{\partial q}(a) = 0, \tag{3.23}$$

where $q = (q_1, \ldots, q_N)$ and $a = (a_1, \ldots, a_N)$. The constant λ can be considered as a Lagrange multiplier; and thus a central configuration is a critical point of the self-potential U restricted to a constant moment of inertia manifold, $I = I_0$. Fixing I_0 fixes the scale.

Let a be a central configuration. Take the dot product of the vector a and Equation (3.23) to get

$$\frac{\partial U}{\partial q}(a) \cdot a + \lambda \frac{\partial I}{\partial q}(a) \cdot a = 0. \tag{3.24}$$

Because U is homogeneous of degree -1, and I is homogeneous of degree 2, Euler's theorem on homogeneous functions gives $-U + 2\lambda I = 0$, or

$$\lambda = \frac{U(a)}{2I(a)} > 0. \tag{3.25}$$

Summing (3.21) on i gives $\sum m_i a_i = 0$, so the center of mass of a c.c. is at the origin. If A is an orthogonal matrix, either 3×3 in general or 2×2 in the planar case, then clearly $Aa = (Aa_1, \ldots, Aa_N)$ is a c.c. also with the same λ. If $\tau \neq 0$, then $(\tau a_1, \tau a_2, \ldots, \tau a_N)$ is a c.c. also with λ replaced by λ/τ^3. Indeed, any configuration similar to a c.c. is a c.c. When counting c.c., one only counts similarity classes.

The Lagrange Central Configuration. Consider the c.c. formula (3.21) for the planar 3-body problem. Then we seek six unknowns, two components each for a_1, a_2, a_3. If we hold the center of mass at the origin, we can eliminate two variables; if we fix the moment of inertia I, we can reduce the dimension by one; and if we identify two configurations that differ by a rotation only, we can reduce the dimension by one again. Thus in theory you can reduce the problem by four dimensions, so that you have a problem of finding critical points of a function on a two-dimensional manifold. This reduction is difficult in general, but there is a trick that works well for the planar 3-body problem.

Let $\rho_{ij} = \|q_i - q_j\|$ denote the distance between the ith and j^{th} particles. Once the center of mass is fixed at the origin and two rotationally equivalent configurations are identified, then the three variables $\rho_{12}, \rho_{23}, \rho_{31}$ are local coordinates near a noncollinear configuration. That is, by specifying the angle between a fixed line and $q_2 - q_1$, the location of the center of mass, and the three variables $\rho_{12}, \rho_{23}, \rho_{31}$, then the configuration of the masses is uniquely specified. The function U is already written in terms of these variables because

$$U = \mathcal{G}\left(\frac{m_1 m_2}{\rho_{12}} + \frac{m_2 m_3}{\rho_{23}} + \frac{m_3 m_1}{\rho_{31}}\right). \tag{3.26}$$

Let M be the total mass, i.e., $M = \sum m_i$, and assume that the center of mass is at the origin; then

$$\sum_i \sum_j m_i m_j \rho_{ij}^2 = \sum_i \sum_j m_i m_j \|q_i - q_j\|^2$$

$$= \sum_i \sum_j m_i m_j \|q_i\|^2 - 2\sum_i \sum_j m_i m_j (q_i, q_j)$$

$$+ \sum_i \sum_j m_i m_j \|q_j\|^2$$

$$= 2MI - 2\sum_i m_i(q_i, \sum_j m_j q_j) + 2MI$$

$$= 4MI.$$

Thus, if the center of mass is fixed at the origin,

$$I = \frac{1}{4M} \sum_i \sum_j m_i m_j \rho_{ij}^2. \tag{3.27}$$

So, I can be written in terms of the mutual distances also. Holding I fixed is the same as holding $I^* = \frac{1}{2}(m_{12}\rho_{12}^2 + m_{23}\rho_{23}^2 + m_{31}\rho_{31}^2)$ fixed. Thus, the conditions for U to have a critical point on the set $I^* = $ constant in these coordinates are

$$-\mathcal{G}\frac{m_i m_j}{\rho_{ij}^2} + \lambda m_i m_j \rho_{ij} = 0, \quad (i,j) = (1,2), (2,3), (3,1), \tag{3.28}$$

which clearly has as its only solution $\rho_{12} = \rho_{23} = \rho_{31} = (\mathcal{G}/\lambda)^{-1/3}$. This solution is an equilateral triangle, and λ is a scale parameter. These solutions are attributed to Lagrange.

Theorem 3.5.1. *For any values of the masses, there are two and only two noncollinear central configurations for the 3-body problem, namely, the three particles are at the vertices of an equilateral triangle. The two solutions correspond to the two orientations of the triangle when labeled by the masses.*

It is trivial to see in these coordinates that the equilateral triangle c.c. is a nondegenerate minimum of the self-potential U.

The above argument would also show that for any values of the masses, there are two and only two noncoplanar c.c. for the 4 body problem, namely, the regular tetrahedron configuration with two orientations.

The Euler-Moulton Collinear Central Configurations. Consider the collinear N-body problem, so $q = (q_1, \ldots, q_N) \in \mathbb{R}^N$. Set $S' = \{q : I(q) = 1\}$, an ellipsoid or topological sphere of dimension $N-1$ in $\mathbb{R}^N$; set $D = \{C(q) = \sum m_i q_i = 0\}$, a plane of dimension $N-1$ in $\mathbb{R}^N$; and $S = S' \cap D$, a sphere of dimension $N-2$ in the plane D. Let $\Delta'_{ij} = \{q : q_i = q_j\}$ and $\Delta' = \cup \Delta'_{ij}$; so U is defined and smooth on $\mathbb{R}^N \backslash \Delta'$. Because Δ' is a union of planes through the origin, it intersects S in spheres of dimension $N-3$, denoted by Δ.

Let $\mathcal{U}$ be the restriction of U to $S \backslash \Delta$, so a critical point of $\mathcal{U}$ is a central configuration. Note that $S \backslash \Delta$ has $N!$ connected components. This is because a component of $S \backslash \Delta$ corresponds to a particular ordering of the q_is. That is, to each connected component there is an ordering $q_{i_1} < q_{i_2} < \cdots < q_{i_N}$ where $(i_1, i_2, \ldots, i_N)$ is a permutation of $1, 2, \ldots, N$. There are $N!$ such permutations. Because $\mathcal{U} \to \infty$ as $q \to \Delta$, the function $\mathcal{U}$ has at least one minimum per connected component. Thus there are at least $N!$ critical points.

Let a be a critical point of $\mathcal{U}$, so a satisfies (3.21) and $\lambda = U(a)/2I(a)$. The derivative of $\mathcal{U}$ at a in the direction $v = (v_1, \ldots, v_N) \in T_a S$ is

$$DU(a)(v) = -\sum \frac{\mathcal{G} m_i m_j (v_j - v_i)}{\|a_j - a_i\|} + \lambda \sum m_i a_i v_i, \qquad (3.29)$$

and the second derivative $D^2 \mathcal{U} = D^2 \mathcal{U}(a)(v, w)$ is

$$D^2 \mathcal{U} = 2 \sum \frac{\mathcal{G} m_i m_j}{\|a_j - a_i\|^3} ((w_j - w_i)(v_j - v_i)) + \lambda \sum m_i w_i v_i. \qquad (3.30)$$

From the above, $D^2 \mathcal{U}(a)(v, v) > 0$ when $v \neq 0$, so the Hessian is positive definite at a critical point and each such critical point is a local minimum of $\mathcal{U}$. Thus there can only be one critical point of $\mathcal{U}$ on each connected component, or there are $N!$ critical points.

In counting the critical points above, we have not removed the symmetry from the problem. The only one-dimensional orthogonal transformation is a reflection in the origin. When we counted a c.c. and its reflection we have counted each c.c. twice. Thus we have the following.

Theorem 3.5.2 (*Euler-Moulton*). *There are exactly $N!/2$ collinear central configurations in the N-body problem, one for each ordering of the masses on the line.*

These c.c. are minima of $\mathcal{U}$ only on the line. It can be shown that they are saddle points in the planar problem.

Euler gave these solutions for $n = 3$ and the complete solution for all n was given in Moulton. The literature on c.c. is vast as one can easily see by search of the web.

3.5.3 Rotating Coordinates

Planar central configurations give rise to equilibria in rotating coordinates called *relative equilibria*. Let

$$K = \begin{bmatrix} 0 & 1 \\ -1 & 0 \end{bmatrix}, \qquad \exp(\omega K t) = \begin{bmatrix} \cos \omega t & \sin \omega t \\ -\sin \omega t & \cos \omega t \end{bmatrix} \tag{3.31}$$

be 2×2 matrices, and consider the planar N-body problem; so, the vectors q_i, p_i in Section 3.1 are 2-vectors. Introduce a set of coordinates that uniformly rotate with frequency ω by

$$u_i = \exp(\omega K t) q_i, \quad v_i = \exp(\omega K t) p_i. \tag{3.32}$$

Because K is skew-symmetric, $\exp(\omega K t)$ is orthogonal for all t, so, the change of variables is symplectic. The remainder function is $-\Sigma \omega u_i^T K v_i$, and so the Hamiltonian of the N-body problem in rotating coordinates is

$$H = \sum_{i=1}^{N} \frac{\|v_i\|^2}{2m_i} - \sum_{i=1}^{N} \omega u_i^T K v_i - U(u), \tag{3.33}$$

and the equations of motion are

$$\dot{u}_i = v_i + \omega K u_i, \qquad \dot{v}_i = \omega K v_i + \frac{\partial U}{\partial u_i}.$$

The remainder term gives rise to extra terms in the equations of motion that are sometimes called Coriolis forces.

To find an equilibria in rotating coordinates solve $\dot{u}_i = \dot{v}_i = 0$ so get

$$\omega^2 u_i + \frac{\partial U}{\partial u_i} = 0,$$

which is the same as (3.23) with $\lambda = \omega^2$.

For the N-body problem in $\mathbb{R}^3$ simply let

$$K = \begin{bmatrix} 0 & 1 & 0 \\ -1 & 0 & 0 \\ 0 & 0 & 0 \end{bmatrix}.$$

3.6 Total Collapse

There is an interesting differential formula relating I and the various energies of the system.

Lemma 3.6.1 (Lagrange–Jacobi formula). *Let I be the moment of inertia, T be the kinetic energy, U the self potential, and h the total energy of the system of N-bodies, then*

$$\ddot{I} = 2T - U = T + h \tag{3.34}$$

Proof. Starting with (3.22) differentiate I twice with respect to t and use (3.3), (3.6), and (3.18) to get

$$\ddot{I} = \sum_{1}^{N} m_i \dot{q}_i \cdot \dot{q}_i + \sum_{1}^{N} m_i q_i \cdot \ddot{q}_i$$

$$= \sum_{1}^{N} m_i \|\dot{q}_i\|^2 + \sum_{1}^{N} q_i \cdot \frac{\partial U}{\partial q_i}$$

$$= 2T - U.$$

This formula and its variations are known as the Lagrange–Jacobi formula and it is used extensively in the studies of the growth and collapse of gravitational systems. We give only one simple, but important application.

First we need another basic result.

Lemma 3.6.2 (Sundman's inequality). *Let $c = \|A\|$ be the magnitude of angular momentum and $h = T - U$ the total energy of the system, then*

$$c^2 \le 4I(\ddot{I} - h). \tag{3.35}$$

Proof. Note

$$c = \|A\| = \|\sum m_i q_i \times \dot{q}_i\|$$

$$\le \sum m_i \|q_i\| \|\dot{q}_i\| = \sum (\sqrt{m_i} \|q_i\|)(\sqrt{m_i} \|\dot{q}_i\|).$$

Now apply Cauchy's inequality to the right side of the above to conclude

$$c^2 \le \sum m_i \|q_i\|^2 \sum m_i \|\dot{q}_i\|^2 = 2I2T.$$

The conclusion follows at once from the Lagrange–Jacobi formula.

Theorem 3.6.1 (Sundman's theorem on total collapse). *If total collapse occurs then angular momentum is zero and it will only take a finite amount of time. That is, if $I(t) \to 0$ as $t \to t_1$ then $t_1 < \infty$ and $A = 0$.*

Proof. Let h be the total energy of the system, so by (3.34) $\ddot{I} = T + h$. Assume $I(t)$ is defined for all $t \geq 0$ and $I \to 0$ as $t \to \infty$. Then $U \to \infty$ and because h is constant $T \to \infty$ also. So there is a $t^* > 0$ such that $\ddot{I} \geq 1$ for $t \geq t^*$. Integrate this inequality to get $I(t) \geq \frac{1}{2}t^2 + at + b$ for $t \geq t^*$ where a and b are constants. But this contradicts total collapse, so total collapse can only take a finite amount of time.

Now suppose that $I \to 0$ as $t \to t_1^- < \infty$ and so as before $U \to \infty$ and $\ddot{I} \to \infty$. Thus, there is a t_2 such that $\ddot{I}(t) > 0$ on $t_2 \leq t < t_1$. Because $I(t) > 0$, $\ddot{I} > 0$ on $t_2 \leq t < t_1$, and $I(t) \to 0$ as $t \to t_1$ it follows that $\dot{I} \leq 0$ on $t_2 \leq t < t_1$.

Now multiply both sides of Sundman's inequality (3.35) by $-\dot{I}I^{-1} > 0$ to get

$$-\frac{1}{4}c^2 \dot{I}I^{-1} \leq h\dot{I} - \dot{I}\ddot{I}.$$

Integrate this inequality to get

$$\frac{1}{4}c^2 \log I^{-1} \leq hI - \frac{1}{2}\dot{I}^2 + K \leq hI + K$$

where K is an integration constant. Thus

$$\frac{1}{4}c^2 \leq \frac{hI + K}{\log I^{-1}}.$$

As $t \to t_1$, $I \to 0$ and so the right side of the above tends to zero. But this implies $c = 0$

3.7 Problems

1. Draw the complete phase portrait of the collinear Kepler problem. Integrate the collinear Kepler problem.
2. Show that $\mu^2(\epsilon^2 - 1) = 2hc$ for the Kepler problem.
3. The area of an ellipse is $\pi a^2 (1-\epsilon^2)^{1/2}$, where a is the semi-major axis. We have seen in Kepler's problem that area is swept out at a constant rate of $c/2$. Prove Kepler's third law: The period p of a particle in a circular or elliptic orbit ($\epsilon < 1$) of the Kepler problem is $p = (2\pi/\sqrt{\mu})a^{3/2}$.
4. Let

$$K = \begin{bmatrix} 0 & 1 \\ -1 & 0 \end{bmatrix};$$

then

$$\exp(Kt) = \begin{bmatrix} \cos t & \sin t \\ -\sin t & \cos t \end{bmatrix}.$$

Find a circular solution of the two-dimensional Kepler problem of the form $q = \exp(Kt)a$ where a is a constant vector.

5. Assume that a particular solution of the N-body problem exists for all $t > 0$ with $h > 0$. Show that $U \to \infty$ as $t \to \infty$. Does this imply that the distance between one pair of particles goes to infinity? (No.)

6. Scale the Hamiltonian of the N-body problem in rotating coordinates, so that ω is 1.

7. There is a general theorem which says that all Lie groups are just closed subgroups of the general linear group $Gl(m, \mathbb{R})$. $\mathbb{R}$ is a Lie group where the group action is addition. This Lie group is the same as the matrix group consisting of all 2×2 matrices of the form

$$\begin{bmatrix} 1 & a \\ 0 & 1 \end{bmatrix}.$$

Use this fact to show that the translational symmetry of the n-body problem can be realized as a Lie group action. Thus give an alternate proof of the conservation of linear momentum.

4. The Restricted Problem

A special case of the 3-body problem is the limiting case in which one of the masses tends to zero. In the traditional derivation of the restricted 3-body problem, one is asked to consider the motion of a particle of infinitesimal mass moving in the plane under the influence of the gravitational attraction of two finite particles that move around each other on a circular orbit of the Kepler problem. Although this description is picturesque, it hardly clarifies the relationship between the restricted 3-body problem and the full problem. Here we shall find the equations of first approximation when one mass tends to zero which then defines the restricted problem.

4.1 Defining

Consider the 3-body problem in rotating coordinates (3.33) with $N = 3$ and $\omega = 1$. Let the third mass be small by setting $m_3 = \epsilon^2$ and considering ϵ as a small positive parameter. Making this substitution into (3.33) and rearranging terms gives

$$H_3 = \frac{\|v_3\|^2}{2\epsilon^2} - u_3^T K v_3 - \sum_{i=1}^{2} \frac{\epsilon^2 m_i}{\|u_i - u_3\|} + H_2.$$

Here H_2 is the Hamiltonian of the 2-body problem in rotating coordinates, i.e., (3.33) with $N = 2$. ϵ is a small parameter that measures the smallness of one mass. A small mass should make a small perturbation on the other particles, thus, we should attempt to make ϵ measure the deviation of the motion of the two finite particles from a circular orbit. That is, ϵ should measure the smallness of the mass and how close the two finite particles' orbits are to circular. To accomplish this we must prepare the Hamiltonian so that one variable represents the deviation from a circular orbit.

© Springer International Publishing AG 2017
K.R. Meyer, D.C. Offin, *Introduction to Hamiltonian Dynamical Systems and the N-Body Problem*, Applied Mathematical Sciences 90, DOI 10.1007/978-3-319-53691-0_4

Let $Z = (u_1, u_2, v_1, v_2)$, so, H_2 is a function of the 8-vector Z. A circular solution of the 2-body problem is a critical point of the Hamiltonian of the 2-body problem in rotating coordinates; i.e., H_2. Let $Z^* = (a_1, a_2, b_1, b_2)$ be such a critical point (later we specify Z^*). By Taylor's theorem

$$H_2(Z) = H_2(Z^*) + \frac{1}{2}(Z - Z^*)^T S(Z - Z^*) + O(\|Z - Z^*\|^3),$$

where S is the Hessian of H_2 at Z^*. Because the equations of motion do not depend on constants, drop the constant term in the above. If the motion of the two finite particles were nearly circular, the $Z - Z^*$ would be small; so this suggests that one should change variables by $Z - Z^* = \epsilon\Psi$, but to make the change of variables symplectic, you must also change coordinates by $u_3 = \xi, v_3 = \epsilon^2\eta$, which gives a symplectic change of variables with multiplier ϵ^{-2}. The Hamiltonian becomes

$$H_3 = \left\{ \frac{\|\eta\|^2}{2} - \xi^T K\eta - \sum_{i=1}^{2} \frac{m_i}{\|\xi - a_i\|} \right\} + \frac{1}{2}\Psi^T S\Psi + O(\epsilon).$$

The quantity in the braces in the above is the Hamiltonian of the restricted 3-body problem, if we take $m_1 = \mu, m_2 = 1 - \mu, a_1 = (1 - \mu, 0)$, and $a_2 = (-\mu, 0)$. The quadratic term above is simply the linearized equations about the circular solutions of the 2-body problem in rotating coordinates. Thus to first order in ϵ the Hamiltonian of the full 3-body problem is the sum of the Hamiltonian for the restricted problem and the Hamiltonian of the linearized equations about the circular solution. In other words the equations of the full 3-body problem decouples into the equations for the restricted problem and the linearized equations about the circular solution.

In Chapter 9, this scaled version of the restricted problem is used to prove that nondegenerate periodic solutions of the restricted problem can be continued into the full 3-body problem for small mass.

Putting it all together defines the (planar, circular) restricted 3-body problem. The two finite particles, called the *primaries*, have mass $\mu > 0$ and $1 - \mu > 0$, $x \in \mathbb{R}^2$ is the coordinate of the *infinitesimal particle* in uniformly rotating coordinate system and $y \in \mathbb{R}^2$ the momentum conjugate to x. The rotating coordinate system is so chosen that the particle of mass μ is always at $(1-\mu, 0)$ and the particle of mass $1-\mu$ is at $(-\mu, 0)$. The Hamiltonian governing the motion of the third (infinitesimal) particle in these coordinates is

$$H = \frac{1}{2}\|y\|^2 - x^T Ky - U, \qquad (4.1)$$

where $x, y \in \mathbb{R}^2$ are conjugate,

$$K = J_2 = \begin{bmatrix} 0 & 1 \\ -1 & 0 \end{bmatrix},$$

and U is the self-potential

$$U = \frac{\mu}{d_1} + \frac{1-\mu}{d_2}, \tag{4.2}$$

with d_i the distance from the infinitesimal body to the ith primary, or

$$d_1^2 = (x_1 - 1 + \mu)^2 + x_2^2, \quad d_2^2 = (x_1 + \mu)^2 + x_2^2. \tag{4.3}$$

The equations of motion are

$$\dot{x} = \frac{\partial H}{\partial y} = y + Kx$$

$$\dot{y} = -\frac{\partial H}{\partial x} = Ky + \frac{\partial U}{\partial x}. \tag{4.4}$$

The term $x^T K y$ in the Hamiltonian H reflects the fact that the coordinate system is not a Newtonian system, but a rotating coordinate system. It gives rise to the Coriolis forces in the equations of motion (4.4). The line joining the masses is known as the line of syzygy.

The proper definition of the restricted 3-body problem is the system of differential equations (4.4) defined by the Hamiltonian in (4.1). It is a two degree of freedom problem that seems simple but has defied integration. It has given rise to an extensive body of research. We return to this problem often in the subsequent chapters.

In much of the literature, the equations of motion for the restricted problem are written as a second-order equation in the position variable x. Eliminating y from Equation (4.4) gives

$$\ddot{x} - 2K\dot{x} - x = \frac{\partial U}{\partial x}, \tag{4.5}$$

and the integral H becomes

$$H = \frac{1}{2}\|\dot{x}\|^2 - \frac{1}{2}\|x\|^2 - U. \tag{4.6}$$

Usually in this case one refers to the Jacobi constant, C, as the integral of motion with $C = -2H + \nu(1-\mu)$, i.e.,

$$C = \|x\|^2 + 2U + (\mu(1-\mu) - \|\dot{x}\|^2, \tag{4.7}$$

and one refers to

$$W = \|x\|^2 + 2U + \mu(1-\mu) \tag{4.8}$$

as the amended potential for the restricted 3-body problem.

The amended potential is positive and it tends to infinity as $x \to \infty$ or when x tends to a primary – see Figure 4.1.[1]

---

[1]This figure was constructed by Patricia Yanguas and Jesús Palacián.

Figure 4.1. The amended potential.

Evidently Euler considered a restricted problem, but it was Jacobi who formulated it as a conservative system and Poincaré who called it the restricted problem. The spatial restricted 3-body problem is essentially the same, but we need to replace $K = J_2$ by

$$K = \begin{bmatrix} 0 & 1 & 0 \\ -1 & 0 & 0 \\ 0 & 0 & 0 \end{bmatrix}$$

throughout. Use

$$d_1^2 = (x_1 - 1 + \mu)^2 + x_2^2 + x_3^2, \quad d_2^2 = (x_1 + \mu)^2 + x_2^2 + x_3^2,$$

in the definition of U, and note that the amended potential becomes

$$W = x_1^2 + x_2^2 + 2U + \mu(1 - \mu),$$

with a corresponding change in the Jacobi constant.

4.2 Discrete Symmetry

The Hamiltonian of the restricted problem (4.1) has the following symmetry

$$H(x_1, x_2, y_1, y_2) = H(x_1, -x_2, -y_1, y_2) \tag{4.9}$$

and such a system is said to be reversible. It follows that to any solution $(x_1(t), x_2(t), y_1(t), y_2(t))$ there corresponds another solution $(x_1(-t), -x_2(-t), -y_1(-t), y_2(-t))$. The second is the reflection of the first in the x_1 axis with time reversed (and vise versa).

An orbit is symmetric if it is its own reflection. That means that somewhere along the orbit there is a point such that $x_2 = y_1 = 0$, i.e., at some time it crosses the x_1 axis perpendicularly. If there are two such points the orbit is periodic. Most studies of periodic orbits have been limited to symmetric periodic orbits, mostly because their numerical computation is shorter and their classification is easier, but also because many simple orbits of interest fall into this category. A systematic and quite complete description of the symmetric periodic orbits of the restricted problem can be found in the book by Hénon (1997).

There is another symmetry involving the parameter μ of the problem. A symmetry with respect to the x_2 axis exchanges μ and $(1 - \mu)$. One can take advantage of this symmetry and restrict the range of the parameter μ to the interval $(0, 1/2]$. Note that for $\mu = 1/2$, there is a second symmetry of the phase space; namely

$$H(x_1, x_2, y_1, y_2) = H(-x_1, x_2, y_1, -y_2).$$

This implies that to any solution $(x_1(t), x_2(t), y_1(t), y_2(t))$ there corresponds another solution $(-x_1(-t), x_2(-t), y_1(-t), -y_2(-t))$. This one can be interpreted as a reflection with respect to the x_2 axis plus a reversal of time.

One obtains a doubly symmetry period solution if a solution starts perpendicular to the x_1 axis and then at a later time, $T > 0$, crosses the y_1 axis perpendicularly – the period is $4T$. One only needs to compute a quarter of an orbit. In the 1920s and 30s many doubly symmetric periodic solutions were computed by human computers at the Copenhagen observatory by Strömgren and followers (see for instance Strömgren (1935)). That is why the restricted problem with $\mu = \frac{1}{2}$ is sometimes called the Copenhagen problem.

4.3 Equilibria of the Restricted Problem

The full 3-body problem has no equilibrium points, but as we have seen there are solutions of the planar problem in which the particles move on uniformly rotating solutions. In particular, there are the solutions in which the particles move along the equilateral triangular solutions of Lagrange, and there are also the collinear solutions of Euler. These solutions would be equilibrium solutions in a rotating coordinates system. Because the restricted 3-body problem is a limiting case in rotating coordinates, we expect to see vestiges of these solutions as equilibria.

From (4.4), an equilibrium solution for the restricted problem would satisfy

$$0 = y + Kx, \qquad 0 = Ky + \frac{\partial U}{\partial x}, \tag{4.10}$$

which implies

$$0 = x + \frac{\partial U}{\partial x} \quad \text{or} \quad 0 = \frac{\partial W}{\partial x}. \tag{4.11}$$

Thus an equilibrium solution is a critical point of the amended potential. By looking at Figure 4.1 we can see three saddle type critical points which are collinear with the primaries and two minimum critical points on either side of the line of primaries in red.

First, seek solutions that do not lie on the line joining the primaries. As in the discussion of the Lagrange c.c., use the distances d_1, d_2 given in (4.3) as coordinates. From (4.3), we obtain the identity

$$x_1^2 + x_2^2 = \mu d_1^2 + (1 - \mu)d_2^2 - \mu(1 - \mu), \tag{4.12}$$

so W can be written

$$W = \mu d_1^2 + (1 - \mu)d_2^2 + \frac{2\mu}{d_1} + \frac{2(1 - \mu)}{d_2}. \tag{4.13}$$

The equation $\partial W / \partial x = 0$ in these variables becomes

$$\mu d_1 - \frac{\mu}{d_1^2} = 0, \qquad (1 - \mu)d_2 - \frac{(1 - \mu)}{d_2^2} = 0, \tag{4.14}$$

which clearly has the unique solution $d_1 = d_2 = 1$. This solution lies at the vertex of an equilateral triangle whose base is the line segment joining the two primaries. Because there are two orientations, there are two such equilibrium solutions: one in the upper half-plane denoted by $\mathcal{L}_4$, and one in the lower half-plane denoted by $\mathcal{L}_5$. The Hessian of W at these equilibria is

$$\frac{\partial^2 W}{\partial d^2} = \begin{bmatrix} 6\mu & 0 \\ 0 & 6(1 - \mu) \end{bmatrix},$$

and so W has a minimum at each equilibrium and takes the minimum value 3.

These solutions are attributed to Lagrange. Lagrange thought that they had no astronomical significance, but in the twentieth century, hundreds of asteroids, the Trojans, were found oscillating around the $\mathcal{L}_4$ position in the sun–Jupiter system and a similar number, the Greeks, were found oscillating about the $\mathcal{L}_5$ position. That is, one group of asteroids, the sun, and Jupiter form an equilateral triangle, approximately, and so does the other group. With better telescopes more and more asteroids have been found.

Now consider equilibria along the line of the primaries where $x_2 = 0$. In this case, the amended potential is a function of x_1, which we denote by x for the present, and so W has the form

$$W = x^2 \pm \frac{2\mu}{(x - 1 + \mu)} \pm \frac{2(1 - \mu)}{(x + \mu)}. \tag{4.15}$$

In the above, one takes the signs so that each term is positive. There are three cases: (i) $x < -\mu$, where the signs are $-$ and $-$; (ii) $-\mu < x < 1 - \mu$, where the signs are $-$ and $+$; and (iii) $1 - \mu < x$, where the signs are $+$

and $+$. Clearly $W \to \infty$ as $x \to \pm\infty$, as $x \to -\mu$, or as $x \to 1 - \mu$, so W has at least one critical point on each of these three intervals. Also

$$\frac{d^2W}{dx^2} = 2 \pm \frac{2\mu}{(x - 1 + \mu)^3} \pm \frac{2(1 - \mu)}{(x + \mu)^3}, \qquad (4.16)$$

where the signs are again taken so that each term is positive, so, W is a convex function. Therefore, W has precisely one critical point in each of

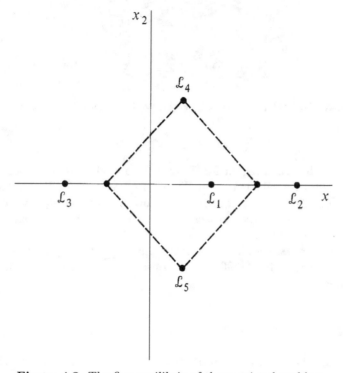

Figure 4.2. The five equilibria of the restricted problem.

these intervals, or three critical points. These three collinear equilibria are attributed to Euler and are denoted by $\mathcal{L}_1, \mathcal{L}_2$, and $\mathcal{L}_3$ as shown in Figure 4.2. In classical celestial mechanics literature such equilibrium points are called libration points, hence the use of the symbol $\mathcal{L}$.vspace*-8pt

4.4 Hill's Regions

The Jacobi constant is $C = W - \|\dot{x}\|$, where W is the amended potential and so $W \geq C$. This inequality places a constraint on the position variable x for each value of C, and if x satisfies this condition, then there is a solution of the restricted problem through x for that value of C. The set

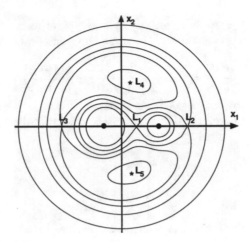

Figure 4.3. Zero-velocity curves for some typical values of the Jacobi constant.

$$\mathfrak{H}(C) = \{x : W(x) \geq C\}$$

is known as the Hill's region for C, and its boundary where equality holds is called a zero velocity curve. In Figure 4.1 level sets of W are the curves where the color changes, but for a less colorful depiction see Figure 4.3.

As seen before, W has critical points at the libration points $\mathcal{L}_i$, $i = 1, \ldots, 5$. Let the collinear points $\mathcal{L}_i$, $i = 1, 2, 3$ are minima of W along the x_1-axis, but they are saddle points in the plane and $\mathcal{L}_i$, $i = 4, 5$ are minima in the plane. See Szebehely (1967) for a complete analysis of the Hill's regions with figures.

4.5 Spectrum at the Equilibrium Points

Here we start the linear analysis of the equilibria by looking at the characteristic equations of the linearized equations at the equilibria. In later chapters we will study the stability of these points and the occurrence of nearby periodic solutions. The presentation given in this section and the next is due to Professor Dieter S. Schmidt with our thanks.

If x_1, x_2 is a critical point of the amended potential (4.8), then x_1, x_2, $y_1 = -x_2$, $y_2 = x_1$ is an equilibrium point. Let ξ_1 and ξ_2 be one of the five critical points. In order to study the motion near this equilibrium point, we translate to new coordinates by

$$u_1 = x_1 - \xi_1, \qquad v_1 = y_1 + \xi_2,$$

$$u_2 = x_2 - \xi_2, \qquad v_2 = y_2 - \xi_1.$$

This translation to the new coordinates (u_1, u_2, v_1, v_2) is obviously symplectic, so, we can perform this change of coordinates in the Hamiltonian (4.1) and preserve its structure. Expanding through second-order terms in the new variables, we obtain

$$H = \frac{1}{2}(v_1^2 + v_2^2) + u_2 v_1 - u_1 v_2 - \frac{1}{2}\left(U_{x_1 x_1} u_1^2 + 2 U_{x_1 x_2} u_1 u_2 + U_{x_2 x_2} u_2^2\right) + \cdots.$$

There are no linear terms because the expansion is performed at an equilibrium and the constant term has been omitted because it contributes nothing in forming the corresponding system of differential equations. The above quadratic Hamiltonian function gives rise to the following Hamiltonian matrix

$$\begin{bmatrix} 0 & 1 & 1 & 0 \\ -1 & 0 & 0 & 1 \\ U_{x_1 x_1} & U_{x_1 x_2} & 0 & 1 \\ U_{x_1 x_2} & U_{x_2 x_2} & -1 & 0 \end{bmatrix}. \tag{4.17}$$

The following computation is a little more compact by introducing

$$V = \frac{1}{2}(x_1^2 + x_2^2) + U(X_1, x_2).$$

The eigenvalues of the matrix (4.17) determine the behavior of the linearized system. The characteristic equation is

$$\lambda^4 + (4 - V_{x_1 x_1} - V_{x_2 x_2})\lambda^2 + V_{x_1 x_1} V_{x_2 x_2} - V_{x_1 x_2}^2 = 0.$$

The partial derivatives are

$$V_{x_1 x_1} = 1 + (1 - \mu)\frac{3(x_1 + \mu)^2 - d_1^2}{d_1^5} + \mu\frac{3(x_1 + \mu - 1)^2 - d_2^2}{d_2^5},$$

$$V_{x_1 x_2} = 3 x_1 x_2\left(\frac{1 - \mu}{d_1^5} + \frac{\mu}{d_2^5}\right),$$

$$V_{x_2 x_2} = 1 + (1 - \mu)\frac{3 x_2^2 - d_1^2}{d_1^5} + \mu\frac{3 x_2^2 - d_2^2}{d_2^5}.$$

They have to be evaluated at the critical points. Thus we have to consider the collinear points and the triangular points separately.

Lemma 4.5.1. *At the collinear points, the matrix (4.17) has two real eigenvalues and two purely imaginary eigenvalues.*

Proof. By direct computation one finds that for the collinear points

$$V_{x_1 x_1} = 1 + 2(1 - \mu)d_1^{-3} + 2\mu d_2^{-3} > 0$$
$$V_{x_1 x_2} = 0$$
$$V_{x_2 x_2} = 1 - (1 - \mu)d_1^{-3}\mu d_2^{-3} < 0.$$

Only the last statement requires some additional work. We present it for $\mathcal{L}_1$ and leave the other cases as exercises.

If $(\xi_1, 0)$ are the coordinates of the Eulerian point $\mathcal{L}_1$, then $d_1 = \xi_1 + \mu$, $d_2 = \xi_1 - 1 + \mu$, and ξ_1 is the real solution of $V_{x_1} = 0$, that is, of a quintic polynomial

$$\xi_1 - (1 - \mu)d_1^{-2} - \mu d_2^{-2} = 0.$$

We use this relationship in the form

$$(1 - \mu)d_1^{-2} = d_1 - \mu d_2^{-2} - \mu$$

when we evaluate the second derivative of V at $(\xi_1, 0)$; that is, we get

$$V_{x_2 x_2} = 1 - \frac{1}{d_1}(d_1 - \mu d_2^{-2} - \mu) - \mu d_2^{-3}$$

$$= \frac{\mu}{d_1}(1 + d_2^{-2} - d_1 d_2^{-3})$$

$$= \frac{\mu}{d_1}(1 - d_2^{-3}) < 0.$$

The last equality follows from $d_1 = 1 + d_2$ and the inequality follows then from the fact that $0 < d_2 < 1$.

Setting $A = 2 - \frac{1}{2}(V_{x_1 x_1} + V_{x_2 x_2})$ and $B = V_{x_1 x_1} V_{x_2 x_2}$ the characteristic equation for the collinear points takes on the form

$$\lambda^4 + 2A\lambda^2 - B = 0$$

with the solutions

$$\lambda^2 = -A \pm \sqrt{A^2 + B}.$$

Because $B > 0$ the statement of the lemma follows. It also means that the collinear points of Euler are unstable. Therefore, some solutions that start near the Euler points will tend away from these points as time tends to infinity.

Lemma 4.5.2. *At the triangular equilibrium points, the matrix (4.17) has purely imaginary eigenvalues for values of the mass ratio μ in the interval $0 < \mu < \mu_1$, where $\mu_1 = \frac{1}{2}(1 - \sqrt{69}/9)$. For $\mu = \mu_1$ the matrix has the repeated eigenvalues $\pm i\sqrt{2}/2$ with nonelementary divisors. For $\mu_1 < \mu \leq \frac{1}{2}$, the eigenvalues are off the imaginary axis. (μ_1 is called Routh's critical mass ratio.)*

Proof. Because the coordinates for the Lagrangian point $\mathcal{L}_4$ have been found to be $\xi_1 = \frac{1}{2} - \mu$ and $\xi_2 = \frac{1}{2}\sqrt{3}$, the second derivatives of V can be computed explicitly. They are

$$V_{x_1 x_1} = \frac{3}{4}, \quad V_{x_1 x_2} = -\frac{3\sqrt{3}}{4}(1 - 2\mu), \quad V_{x_2 x_2} = \frac{9}{4}.$$

The characteristic equation for (4.17) is then

$$\lambda^4 + \lambda^2 + \frac{27}{4}\mu(1 - \mu) = 0. \tag{4.18}$$

It has the roots

$$\lambda^2 = \frac{1}{2}\{-1 \pm \sqrt{1 - 27\mu(1 - \mu)}\}. \tag{4.19}$$

When the above square root is zero, we have the double eigenvalues $\pm i\sqrt{2}/2$. This occurs for $\mu = \mu_1 = \frac{1}{2}(1 - \sqrt{69}/9)$, that is, for Routh's critical mass ratio (and due to symmetry also for $1 - \mu_1$). It can be seen that the matrix (4.17) has nonsimple elementary divisors, which means it is not diagonalizable. We return to this case later on.

For $\mu_1 < \mu < 1 - \mu_1$, the square root in (4.19) produces imaginary values, and so λ will be complex with nonzero real part. The eigenvalues of (4.17) lie off the imaginary axis, and the triangular Lagrangian points cannot be stable. In this case the equilibrium is said to be hyperbolic.

This leaves the interval $0 < \mu < \mu_1$ (and $1 - \mu_1 < \mu < 1$) where the matrix (4.17) has purely imaginary eigenvalues of the form $\pm i\omega_1$ and $\pm i\omega_2$. We adopt the convention that ω_1 will be the larger of the two values so that ω_1 and ω_2 are uniquely defined by the conditions that follow from (4.18),

$$0 < \omega_2 < \frac{\sqrt{2}}{2} < \omega_1,$$

$$\omega_1^2 + \omega_2^2 = 1, \tag{4.20}$$

$$\omega_1^2 \omega_2^2 = \frac{27\mu(1 - \mu)}{4}.$$

4.6 Mass Ratios

As we shall see later other particular values of the mass-ratio μ are of interest. They are the $\mu_{\alpha/\beta}$ for which the frequencies at the equilibrium are in (α, β) resonance (i.e., $\beta\omega_1 = \alpha\omega_2$). Of special interest are the μ_α, when $\beta = 1$. Three other special mass ratios μ_d, μ^* and μ^{**} will also be considered. These values are listed in the following table together with the value μ_{EM} of the mass-ratio of the system Earth-Moon and the value μ_{SJ} of the system Sun-Jupiter. Except for the couple Pluto-Charon, with a mass-ratio of approximately 0.22, all the other systems "Sun-Planet" or "Planet-Satellite" in the Solar system have mass-ratios smaller (and usually much smaller) than μ_{15} (Table 4.1).

In the following chapters we will investigate in detail the classification of orbits in the vicinity of the L_4 and L_5 points, particularly for $\mu \le \mu_1$.

Table 4.1. Critical values of the mass-ratio μ.

μ_1	$= 0.038520896504551$	μ_7	$= 0.002912184522396$
μ_2	$= 0.024293897142052$	μ_8	$= 0.002249196513710$
μ^{**}	$= 0.02072$	μ_9	$=0.001787848394744$
μ_3	$= 0.013516016022453$	μ_{10}	$=0.001454405739621$
μ_{EM}	$= 0.01215002$	μ_{11}	$=0.001205829591109$
μ^*	≈ 0.01228	μ_{12}	$= 0.001015696721082$
μ_d	$= 0.010913667677201$	μ_{SJ}	$=0.000953875$
μ_4	$= 0.008270372663897$	μ_{13}	$=0.000867085298404$
μ_5	$= 0.005509202949840$	μ_{14}	$=0.000748764338855$
μ_6	$= 0.003911084259658$	μ_{15}	$=0.000653048708761$

4.7 Canonical Forms for the Matrix at $\mathcal{L}_4$

We now turn our attention to finding the Hamiltonian normal form for the coefficient matrix of the linearized equations at $\mathcal{L}_4$, $\mathcal{L}_5$ when the mass ratio μ is smaller than Routh's critical value μ_1. See Section 5.4 for a general discussion of normal forms for Hamiltonian matrices.

The quadratic part of the Hamiltonian function at $\mathcal{L}_4$ is

$$Q = \frac{1}{2}(v_1^2 + v_2^2) + u_2 v_1 - u_1 v_2 + \frac{1}{8}u_1^2 - \frac{3\sqrt{3}}{4}\mu(1-\mu)u_1 u_2 - \frac{5}{8}u_2^2.$$

We construct the symplectic linear transformation which brings this Hamiltonian function into its normal form. In terms of complex coordinates, this normal form turns out to be

$$K = -i\omega_1 z_1 \bar{z}_1 + i\omega_2 z_2 \bar{z}_2.$$

It is the Hamiltonian function for two harmonic oscillators with frequencies ω_1 and ω_2. Because the original Hamiltonian was indefinite, the two terms do not have the same sign.

When we perform these calculations it is not very convenient to work with the parameter μ. It hides the symmetry of the problem with respect to $\mu = \frac{1}{2}$. The calculations are simpler if we use $1 - 2\mu$ as a parameter instead of μ. At the same time we can simplify the calculations further by absorbing the factor $3\sqrt{3}$ into this parameter. We thus introduce

$$\gamma = 3\sqrt{3}(1 - 2\mu).$$

The other difficulty in performing the calculations by hand and even more so by machine has to do with the fact that the expressions for ω_1 and ω_2 are rather lengthy and it is easier to express everything in terms of these variables instead of μ. But ω_1 and ω_2 are not independent as (4.20) shows. Indeed, in order to simplify intermediate results, one has to use these relationships. One strategy is to replace ω_2^2 by $1 - \omega_1^2$ whenever it occurs and thus restrict the exponents of w_2 to 0 and 1. But most expressions are shorter if the symmetry between the frequencies w_1 and w_2 is preserved within the formulas.

Our approach reduces the problem so that it has the minimum number of essential parameters. We divide the Hamiltonian function by w_1, and set $w = w_2/w_1$. Due to our previous convention for w_1 and w_2 one sees that w lies in $0 < w < 1$. We use the second formula in (4.19) to express the terms containing w_1 and w_2 as a function of w. The third relationship in (4.19) then reads

$$\frac{16w^2}{(1+w^2)^2} = 27 - \gamma^2$$

or

$$\gamma^2 - \frac{27 + 38w^2 + 27w^4}{(1+w^2)^2}.$$

The last form is used to limit the exponent of γ to 0 and 1 in all intermediate expressions.

The Hamiltonian matrix derived from (4.17) is

$$A = \frac{1}{w_1} \begin{bmatrix} 0 & 1 & 1 & 0 \\ -1 & 0 & 0 & 1 \\ -1/4 & \gamma/4 & 0 & 1 \\ \gamma/4 & 5/4 & -1 & 0 \end{bmatrix}. \tag{4.21}$$

Its eigenvalues are $\pm i$ and $\pm iw$. The eigenvectors belonging to $+i$ and to $+iw$ are denoted by α_1 and α_2, respectively. They are given by

$$\alpha_1 = \begin{bmatrix} 1 \\ \dfrac{-(w^2+1)\gamma + 8i\sqrt{w^2+1}}{9w^2+13} \\ \dfrac{(w^2+1)\gamma + i(w^2+5)/\sqrt{w^2+1}}{9w^2+13} \\ \dfrac{9w^2+5 - i\gamma\sqrt{w^2+1}}{9w^2+13} \end{bmatrix},$$

$$\alpha_2 = \begin{bmatrix} 1 \\[2mm] \dfrac{-(\omega^2+1)\gamma + 8i\omega\sqrt{\omega^2+1}}{13\omega^2+9} \\[4mm] \dfrac{(\omega^2+1)\gamma + i\omega(5\omega^2+1)/\sqrt{\omega^2+1}}{13\omega^2+9} \\[4mm] \dfrac{5\omega^2+9 - i\gamma\omega\sqrt{\omega^2+1}}{13\omega^2+9} \end{bmatrix}.$$

Because $\alpha_1^T J \bar{\alpha}_1 = -ir_1^2/2$ and $\alpha_2^T J \bar{\alpha}_2 = ir_2^2/2$, where r_1 and r_2 are the positive real roots of

$$r_1^2 = \frac{16(1-\omega^2)}{\sqrt{\omega^2+1}(9\omega^2+13)} \quad \text{and} \quad r_2^2 = \frac{16(1-\omega^2)}{\sqrt{\omega^2+1}(13\omega^2+9)},$$

respectively, we create the transformation matrix T to the new set of complex-valued variables $(z_1, z_2, \bar{z}_1, \bar{z}_2)$ by

$$T = (\bar{\alpha}_1/r_1, \alpha_2/r_2, \alpha_1/r_1, \bar{\alpha}_2/r_2).$$

Because we have $T^T J T = \frac{1}{2}iJ$, the transformation is symplectic with multiplier $\frac{1}{2}i$. The old and new Hamiltonians are related by

$$K(z_1, z_2, \bar{z}_1, \bar{z}_2) = -2iQ(u_1, u_2, v_1, v_2),$$

which leads to

$$K = -iz_1\bar{z}_1 + i\omega z_2\bar{z}_2. \tag{4.22}$$

We remark in passing that T is not the only symplectic matrix that accomplishes the transformation to this complex normal form. The matrix

$$(\bar{\alpha}_1/r_1^2, \alpha_2/r_2^2, \alpha_1, \bar{\alpha}_2)$$

would do the same and at the same time has a simpler form than T. On the other hand, the reality conditions for it are more complicated. The advantage of a simpler form is lost when we want to go back to real coordinates.

Therefore, we stay with the above form for T and introduce a new set of real variables $(\xi_1, \xi_2, \eta_1, \eta_2)$ by $z_j = \xi_j + i\eta_j$, $j = 1, 2$. It is a symplectic transformation with multiplier $-2i$, and the transformed Hamiltonian becomes

$$\mathcal{H} = \frac{1}{2}(\xi_1^2 + \eta_1^2) - \frac{\omega}{2}(\xi_2^2 + \eta_2^2). \tag{4.23}$$

The transformation from the original coordinates to these new coordinates is then given by

$$\begin{bmatrix} x_1 \\ x_2 \\ y_1 \\ y_2 \end{bmatrix} = \frac{1}{2\sqrt{1-\omega^2}} RS \begin{bmatrix} \xi_1 \\ \xi_2 \\ \eta_1 \\ \eta_2 \end{bmatrix},$$

where R is the matrix

$$R = \begin{bmatrix} 9\omega^2 + 13 & 13\omega^2 + 9 & 0 & 0 \\ -\gamma(\omega^2 + 1) & -\gamma(\omega^2 + 1) & 8(\omega^2 + 1) & -8(\omega^2 + 1) \\ \gamma(\omega^2 + 1) & \gamma(\omega^2 + 1) & \omega^2 + 5 & -5(\omega^2 - 1) \\ 9\omega^2 + 5 & 5\omega^2 + 9 & -\gamma(\omega^2 + 1) & \gamma(\omega^2 + 1) \end{bmatrix}$$

and S is the diagonal matrix

$$S = \mathrm{diag} \left(\frac{\sqrt[4]{\omega^2 + 1}}{\sqrt{9\omega^2 + 13}}, \frac{\sqrt[4]{\omega^2 + 1}}{\sqrt{\omega(13\omega^2 + 9)}}, \right.$$

$$\left. \frac{1}{\sqrt[4]{\omega^2 + 1}\sqrt{9\omega^2 + 13}}, \frac{\sqrt{\omega}}{\sqrt[4]{\omega^2 + 1}\sqrt{9\omega^2 + 13}} \right).$$

The matrix A in (4.21) and the subsequent Hamiltonian $\mathcal{H}$ in (4.23) have been scaled. The true matrix of the restricted problem at $\mathcal{L}_4$ is $\omega_1 A$. The transformations given above will diagonalize $\omega_1 A$ also. In fact K in (4.22) becomes $K = -i\omega_1 z_1 \bar{z}_1 + i\omega_2 z_2 \bar{z}_2$, and $\mathcal{H}$ in (4.23) becomes $\mathcal{H} = (\omega_1/2)(\xi_1^2 + \eta_1^2) - (\omega_2/2)(\xi_2^2 + \eta_2^2)$.

The above transformation becomes singular when $\omega = 1$. This is due to the fact that the Hamiltonian matrix (4.21) is not diagonalizable when $\gamma = \sqrt{23}$ or when $\mu = \mu_1$. We construct the linear transformation which brings

$$A = \sqrt{2} \begin{bmatrix} 0 & 1 & 1 & 0 \\ -1 & 0 & 0 & 1 \\ -1/4 & \sqrt{23}/4 & 0 & 1 \\ \sqrt{23}/4 & 5/4 & -1 & 0 \end{bmatrix} \tag{4.24}$$

into its complex normal form

$$C = \begin{bmatrix} -i & 0 & 0 & 0 \\ 0 & i & 0 & 0 \\ 0 & -1 & i & 0 \\ -1 & 0 & 0 & -i \end{bmatrix} \tag{4.25}$$

and afterwards we convert it into the corresponding real normal form

$$
C = \begin{bmatrix}
0 & 1 & 0 & 0 \\
-1 & 0 & 0 & 0 \\
-1 & 0 & 0 & 1 \\
0 & -1 & -1 & 0
\end{bmatrix}.
\tag{4.26}
$$

For the eigenvalue $+i$ of the matrix A, we calculate the eigenvector α and the generalized eigenvector β. They are given by

$$
\alpha = r \begin{bmatrix}
2\sqrt{23} + 8i\sqrt{2} \\
-10 \\
2 + i\sqrt{46} \\
2\sqrt{23} + 3i\sqrt{2}
\end{bmatrix},
\quad
\beta = s\alpha + \frac{r}{5} \begin{bmatrix}
-8\sqrt{2} - 8i\sqrt{23} \\
0 \\
-\sqrt{46} - 48i \\
17\sqrt{2} - 8i\sqrt{23}
\end{bmatrix},
$$

where r and s are complex-valued constants that have to be determined so that the transformation is symplectic. Due to the form of C the transformation matrix T from real coordinates to the new complex coordinates has to be $T = (\bar{\beta}, \beta, \alpha, \bar{\alpha})$.

The only terms that are nonzero in $T^T J T$ are $\beta^T J \bar{\beta}$, $\beta^T J \bar{\alpha}$, and those directly related to them. We compute

$$
\beta^T J \bar{\alpha} = (80\sqrt{2}) r \bar{r} \quad \text{and} \quad \beta^T J \bar{\beta} = i16\sqrt{2}\{10\Im(r\bar{s}) - r\bar{r}\}.
$$

In order to get a symplectic transformation to the new complex coordinates (z_1, z_2, z_3, z_4), we set $r = 1/\sqrt{80\sqrt{2}}$ and $s = -ir/10$. From the form of the matrix C, it also follows that the reality conditions have to be $z_1 = \bar{z}_2$ and $z_3 = \bar{z}_4$. It requires that the transformation to real position coordinates ξ_1, ξ_2 and their conjugate momenta η_1, η_2 has to be set up in the following special way

$$
z_1 = \xi_1 + i\xi_2,
$$

$$
z_2 = \xi_1 - i\xi_2,
$$

$$
z_3 = \eta_1 - i\eta_2,
$$

$$
z_4 = \eta_1 + i\eta_2.
$$

This form is forced upon us if we want to preserve the two-form; that is, $dz_1 \wedge dz_3 + dz_2 \wedge dz_4 = 2(d\xi_1 \wedge d\eta_1 + d\xi_1 \wedge d\eta_2)$. (It is shown in Chapter 2.6 that this ensures the transformation is symplectic.)

Summarizing, we first transformed the original Hamiltonian function $\sqrt{2}H$ into the complex normal form

$$K = -iz_1z_3 + iz_2z_4 + z_1z_2,$$

which we then transformed into the real normal form

$$\mathcal{H} = -\xi_1\eta_2 + \xi_2\eta_1 + \frac{1}{2}(\xi_1^2 + \xi_2^2).$$

The composite transformation from the original to the new real coordinates is given by

$$
\begin{bmatrix} x_1 \\ x_2 \\ y_1 \\ y_2 \end{bmatrix}
= \frac{\sqrt{5\sqrt{2}}}{100}
\begin{bmatrix}
4\sqrt{2} & 9\sqrt{23} & -10\sqrt{23} & -40\sqrt{2} \\
0 & -5 & 50 & 0 \\
\sqrt{46}/2 & 49 & -10 & -5\sqrt{46} \\
-37\sqrt{2}/2 & 9\sqrt{23} & -10\sqrt{23} & -15\sqrt{2}
\end{bmatrix}
\begin{bmatrix} \xi_1 \\ \xi_2 \\ \eta_1 \\ \eta_2 \end{bmatrix}.
\tag{4.27}
$$

The transformations given above take the matrix A in (4.24) to its normal form but $A = \sqrt{2}B$ where B is the true matrix at $\mathcal{L}_4$. Similarly the transformations take $\sqrt{2}Q$ to normal form where Q is the true quadratic Hamiltonian at $\mathcal{L}_4$. The transformations take Q to $K = (\sqrt{2}/2)\{-iz_1z_3 + iz_2z_4 + z_1z_2\}$ and $\mathcal{H} = (\sqrt{2}/2)\{-\xi_1\eta_2 + \xi_2\eta_1 + (\xi_1^2 + \xi_2^2)/2\}$. In order to get Q into its true normal form, one additional scaling, $\xi_i \to \sqrt[4]{2}\xi_i$, $\eta_i \to 1/\sqrt[4]{2}\eta_i$, is required. This scaling is symplectic, and the Q becomes $(\sqrt{2}/2)\{-\xi_1\eta_2 + \xi_2\eta_1\} + (\xi_1^2 + \xi_2^2)/2$.

4.8 Other Restricted Problems

There are a multitude of restricted problems, but here we give just two. An occasional reference to them or problem about them will be given in subsequent chapters. A more substantial introduction with similar notation can be found in Meyer (1999).

4.8.1 Hill's Lunar Equations

In a popular description of Hill's lunar equations, one is asked to consider the motion of an infinitesimal body (the moon) which is attracted to a body (the earth) fixed at the origin. The infinitesimal body moves in a rotating coordinate system: the system rotates so that the positive x-axis points to an infinite body (the sun) infinitely far away. The ratio of the two infinite quantities is taken so that the gravitational attraction of the sun on the moon is finite. Here we will give a derivation of Hill's problem similar to the derivation of the restricted problem given in Section 4.1.

One of Hill's major contributions to celestial mechanics was his reformulation of the main problem of lunar theory: he gave a new definition for the equations of the first approximation for the motion of the moon. Because his equations of the first approximation contained more terms than the older first approximations, the perturbations were smaller and he was able to obtain series representations for the position of the moon that converge more rapidly than the previously obtained series. Indeed, for many years lunar ephemerides were computed from the series developed by Brown, who used the main problem as defined by Hill. Even today, most of the searchers for more accurate series solutions for the motion of the moon use Hill's definition of the first approximation.

Before Hill, the first approximation consisted of two Kepler problems: one describing the motion of the earth and moon about their center of mass, and the other describing the motion of the sun and the center of mass of the earth-moon system. The coupling terms between the two Kepler problems are neglected at the first approximation. Delaunay used this definition of the first approximation for his solution of the lunar problem, but after 20 years of computation he was unable to meet the observational accuracy of his time.

In Hill's definition of the main problem, the sun and the center of mass of the earth-moon system still satisfy a Kepler problem, but the motion of the moon is described by a different system of equations known as Hill's lunar equations. Using heuristic arguments about the relative sizes of various physical constants, he concluded that certain other terms were sufficiently large that they should be incorporated into the main problem. This heuristic grouping of terms does not lead to a precise description of the relationship between the equations of the first approximation and the full problem.

Here we give a derivation of Hill's lunar problem as a limit of the restricted problem following these heuristics. To see this we make a sequence of symplectic coordinate changes and scaling.

Start with the restricted problem given in Section (4) in x, y coordinates and move one primary to the origin by the change of coordinates

$$x_1 \to x_1 + 1 - \mu, \qquad x_2 \to x_2,$$
$$y_1 \to y_1, \qquad y_2 \to y_2 + 1 - \mu,$$

so that the Hamiltonian becomes

$$H = \frac{1}{2}\left(y_1^2 + y_2^2\right) - y_2 x_1 + y_1 x_2 - \frac{1-\mu}{r_1} - \frac{\mu}{r_2} - (1-\mu)x_1, \qquad (4.28)$$

where

$$r_1^2 = (x_1 + 1)^2 + x_2^2, \qquad r_2^2 = x_1^2 + x_2^2.$$

(Here we drop all constants from the Hamiltonian.) By Newton's binomial series

$$\{1+u\}^{-1/2} = 1 - \frac{1}{2}u + \frac{3}{8}u^2 + \cdots$$

so

$$-\frac{1-\mu}{\sqrt{(x_1+1)^2+x_2^2}} = -(1-\mu)\{1 - x_1 + x_1^2 - \frac{1}{2}x_2^2 + \cdots\}$$

and the Hamiltonian becomes

$$H = \frac{1}{2}\left(y_1^2 + y_2^2\right) - y_2 x_1 + y_1 x_2 - \frac{\mu}{r_2} - (1-\mu)\{x_1^2 - \frac{1}{2}x_2^2 + \cdots\}. \qquad (4.29)$$

We consider the mass μ as a small parameter and distance to the primary to be small by scaling

$$x \to \mu^{1/3}x, \qquad y \to \mu^{1/3}y$$

which is symplectic with multiplier $\mu^{-2/3}$ so the Hamiltonian becomes

$$H = L + O(\mu^{1/3}), \qquad (4.30)$$

where L is the Hamiltonian of Hill's lunar problem

$$L = \frac{1}{2}\left(y_1^2 + y_2^2\right) - y_2 x_1 + y_1 x_2 - \frac{1}{\|x\|} - x_1^2 + \frac{1}{2}x_2^2. \qquad (4.31)$$

The Hamiltonian L has an equilibrium point at

$$(x_1, x_2, y_1, y_2) = (-3^{-1/3}, 0, 0, -3^{-1/3}),$$

which is the limit of the equilibrium point $\mathcal{L}_2$ as $\mu \to 0$ in the scaling given above. The exponents at this equilibrium point are $\pm i\sqrt{2\sqrt{7}-1}, \sqrt{2\sqrt{7}+1}$.

4.8.2 Elliptic Restricted Problem

Previously we gave the Hamiltonian of the circular restricted three-body problem. In that problem, it is assumed that the two primaries move on circular orbits of the two-body problem. If one assumes instead that the primaries move on elliptical orbits of the two-body problem, then one has the elliptical restricted problem. It is no longer autonomous (time-independent) but rather periodic in time. It also contains a parameter e, denoting the eccentricity of the primaries' orbit. In Section 8.10.1, we give a detailed derivation of these equations and the special coordinates used to study this problem; for now we will just give the Hamiltonian.

The Hamiltonian governing the motion of the third (infinitesimal) particle in the elliptic restricted problem is

$$H = \frac{\|y\|^2}{2} - x^T J y - r(t)U + \left(\frac{1-r(t)}{2}\right)y^T x,$$

where $x, y \in \mathbb{R}^2$ are conjugate, J is J_2, U is the self-potential

$$U = \frac{\mu}{d_1} + \frac{1-\mu}{d_2},$$

d_i is the distance from the infinitesimal body to the ith primary, or

$$d_1^2 = (x_1 - 1 + \mu)^2 + x_2^2, \qquad d_2^2 = (x_1 + \mu)^2 + x_2^2,$$

and $r(t) = 1/(1 + e \cos t)$. These are special coordinates that keep the primaries at a fixed position.

In all the other examples of this chapter, the Hamiltonian is independent of time t, so it is a constant of the motion or an integral. However, in the elliptic case, the Hamiltonian is not independent of time and so is not an integral.

4.9 Problems

1. To investigate solutions near ∞ in the restricted problem, scale by $x \to \epsilon^{-2}x$, $y \to \epsilon y$. Show that the Hamiltonian becomes $H = -x^T K y + \epsilon^3 \{\|y\|^2/2 - 1/\|x\|\} + O(\epsilon^2)$. Justify this result on physical grounds.
2. To investigate solutions near one of the primaries in the restricted problem, first shift the origin to one primary as was done in Section 4.8.1. Then scale by $x \to \epsilon^2 x$, $y \to \epsilon^{-1}y$, $t \to \epsilon^3 t$.
3. Consider Hill's lunar problem (4.31).
   a) Write the equations of motion.
   b) Show that there are two equilibrium points on the x_1-axis.
   c) Show that the linearized system at these equilibrium points are saddle-centers, i.e., it has one pair of real eigenvalues and one pair of imaginary eigenvalues.
   d) Sketch the Hill's regions for Hill's lunar problem.
   e) Why did Hill say that the motion of the moon was bounded? (He had the Earth at the origin, and an infinite sun infinitely far away and x was the position of the moon in this ideal system. What can you say if x and y are small?
4. Show that the elliptic restricted three-body problem has five equilibrium points — two at the vertices of an equilateral triangle and three collinear.

5. Topics in Linear Theory

This chapter contains various special topics in the linear theory of Hamiltonian systems. Therefore, the chapter can be skipped on first reading and referred back to when the need arises. Sections 5.1, 5.5, and 5.6 are independent of each other.

5.1 Parametric Stability

Stability questions for Hamiltonian systems have been studied since the time of Newton. Is the solar system stable? This is an easy question to ask with obvious consequences, but it is difficult to answer. We have seen some simple results for linear systems, some positive and some negative. A satisfactory stability theory for Hamiltonian systems exists for linear autonomous and periodic systems only. The richness of this theory in this simplest of all cases foreshadows the complexity of the nonlinear problem discussed in Chapter 12. We present all the essential features of this theory and the reader will find an extensive discussion of periodic systems with examples in the two-volume set by Yakubovich and Starzhinskii (1975).

Consider a periodic Hamiltonian system

$$\dot{z} = P(t)z, \qquad P(t+T) \equiv P(t). \tag{5.1}$$

Recall that if $Z(t)$ is the fundamental matrix solution of (5.1), then $Z(t)$ is symplectic, $Z(T)$ is called the monodromy matrix of the system, and the eigenvalues of $Z(T)$ are called the (characteristic) multipliers of the system. By the Floquet–Lyapunov theory and Corollary 2.5.1 there is a periodic symplectic change of variables $z = Q(t)w$ which reduces the periodic Hamiltonian system (5.1) to the constant system

$$\dot{w} = Aw. \tag{5.2}$$

The period of Q is either T or $2T$ and either

$$e^{AT} = Z(T) \quad \text{or} \quad e^{2AT} = Z(2T) = Z(T)^2.$$

© Springer International Publishing AG 2017
K.R. Meyer, D.C. Offin, *Introduction to Hamiltonian Dynamical Systems and the N-Body Problem*, Applied Mathematical Sciences 90,
DOI 10.1007/978-3-319-53691-0_5

(Recall, this result depends on the existence of a Hamiltonian logarithm of a symplectic matrix which was only established in the case when the symplectic matrix was diagonalizable in Chapter 2. We do not need the full result here. A complete treatment of the logarithm is given in Section 5.2.)

The eigenvalues of A are called the (characteristic) exponents of the systems (5.1) and (5.2). The exponents of a periodic system are defined modulo $2\pi i$ because in this case the matrix A is the logarithm of the monodromy matrix.

A linear periodic Hamiltonian system is stable if all solutions are bounded for all $t \in \mathbb{R}$. (For linear Hamiltonian systems stability is equivalent to the origin being positively and negatively stable; see the Problems).

If Equation (5.1) is the mathematical model of a physical problem, then the coefficients in the equation, i.e., the matrix $P(t)$, may not be known exactly. Is the question of stability sensitive to small changes in the Hamiltonian matrix $P(t)$? This question gives rise to the following concept. A linear periodic Hamiltonian system is parametrically stable or strongly stable if it and all sufficiently small periodic linear Hamiltonian perturbations of it are stable. That is, (5.1) is parametrically stable if there is an $\epsilon > 0$ such that $\dot{z} = R(t)z$ is stable, where $R(t)$ is any periodic Hamiltonian matrix with the same period such that $\|Q(t) - R(t)\| < \epsilon$ for all t. For a constant system the definition is the same except that the perturbations remain in the constant class.

Autonomous Systems. Now consider the constant system (5.2) only. Solutions of the constant system (5.2) are linear combinations of the basic solutions of the form $t^k \exp(\lambda t)a$, where k is a nonnegative integer, a is a constant vector, and λ is an eigenvalue of A. All solutions of (5.2) will tend to 0 as $t \to \infty$ (the origin is asymptotically stable) if and only if all the eigenvalues of A have negative real parts. By Proposition 2.3.1 this never happens for a Hamiltonian system. All solutions of (5.2) are bounded for $t > 0$ if and only if (i) all the eigenvalues of A have nonpositive real parts and (ii) if λ is an eigenvalue of A with zero real part (pure imaginary), then the k in the basic solutions, $t^k \exp(\lambda t)a$, is zero. This last condition, (ii), is equivalent to the condition that the Jordan blocks for all the pure imaginary eigenvalues of A in the Jordan canonical form for A are diagonal. That is, there are no off-diagonal terms in the Jordan blocks for pure imaginary eigenvalues of A. For Hamiltonian systems by Proposition 2.3.1 if all the eigenvalues have nonpositive real parts, then they must be pure imaginary. Thus if a Hamiltonian system has all solutions bounded for $t > 0$, then all solutions are bounded for all time. This is why for linear Hamiltonian systems, the meaningful concept of stability is that all solutions are bounded for all $t \in \mathbb{R}$.

Proposition 5.1.1. *The linear constant Hamiltonian system (5.2) is stable if and only if (i) A has only pure imaginary eigenvalues and (ii) A is diagonalizable (over the complex numbers).*

Let $A = JS$, where S is a $2n \times 2n$ constant symmetric matrix, and $H(z) = \frac{1}{2}z^T S z$ is the Hamiltonian.

Lemma 5.1.1. *If the Hamiltonian H is positive (or negative) definite, then the system* (5.2) *is parametrically stable.*

Proof. Let H be positive definite. Because H is positive definite, the level set $H = h$ where h is a positive constant is an ellipsoid in $\mathbb{R}^{2n}$ and hence a bounded set. Because H is an integral, any solution that starts on $H = h$ remains on $H = h$ and so is bounded. So H being positive definite implies (5.2) is stable. (This is just a special case of Dirichlet's Theorem 1.2.2.)

Any sufficiently small perturbation of a positive definite matrix is positive definite, and so any sufficiently small perturbation of (5.2) is stable also.

Lemma 5.1.2. *If* (5.2) *is parametrically stable, then the eigenvalues of A must be pure imaginary, and A must be diagonalizable.*

Proof. If (5.2) is parametrically stable, then it is stable.

Recall that $\eta(\lambda)$ denotes the eigenspace and $\eta^\dagger(\lambda)$ denotes the generalized eigenspace of A corresponding to the eigenvalue λ.

Lemma 5.1.3. *If* (5.2) *is parametrically stable, then zero is not an eigenvalue of A.*

Proof. Assume not, so, A is parametrically stable and $\eta(0) = \eta^\dagger(0)$ is not trivial. By the discussion in Section 2.5, the subspace $\eta(0)$ is an A-invariant symplectic subspace, so, A restricted to this subspace, denoted by A', is Hamiltonian. Because A is diagonalizable so is A'. But a diagonalizable matrix all of whose eigenvalues are zero is the zero matrix, i.e., $A' = 0$. Let B be a Hamiltonian matrix of the same size as A' with real eigenvalues ± 1; then ϵB is a small perturbation of $A' = 0$ for small ϵ and has eigenvalues $\pm \epsilon$. Thus by perturbing A along the subspace $\eta^\dagger(0)$ by ϵB and leaving A fixed on the other subspaces gives a small Hamiltonian perturbation that is not stable.

Let system (5.2) be stable and let A have distinct eigenvalues

$$\pm \beta_1 i, \ldots, \pm \beta_s i,$$

with $\beta_j \neq 0$. The space $\eta(+\beta_j i) \oplus \eta(-\beta_j i)$ is the complexification of a real space $\mathbb{V}_j$ of dimension $2n_j$ and A restricted to $\mathbb{V}_j$ is denoted by A_j. $\mathbb{V}_j$ is a symplectic linear space, and A_j is invariant. A_j is a real diagonalizable Hamiltonian matrix with eigenvalues $\pm \beta_j i$. Define the symmetric matrix S_j by $A_j = JS_j$ and H_j the restriction of H to $\mathbb{V}_j$.

Lemma 5.1.4. *The system* (5.2) *is parametrically stable if and only if the restriction of* (5.2) *to each $\mathbb{V}_j$ is parametrically stable, i.e., if and only if each Hamiltonian system with Hamiltonian H_j and coefficient matrix A_j is parametrically stable.*

Theorem 5.1.1 (Krein–Gel'fand). *Using the notation given above, the system* (5.2) *is parametrically stable if and only if*

1. *All the eigenvalues of A are pure imaginary.*
2. *A is nonsingular.*
3. *The matrix A is diagonalizable over the complex numbers.*
4. *The Hamiltonian H_j is positive or negative definite for each j.*

Thus for example, the systems defined by the Hamiltonians

$$2H = (x_1^2 + y_1^2) - 4(x_2^2 + y_2^2) \quad \text{and} \quad 2H = (x_1^2 + y_1^2) + (x_2^2 + y_2^2)$$

are parametrically stable, whereas the system defined by the Hamiltonian

$$2H = (x_1^2 + y_1^2) - (x_2^2 + y_2^2)$$

is not parametrically stable.

Proof. First, the if part. Given A let $\mathbb{V}$ be the decomposition into the invariant symplectic subspaces $\mathbb{V}_1, \ldots, \mathbb{V}_s$, as defined above. Then there is an ϵ so small that if B is any Hamiltonian ϵ-perturbation of A, then there are B-invariant symplectic spaces $\mathbb{W}_1, \ldots, \mathbb{W}_s$ with $\mathbb{W}_j$ close to $\mathbb{V}_j$. Moreover, $\dim \mathbb{V}_j = \dim \mathbb{W}_j$, the eigenvalues of B restricted to $\mathbb{W}_j$ are close to $\pm \beta_j i$, and the Hamiltonian $\tilde{H}_j$ of B restricted to $\mathbb{W}_j$ is positive or negative definite. Because $\tilde{H}_j$ is positive or negative definite on each $\mathbb{W}_j$ all the solutions of the system with coefficient matrix B are bounded, and hence the system is stable.

Second, the only if part. What we need to show is that if the Hamiltonian is not definite on one of the spaces A_j, then some perturbation will be unstable. We know that A_j is diagonalizable and all its eigenvalues are $\pm \beta_j$ thus by a linear symplectic change of variables

$$H_j = \frac{1}{2} \sum_{s=1}^{n_j} \pm \beta_j (x_s^2 + y_s^2).$$

We must show that if H_j is not positive or negative definite then the system is not parametrically stable. So there must be one plus sign and one minus sign in the form for H_j.

Without loss of generality we may assume $\beta_j = 1$ and the first term is positive and the second term is negative and forget all the other terms. That is, there is no loss in generality in considering the Hamiltonian of two harmonic oscillators with equal frequencies; namely

$$2H = (x_1^2 + y_1^2) - (x_2^2 + y_2^2).$$

Then the perturbation

$$2H_\epsilon = (x_1^2 + y_1^2) - (x_2^2 + y_2^2) + \epsilon y_1 y_2$$

is unstable for small ϵ, because the characteristic equation of the above system is $(\lambda^2 + 1)^2 + \epsilon^2$; and so, the eigenvalues are $\pm \sqrt{(-1 \pm \epsilon i)}$, which has real part nonzero for $\epsilon \neq 0$.

Periodic Systems. One way to reduce the parametric stability questions of the periodic system (5.1) to the corresponding question for the constant system is to use the Floquet–Lyapunov theorem which states that there is a T or $2T$-periodic (hence bounded) change of variables that takes the periodic system (5.1) to the constant system (5.2). The system (5.1) is parametrically stable if and only if the system (5.2) is parametrically stable. This approach requires a detailed analysis of the logarithm function applied to symplectic matrices given later in this chapter. A simpler and more direct approach using a simple Möbius transform is given here.

Consider the periodic system (5.1) with fundamental matrix solution $Y(t)$ and monodromy matrix $M = Y(T)$.

Lemma 5.1.5. $Y(t)$ *is bounded for all t if and only if M^k is bounded for all integers k. That is, the periodic system (5.1) is stable if and only if M^k is bounded for all k.*

Proof. Both $Y(kT + t)$ and $Y(t)M^k$ satisfy Equation (5.1) as functions of t and they satisfy the same initial condition when $t = 0$, so by the uniqueness theorem for differential equations $Y(kT + t) = Y(t)M^k$. Because $Y(t)$ is bounded for $0 \leq t \leq T$ the result follows from this identity. Thus the stability analysis is reduced to a study of the matrix M under iteration.

The next two results are proved in a manner similar to the corresponding results for the constant coefficient case.

Proposition 5.1.2. *The periodic Hamiltonian system (5.1) is stable if and only if (i) the monodromy matrix $Y(T)$ has only eigenvalues of unit modulus and (ii) $Y(T)$ is diagonalizable (over the complex numbers).*

Lemma 5.1.6. *If (5.1) is parametrically stable then ± 1 are not multipliers of the monodromy matrix $Y(T)$.*

The particular Möbius transformation

$$\phi : z \to w = (z - 1)(z + 1)^{-1}, \qquad \phi^{-1} : w \to z = (1 + w)(1 - w)^{-1}$$

is known as the Cayley transformation. One checks that $\phi(1) = 0, \phi(i) = i, \phi(-1) = \infty$ and so ϕ takes the unit circle in the z-plane to the imaginary axis in the w-plane, the interior of the unit circle in the z-plane to the left half w-plane, etc. ϕ can be applied to any matrix B that does not have -1 as an eigenvalue and if λ is an eigenvalue of B then $\phi(\lambda)$ is an eigenvalue of $\phi(B)$.

Lemma 5.1.7. *Let M be a symplectic matrix that does not have the eigenvalue -1; then $C = \phi(M)$ is a Hamiltonian matrix. Moreover, if M has only eigenvalues of unit modulus and is diagonalizable, then $C = \phi(M)$ has only pure imaginary eigenvalues and is diagonalizable.*

Proof. Because M is symplectic $M^T J M = M J M^T = J$ and $C = \phi(M) = (M-I)(M+I)^{-1} = (M+I)^{-1}(M-I)$. We must show that JC is symmetric.

$$
\begin{aligned}
(JC)^T &= (M+I)^{-T}(M-I)^T J^T \\
&= -(M^T+I)^{-1}(M^T-I)J \\
&= -(JM^{-1}J^{-1} + JJ^{-1})^{-1}(JM^{-1}J^{-1} - JJ^{-1})J \\
&= -J(M^{-1}+I)^{-1}J^{-1}J(M^{-1}-I)J^{-1}J \\
&= -J(M^{-1}+I)^{-1}(M^{-1}-I) \\
&= -J(I+M)^{-1}(I-M) = JC.
\end{aligned}
$$

Assume the monodromy matrix $M = Y(T)$ is diagonalizable and ± 1 are not eigenvalues. Then $C = \phi(M)$ is defined and Hamiltonian. Let C have distinct eigenvalues

$$\pm \beta_1 i, \ldots, \pm \beta_s i,$$

with $\beta_j \neq 0$. The space $\eta(+\beta_j i) \oplus \eta(-\beta_j i)$ is the complexification of a real space $\mathbb{V}_j$ of dimension $2n_j$ and C restricted to $\mathbb{V}_j$ is denoted by C_j. $\mathbb{V}_j$ is a symplectic linear space, and C_j is invariant. C_j is a real diagonalizable Hamiltonian matrix with eigenvalues $\pm \beta i$. Define the symmetric matrix S_j by $C_j = JS_j$ and $H_j(u) = \frac{1}{2}u^T S_j u$ where $u \in \mathbb{V}_j$.

Theorem 5.1.2 (Krein–Gel'fand). *Using the notation given above, the system* (5.1) *is parametrically stable if and only if*

1. *All the multipliers have unit modulus.*
2. *± 1 are not multipliers.*
3. *The monodromy matrix $M = Y(T)$ is diagonalizable over the complex numbers.*
4. *The Hamiltonian H_j is positive or negative definite for each j.*

Problem

1. Let $\phi(t, \zeta)$ be the solution to the linear periodic Hamiltonian system (5.1) such that $\phi(0, \zeta) = \zeta$. The zero solution (the origin) for (5.1) is stable if for every $\epsilon > 0$ there is a $\delta > 0$ such that $\|\zeta\| < \delta$ implies $\|\phi(t, \zeta)\| < \epsilon$ for all $t \in \mathbb{R}$. Prove that all solutions of (5.1) are bounded (i.e., the system is stable) if and only if the origin is stable.

5.2 Logarithm of a Symplectic Matrix

The simplest proof that a symplectic matrix has a Hamiltonian logarithm uses the theory of analytic functions of a matrix. This theory is not widely known, therefore we give a cursory introduction here. This proof and some of the background material are found in Yakubovich and Starzhinskii (1975). See Sibuya (1960) for a more algebraic, but not necessarily simpler proof.

5.2.1 Functions of a Matrix

Let A be any square matrix and $f(z) = \sum_0^\infty a_k z^k$ a convergent power series. We define $f(A)$ by

$$f(A) = \sum_{k=0}^{\infty} a_k A^k.$$

Use any norm on matrices you like, e.g., $\|A\| = \sup |a_{ij}|$, and you will find that $f(A)$ has essentially the same radius of convergence as $f(z)$. The theory of analytic functions of a matrix proceeds just as the standard theory: analytic continuation, persistence of formulas, integral formulas, etc. For example

$$e^A e^A = e^{2A}, \qquad (\sin \Lambda)^2 + (\cos A)^2 = I$$

still hold, but there is one major cavitate. Namely, formulas such as

$$e^A e^B = e^{A+B}, \qquad \sin(A + B) = \sin A \cos B + \sin B \cos A$$

only hold if A and B commute. Any formula that can be proved by rearranging series is valid as long as the matrices involved commute.

The definition behaves well with respect to similarity transformations. Because

$$(P^{-1}AP)^k = P^{-1}APP^{-1}AP \cdots P^{-1}AP = P^{-1}A^k P$$

it follows that

$$f(P^{-1}AP) = P^{-1}f(A)P.$$

Hence, if λ is an eigenvalue of A then $f(\lambda)$ is an eigenvalue of $f(A)$.

If N is a Jordan block then $f(N)$ has a triangular form:

$$N = \begin{bmatrix} \lambda & 0 & \cdots & 0 & 0 \\ 1 & \lambda & \cdots & 0 & 0 \\ \vdots & \vdots & & \vdots & \vdots \\ 0 & 0 & \cdots & \lambda & 0 \\ 0 & 0 & \cdots & 1 & \lambda \end{bmatrix}, \quad f(N) = \begin{bmatrix} f(\lambda) & 0 & \cdots & 0 & 0 \\ f'(\lambda) & f(\lambda) & \cdots & 0 & 0 \\ \vdots & \vdots & & \vdots & \vdots \\ \dfrac{f^{(k-2)}(\lambda)}{(k-2)!} & \dfrac{f^{(k-3)}(\lambda)}{(k-3)!} & \cdots & f(\lambda) & 0 \\ \dfrac{f^{(k-1)}(\lambda)}{(k-1)!} & \dfrac{f^{(k-2)}(\lambda)}{(k-2)!} & \cdots & f'(\lambda) & f(\lambda) \end{bmatrix}.$$

The integral calculus extends also. The integral formula

$$f(A) = \frac{1}{2\pi i} \oint_\Gamma (\zeta I - A)^{-1} f(\zeta) d\zeta, \tag{5.3}$$

holds provided the domain of analyticity of f contains the spectrum of A in its interior and Γ encloses the spectrum of A. One checks that the two definitions agree using residue calculus.

In particular

$$I = \frac{1}{2\pi i} \oint_\Gamma (\zeta I - A)^{-1} d\zeta, \qquad A = \frac{1}{2\pi i} \oint_\Gamma (\zeta I - A)^{-1} \zeta d\zeta,$$

which seem uninteresting until one writes $\Gamma = \sum \Gamma_j$, where Γ_j is a contour encircling just one eigenvalue λ_j. Then $I = \sum P_j$ and $A = \sum A_j$ where

$$P_j = \frac{1}{2\pi i} \oint_{\Gamma_j} (\zeta I - A)^{-1} d\zeta, \qquad A_j = \frac{1}{2\pi i} \oint_{\Gamma_j} (\zeta I - A)^{-1} \zeta d\zeta.$$

P_j is the projection on the generalized eigenspace corresponding to λ_j and $A_j = A P_j$ is the restriction of A to this generalized eigenspace.

5.2.2 Logarithm of a Matrix

Let A be nonsingular and let log be any branch of the logarithm function that contains the spectrum of A in the interior of its domain. Then

$$B = \log(A) = \frac{1}{2\pi i} \oint_\Gamma (\zeta I - A)^{-1} \log(\zeta) d\zeta \tag{5.4}$$

is a logarithm of A. Of course this logarithm may not be real even if A is real. The logarithm is not unique in general, even in the real case, in as much as $I = \exp O = \exp 2\pi J$.

Lemma 5.2.1. *Let A be a real nonsingular matrix with no negative eigenvalues. Then A has a real logarithm.*

Proof. Let A have distinct eigenvalues $\lambda_1, \ldots, \lambda_k$, with λ_i not a negative number for all i. The set of eigenvalues of A is symmetric with respect to the real axis. Let $\Gamma_1, \ldots, \Gamma_k$ be small nonintersecting circles in the complex plane centered at $\lambda_1, \ldots, \lambda_k$, respectively, which are symmetric with respect to the real axis. Thus conjugation, $z \to \bar{z}$, takes the set of circles $\Gamma_1, \ldots, \Gamma_k$ into itself (possibly permuting the order).

Let Log be the branch of the logarithm function defined by slitting the complex plane along the negative real axis and $-\pi < \arg(\text{Log } z) < \pi$. Then a logarithm of A is given by

$$B = \log A = \frac{1}{2\pi i} \sum_{j=1}^k \oint_{\Gamma_j} (\zeta I - T)^{-1} \text{Log } \zeta d\zeta. \tag{5.5}$$

Let conjugation take Γ_j to $-\Gamma_s = \bar{\Gamma}_j$ (the minus indicates that conjugation reverses orientation). Then

$$\overline{\frac{1}{2\pi i} \oint_{\Gamma_j} (\zeta I - T)^{-1} \text{Log } \zeta d\zeta} = -\frac{1}{2\pi i} \oint_{\bar{\Gamma}_j} (\bar{\zeta} I - T)^{-1} \text{Log } \bar{\zeta} d\bar{\zeta}$$

$$= -\frac{1}{2\pi i} \oint_{-\Gamma_s} (\zeta I - T)^{-1} \text{Log } \zeta d\zeta \qquad (5.6)$$

$$= \frac{1}{2\pi i} \oint_{\Gamma_s} (\zeta I - T)^{-1} \text{Log } \zeta d\zeta$$

So conjugation takes each term in (5.5) into another, which implies that B is real.

If A has a real logarithm, $A = \exp B$, then A is nonsingular and has a real square root $A^{1/2} = \exp(B/2)$. A straightforward computation shows that the matrix

$$R = \begin{bmatrix} -1 & 1 \\ 0 & -1 \end{bmatrix}$$

has no real square root and hence no real logarithm. But, the matrix

$$S = \begin{bmatrix} -1 & 0 \\ 0 & -1 \end{bmatrix} = \begin{bmatrix} \cos \pi & \sin \pi \\ -\sin \pi & \cos \pi \end{bmatrix} = e^{\begin{bmatrix} 0 & \pi \\ -\pi & 0 \end{bmatrix}}$$

has a real logarithm even though it has negative eigenvalues. This can be generalized.

We say that a real nonsingular matrix C has negative eigenvalues in pairs if it is similar to a matrix of the form $\text{diag}(A, D, D)$ where A is real and has no negative eigenvalues and D is a matrix with only negative eigenvalues. That is, the number of Jordan blocks for C of any particular size for a negative eigenvalue must be even.

Theorem 5.2.1. *A nonsingular matrix C that has negative eigenvalues in pairs has a real logarithm.*

Proof. The logarithm of A is given by Lemma 5.2.1, so we need to consider the case when $C = \text{diag}(D, D)$. The matrix $-D$ has only positive eigenvalues, so $\log(-D)$ exists as a real matrix by Lemma 5.2.1. Let

$$E = \begin{bmatrix} \log(-D) & \pi I \\ -\pi I & \log(-D) \end{bmatrix} = \begin{bmatrix} 0 & \pi I \\ -\pi I & 0 \end{bmatrix} + \begin{bmatrix} \log(-D) & 0 \\ 0 & \log(-D) \end{bmatrix}.$$

Notice that in the above E is the sum of two commuting matrices. So

$$e^E = e^{\begin{bmatrix} 0 & \pi I \\ -\pi I & 0 \end{bmatrix}} e^{\begin{bmatrix} \log(-D) & 0 \\ 0 & \log(-D) \end{bmatrix}} = \begin{bmatrix} D & 0 \\ 0 & D \end{bmatrix}.$$

5.2.3 Symplectic Logarithm

Turn now to the symplectic case.

Lemma 5.2.2. *Let A be a real symplectic matrix with no negative eigenvalues; then A has a real Hamiltonian logarithm.*

Proof. Let A have distinct eigenvalues $\lambda_1, \ldots, \lambda_{2k}$, with λ_i not a negative number for all i. The set of eigenvalues of A is symmetric with respect to the real axis and the unit circle by Proposition 2.3.1. Let $\Gamma_1, \ldots, \Gamma_{2k}$ be small nonintersecting circles in the complex plane centered at $\lambda_1, \ldots, \lambda_{2k}$, respectively, which are symmetric with respect to the real axis and the unit circle. Thus conjugation, $z \to \bar{z}$, and inversion, $z \to 1/z$, take the set of circles $\Gamma_1, \ldots, \Gamma_{2k}$ into itself (possibly permuting the order).

As before let Log be the branch of the logarithm function defined by slitting the complex plane along the negative real axis and $-\pi < \arg(\text{Log } z) < \pi$. Then a logarithm of A is given by

$$B = \log A = \frac{1}{2\pi i} \sum_{j=1}^{2k} \oint_{\Gamma_j} (\zeta I - A)^{-1} \text{Log } \zeta d\zeta. \qquad (5.7)$$

The matrix B is real by the argument in the proof of Lemma 5.2.1, so it remains to show that B is Hamiltonian.

Let inversion take Γ_j into Γ_s (inversion is orientation-preserving). Make the change of variables $\zeta = 1/\xi$ in the integrals in (5.7) and recall that $A^{-1} = -JA^T J$. Then

$$
\begin{aligned}
(\xi - A)^{-1} \text{Log } \zeta d\xi &= \{(1/\xi)I - A\}^{-1}(-\text{Log } \xi)(-d\xi/\xi^2) \\
&= (I - \xi A)^{-1}\xi^{-1}\text{Log } \xi d\xi \\
&= \{A(I - \xi A)^{-1} + \xi^{-1}I\}\text{Log } \xi d\xi \\
&= \{(A^{-1} - \xi I)^{-1} + \xi^{-1}I\}\text{Log } \xi d\xi \\
&= \{(-JA^T J - \xi I)^{-1} + \xi^{-1}I\}\text{Log } \xi d\xi \\
&= -J(A^T - \xi I)^{-1}J\text{Log } \xi d\xi + \xi^{-1}\text{Log } \xi d\xi.
\end{aligned}
\qquad (5.8)
$$

The circle Γ_j does not enclose the origin, thus $\oint \xi^{-1}\text{Log } \xi d\xi = 0$ on Γ_j for all j. Making the substitution $\zeta = 1/\xi$ in (5.7) and using (5.8) shows that $B = JB^T J$ or $JB^T + BJ = 0$. Thus B is Hamiltonian.

We say that a symplectic matrix G has negative eigenvalues in pairs if it is symplectically similar to a matrix of the form $\text{diag}(A, C, C)$ where A is symplectic and has no negative eigenvalues and C is symplectic and has only negative eigenvalues. The symplectic matrix J is replaced by $\text{diag}(J, J, J)$.

Theorem 5.2.2. *A real symplectic matrix that has negative eigenvalues in pairs has a real Hamiltonian logarithm.*[1]

Proof. The proof proceeds just as the proof of Theorem 5.2.1.

Problem: Discuss the question of finding a real, skew-symmetric logarithm of an orthogonal matrix

5.3 Spectral Decomposition

In this section we complete the spectral analysis of a Hamiltonian matrix A and a symplectic matrix T. Here we drop the assumption that these matrices are diagonalizable. The material in this section is be used in Sections 5.4 and 5.2 and to some extent in Sections 5.1 and 4.5. The main references on Hamiltonian matrices discussed in this section and in Section 5.4 are still Williamson (1936, 1937, 1939), but we follow the presentation in Laub and Meyer (1974).

Recall the definition of the generalized eigenspace. Let λ be an eigenvalue of A, and define subspaces of $\mathbb{C}^{2n}$ by $\eta_k(\lambda) = \text{kernel}\,(A - \lambda I)^k$. From elementary linear algebra $\eta_k(\lambda) \subset \eta_{k+1}(\lambda)$ and for each eigenvalue λ there is a smallest k', $1 \leq k' < 2n$, such that

$$\eta_{k'}(\lambda) = \eta_{k'+1}(\lambda) = \eta_{k'+2}(\lambda) = \cdots.$$

Let $\eta^\dagger(\lambda) = \eta_{k'}(\lambda)$. The eigenspace of A corresponding to the eigenvalue λ is $\eta(\lambda) = \eta_1(\lambda)$, and the generalized eigenspace is $\eta^\dagger(\lambda)$. The same definitions hold for any matrix including T.

Also recall that if $\{x, y\} = x^T J y = 0$, then x and y are J-orthogonal.

Proposition 5.3.1. *Let A (respectively T) be any $2n \times 2n$ real Hamiltonian (respectively, symplectic) matrix with eigenvalues λ and μ, $\lambda + \mu \neq 0$ (respectively, $\lambda\mu \neq 1$). Then $\{\eta^\dagger(\lambda), \eta^\dagger(\mu)\} = 0$; i.e., the generalized eigenspaces are J-orthogonal.*

Proof. Let A be Hamiltonian with eigenvalues λ and μ, $\lambda + \mu \neq 0$. By Lemma 2.3.1, $\{\eta_1(\lambda), \eta_1(\mu)\} = 0$; so, make the induction assumption that $\{\eta_k(\lambda), \eta_k(\mu)\} = 0$. We first show that $\{\eta_{k+1}(\lambda), \eta_k(\mu)\} = 0$. Recall that $\{Ax, y\} = -\{x, Ay\}$ for all $x, y \in \mathbb{V}$. Let $u \in \eta_{k+1}(\lambda)$ and $v \in \eta_k(\mu)$; so, $(A - \lambda I)^{k+1}u = 0$ and $(A - \mu I)^k v = 0$. Then

---

[1]The statement of the theorem on logarithms of symplectic matrices in Meyer and Hall (1991) is wrong.

$$0 = \{u, (A - \mu I)^k v\}$$

$$= \{u, (A + \lambda I + [-\lambda - \mu]I)^k v\}$$

$$= \sum_{j=0}^{k} \binom{k}{j} (-\lambda - \mu)^{k-j} \{u, (A + \lambda I)^j v\}$$

$$= \sum_{j=0}^{k} \binom{k}{j} (-\lambda - \mu)^{k-j} \{(-A + \lambda I)^j u, v\}.$$

Because $u \in \eta_{k+1}(\lambda)$, we have $(-A + \lambda I)^j u \in \eta_k(\lambda)$ for $j = 1, 2, \ldots$; so all the terms in the last sum above are zero except the term with $j = 0$. Therefore $(-\lambda - \mu)^k \{u, v\} = 0$. This proves $\{\eta_{k+1}(\lambda), \eta_k(\mu)\} = 0$. A similar argument shows that $\{\eta_{k+1}(\lambda), \eta_k(\mu)\} = 0$ implies $\{\eta_{k+1}(\lambda), \eta_{k+1}(\mu)\} = 0$; thus by induction the lemma holds for Hamiltonian matrices.

Let T be symplectic and $Tx = \lambda x$, $Ty = \mu y$, where $x, y \neq 0$ and $\lambda\mu \neq 0$. Then

$$\lambda\{x, y\} = \{Tx, y\} = x^T T^T J y = x^T J T^{-1} y = x^T J \mu^{-1} y = \mu^{-1}\{x, y\},$$

so $\{x, y\} = 0$ or $\{\eta(\lambda), \eta(\mu)\} = 0$. This is the initial step in the induction which proceeds just as in the Hamiltonian case. See Laub and Meyer (1974).

Let A be a real Hamiltonian matrix. The eigenvalues of A fall into four groups:

(i)   The eigenvalue 0
(ii)   The real eigenvalues $\pm\alpha_1, \ldots, \pm\alpha_s$
(iii) The pure imaginary $\pm\beta_1 i, \ldots, \pm\beta_r i$
(iv)   The truly complex $\pm\gamma_1 \pm \delta_1 i, \ldots, \pm\gamma_t \pm \delta_t i$

This defines a direct sum decomposition

$$\mathbb{V} = \mathbb{X} \oplus (\oplus_j \mathbb{U}_j) \oplus (\oplus_j \mathbb{W}_j) \oplus (\oplus_j \mathbb{Z}_j),$$

where

$$\mathbb{X} = \eta^\dagger(0),$$

$$\mathbb{U}_j = \eta^\dagger(\alpha_j) \oplus \eta^\dagger(-\alpha_j),$$

$$\mathbb{W}_j = \eta^\dagger(\beta_j i) \oplus \eta^\dagger(-\beta_j i),$$

$$\mathbb{Z}_j = \{\eta^\dagger(\gamma_j + \delta_j i) \oplus \eta^\dagger(\gamma_j - \delta_j i)\} \oplus \{\eta^\dagger(-\gamma_j - \delta_j i) \oplus \eta^\dagger(-\gamma_j + \delta_j i)\}.$$

Each of the spaces given above is an invariant subspace for A. By Lemma 5.3.1 each space is J-orthogonal to every other, and so by Proposition 2.2.1, each space must be a symplectic subspace. Because each subspace is invariant under complex conjugation, each is the complexification of a real space. Thus we can choose symplectic coordinates for each of the spaces so that A in these coordinates is block diagonal. Therefore, the next task would be to consider each space separately.

For the discussion assume that the α_j and γ_j are positive. Regroup the spaces as follows.

$$\mathbb{V} = \mathbb{I} \oplus \mathbb{N} \oplus \mathbb{P},$$

$$\mathbb{I} = \mathbb{X} \oplus (\oplus_j \mathbb{W}_j),$$

$$\mathbb{N} = (\oplus_j \eta^|(-\alpha_j)) \oplus (\oplus_j \{\eta^\dagger(-\gamma_j - \delta_j i) \oplus \eta^\dagger(-\gamma_j + \delta_j i)\})$$

$$\mathbb{P} = (\oplus_j \eta^\dagger(+\alpha_j)) \oplus (\oplus_j \{\eta^\dagger(+\gamma_j - \delta_j i) \oplus \eta^\dagger(+\gamma_j + \delta_j i)\}).$$

(5.9)

All these spaces are real and A invariant. $\mathbb{I}$ is symplectic and the restriction of A to $\mathbb{I}$ has only pure imaginary eigenvalues. All the eigenvalues of A restricted to $\mathbb{N}$ (respectively, $\mathbb{P}$) have negative (respectively, positive) real parts. $\mathbb{N} \oplus \mathbb{P}$ is symplectic and the splitting is Lagrangian. The following lemma is a direct result of Lemma 2.2.5.

Proposition 5.3.2. *Let A be a real, $2n \times 2n$, Hamiltonian matrix all of whose eigenvalues have nonzero real parts, i.e., $\mathbb{I} = 0$. Then there exists a real $2n \times 2n$ symplectic matrix P such that $P^{-1}AP = diag(B^T, -B)$, where B is a real $n \times n$ matrix, all of whose eigenvalues have negative real parts. In particular, B could be taken in real Jordan form.*

It remains to consider the restriction of A to $\mathbb{I}$. The detailed and lengthy discussion of these normal forms is given in Section 5.4.

Let T be a real symplectic matrix. The eigenvalues of T fall into five groups:

(i) $+1$
(ii) -1
(iii)   The real eigenvalues $\mu_1^{\pm 1}, \ldots, \mu_s^{\pm 1}$
(iv)   The eigenvalues of unit modulus $\alpha_1 \pm \beta_1 i, \ldots, \alpha_r \pm \beta_r i$
(v)   The eigenvalues of modulus different from one
       $(\gamma_1 \pm \delta_1 i)^{\pm 1}, \ldots, (\gamma_t \pm \delta_t i)^{\pm 1}$.

This defines a direct sum decomposition

$$\mathbb{V} = \mathbb{X} \oplus \mathbb{Y} \oplus (\oplus_j \mathbb{U}_j) \oplus (\oplus_j \mathbb{W}_j) \oplus (\oplus_j \mathbb{Z}_j) \tag{5.10}$$

where

$$\mathbb{X} = \eta^\dagger(+1), \qquad \mathbb{Y} = \eta^\dagger(-1),$$

$$\mathbb{U}_j = \eta^\dagger(\nu_j) \oplus \eta^\dagger(\nu_j^{-1}),$$

$$\mathbb{W}_j = \eta^\dagger(\alpha_j + \beta_j i) \oplus \eta^\dagger(\alpha_j - \beta_j i),$$

$$\mathbb{Z}_j = \{\eta^\dagger(\gamma_j + \delta_j i) \oplus \eta^\dagger(\gamma_j - \delta_j i)\} \oplus \{\eta^\dagger((\gamma_j - \delta_j i)^{-1}) \oplus \eta^\dagger((\gamma_j + \delta_j i)^{-1})\}.$$

Each of the summands in the above is an invariant subspace for T. By Lemma 5.3.1, each space is J-orthogonal to every other, and so each space must be a symplectic subspace. Because each subspace is invariant under complex conjugation, each is the complexification of a real space. Thus we can choose symplectic coordinates for each of the spaces so that T in these coordinates is block diagonal. Because each of the spaces in the above is symplectic it must be even-dimensional and so, the multiplicity of the eigenvalue -1 is even. The restriction of T to $\eta^\dagger(-1)$ has determinant $+1$. This gives another proof of Theorem 2.1.7.

Corollary 5.3.1. *The determinant of a symplectic matrix is $+1$.*

Again we can group these spaces together, as follows.

$$\mathbb{V} = \mathbb{I} \oplus \mathbb{N} \oplus \mathbb{P},$$

$$\mathbb{I} = \mathbb{X} \oplus \mathbb{Y} \oplus (\oplus_j \mathbb{W}_j),$$

$$\mathbb{N} = \{\eta^\dagger(\gamma_j + \delta_j i) \oplus \eta^\dagger(\gamma_j - \delta_j i)\},$$

$$\mathbb{P} = \{\eta^\dagger((\gamma_j + \delta_j i)^{-1} \oplus \eta^\dagger((\gamma_j - \delta_j i)^{-1})\}.$$

All these spaces are real and T invariant. $\mathbb{I}$ is symplectic and the restriction of T to $\mathbb{I}$ has only eigenvalues with unit modulus. All the eigenvalues of T restricted to $\mathbb{N}$ (respectively, $\mathbb{P}$) have modulus less than (respectively, greater than) one. $\mathbb{N} \oplus \mathbb{P}$ is symplectic and the splitting is Lagrangian. The following lemma is a direct result of Lemma 2.2.5.

Proposition 5.3.3. *Let T be a real, $2n \times 2n$, symplectic matrix all of whose eigenvalues have modulus different from one, i.e., $\mathbb{I} = 0$. Then there exists a real $2n \times 2n$ symplectic matrix P such that $P^{-1}TP = \mathrm{diag}(B^{\mathrm{T}}, B^{-1})$, where B is a real $n \times n$ matrix, all of whose eigenvalues have negative real parts. In particular, B could be taken in real Jordan form.*

It remains to consider the restriction of T to $\mathbb{I}$. For this discussion we refer the reader to Laub and Meyer (1974).

Problem: Prove Lemma 2.5.1 for the symplectic matrix T by using induction on the formula $\{\eta_k(\lambda), \eta_k(\mu)\} = 0$, where $\eta_k(\lambda) = \mathrm{kernel}\,(T^k - \lambda I)$.

5.4 Normal Forms for Hamiltonian Matrices

By the discussion of Section 5.3 it remains to study a Hamiltonian matrix where all the eigenvalues have zero real part. We further subdivide this into the case when the matrix has only the eigenvalue zero and to the case when the matrix has a pair of pure imaginary eigenvalues. These two cases are covered in the next two subsections.

5.4.1 Zero Eigenvalue

Throughout this subsection $\mathbb{V}$ is a real symplectic linear space with Poisson bracket $\{\cdot, \cdot\}$ ($\{x, y\} = x^T J y$), and $A : \mathbb{V} \to \mathbb{V}$ is a real Hamiltonian linear operator (or matrix) whose sole eigenvalue is zero, i.e., A is nilpotent.

Theorem 5.4.1. $\mathbb{V} = \oplus_j U_j$ where U_j is an A-invariant symplectic subspace and there is a special symplectic basis for U_j. Let the dimension of U_j be $2s \times 2s$ and B_δ the matrix of the restriction of A to U_j in this basis. Then

$$B_\delta = \begin{bmatrix} N & 0 \\ D & -N^T \end{bmatrix},$$

where

$$N = \begin{bmatrix} 0\,0\,0\,\cdots\,0\,0 \\ 1\,0\,0\,\cdots\,0\,0 \\ 0\,1\,0\,\cdots\,0\,0 \\ \cdots \\ 0\,0\,0\,\cdots\,0\,0 \\ 0\,0\,0\,\cdots\,1\,0 \end{bmatrix}, \quad D = \begin{bmatrix} 0\,0\,0\,\cdots\,0\,0 \\ 0\,0\,0\,\cdots\,0\,0 \\ \cdots \\ 0\,0\,0\,\cdots\,0\,0 \\ 0\,0\,0\,\cdots\,0\,\delta \end{bmatrix},$$

and $\delta = 0$ or 1 or -1.
The Hamiltonian is

$$H_\delta = \sum_{j=1}^{s-1} x_j y_{j+1} - \frac{\delta}{2} x_s^2.$$

Remarks: B_0 has rank $2s - 2$ whereas $B_{\pm 1}$ has rank $2s - 1$ so they are not similar. The Hessian of H_{+1} (respectively. of H_{-1}) has index s (respectively, $s - 1$) so B_{+1} and B_{-1} are not symplectically similar.
B_0 is the coefficient matrix of the pair of nth-order equations

$$\frac{d^s x}{dt^s} = 0, \quad \frac{d^s y}{dt^s} = 0$$

written as a system and $B_{\pm 1}$ is the coefficient matrix of the $2n^{th}$-order equations

$$\frac{d^{2s}x}{dt^{2s}} = 0$$

written as a system. The exponential of B_δ is easy to compute because the series expansion terminates due to the fact that B_δ is nilpotent. For example, if $s = 2$ we compute

$$e^{B_\delta t} = \begin{bmatrix} 1 & 0 & 0 & 0 \\ t & 1 & 0 & 0 \\ -\delta t^3/3! & -\delta t^2/2! & 1 & -t \\ \delta t^2/2! & \delta t & 0 & 1 \end{bmatrix}.$$

The proof of this proposition requires a series of lemmas and some definitions. The general ideas are used over and over. Let A have nilpotent index $k + 1$, i.e., $A^k \neq 0$, $A^{k+1} = 0$. By the Jordan canonical form theorem, $\mathbb{V}$ has a basis of the form

$$v_1, \; Av_1, \; \ldots, \; A^{s_1}v_1$$
$$v_2, \; Av_2, \; \ldots, \; A^{s_2}v_2$$
$$\ldots$$
$$v_r, \; Av_r, \; \ldots, \; A^{s_r}v_r,$$

where $A^{s_i+1}v_i = 0$, $i = 1, \ldots, r$. However, this basis is not symplectic.

Let $\mathcal{A}$ be the commutative algebra,

$$\mathcal{A} = \{\alpha_0 I + \alpha_1 A + \alpha_2 A^2 + \cdots + \alpha_k A^k : \alpha_i \in \mathbb{R}\}$$

and $\mathcal{V}$ be $\mathbb{V}$ considered as a module over $\mathcal{A}$. Note that $v_1, \ldots, v_r$ is a basis for $\mathcal{V}$. Given a set of vectors S let $\mathcal{L}(S)$ denote the linear span of this set in the module sense.

Let $\Phi = \alpha_0 I + \alpha_1 A + \alpha_2 A^2 + \cdots + \alpha_k A^k$ and define

$$\Phi^* = \alpha_0 I - \alpha_1 A + \alpha_2 A^2 - \cdots + (-1)^k \alpha_k A^k,$$
$$\Lambda(\Phi) = \alpha_k,$$
$$\Omega(x, y) = \{A^k x, y\}I + \{A^{k-1}x, y\}A + \cdots + \{x, y\}A^k.$$

Lemma 5.4.1. *For all $\beta_1, \beta_2 \in \mathbb{R}$, $\Phi, \Phi_1, \Phi_2 \in \mathcal{A}$, and $x, x_1, x_2, y, y_1, y_2 \in \mathbb{V}$,*

1. $\Omega(x, y) = (-1)^{k+1}\Omega(y, x)^*$,
2. $\Omega(\beta_1 \Phi_1 x_1 + \beta_2 \Phi_2 x_2, y) = \beta_1 \Phi_1 \Omega(x_1, y) + \beta_2 \Phi_2 \Omega(x_2, y)$,
3. $\Omega(x, \beta_1 \Phi_1 y_1 + \beta_2 \Phi_2 y_2) = \beta_1 \Phi_1^* \Omega(x, y_1) + \beta_2 \Phi_2^* \Omega(x, y_2)$,
4. $\Omega(x, y) = 0$ *for all y implies $x = 0$,*
5. $\{\Phi x, y\} = \Lambda(\Phi \Omega(x, y))$.

Proof. A is Hamiltonian so $\{Ax, y\} = -\{x, Ay\}$ which yields the first statement. By the linearity of the Poisson bracket $\Omega(\beta_1 x_1 + \beta_2 x_2, y_1) = \beta_1 \Omega(x_1, y_1) + \beta_2 \Omega(x_2, y_1)$. Note that

$$A\Omega(x, y) = A\{0 + \{A^k x, y\}I + \cdots + \{Ax, y\}A^{k-1} + \{x, y\}A^k\}$$
$$= 0 + \{A^k x, y\}A + \cdots + \{Ax, y\}A^k + \{x, y\}A^{k+1}$$
$$= \{A^k Ax, y\}I + \{A^{k-1}Ax, y\} + \cdots + \{Ax, y\}A^k + 0$$
$$= \Omega(Ax, y).$$

Thus $A\Omega(x, y) = \Omega(Ax, y)$ and with the linearity this implies the second statement. The third statement follows from the first two. The nondegeneracy of Ω follows at once from the nondegeneracy of the Poisson bracket. The last statement is just a computation.

Lemma 5.4.2. $\Phi = \alpha_0 I + \alpha_1 A + \alpha_2 A^2 + \cdots + \alpha_k A^k$ is nonsingular if and only if $\alpha_0 \neq 0$. If Φ is nonsingular then it has a nonsingular square root Ψ such that $\Psi^2 = \mathrm{sign}\,(\alpha_0)\Phi$.

Proof. Let $\Psi = \sum_{j=0}^{k} \beta_k A^k$. Then $\Psi\Phi = \sum_{l=0}^{k}\{\sum_{j=0}^{l} \alpha_j \beta_{l-j}\}A^l$ so to find an inverse we must solve

$$\alpha_0 \beta_0 = 1 \quad \text{and} \quad \sum_{j=0}^{l} \alpha_j \beta_{l-j} = 0 \quad \text{for} \quad l = 1, \ldots, k.$$

The first equation has as its solution $\beta_0 = 1/\alpha_0$. The remaining equation can be solved recursively for β_j, $j = 1, \ldots, k$.

Assume $\alpha_0 > 0$, then to solve $\Psi^2 = \Phi$ leads to $\alpha_l = \sum_{j=0}^{l} \beta_j \beta_{l-j}$. For $l = 0$ the formula is $\alpha_0 = \beta_0^2$ which has a solution $\beta_0 = \sqrt{\alpha_0}$. The remaining β_j are again found by recursion. If $\alpha_0 < 0$ then solve $\Psi^2 = -\Phi$ as above.

Lemma 5.4.3. *Let $\mathbb{W} \subset \mathbb{V}$ be an A-invariant subspace and let the subscript W indicate the restriction of all the various objects to this subspace. That is, $A_W = A|\mathbb{W}$; the nilpotent index of A_W is $k_W + 1 \leq k + 1$; $\mathcal{A}_W$ is the algebra generated by A_W; $\Omega_W = \Omega|(W \times W)$; and let $\mathcal{W}$ denote W considered as a module over $\mathcal{A}_W$.*

Let $\xi_1, \ldots, \xi_\gamma$ be a basis for $\mathcal{W}$. By relabeling if necessary suppose that $\Omega(\xi_1, \xi_1)$ is nonsingular.

Then there exist a basis $e_W, \xi_2', \ldots, \xi_\gamma'$ for $\mathcal{W}$ with $\Omega_W(e_W, e_W) = \pm I$, $\Omega_W(e_W, \xi_j') = 0$ for $j = 2, \ldots, \gamma$, and $\mathcal{W} = \mathcal{L}(e_W) \oplus \mathcal{L}(\xi_2', \ldots, \xi_\gamma')$. Also $e_W, A_W e_W, \ldots, A_W^{k_W} e_W$ is a basis for $\mathcal{L}(e_W)$ as a vector space over $\mathbb{R}$.

Proof. For notional convenience in this proof we drop the subscript W. Note that k must be odd since from Lemma 5.4.1 we have $\Omega(\xi_1, \xi_1) = (-1)^{k+1}\Omega(\xi_1, \xi_1)^*$. The coefficient α_0 of I in $\Omega(\xi_1, \xi_1)$ must be nonzero by Lemma 5.4.2 and is invariant under the $*$-operation so $\alpha_0 = (-1)^{k+1}\alpha_0 \neq 0$ because k is odd. Thus $\Phi = \Omega(\xi_1, \xi_1) = \Omega(\xi_1, \xi_1)^*$ so that Φ must be of the form $\alpha_0 I + \alpha_2 A^2 + \cdots + \alpha_{k-1} A^{k-1}$. We say that Φ is even.

Because $\Phi = \Phi^*$ is nonsingular, by Lemma 5.4.2 there exists a square root Ψ such that $\Psi^2 = \mathrm{sign}\,(\alpha_0)\Phi$. Moreover, from the proof of Lemma 5.4.2, it is clear that Ψ is even, i.e., $\Psi = \Psi^*$. Suppose that $\alpha_0 > 0$. (A precisely analogous proof works when $\alpha_0 < 0$.) Let e be defined by $\xi_1 = \Psi e$ (or $e = \Psi^{-1}\xi_1$). Then

$$\Psi\Psi^* = \Psi^2 = \Phi = \Omega(\xi_1, \xi_1) = \Omega(\Psi e, \Psi e) = \Psi\Psi^*\Omega(e, e).$$

Thus $\Omega(e, e) = I$ (or $\Omega(e, e) = -I$ when $\alpha_0 < 0$).

Now, for $j = 2, \ldots, \gamma$, let $\xi'_j = \xi_j - \Omega(e, \xi_j)^* e$. Then

$$
\begin{aligned}
\Omega(e, \xi'_j) &= \Omega(e, \xi_j) - \Omega(e, \Omega(e, \xi_i)^* e) \\
&= \Omega(e, \xi_j) - \Omega(e, \xi_j)\Omega(e, e) \\
&= \Omega(e, \xi_j) - \Omega(e, \xi_j) \\
&= 0.
\end{aligned}
$$

The transformation from the basis $\xi_1, \xi_2, \ldots, \xi_\gamma$ to $e, \xi'_2, \ldots, \xi'_\gamma$ is invertible, thus the latter set of vectors is also a basis for $\mathcal{W}$.

Lemma 5.4.4. *Let $W, \mathcal{W}, A_W, \Omega_W$ be as in Lemma 5.4.3 (except now A_W is nilpotent of index $m_W + 1 \leq k+1$). Suppose again that $\xi_1, \ldots, \xi_\gamma$ is a basis for $\mathcal{W}$, but now $\Omega_W(\xi_j, \xi_j)$ is singular for all $j = 1, \ldots, \gamma$ and by relabeling if necessary $\Omega_W(\xi_1, \xi_2)$ is nonsingular.*

Then there exists a basis $f_W, g_W, \xi'_3, \ldots, \xi'_\gamma$ for $\mathcal{W}$ such that

$$
\Omega_W(f_W, g_W) = I,
$$

$$
\Omega_W(f_W, f_W) = \Omega_W(g_W, g_W) = \Omega_W(f_W, \xi'_j) = \Omega_W(g_W, \xi'_j) = 0,
$$

and $\mathcal{W} = \mathcal{L}(f_W, g_W) \oplus \mathcal{L}(\xi'_3, \ldots, \xi'_\gamma)$ where $\mathcal{L}(f_W, g_W)$ has a vector space basis $f_W, \ldots, A_W^{m_W} f_W, g_W, \ldots, A_W^{m_W} g_W$.

Proof. Again for notational convenience we drop the W subscript. Also we may assume that $\Omega(\xi_1, \xi_2) = I$; if not, simply make a change of variables. There are two cases to consider.

Case 1: m is even. By Lemma 5.4.1, $\Omega(\xi_j, \xi_j) = (-1)^{m+1}\Omega(\xi_j, \xi_j)^* = -\Omega(\xi_j, \xi_j)^*$ thus $\Omega(\xi_j, \xi_j)$ is of the form $\alpha_1 A + \alpha_3 A^3 + \cdots + \alpha_{m-1}A^{m-1}$ which we call odd. Also by Lemma 5.4.1, $\Omega(\xi_2, \xi_1) = -I$. Let $f = \xi_1 + \Phi\xi_2$ where Φ is to be found so that $\Omega(f, f) = 0$ and Φ is odd.

$$
\begin{aligned}
0 = \Omega(f, f) &= \Omega(\xi_1, \xi_1) + \Omega(\xi_1, \Phi\xi_2) + \Omega(\Phi\xi_2, \xi_1) + \Omega(\Phi\xi_2, \Phi\xi_2) \\
&= \Omega(\xi_1, \xi_1) + \Phi^* - \Phi + \Phi\Phi^*\Omega(\xi_2, \xi_2).
\end{aligned}
$$

Because we want $\Phi = -\Phi^*$, we need to solve

$$
\Phi = \frac{1}{2}[\Omega(\xi_1, \xi_1) + \Phi^2 \Omega(\xi_2, \xi_2)].
$$

Notice that the product of three odd terms is again odd, so the right-hand side is odd. Clearly, we may solve recursively for the coefficients of Φ starting with the coefficient of A in Φ.

Case 2: m is odd. By Lemma 5.4.1 $\Omega(\xi_i, \xi_i)$ is even and because it is singular $\alpha_0 = 0$. Also $\Omega(\xi_2, \xi_1) = I$. Set $f = \xi_1 + \Phi\xi_2$ and determine an even Φ so that $\Omega(f, f) = 0$. We need to solve

$$
\Phi = \frac{1}{2}[\Omega(\xi_1, \xi_1) + \Phi^2 \Omega(\xi_2, \xi_2)].
$$

The right-hand side is even and we can solve the equation recursively starting with the coefficient of A^2.

In either case $f, \xi_2, \ldots, \xi_\gamma$ is a basis and we may assume that $\Omega(f, \xi_2) = I$ (and $\Omega(\xi_2, f) = \pm I$ accordingly as m is odd or even). Let $h = \xi_2 - \frac{1}{2}\Omega(\xi_2, \xi_2)f$. One can check that $\Omega(h, h) = 0$ whether m is even or odd. Let g be defined by $h = \Omega(f, h)^*g$. Then $\Omega(g, g) = 0$ and $\Omega(f, h) = \Omega(f, h)\Omega(f, g)$ so $\Omega(f, g) = I$. Finally, note that $\Omega(g, h) = I$ if m is odd and $-I$ if m is even. Let $\xi'_j = \xi_j \mp \Omega(g, \xi_j)^*f - \Omega(f, \xi_j)^*g$ (minus sign when m is odd and plus sign when m is even). One checks that $\Omega(f, \xi'_j) = \Omega(g, \xi'_j) = 0$ for $j = 3, \ldots, \gamma$.

Proposition 5.4.1. *Let $A : V \to V$ be nilpotent, then $\mathbb{V}$ has a symplectic decomposition*

$$\mathbb{V} = U_1 \oplus \cdots \oplus U_\alpha \oplus V_1 \oplus \cdots \oplus V_\beta$$

into A-invariant subspaces. Furthermore, U_j has a basis $e_j, \ldots, A^{k_j}e_j$ ($A|U_j$ is nilpotent with index $k_j + 1 \le k + 1$) with

$$\{A^s e_j, e_j\} = \begin{cases} \pm 1 & \text{if } s = k_j \\ 0 & \text{otherwise,} \end{cases}$$

and V_j has a basis $f_j, Af_j, \ldots, A^{m_j}f_j, g_j, Ag_j, \ldots, A^{m_j}g_j$ ($A|V_j$ is nilpotent of index $m_j + 1 \le k + 1$) with

$$\{A^s f_j, g_j\} = \begin{cases} 1 & \text{if } s = m_j \\ 0 & \text{otherwise} \end{cases}$$

and

$$\{A^s f_j, f_j\} = \{A^s g_j, g_j\} = 0 \quad \text{for all } s.$$

Proof. Let $\xi_1, \ldots, \xi_\gamma$ be a basis for $\mathcal{V}$. First we need to show that $\Omega(\xi_i, \xi_j)$ is nonsingular for some i and j (possibly equal). Suppose to the contrary that $\Omega(\xi_i, \xi_j)$ is singular for all i and j. By Lemma 5.4.2, the coefficient of I must be 0, i.e., $\{A^k \xi_i, \xi_j\} = 0$. Furthermore, $\{A^{k+l_1+l_2}\xi_i, \xi_j\} = 0$ for all nonnegative integers l_1, l_2 by the nilpotence of A. Fix l_1. Then $\{A^k(A^{l_1}\xi_j), A^{l_2}\xi_j\} = 0$ for all $l_2 \ge 0$. Because $\{A^{l_2}\xi_j : l_2 \ge 0\}$ forms a basis for $\mathbb{V}$ and $\{\cdot, \cdot\}$ is nondegenerate, we conclude that $A^k(A^{l_1}\xi_j) = 0$. But l_1 is arbitrary so $A^k = 0$, which contradicts the hypothesis of $k + 1$ as the index of nilpotence of A.

By relabeling if necessary, we may assume that either $\Omega(\xi_1, \xi_1)$ or $\Omega(\xi_1, \xi_2)$ is nonsingular. In the first case apply Lemma 5.4.3 and in the second case Lemma 5.4.4. We have either $\mathbb{V} = \mathcal{L}(e_W) \oplus \mathbb{Z}$, where $\mathbb{Z} = \mathcal{L}(\xi'_2, \ldots, \xi'_\gamma)$ or $\mathbb{V} = \mathcal{L}(f_W, g_W) \oplus \mathbb{Z}$ where $\mathbb{Z} = \mathcal{L}(\xi'_2, \ldots, \xi'_\gamma)$. In the first case call $U_1 = \mathcal{L}(e_W)$; in the second case call $V_1 = \mathcal{L}(f_W, g_W)$. In either case repeat the process on $\mathbb{Z}$.

Proof (Theorem 5.4.1). All we need to do is construct a symplectic basis for each of the subspaces in Proposition 5.4.1.

Case 1: $A|U_j$. Let $A|U_j$ be denoted by A, U denote U_j and $k + 1$ the index of nilpotence of A. Then there is an $e \in U$ such that $\{A^k e, e\} = \pm 1$

and $\{A^s e, e\} = 0$ for $s \neq k$. Consider the case when $\{A^k e, e\} = +1$ first and recall that k must be odd. Let $l = (k+1)/2$. Define

$$q_j = A^{j-1} e, \quad p_j = (-1)^{k+1-j} A^{k+1-j} e \quad \text{for } j = 1, \ldots, l.$$

Then for $i, j = 1, \ldots, l$,

$$\begin{aligned}
\{q_i, q_j\} &= \{A^{i-1} e, A^{j-1}\} \\
&= (-1)^{j-1} \{A^{i+j-2} e, e\} \\
&= 0 \text{ because } i + j - 2 \leq k - 1,
\end{aligned}$$

$$\begin{aligned}
\{p_i, p_j\} &= \{(-1)^{k+1-i} A^{k+1-i} e, (-1)^{k+1-j} A^{k+1-j} e\} \\
&= (-1)^{k+1-i} \{A^{2k-i-j} e, e\} \\
&= 0 \text{ because } 2(k+1) - i - j > k + 1,
\end{aligned}$$

$$\begin{aligned}
\{q_i, p_j\} &= \{A^{i-1} e, (-1)^{k+1-j} A^{k+1-j} e\} \\
&= \{A^{k+i-j} e, e\} \\
&= \begin{cases} 1 \text{ if } i = j \\ 0 \text{ if } i \neq j. \end{cases}
\end{aligned}$$

Thus $q_1, \ldots, p_l$ is a symplectic basis. With respect to this basis A has the matrix form B_δ given in Proposition 5.4.1 with $\delta = (-1)^l$.

In the case $\{A^k e, e\} = -1$ define

$$q_j = A^{j-1} e, \quad p_j = (-1)^{k-j} A^{k+1-j} e \quad \text{for } j = 1, \ldots, l.$$

to find that A in these coordinates is B_δ with $\delta = (-1)^{l+1}$.

Case 2: $A|V_j$. Let $A|V_j$ be denoted by A, V denote V_j and $m+1$ the index of nilpotence of A. Then there are $f, g \in V$ such that $\{A^s f, f\} = \{A^s g, g\} = 0$, $\{A^m f, g\} = 1$, and $\{A^s f, g\} = 0$ for $s \neq m$. Define

$$q_j = A^{j-1} f, \qquad p_j = (-1)^{m+1-j} A^{m+1-j} g,$$

and check that $q_1, \ldots, p_m$ is a symplectic basis for U. The matrix representation of A in this basis is the B_0 of Theorem 5.4.1.

5.4.2 Pure Imaginary Eigenvalues

Throughout this section let $A : \mathbb{V} \to \mathbb{V}$ be a real Hamiltonian linear operator (or matrix) that has a single pair of pure imaginary eigenvalues $\pm i\nu$, $\nu \neq 0$. It is necessary to consider $\mathbb{V}$ as a vector field over $\mathbb{C}$ the complex numbers so that we may write $\mathbb{V} = \eta^\dagger(i\nu) \oplus \eta^\dagger(-i\nu)$.

Theorem 5.4.2. $\mathbb{V} = \oplus_j W_j$ where W_j is an A-invariant symplectic subspace and there is a special symplectic basis for W_j. If C is the matrix of the restriction of A to W_j in this basis, then C has one of the complex forms (5.11), (5.16) or one of the real forms (5.13), (5.14),(5.15), or (5.17).

This case is analogous to the nilpotent case. Let $B = A|\eta^\dagger(i\nu) - i\nu I$ and suppose that B is nilpotent of index $k + 1 \leq n$. Let $\mathcal{A}$ be the algebra generated by B, i.e., $\mathcal{A} = \{\alpha_0 I + \alpha_1 B + \cdots + \alpha_k B^k : \alpha_j \in \mathbb{C}\}$ and let $\mathcal{V}$ be $\eta^\dagger(i\nu)$ considered as a module over $\mathcal{A}$.

Let $\Phi = \alpha_0 I + \alpha_1 B + \alpha_2 B^2 + \cdots + \alpha_k B^k$ and define

$$\Phi^* = \bar{\alpha}_0 I - \bar{\alpha}_1 B + \bar{\alpha}_2 B^2 - \cdots + (-1)^k \bar{\alpha}_k B^k,$$
$$\Lambda(\phi) = \alpha_k,$$
$$\Omega(x, y) = \{B^k x, \bar{y}\} I + \{B^{k-1} x, \bar{y}\} B + \cdots + \{x, \bar{y}\} B^k.$$

The next three lemmas are proved like the analogous lemmas for the nilpotent case.

Lemma 5.4.5. *For all $\beta_1, \beta_2 \in \mathbb{C}$, $\Phi, \Phi_1, \Phi_2 \in \mathcal{A}$, and $x, x_1, x_2, y, y_1, y_2 \in \mathbb{V}$ we have*

1. $\Omega(x, y) = (-1)^{k+1} \Omega(y, x)^*$,
2. $\Omega(\beta_1 \Phi_1 x_1 + \beta_2 \Phi_2 x_2, y) = \beta_1 \Phi_1 \Omega(x_1, y) + \beta_2 \Phi_2 \Omega(x_2, y)$,
3. $\Omega(x, \beta_1 \Phi_1 y_1 + \beta_2 \Phi_2 y_2) = \bar{\beta}_1 \Phi_1^* \Omega(x, y_1) + \bar{\beta}_2 \Phi_2^* \Omega(x, y_2)$.
4. $\Omega(x, y) = 0$ *for all y implies $x = 0$,*
5. $\{\Phi x, \bar{y}\} = \Lambda(\Phi \Omega(x, y))$.

Lemma 5.4.6. $\Phi = \alpha_0 I + \alpha_1 B + \alpha_2 B^2 + \cdots + \alpha_k B^k$ *is nonsingular if and only if $\alpha_0 \neq 0$.*

Lemma 5.4.7. *Let $\Phi = \alpha_0 I + \alpha_1 B + \alpha_2 B^2 + \cdots + \alpha_k B^k$ be nonsingular and satisfy $\Phi = (-1)^{k+1} \Phi^*$. If k is even (respectively, odd), then Φ has a nonsingular square root Ψ such that $\Psi\Psi^* = i \operatorname{sign}(\alpha_0/i)\Phi$ (respectively, $\Psi\Psi^* = \operatorname{sign}(\alpha_0)\Phi$). Moreover, $\Psi = (-1)^{k+1}\Psi^*$.*

Proposition 5.4.2. *Let $A : \mathbb{V} \to \mathbb{V}$ have only the pure imaginary eigenvalues $\pm i\nu$ and $B = A - i\nu I$. Then $\mathbb{V}$ has a symplectic decomposition into the A-invariant of the form*

$$\mathbb{V} = (U_1 \oplus \bar{U}_1) \oplus \cdots \oplus (U_\alpha \oplus \bar{U}_\alpha) \oplus (Y_1 \oplus \bar{Y}_1) \oplus \cdots \oplus (Y_\beta \oplus \bar{Y}_\beta)$$

where U_j and Y_j are subspaces of $\eta^\dagger(i\nu)$.

U_j *has a basis $e_j, Be_j, \ldots, B^{k_j} e_j$ where $B|U_j$ is nilpotent of index $k_j + 1$, $\bar{U}_j$ has a basis $\bar{e}_j, \bar{B}\bar{e}_j, \ldots, \bar{B}^{k_j}\bar{e}_j$, and*

$$\{B^s e_j, \bar{e}_j\} = \begin{cases} \pm 1 & \text{if } s = k_j, \ k_j \text{ odd}, \\ \pm i & \text{if } s = k_j, \ k_j \text{ even}, \\ 0 & \text{otherwise}. \end{cases}$$

Y_j *has a basis $f_j, Bf_j, \ldots, B^{k_j} f_j, g_j, Bg_j, \ldots, B^{k_j} g_j$ where $B|Y_j$ is nilpotent of index $k_j + 1$, $\bar{Y}_j$ has a basis $\bar{f}_j, \ldots, \bar{B}^{k_j}\bar{g}_j, \ldots, \bar{B}^{k_j}\bar{g}_j$, and*

$$\{B^s f_j, \bar{g}_j\} = \begin{cases} 1 & \text{if } s = k_j, \\ 0 & \text{otherwise} \end{cases},$$
$$\{B^s f_j, \bar{f}_j\} = 0 \text{ all } s,$$
$$\{B^s g_j, \bar{g}_j\} = 0 \text{ all } s.$$

Proof. The proof of this proposition is essentially the same as the proof of Proposition 5.4.1 and depends on extensions of lemmas like Lemma 5.4.3 and Lemma 5.4.4. See Theorem 14 and Lemmas 15 and 16 of Laub and Meyer (1974).

Proof (Theorem 5.4.2). The real spaces $\mathbb{W}_j$ of the theorem have as their complexification either one of $(U_j \oplus \bar{U}_j)$ or one of $(Y_j \oplus \bar{Y}_j)$. Using Lemma 5.4.2 we construct a symplectic basis for each of these subspaces.

   Case 1: Let us consider $(U_j \oplus \bar{U}_j)$ first and drop the subscript. Indeed, for the moment let $\mathbb{V} = (U_j \oplus \bar{U}_j)$ and $A = A|(U_j \oplus \bar{U}_j)$, etc. Then there is a complex basis $e, Be, \ldots, B^k e, \bar{e}, \bar{B}\bar{e}, \ldots, \bar{B}^k \bar{e}$ and $\{B^k e, \bar{e}\} = a$ where $a = \pm 1, \pm i$. Consider the complex basis

$$u_j = B^{j-1} e, \quad v_j = a^{-1}(-1)^{k-j+1} \bar{B}^{k-j+1} \bar{e}, \quad j = 1, \ldots, k+1.$$

Because $(U_j \oplus \bar{U}_j)$ is a Lagrangian splitting $\{u_j, u_s\} = \{v_j, v_s\} = 0$.

$$\begin{aligned}
\{u_j, v_s\} &= \{B^{j-1} e, a^{-1}(-1)^{k-s+1} \bar{B}^{k-s+1} \bar{e}\} \\
&= a^{-1}\{B^{k+j-s} e, \bar{e}\} \\
&= \begin{cases} 1 & \text{if } j = s, \\ 0 & \text{otherwise} \end{cases}
\end{aligned}$$

Thus the basis is symplectic and in this basis A has the form

$$\begin{bmatrix} N & 0 \\ 0 & -N^T \end{bmatrix}, \quad \text{where} \quad N = \begin{bmatrix} i\nu & 0 & 0 & \cdots & 0 & 0 \\ 1 & i\nu & 0 & \cdots & 0 & 0 \\ 0 & 1 & i\nu & \cdots & 0 & 0 \\ & & & \cdots & & \\ 0 & 0 & 0 & \cdots & i\nu & 0 \\ 0 & 0 & 0 & \cdots & 1 & i\nu \end{bmatrix}, \tag{5.11}$$

and the reality condition is $u_j = \bar{a}(-1)^{k-j+2} \bar{v}_{k-j+2}$. The real normal forms depend on the parity of k.

   Case k odd: The reality condition is $u_j = a(-1)^{j-1} \bar{v}_{k-j+2}$ where $a = \pm 1$. Consider the following real basis

$$q_j = \begin{cases} \sqrt{2}\Re u_j = \dfrac{1}{\sqrt{2}}(u_j + a(-1)^{j-1} v_{k-j+2}), & j \text{ odd}, \\[2mm] \sqrt{2}\Im u_j = \dfrac{1}{\sqrt{2i}}(u_j - a(-1)^{j-1} v_{k-j+2}), & j \text{ even}, \end{cases}$$

$$\tag{5.12}$$

$$p_j = \begin{cases} \sqrt{2}\Re v_j = \dfrac{1}{\sqrt{2}}(v_j - a(-1)^{j-1} u_{k-j+2}), & j \text{ odd}, \\[2mm] -\sqrt{2}\Im v_j = \dfrac{1}{\sqrt{2i}}(-v_j - a(-1)^{j-1} u_{k-j+2}), & j \text{ even}. \end{cases}$$

A direct computation verifies that $\{q_j, q_s\} = \{p_j, p_j\} = 0$ for $j, s = 1, \ldots, n$ and $\{q_j, p_s\} = 0$ for $j, s = 1, \ldots, n$ and $j \neq s$. If j is odd

$$\{q_j, p_j\} = \tfrac{1}{2}\{u_j + a(-1)^{j-1}v_{k-j+2}, v_j - a(-1)^{j-1}u_{k-j+2}\}$$

$$= \tfrac{1}{2}[\{u_j, v_j\} - \{v_{k-j+2}, u_{k-j+2}\}] = 1$$

and if j is even

$$\{q_j, p_j\} = -\tfrac{1}{2}\{u_j - a(-1)^{j-1}v_{k-j+2}, -v_j - a(-1)^{j-1}u_{k-j+2}\}$$

$$= -\tfrac{1}{2}[-\{u_j, v_j\} + \{v_{k-j+2}, u_{k-j+2}\}] = 1.$$

Thus (5.12) defines a real symplectic basis and the matrix A in this basis is

$$\begin{bmatrix} 0 & aN \\ -aN^T & 0 \end{bmatrix}, \quad \text{where} \quad N = \begin{bmatrix} 0 & 0 & 0 & \cdots & 0 & 0 & \nu \\ 0 & 0 & 0 & \cdots & 0 & \nu & 1 \\ 0 & 0 & 0 & \cdots & \nu & -1 & 0 \\ & \vdots & & & & \vdots & \\ 0 & 0 & \nu & \cdots & 0 & 0 & 0 \\ 0 & \nu & -1 & \cdots & 0 & 0 & 0 \\ \nu & 1 & 0 & \cdots & 0 & 0 & 0 \end{bmatrix} \qquad (5.13)$$

For example, when $k = 1$ and $a = 1$,

$$A = \begin{bmatrix} 0 & 0 & 0 & \nu \\ 0 & 0 & \nu & 1 \\ -1 & -\nu & 0 & 0 \\ -\nu & 0 & 0 & 0 \end{bmatrix}.$$

The Hamiltonian is

$$H = \nu(x_1 x_2 + y_1 y_2) + \frac{1}{2}(x_1^2 + y_2^2).$$

Noting that A is the sum of two commuting matrices, one semisimple and one nilpotent, we have

$$a^{At} = \begin{bmatrix} 1 & 0 & 0 & 0 \\ 0 & 1 & 0 & t \\ -t & 0 & 1 & 0 \\ 0 & 0 & 0 & 1 \end{bmatrix} \begin{bmatrix} \cos \nu t & 0 & 0 & \sin \nu t \\ 0 & \cos \nu t & \sin \nu t & 0 \\ 0 & -\sin \nu t & \cos \nu t & 0 \\ -\sin \nu t & 0 & 0 & \cos \nu t \end{bmatrix}$$

$$= \begin{bmatrix} \cos \nu t & 0 & 0 & \sin \nu t \\ -t \sin \nu t & \cos \nu t & \sin \nu t & t \cos \nu t \\ -t \cos \nu t & -\sin \nu t & \cos \nu t & -t \sin \nu t \\ -\sin \nu t & 0 & 0 & \cos \nu t \end{bmatrix}.$$

Normal forms are not unique. Here is another normal form when $k = 1$; consider the following basis

$$\begin{aligned}
q_1 &= \sqrt{2}\Re u_1 = \frac{1}{\sqrt{2}}(u_1 + \bar{u}_1) = \tfrac{1}{\sqrt{2}}(u_1 + av_2)\\
q_2 &= \sqrt{2}\Im u_1 = \frac{1}{\sqrt{2}i}(u_1 - \bar{u}_1) = \tfrac{1}{\sqrt{2}i}(u_1 - av_2)\\
p_1 &= \sqrt{2}\Re v_1 = \frac{1}{\sqrt{2}}(v_1 + \bar{v}_1) = -\tfrac{1}{\sqrt{2}}(v_1 - au_2)\\
p_2 &= -\sqrt{2}\Im v_1 = \frac{-1}{\sqrt{2}i}(v_1 - \bar{v}_1) = \tfrac{1}{\sqrt{2}i}(v_1 + au_2).
\end{aligned}$$

This is a symplectic basis. In this basis

$$A = \begin{bmatrix} 0 & \nu & 0 & 0 \\ -\nu & 0 & 0 & 0 \\ a & 0 & 0 & \nu \\ 0 & a & -\nu & 0 \end{bmatrix}, \tag{5.14}$$

the Hamiltonian is

$$H = \nu(x_2 y_1 - x_1 y_2) - \frac{a}{2}(x_1^2 + x_2^2),$$

and the exponential is

$$e^{At} = \begin{bmatrix} \cos\nu t & \sin\nu t & 0 & 0 \\ -\sin\nu t & \cos\nu t & 0 & 0 \\ at\cos\nu t & at\sin\nu t & \cos\nu t & \sin\nu t \\ -at\sin\nu t & at\cos\nu t & -\sin\nu t & \cos\nu t \end{bmatrix},$$

where $a = \pm 1$.

Case k even: In this case $a = \pm i$ and the reality condition is $u_j = \bar{a}(-1)^j \bar{v}_{k-j+2}$. Consider the following real basis

$$q_j = \sqrt{2}\Re u_j = \frac{1}{\sqrt{2}}(u_j + a(-1)^j v_{k-j+2}),$$

$$p_j = (-1)^j \sqrt{2}\Im u_j = \frac{1}{\sqrt{2}}(v_j + \bar{a}(-1)^{j+1} u_{k-j+2}),$$

for $j = 1, \ldots, k+1$. By inspection $\{q_j, q_s\} = \{p_j, p_s\} = 0$, $\{q_j, p_s\} = 0$ for $j \neq s$ and

$$\{q_j, p_j\} = \tfrac{1}{2}\{u_j + a(-1)^j v_{k-j+2}, \bar{a}(-1)^{j+1} u_{k-j+2} + v_j\}$$

$$= \tfrac{1}{2}(\{u_j, v_j\} - a\bar{a}\{v_{k-j+2}, u_{k-j+2}\}) = 1.$$

Thus the basis is symplectic and A in this basis is

$$A = \begin{bmatrix} N & M \\ -M & -N^T \end{bmatrix}, \tag{5.15}$$

where

$$
N = \begin{bmatrix} 0\,0\,0 \cdots 0\,0\,0 \\ 1\,0\,0 \cdots 0\,0\,0 \\ 0\,1\,0 \cdots 0\,0\,0 \\ 0\,0\,1 \cdots 0\,0\,0 \\ \cdots \\ 0\,0\,0 \cdots 1\,0\,0 \\ 0\,0\,0 \cdots 0\,1\,0 \end{bmatrix}, \qquad
M = \pm \begin{bmatrix} 0 & 0 & 0 & \cdots & 0 & 0 & -\nu \\ 0 & 0 & 0 & \cdots & 0 & \nu & 0 \\ 0 & 0 & 0 & \cdots & -\nu & 0 & 0 \\ & & & \cdots & & & \\ 0 & 0 & -\nu & \cdots & 0 & 0 & 0 \\ 0 & \nu & 0 & \cdots & 0 & 0 & 0 \\ -\nu & 0 & 0 & \cdots & 0 & 0 & 0 \end{bmatrix},
$$

and the Hamiltonian is

$$
H - \pm\nu \sum_{j=1}^{k+1} (-1)^j (x_j x_{k-j+1} + y_j y_{k-j+1}) + \sum_{j=1}^{k} (x_j y_{j+1} + x_{j+1} y_j).
$$

The matrix A is the sum of commuting matrices

$$
A = B + C, \qquad B = \begin{bmatrix} N & 0 \\ 0 & -N^T \end{bmatrix}, \qquad C = \begin{bmatrix} 0 & M \\ -M & 0 \end{bmatrix},
$$

so $e^{At} = e^{Bt} e^{Ct}$ where

$$
e^{Bt} = B + Bt + B^2 \frac{t^2}{2} + \cdots + B^k \frac{t^k}{k!}, \qquad e^{Ct} = \begin{bmatrix} \cos Mt & \pm \sin Ml \\ + \sin Mt & \cos Mt \end{bmatrix}.
$$

In particular when $k = 2$

$$
e^{Bt} = \begin{bmatrix} 1 & 0 & 0 & 0 & 0 & 0 \\ t & 1 & 0 & 0 & 0 & 0 \\ \frac{t^2}{2} & t & 1 & 0 & 0 & 0 \\ 0 & 0 & 0 & 1 & -t & \frac{t^2}{2} \\ 0 & 0 & 0 & 0 & 1 & -t \\ 0 & 0 & 0 & 0 & 0 & 1 \end{bmatrix},
$$

$$
e^{Ct} = \begin{bmatrix} \cos \nu t & 0 & 0 & 0 & 0 & \pm \sin \nu t \\ 0 & \cos \nu t & 0 & 0 & \mp \sin \nu t & 0 \\ 0 & 0 & \cos \nu t & \pm \sin \nu t & 0 & 0 \\ 0 & 0 & \mp \sin \nu t & \cos \nu t & 0 & 0 \\ 0 & \pm \sin \nu t & 0 & 0 & \cos \nu t & 0 \\ \mp \sin \nu t & 0 & 0 & 0 & 0 & \cos \nu t \end{bmatrix}.
$$

Case 2: Now let us consider $(Y_j \oplus \bar{Y}_j)$ and drop the subscript. Because $\Omega(f, \bar{f}) = \Omega(g, \bar{g}) = 0$ and $\Omega(f, \bar{g}) = I$, by definition of Ω we have

$$
\{B^s f, \bar{f}\} = \{B^s g, \bar{g}\} = 0, \text{ for all } s, \qquad \{B^s f, \bar{g}\} = \begin{cases} 1 & \text{if } s = k \\ 0 & \text{otherwise} \end{cases}
$$

Let

$$u_j = \begin{cases} B^{j-1}f & \text{for } j = 1, \ldots, k+1 \\ (-1)^{j-k-1}B^{j-k-2}g & \text{for } j = k+2, \ldots, 2k+2 \end{cases}$$

and

$$v_j = \begin{cases} (-1)^{k-j+1}\bar{B}^{k-j+1}\bar{g} & \text{for } j = 1, \ldots, k+1 \\ \bar{B}^{2k-j+2}\bar{f} & \text{for } j = k+2, \ldots, 2k+2. \end{cases}$$

One can verify that this is a symplectic basis and with respect to this basis

$$A = \begin{bmatrix} N & 0 & 0 & 0 \\ 0 & -\bar{N} & 0 & 0 \\ 0 & 0 & -N^T & 0 \\ 0 & 0 & 0 & \bar{N}^T \end{bmatrix}, \quad \text{where } N = \begin{bmatrix} i\nu & 0 & 0 & \cdots & 0 & 0 \\ 1 & i\nu & 0 & \cdots & 0 & 0 \\ 0 & 1 & i\nu & \cdots & 0 & 0 \\ \vdots & \vdots & \vdots & & \vdots & \vdots \\ 0 & 0 & 0 & \cdots & i\nu & 0 \\ 0 & 0 & 0 & \cdots & 1 & i\nu \end{bmatrix} \tag{5.16}$$

with the reality conditions

$$u_j = \begin{cases} \bar{v}_{2k-j+3} & \text{for } j = 1, \ldots, k+1 \\ -\bar{v}_{2k-j+3} & \text{for } j = k+2, \ldots, 2k+2. \end{cases}$$

Define the following real symplectic basis

$$q_j = \begin{cases} \sqrt{2}\Re u_j & \text{for } j = 1, \ldots, k+1 \\ -\sqrt{2}\Im u_{2k-j+3} & \text{for } j = k+2, \ldots, 2k+2 \end{cases}$$

$$p_j = \begin{cases} \pm\sqrt{2}\Re u_{2k-j+3} & \text{for } j = 1, \ldots, k+1 \\ -\sqrt{2}\Im u_j & \text{for } j = k+2, \ldots, 2k+2 \end{cases}$$

with respect to this basis

$$A = \begin{bmatrix} N & -M & 0 & 0 \\ M & N^T & 0 & 0 \\ 0 & 0 & -N^T & -M \\ 0 & 0 & M & -N \end{bmatrix}, \tag{5.17}$$

where

$$N = \begin{bmatrix} 0 & 0 & 0 & \cdots & 0 & 0 \\ 1 & 0 & 0 & \cdots & 0 & 0 \\ 0 & 1 & 0 & \cdots & 0 & 0 \\ \vdots & \vdots & \vdots & & \vdots & \vdots \\ 0 & 0 & 0 & \cdots & 1 & 0 \end{bmatrix}, \quad M = \begin{bmatrix} 0 & 0 & 0 & \cdots & 0 & \nu \\ 0 & 0 & 0 & \cdots & \nu & 0 \\ \vdots & \vdots & \vdots & & \vdots & \vdots \\ 0 & 0 & \nu & \cdots & 0 & 0 \\ 0 & \nu & 0 & \cdots & 0 & 0 \\ \nu & 0 & 0 & \cdots & 0 & 0 \end{bmatrix}. \tag{5.18}$$

The matrix $A = B + C$ where B and C are the commuting matrices

$$B = \begin{bmatrix} N & 0 & 0 & 0 \\ 0 & N^T & 0 & 0 \\ 0 & 0 & -N^T & 0 \\ 0 & 0 & 0 & -N \end{bmatrix}, \quad C = \begin{bmatrix} 0 & -M & 0 & 0 \\ M & 0 & 0 & 0 \\ 0 & 0 & 0 & -M \\ 0 & 0 & M & 0 \end{bmatrix}$$

so $e^{At} = e^{Bt} e^{Ct}$ where

$$e^{Bt} = B + Bt + B^2 \frac{t^2}{2} + \cdots + B^k \frac{t^k}{k!},$$

$$e^{Ct} = \begin{bmatrix} \cos Mt & -\sin Mt & 0 & 0 \\ \sin Mt & \cos Mt & 0 & 0 \\ 0 & 0 & \cos Mt & -\sin Mt \\ 0 & 0 & \sin Mt & \cos Mt \end{bmatrix}.$$

When $k = 1$ the Hamiltonian is

$$H = \nu(x_1 y_4 + x_2 y_3 - x_3 y_2 - x_4 y_1) + (x_1 y_2 + x_4 y_3)$$

5.5 Topology of $Sp(2n, \mathbb{R})$

The group $Sp(2n, \mathbb{R})$ is a manifold, so in this section we discuss some of its topology following the class notes of Larry Markus. The main results are the definition of a function, Θ, that assigns an angle to each symplectic matrix and Theorem 5.5.1 which says the fundamental group of the symplectic group, $\pi_1(Sp(n, \mathbb{R}))$, is $\mathbb{Z}$. That fact should bring the words degree and index to mind. Indeed the next topic is the Maslov and Conley-Zehnder indexes already introduced in Section 2.4 for $n = 2$.

5.5.1 Angle Function

A crucial element in this analysis is the existence of the polar decomposition of a symplectic matrix found in Theorem 2.1.6. In particular this theorem says that

$$Sp(2n, \mathbb{R}) = PSp(2n, \mathbb{R}) \times OSp(2n, \mathbb{R})$$

as manifolds (not groups) where $PSp(2n, \mathbb{R})$ is the set of all positive definite, symmetric, symplectic matrices and $OSp(2n, \mathbb{R}) = Sp(2n, \mathbb{R}) \cap O(2n, \mathbb{R})$ is the group of orthogonal symplectic matrices.

Proposition 5.5.1. $PSp(2n, \mathbb{R})$ *is diffeomorphic to* $\mathbb{R}^{n(n+1)}$. $OSp(2n, \mathbb{R})$ *is a strong deformation retract of* $Sp(2n, \mathbb{R})$.

Proof. Let $psp(2n, \mathbb{R})$ be the set of all symmetric Hamiltonian matrices. If $A \in psp(2n, \mathbb{R})$ then e^A is symmetric and symplectic. Because A is symmetric its eigenvalues are real and e^A has positive eigenvalues so e^A is positive definite. Any $T \in PSp(2n, \mathbb{R})$ has a real Hamiltonian logarithms, so the map $\Phi : psp(2n, \mathbb{R}) \to PSp(2n, \mathbb{R}) : A \mapsto e^A$ is a global diffeomorphism.

It is easy to see that $M \in psp(2n, \mathbb{R})$ if and only if

$$M = \begin{bmatrix} A & B \\ B & -A \end{bmatrix},$$

where A and B are symmetric $n \times n$ matrices. So $PDSp(2n, \mathbb{R})$ is diffeomorphic to $\mathbb{R}^{n(n+1)}$.

Proposition 5.5.2. *$OSp(2n, \mathbb{R})$ is isomorphic to $U(n, \mathbb{C})$ the group of $n \times n$ unitary matrices.*

Proof. If $T \in Sp(2n, \mathbb{R})$ then in block form

$$T = \begin{bmatrix} a & b \\ c & d \end{bmatrix}, \qquad T^{-1} = \begin{bmatrix} d^T & -b^T \\ -c^T & a^T \end{bmatrix},$$

with $a^T d - c^T b = I$ and $a^T c$ and $b^T d$ both symmetric; see Section 2.1.

If $T \in OSp(2n, \mathbb{R})$, then by the equation $T^{-1} = T^T$ we have that

$$T = \begin{bmatrix} a & b \\ -b & a \end{bmatrix}$$

with $a^T a + b^T b = I$ and $a^T b$ symmetric.

The map $\Phi : OSp(2n, \mathbb{R}) \to U(n, \mathbb{C})$ given by

$$\Phi : \begin{bmatrix} a & b \\ -b & a \end{bmatrix} \longmapsto a + bi$$

is the desired isomorphism.

Because $O(1, \mathbb{C})$ is just the set of complex numbers of unit modulus, we have

Corollary 5.5.1. *$Sp(2, \mathbb{R})$ is diffeomorphic to $S^1 \times \mathbb{R}^2$.*

The determinant of a unitary matrix has absolute value 1, so is of the form $e^{i\theta}$ where θ is a angle defined modulo 2π. Thus we have a well-defined mapping $\Theta : SP(n, \mathbb{R}) \to S^1 : T \mapsto e^{i\theta}$, in other words to each symplectic matrix there is a well-defined angle θ defined modulo 2π.

It is enlightening to look at two examples

$$T_1 = \begin{bmatrix} \alpha_1 & 0 & \beta_1 & 0 \\ 0 & \alpha_2 & 0 & \beta_2 \\ -\beta_1 & 0 & \alpha_1 & 0 \\ 0 & -\beta_2 & 0 & \alpha_2 \end{bmatrix}, \qquad T_2 = \begin{bmatrix} \mu_1 & 0 & 0 & 0 \\ 0 & \mu_2 & 0 & 0 \\ 0 & 0 & \mu_1^{-1} & 0 \\ 0 & 0 & 0 & \mu_2^{-1} \end{bmatrix},$$

with $\alpha_1^2 + \beta_1^2 = \alpha_2^2 + \beta_2^2 = 1$ and μ_1, μ_2 nonzero real. Both T_1 and T_2 are symplectic and in normal form. T_1 is orthogonal also so it is its own polar decomposition. It is two rotations about two planes in $\mathbb{R}^4$. The eigenvalues of T_1 are $\alpha_1 \pm i\beta_1, \alpha_2 \pm i\beta_2$. So

$$\Theta(T_1) = \det \begin{bmatrix} \alpha_1 + i\beta_1 & 0 \\ 0 & \alpha_2 + i\beta_2 \end{bmatrix} = (\alpha_1 + i\beta_1)(\alpha_2 + i\beta_2),$$

which gives the algebraic sum of two angles using the sign convention given in Section 2.4.

The eigenvalues T_2 are $\mu_1, \mu_2, \mu_1^{-1}, \mu_2^{-1}$. The polar decomposition of B is

$$T_2 = \begin{bmatrix} |\mu_1| & 0 & 0 & 0 \\ 0 & |\mu_2| & 0 & 0 \\ 0 & 0 & |\mu_1^{-1}| & 0 \\ 0 & 0 & 0 & |\mu_2^{-1}| \end{bmatrix} \begin{bmatrix} \text{sign}\,(\mu_1) & 0 & 0 & 0 \\ 0 & \text{sign}\,(\mu_2) & 0 & 0 \\ 0 & 0 & \text{sign}\,(\mu_1) & 0 \\ 0 & 0 & 0 & \text{sign}\,(\mu_2) \end{bmatrix}$$

Also

$$\Theta(T_2) = \det \begin{bmatrix} \text{sign}\,(\mu_1) & 0 \\ 0 & \text{sign}\,(\mu_2) \end{bmatrix} = \text{sign}\,(\mu_1 \mu_2)$$

which is ± 1 so an angle of 0 or π.

Now let A_1 and A_2 be symplectic matrices which are symplectically similar, so $A_1 = C^{-1} A_2 C$ where C is symplectic. Let $A_1 = P_1 O_1$ and $A_2 = P_2 O_2$ be their polar decomposition, then

$$A_1 = P_1 O_1 = C^{-1} A_2 C = C^{-1} P_2 O_2 C = C^{-1} P_2 C C^{-1} O_2 C.$$

5.5.2 Fundamental Group

Let us turn to the topology of $U(n, \mathbb{C})$, for which we follow Chevally (1946).

Proposition 5.5.3. $U(n, \mathbb{C})$ is homeomorphic to $S^1 \times SU(n, \mathbb{C})$.

Proof. Let $\mathcal{G}$ be the subgroup of $U(n, \mathbb{C})$ of matrices of the form $G(\phi) = \text{diag}(e^{i\phi}, 1, 1, \ldots, 1)$ where ϕ is just an angle defined mod 2π. Clearly, $\mathcal{G}$ is homeomorphic to S^1.

Let $P \in U(n, \mathbb{C})$ and $\det P = e^{i\phi}$; then $P = G(\phi)Q$ where $Q \in SU(n, \mathbb{C})$. Because $\mathcal{G} \cap SU(n, \mathbb{C}) = \{I\}$ the representation is unique. Thus the map $\mathcal{G} \times SU(n, \mathbb{C}) \to U(n, \mathbb{C}) : (G, Q) \mapsto GQ$ is continuous, one-to-one and onto a compact space so it is a homeomorphism.

Lemma 5.5.1. Let $\mathcal{H} \subset \mathcal{G}$ be a closed subgroup of a topological group $\mathcal{G}$. If $\mathcal{H}$ and the quotient space $\mathcal{G}/\mathcal{H}$ are connected then so is $\mathcal{G}$.

Proof. Let $\mathcal{G} = U \cup V$ where U and V are nonempty open sets. Then $\pi : \mathcal{G} \to \mathcal{G}/\mathcal{H}$ maps U and V onto open sets U' and V' of $\mathcal{G}/\mathcal{H}$ and $\mathcal{G}/\mathcal{H} = U' \cup V'$.

$\mathcal{G}/\mathcal{H}$ is connected so there is a $g\mathcal{H} \in U' \cap V'$. $g\mathcal{H} = (g\mathcal{H} \cap U) \cup (g\mathcal{H} \cap V)$, but because $\mathcal{H}$ is connected so is $g\mathcal{H}$. Therefore there is a point common to $(g\mathcal{H} \cap U) \cup (g\mathcal{H} \cap V)$ and hence to U and V. Thus $\mathcal{G}$ is connected.

Lemma 5.5.2. $U(n,\mathbb{C})/U(n-1,\mathbb{C})$ and $SU(n,\mathbb{C})/SU(n-1,\mathbb{C})$ are homeomorphic to S^{2n-1}.

Proof. Let $\mathcal{H}$ be the subgroup of $U(n,\mathbb{C})$ of matrices of the form $H = \text{diag}(1,H')$ where H' is a $(n-1) \times (n-1)$ unitary matrix. This H has a 1 in the 1,1 position and 0 in the rest of the first row and column. Clearly $\mathcal{H}$ is isomorphic to $U(n-1,\mathbb{C})$.

Let $\phi : U(n,\mathbb{C}) \to \mathbb{C}^n : Q \mapsto$ (the first column of Q). ϕ is continuous and because Q is unitary its columns are of unit length so ϕ maps onto S^{2n-1}. If $\phi(Q) = \phi(P)$ then $Q = PH$ where $H \in \mathcal{H}$. ϕ is constant on the cosets and so

$$\tilde{\phi} : U(n,\mathbb{C})/\mathcal{H} \to S^{2n-1} : (Q\mathcal{H}) \mapsto \phi(Q)$$

is well defined, continuous, one-to-one and onto. The spaces involved are compact thus $\tilde{\phi}$ is the desired homeomorphism. The same proof works for $SU(n,\mathbb{C})/SU(n-1,\mathbb{C})$.

Proposition 5.5.4. *The spaces $SU(n,\mathbb{C})$, $U(n,\mathbb{C})$, and $Sp(2n,\mathbb{R})$ are connected topological spaces.*

Proof. $SU(1,\mathbb{C})$, $U(1,\mathbb{C})$ are, respectively, a singleton and a circle so connected. By Lemmas 5.5.1 and 5.5.2, $SU(2,\mathbb{C})$, $U(2,\mathbb{C})$ are connected. Proceed with the induction to conclude that $SU(n,\mathbb{C})$, $U(n,\mathbb{C})$ are connected.

Propositions 5.5.1 and 5.5.2 imply that $Sp(2n,\mathbb{R})$ is connected.

Corollary 5.5.2. *The determinant of a symplectic matrix is $+1$.*

Proof. From $T^T J T = J$ follows $\det T = \pm 1$. The corollary follows from the proposition because det is continuous.

Theorem 5.5.1. $SU(n,\mathbb{C})$ *is simply connected. The fundamental groups $\pi_1(U(n,\mathbb{C}))$ and $\pi_1(Sp(2n,\mathbb{R}))$ are isomorphic to $\mathbb{Z}$.*

Proof. To prove $SU(n,\mathbb{C})$ is simply connected, we use induction on n. $SU(1,\mathbb{C}) = \{1\}$ and so is simply connected which starts the induction. Assume $n > 1$ and $SU(n-1,\mathbb{C})$ is simply connected. Using the notation of Lemma 5.5.2, $\phi : SU(n,\mathbb{C}) \to S^{2n-1}$ and $\phi^{-1}(p) = H$ where $p \in S^{2n-1}$ and H is homeomorphic to $SU(n-1,\mathbb{C})$ and thus simply connected.

Let $K_1 = \{c \in \mathbb{C}^n : \Re c_1 \geq 0\}$, $K_2 = \{c \in \mathbb{C}^n : \Re c_1 \leq 0\}$, and $K_3 = \{c \in \mathbb{C}^n : \Re c_1 = 0\}$ (think northern hemisphere, southern hemisphere, and equator). K_1 and K_2 are $2n-1$ balls and K_{12} is a $2n-2$ sphere. We write

$$SU(n,\mathbb{C}) = X_1 \cup X_2,$$

where $X_i = \{\phi^{-1}(p) : p \in K_i\}$. X_1 and X_2 are bundles over balls and thus products; i.e., $X_i = K_1 \times H$ for $i = 1, 2$. Therefore by the induction hypothesis they are simply connected. $X_1, X_2, X_3 = X_1 \cap X_2$ are all connected. So by van Kampen's theorem $X_1 \cup X_2 = SU(n,\mathbb{C})$ is simply connected. See Crowell and Fox (1963).

That the fundamental groups $\pi_1(U(n,\mathbb{C}))$ and $\pi_1(Sp(2n,\mathbb{R}))$ are isomorphic to $\mathbb{Z}$ follows from Propositions 5.5.1, 5.5.2, and 5.5.3.

Problem: Prove that if u and v be any two nonzero vectors in $\mathbb{R}^{2n}$ there is a $2n \times 2n$ symplectic matrix A such that $Au = v$. So the symplectic group acts transitively on $\mathbb{R}^{2n} \setminus \{0\}$.

5.6 Maslov Index

We have seen earlier how the dynamics of subspaces of initial conditions of Lagrangian type for the Hamiltonian flow may be important and useful to discuss the full dynamics of the linear vector field X_H. We look at another facet of such subspaces here when we consider the collection of all Lagrangian subspaces of a given symplectic space. Our goal is to explain recent applications to the study of stability for periodic integral curves of X_H. Further aspects of those ideas we touch upon here may be found in Arnold (1985, 1990), Cabral and Offin (2008), Conley and Zehnder (1984), Contreras et al. (2003), Duistermaat (1976), Morse (1973), and Offin (2000). We work initially in the nonlinear case to describe a geometric setting for these ideas before specializing our computations in the case of linear vector fields along periodic solutions of X_H.

The symplectic form ω on $\mathbb{R}^{2n}$ is a closed nondegenerate two form

$$\omega = \sum_{i=1}^{n} dq_i \wedge dp_i.$$

A Lagrange plane in a symplectic vector space such as $\mathbb{R}^{2n}$ is a maximal isotropic subspace, therefore an n-dimensional subspace $\lambda \subset \mathbb{R}^{2n}$ with $\omega|_\lambda = 0$ is a Lagrangian subspace. For example

$$\lambda = \{(0, q_2, \ldots, q_n, p_1, 0, \ldots, 0)\}$$

is a Lagrange plane in $\mathbb{R}^{2n}$. Note that a symplectic transformations $P \in Sp(2n)$ map Lagrange planes to Lagrange planes because $\omega|_{P\lambda} = 0$ whenever $\omega|_\lambda = 0$.

Let M denote a manifold with metric tensor $\langle \cdot, \cdot \rangle$, and $H : T^*M \to \mathbb{R}$ a smooth function convex on the fibers. We denote local coordinates $(q_1, \ldots, q_n)$ on M, and $(p_1, \ldots, p_n)$ on the fiber T_q^*M. An n-dimensional submanifold $i : \mathcal{L} \to T^*M$ is called Lagrangian if $i^*\omega = 0$. An interesting and important example of a Lagrangian manifold is the graph of $\nabla S(x)$, $x \in \mathbb{R}^n$. This is the manifold $\mathcal{L} = \{(x, \nabla S(x)) | x \in \mathbb{R}^n\}$, for which we have the property that the line integral with the canonical one form $\sum p_i dq_i$ on a closed loop γ in $\mathcal{L}$ is zero. This follows easily from the fact that the path integral of a gradient in $\mathbb{R}^n$ is path independent. An application of Stokes theorem then implies that $\mathcal{L}$ is a Lagrangian manifold because $\int_\sigma \omega = 0$ for any surface $\sigma \subset \mathcal{L}$ spanned by γ. For this example, that $\mathcal{L}$ is a Lagrangian manifold may be seen in an equivalent way, using the symmetric Hessian of the function $S(x)$, as a direct computation in the tangent plane to $\mathcal{L}$ by showing that the symplectic form ω vanishes on an arbitrary pair of vectors which are tangent to $\mathcal{L}$.

In applications, the Lagrange manifolds considered will belong to an invariant energy surface, although it is not necessary to make this restriction. Given the Hamiltonian vector field X_H on T^*M, we consider the energy surface $E_h = H^{-1}(h)$ which is invariant under the flow of X_H. The canonical projection $\pi : H^{-1}(h) \to M$ has Lagrangian singularities on the Lagrangian submanifold $i : \mathcal{L} \to H^{-1}(h)$ when $d(i^*\pi)$ is not surjective. Notice that the mapping $i^*\pi$ is a smooth mapping of manifolds of the same dimension. At a nonsingular point, this mapping is a local diffeomorphism. Thus Lagrangian singularities develop when the rank of this mapping drops below n. As an example, we observe that the graph of $\bigtriangledown S(x)$ denoted $\mathcal{L}$ above has no Lagrangian singularities, the projection π is always surjective on $\mathcal{L}$. On the other hand if $\lambda = \{(0, q_2, \ldots, q_n, p_1, 0, \ldots, 0)\}$ and $\gamma = \{(q_1, 0, \ldots, 0, 0, p_2, \ldots, p_n)\}$ then $\mathcal{L} = \text{graph } B$, is a Lagrange plane when $B : \lambda \to \gamma$, and $\omega(\lambda, B\lambda)$ is a symmetric quadratic form on the Lagrange plane λ. Moreover the singular set on $\mathcal{L}$ consists exactly of the codimension one two-sided surface $\partial q_1/\partial p_1 = 0$. The fact that this surface is two sided comes from the fact that as a curve crosses the singular set where $\partial q_1/\partial p_1 = 0$, the derivative changes sign from positive to negative or vice versa. This is the typical case for singular sets of the projection map π. Another example is afforded by the embedded Lagrangian torus $\mathcal{T}^n$ which will almost always develop Lagrangian singularities when projected into the configuration manifold M. The following result of Bialy (1991) is easy to state and illustrates several important facts mentioned above

$$H : T^*M \times S^1 \to \mathbb{R}, \quad M = \mathbb{R}, \text{ or } M = S^1$$

H is smooth and convex on the fibers of the bundle

$$\pi : T^*M \times S^1 \to M \times S^1.$$

Let $\lambda \to T^*M \times S^1$ be an embedded invariant 2-torus without closed orbits, and such that $\pi|_\lambda$ is not a diffeomorphism. Then the set of all singular points of $\pi|_\lambda$ consists of exactly two different smooth nonintersecting simple closed curves not null homotopic on λ. This result illustrates the general fact that the singular set in $\mathcal{L}$ is a codimension one manifold without boundary. This singular set is called the Maslov cycle. We will develop the theory of this counting of Lagrangian singularities along a given curve in $\mathcal{L}$, which is also called the Maslov index. This counting argument can be seen clearly in the case of the embedded torus whose singular set consists of two closed curves. A given closed curve γ on $\mathcal{L}$ intersects the Maslov cycle in a particular way, with a counting of these intersections independent of the homology class which γ belongs to. The Maslov index then is the value of a cohomology class on a closed curve γ within a given homology class.

In addition this result indicates that the kind of Lagrangian singularities developed may be a topological invariant for the Lagrangian submanifold $\mathcal{L}$

which while true will not be pursued here. This idea is at the root of the method to determine stability type of periodic orbits utilizing the counting of Lagrangian singularities along a given closed integral curve Γ.

As a first step, we mention that in general the singular set on $\mathcal{L}$ consists of points where the rank of the mapping $i^*\pi$ is less than n. Thus the singular set can be decomposed into subsets where the rank is constant and less than n. The simple component of the singular set corresponds to the case where rank $= n - 1$. The boundary of this set consists of points where rank $<$ $n - 1$. The singular set is a manifold of codimension 1, whose boundary has codimension bigger or equal to 3. Thus the generic part of the singular set is a two-sided codimension one variety whose topological boundary is empty. A given closed phase curve $z(t) \subset \mathcal{L}$ intersects this cycle transversely, and we may therefore count the algebraic intersection number of such a closed curve with positive contributions as the curve crosses the singular set from negative to positive and negative contributions coming as the curve crosses from positive to negative. This counting is known as the Maslov index of $z(t)$, $0 \leq t \leq T$, where T denotes the period of $z(t)$.

To classify these singularities, it is sometimes helpful to project the locus of singular points into the configuration manifold M. This geometric notion allows us to define caustic singularities which are the locus of projected singular points $i^*\pi : \mathcal{L} \to M$. Suppose that $\mathcal{L}$ is an invariant Lagrangian submanifold, and that $\Gamma \subset \mathcal{L}$ is a region which is foliated by phase curves of the vector field X_H. Caustics occur along envelopes of projected extremals $\gamma_e = \pi\Gamma_e$ which foliate $\mathcal{L}$.

Caustic singularities have been studied in geometric optics for a long time. The analogy here is that integral curves of X_H correspond to light rays in some medium. Now the focusing of projected integral curves on manifold M corresponds to the focusing of light rays. A caustic is the envelope of rays reflected or refracted by a given curve. They are curves of light of infinite brightness consisting of points through which infinitely many reflected or refracted light rays pass. In reality they often can be observed as a pattern of pieces of very bright curves; e.g., on a sunny day at the seashore on the bottom beneath a bit of wavy water.

The caustic singularities play an important role in the evaluation of the Morse index for the second variation of the action functional $\mathcal{A}(q) = \int_0^T L(q, \dot{q})dt$, $q(t) = \pi z(t)$ which we discussed earlier with various types of boundary conditions. The Morse index turns out to be a special case of the Maslov index, which is an important technique for evaluation of the Maslov index. We shall discuss these details more fully below and in the Chapter 13.

In the discussion above, the Lagrangian singularities arise from the singularities of the Lagrangian map $i^*\pi$. In this setting, the vertical distribution $V_z =$ kernel $d\pi(z)$, $z \in T^*M$ plays an important role. In effect, the Maslov index as described above may be calculated by counting the number of intersections of the tangent plane $T_{z(t)}\mathcal{L}$ with the vertical $V_{z(t)}$ along a curve

$z(t) \in \mathcal{L}$. In the following we abstract this and consider the intersections with an arbitrary Lagrange plane λ; however the case above is the setting for our applications. To make precise the considerations above, we turn to the linear theory of Lagrange planes in a given symplectic vector space such as $\mathbb{R}^{2n}$. The following result in Duistermaat (1976) is very useful for understanding Lagrange planes in the neighborhood of a given one.

Lemma 5.6.1. *If λ and γ are transverse Lagrange planes in $\mathbb{R}^{2n}$ so that $\lambda \cap \gamma = 0$, and $B : \lambda \to \gamma$ is linear, then $\alpha = $ graph $B = \{l + Bl | l \in \lambda\}$ is a Lagrange plane if and only if $\omega(\lambda, B\lambda)$ is a symmetric quadratic form on λ.*

Proof. First assume that α is a Lagrange plane. Then for a pair of vectors l_1, l_2 in λ, we have $\omega(l_1 + Bl_1, l_2 + Bl_2) = 0$ and

$$
\begin{aligned}
\omega(l_1, Bl_2) - \omega(l_2, Bl_1) &= \omega(l_1, Bl_2) + \omega(Bl_1, l_2) \\
&= \omega(l_1 + Bl_1, Bl_2) + \omega(Bl_1, l_2) \\
&= \omega(l_1 + Bl_1, l_2 + Bl_2) - \omega(l_1 + Bl_1, l_2) + \omega(Bl_1, l_2) \\
&= -\omega(Bl_1, l_2) + \omega(Bl_1, l_2) \\
&= 0.
\end{aligned}
$$

On the other hand, if $\omega(\lambda, B\lambda)$ is a symmetric quadratic form, then the computation above shows that $\omega(l_1 + Bl_1, l_2 + Bl_2) = 0$ for every pair $l_1, l_2 \in \lambda$. Therefore α is a Lagrange plane.

If we denote by Λ_n the topological space of all Lagrange planes in $\mathbb{R}^{2n}$, called the Lagrangian Grassmannian, then the set of Lagrange planes $\Lambda^0(\gamma)$ which are transverse to a given one γ, $\Lambda^0(\gamma) = \{\lambda \in \Lambda_n | \lambda \cap \gamma = 0\}$, is an open set in Λ_n. The Lemma above gives a parameterization for all Lagrange planes in the open set $\Lambda^0(\gamma)$, in terms of symmetric forms on a given subspace $\lambda \in \Lambda^0(\gamma)$. Moreover, the set of Lagrange planes in a neighborhood of λ which do not intersect λ is diffeomorphic with the space of nondegenerate quadratic forms on λ. We can think of this map $\alpha \mapsto \omega(\lambda, B\lambda)$, where $\alpha = $ graph B, as a coordinate mapping on the chart $\Lambda^0(\gamma)$, and the dimension of the kernel $\ker \omega(\lambda, B\lambda)$ equals the dimension of the intersection graph$B \cap \lambda$.

Definition 5.6.1. *If $\alpha, \lambda \in \Lambda^0(\gamma)$ and $\alpha = $ graph B, then the symmetric form on λ associated with α is denoted $Q(\lambda, \gamma; \alpha) = \omega(\lambda, B\lambda)$.*

From these considerations it is easy to see that the dimension of Λ_n is $\frac{1}{2}n(n+1)$. Moreover, it is also easy to understand why the Maslov cycle of a Lagrange plane λ is a topological cycle, and that its codimension is one on its relative interior where it is a smooth submanifold. Recall that the Maslov cycle consists of the Lagrange planes which intersect the given Lagrangian plane λ in a subspace with dimension larger than or equal to 1 (Arnold refers to this set as the train of the Lagrange plane λ). Therefore if $\Lambda^k(\gamma) = \{\beta \in \Lambda(n) | \dim \beta \cap \lambda = k\}$, then the Maslov cycle of λ is

$$\mathcal{M}(\lambda) = \bigcup_{k \geq 1} \Lambda^k(\lambda)$$

However we can compute the codimension of $\mathcal{M}(\lambda)$ quite nicely using the quadratic forms $Q(\lambda, \gamma; \beta)$, where $\beta \in \Lambda^k(\lambda)$. It is clear on a moments reflection that if dimension of the kernel of $Q(\lambda, \gamma; \beta) = k$, then $Q(\lambda, \gamma; \beta)$ must have a k-dimensional 0 block, and therefore that the submanifold $\Lambda^k(\lambda)$ is codimension $\frac{1}{2}k(k + 1)$ in Λ_n. In conclusion, $\Lambda^1(\lambda)$ is an open submanifold of codimension one, and its boundary consists of the closure of $\Lambda^2(\lambda)$ which has codimension 3 in Λ_n. Thus the closure of the submanifold $\Lambda^1(\lambda)$ is an oriented, codimension one topological cycle.

To describe the index of a path of Lagrange planes λ_t in a neighborhood of a given plane λ, when the endpoints of λ_t are transverse to λ we observe that the space of nondegenerate quadratic forms is partitioned into $n + 1$ regions depending on the number of positive eigenvalues of the quadratic form $\omega(\lambda, B\lambda)$, the so-called positive inertia index. The singular set of Lagrange planes $\Sigma(\lambda)$ which intersect λ is a codimension one manifold in Λ_N given by the condition $\det \omega(\lambda, B\lambda) = 0$. We map the curve $\lambda_t = \text{graph } B_t$ to the corresponding curve of symmetric operators on λ denoted $Q(\lambda, \gamma; \lambda_t)$. This curve has real eigenvalues which are continuous functions of the parameter t. These eigenvalues may cross the zero eigenvalue which signals an intersection of the curve λ_t with the fixed plane λ. Such crossings may occur either from $-$ to $+$ or from $+$ to $-$. We count the algebraic number of crossings with multiplicity with the crossing from $-$ to $+$ as a positive contribution, and from $+$ to $-$ as a negative contribution. The Maslov index then for a curve λ_t whose endpoints do not intersect λ, denoted $[\lambda_t; \lambda]$, is the sum of the positive contributions minus the negative contributions. If on the other hand the endpoints of the curve λ_t are not transverse to λ we use the convention specified by Arnold (1985) and stipulate the Maslov index in this case is obtained by considering a nearby path in Λ_N, say α_t, so that the corresponding eigenvalues of $Q(\lambda, \gamma; \alpha_t)$ which are perturbations of the eigenvalues of $Q(\lambda, \gamma; \lambda_t)$ are nonzero at the endpoints. Moreover on the right-hand endpoint, all zero eigenvalues are moved to the right of zero (positive domain), while at the left-hand endpoint, all zero eigenvalues are moved to the left.

It should be mentioned explicitly that the Maslov index as here described is a homotopy invariant, with fixed endpoints for the path λ_t due to the fact that $\mathcal{M}(\lambda)$ is a cycle in Λ_n. This remark implies in particular that the Maslov index of a curve λ_t can be computed from a nearby path which crosses the Maslov cycle $\mathcal{M}(\lambda)$ transversely, and only intersects the simple part of the cycle, where the intersection number is $+1$ or -1.

There are several translations of the Maslov index which are important in applications. Two of these are the Conley–Zehnder index, and the Morse index which we alluded to above. Roughly, the Maslov index of the vertical space is equivalent to the Morse index when the curve λ_t is a positive curve, in the sense that it intersects the Maslov cycle of the vertical space transversely and only in the positive sense. An important case arises when the curve $\lambda_t = d_z\phi_t\lambda$, where ϕ_t denotes the flow of a Hamiltonian vector field X_H in $\mathbb{R}^{2n}$ with the Hamiltonian convex in the momenta $\partial^2 H/\partial p^2 > 0$. Then the curve $\lambda_t \in \Lambda_n$ intersects the vertical distribution only in the positive sense and this condition also implies that the flow direction is transverse to the vertical distribution Duistermaat (1976), Offin (2000). In the following discussion we let ϕ_t denote the flow of the Hamiltonian vector field X_H with convex Hamiltonian $\partial^2 H/\partial p^2 > 0$. Recall that the action functional with fixed boundary conditions

$$F(q) = \int_0^T L(q(t), \dot{q}(t))dt, \quad q(0) = q_0, \quad q(T) = q_1$$

leads to consideration of the critical curves $q(t) = \pi z(t), \; 0 \le t \le T$ where $z(t)$ is an integral curve of the Hamiltonian vector field X_H such that $q(0) = q_0, \; q(T) = q_1$. For such a critical curve $q(t)$, the second variation is the quadratic form $d^2 F(q)\cdot\xi$, where $\xi(t)$ is a variation vector field along $q(t)$ which satisfies the boundary conditions $\xi(0) = \xi(T) = 0$. These boundary conditions arise as natural with the fixed endpoint problem discussed earlier. The second variation measures second order variations in the action along the tangent directions given by the variation vector field $\xi(t)$. It is shown in textbooks on the calculus of variations that the number of negative eigenvalues of the second variation is a finite number provided that the Legendre condition holds $\partial^2 L/\partial v^2 > 0$, which is equivalent to the condition of convexity of the Hamiltonian; see Hestenes (1966). This finite number is the Morse index of the critical curve $q(t)$ for the fixed endpoint problem. It is known that this index depends crucially on the boundary conditions but in the case of fixed endpoint it is the number of conjugate points (counted with multiplicity) along the critical arc $q(t)$.

A time value t_0 is said to be conjugate to 0 if there is a solution $\zeta(t) = d_{z(t)}\phi_t\zeta(0), \; 0 \le t \le T$ of the linearized Hamiltonian equations along $z(t)$, such that $\pi\zeta(0) = 0 = \pi\zeta(t_0)$. Of course the projected vector field $\pi\zeta(t)$ is just one example of a variation vector field $\xi(t)$ along $q(t)$, however it plays a crucial role in determining the Morse index. We observe that the condition for existence of conjugate points can be equivalently described by allowing the vertical plane $V(z)$ at $t = 0$ to move with the linearized flow $d_z\phi_t$, and to watch for the intersections of this curve of Lagrange planes with the fixed vertical plane $V(z(t_0))$ at $z(t_0)$. The Morse index formula for the fixed endpoint problem can be now stated

$$\text{conjugate index} \; = \; \sum_{0 < t \le T} \dim\,[d_z\phi V(z) \cap V(\phi_t z)]$$

which is a special case of the intersection number for the curve of Lagrangian planes $d_z\phi_t V(z)$ with the fixed distribution $V(\phi_t z)$.

An important generalization of the fixed endpoint example is the case of periodic boundary conditions which will play an important role in the applications to periodic solutions of the N-body problem

$$F(q) = \int_0^T L(q(t), \dot{q}(t))dt, \quad q(0) = q(T). \tag{5.19}$$

The second variation of the action functional for critical curves $q(t) = \pi z(t)$ may be calculated easily from equation (1.31) by differentiating the first variation along the variation vector field $\xi(t)$. We will denote the second variation by $d^2 F(q(t)) \cdot \xi$ whose domain consists of absolutely continuous variation vector fields $\xi(t)$ which satisfy the same periodic boundary conditions $\xi(T) = \xi(0)$. In this case the Maslov–Morse index is modified slightly from the fixed endpoint case to include the case of periodic boundary conditions for variation vector fields.

$$\text{index } d^2 F(q(t)) = \sum_{0 < t \leq T} \dim \left[d_z \phi V(z) \cap V(\phi_t z) \right] + \text{index } d^2 F(q(t))|_W,$$

$$\tag{5.20}$$

where W denotes the subspace of variation vector fields $\zeta(t) = d\phi_t \zeta(0)$ along $z(t)$ which are periodic in the configuration component

$$W = \left\{ \zeta = \xi \frac{\partial}{\partial q} + \eta \frac{\partial}{\partial p} | \pi \zeta(T) = \pi \zeta(0) \right\}. \tag{5.21}$$

We refer to the variation vector fields $\xi(t) = \pi \zeta(t)$, $\zeta(t) = d\phi_t \zeta(0)$ as Jacobi fields along the phase curve $z(t) = \phi_t z$. The second variation restricted to the subspace of Jacobi fields, periodic in the configuration component has a simpler form due to the fact that they are integral curves of the linear vector field $DX_H(z(t))$, $0 \leq t \leq T$, the integral term in the second variation vanishes, leaving only the endpoint contribution. Using the periodic boundary conditions of the Jacobi fields associated with the subspace W yields

$$\begin{aligned} d^2 F(q(t))|_W &= \langle \zeta(T), \xi(T) \rangle - \langle \eta(0), \xi(0) \rangle \\ &= \langle \zeta(T), \xi(0) \rangle - \langle \eta(0), \xi(T) \rangle \\ &= \omega(\zeta(0), \zeta(T)) \\ &= \omega(W, d\phi_T W). \end{aligned}$$

An application of this formula to the case when $n = 1$ yields the case studied by Morse for scalar variational problems with periodic boundary conditions

$$\text{index } d^2 F(q(t)) = \text{conjugate index} + \begin{cases} 0 \text{ if } (\eta(T) - \eta(0))\xi(0) \geq 0 \\ 1 \text{ if } (\eta(T) - \eta(0))\xi(0) < 0, \end{cases}$$

where $\eta(t)$ denotes the Jacobi field which satisfies the periodic boundary condition $\xi(T) = \xi(0)$. The correction term to the conjugate index was called the concavity by Morse (1973). Incidentally, it is easy to see that the subspace W always contains nonzero terms, because we have the useful formula

$$(d_z\phi_T - \mathrm{id})W = V(z),$$

where $V(z)$ denotes the vertical space at z. This implies in particular that if the linearized flow along $z(t)$ has no T-periodic solutions, then W has dimension n.

To tie some of these ideas together, we mention a theorem of Bondarchuk (1984) which describes a typical case of the Maslov index for hyperbolic periodic orbits of a Hamiltonian vector field X_H with convex Hamiltonian H. It is easy to see that the stable and unstable manifolds of $z(t)$ are Lagrangian submanifolds because the symplectic form must vanish for any pair of tangent vectors along $z(t)$. The Lagrangian singularities of the stable or unstable manifold relative to the canonical projection $\pi : H^{-1}(h) \to M$ occur at those moments along $z(t)$ when the tangent space to the stable or unstable manifold becomes vertical. If λ denotes the tangent plane to the stable manifold for example at the point $z(0)$, then the t parameter values of the singularities of the moving plane $d_z\phi_t\lambda$ are called focal points for the Lagrange plane λ. Counting these singularities along $z(t)$ for one period will yield the Maslov index of the stable or unstable manifolds in $H^{-1}(h)$.

Theorem 5.6.1. *If $z(t)$ is T-periodic and hyperbolic on its invariant energy surface $H^{-1}(h)$ (the Poincaré map on $H^{-1}(h)$ has eigenvalues which lie off the unit circle), then the Morse index of the second variation $d^2F(\pi z(t))$ with periodic boundary conditions is equal to the Maslov index of the local stable or unstable manifolds $W^s_{loc}(z(t)), W^u_{loc}(z(t))$. In particular, if $[d_z\phi_t\lambda; V(z(t)] = k \in \mathbb{Z}^+$ where $0 \le t \le T$ then $[d_z\phi_t\lambda; V(z(t)] = mk$ over m covers of the periodic phase curve $0 \le t \le mT$.*

In special circumstances, the subspace W (see equation (5.21)) is Lagrangian: for example in the case described above in Bondarchuk's theorem or in the time reversing case Offin (2000), or in case the orbit is symmetric under an Abelian group of symplectic symmetries Cabral and Offin (2008), or in case the periodic orbit lies on an invariant energy surface $H^{-1}(h)$ for a two degree of freedom Hamiltonian system. This last case is equivalent to the one-dimensional case described above for the periodic problem of the calculus of variations. This Lagrangian property of the subspace W will be shown to lead to an interesting possibility to analyze the orbits for qualitative behavior based only the values of the Maslov index. This opens the door to predicting analytically the stability type of the periodic orbit, without recourse to numerical techniques. It has been shown that in the one-dimensional case considered by Morse, the stability type of an underlying periodic orbit can be deduced solely from the parity of the Maslov index Offin (2001). We shall use these formulas for the Maslov–Morse index of the second variation in applications, and exploit the properties of the iterated Maslov index when we can show that the subspace W is Lagrange.

The other important translation of the Maslov index is the Conley–Zehnder (1984), which concerns a curve of symplectic matrices in $Sp(2n)$ rather than a curve of Lagrange planes. Because the topology of $Sp(2n)$ is similar to the topology of Λ_n, it is not surprising that the two theories are by and large equivalent. The role of the singular cycle is played in the Conley–Zehnder theory by codimension one cycles which are essentially submanifolds of symplectic matrices having eigenvalues ± 1. The Conley–Zehnder index has been extensively studied by Long (2002) and we will refer the reader to this source to investigate further the Maslov index in this case. We shall detail how to compute the Maslov index for the case of critical curves of variational problems further in the chapter on applications.

5.7 Problems

1. Prove Lemma 2.5.1 for the symplectic matrix T by using induction on the formula $\{\eta_k(\lambda), \eta_k(\mu)\} = 0$, where $\eta_k(\lambda) = \mathrm{kernel}\,(T^k - \lambda I)$. (See Laub and Meyer (1974).)
2. Write the 4th-order equation $x^{(4)} = 0$ as a Hamiltonian system. (Hint: See the canonical forms in Section 5.3.)
3. Compute $\exp At$ for each canonical form given in Section 5.4.

6. Local Geometric Theory

This chapter gives a brief introduction to the geometric theory of autonomous Hamiltonian systems by studying some local questions about the nature of the solutions in a neighborhood of a point or a periodic solution. The dependences of periodic solutions on parameters are also presented in the case when no drastic changes occur, i.e., when there are no bifurcations. Bifurcations are addressed in detail in Chapter 11.

The geometric theory of Hamiltonian systems is vast and far from complete. Some of the basic definitions and results from the theory of dynamical systems are given to put the topic in context. In most cases, the background theory for ordinary (non-Hamiltonian) equations is given first. The non-Hamiltonian theory is fairly well documented in the literature, therefore the more lengthy proofs are given by referral. See, for example, Chicone (1999) or Robinson (1999).

6.1 The Dynamical System Point of View

Consider an autonomous system of ordinary differential equations of the form

$$\dot{x} = f(x), \tag{6.1}$$

where $f : \mathcal{O} \to \mathbb{R}^m$ is smooth and $\mathcal{O}$ is an open set in $\mathbb{R}^m$. Let $\psi(t)$ be a solution of (6.1) defined for $t \in (\alpha, \omega)$. A geometric representation of a solution (for a nonautonomous as well as an autonomous system) is the graph of ψ, $\{(t, \psi(t)) : t \in (\alpha, \omega)\}$, in $\mathcal{O} \times (\alpha, \omega) \subset \mathbb{R}^{m+1}$, position–time space. See Figure 6.1. The fundamental existence and uniqueness theorem for differential equations asserts that there is one and only one solution through a point $\xi \in \mathcal{O}$ when $t = t_0$, so, there is one and only one graph of a solution through a point $(\xi, t_0) \in \mathcal{O} \times \mathbb{R}$.

Because (6.1) is independent of any t, the translate of a solution, $\psi(t - \tau)$, is a solution also. (There is no clock for an autonomous equation and so no initial epoch.) If one thinks of ψ as a curve in $\mathcal{O} \subset \mathbb{R}^m$, then alltranslates of

© Springer International Publishing AG 2017
K.R. Meyer, D.C. Offin, *Introduction to Hamiltonian Dynamical Systems and the N-Body Problem*, Applied Mathematical Sciences 90,
DOI 10.1007/978-3-319-53691-0_6

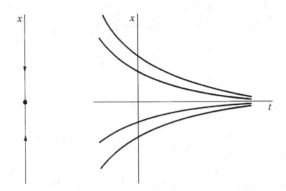

Figure 6.1. Solutions and orbits of $\dot{x} = -x$.

the solution ψ give the same curve in $\mathbb{R}^m$. The parameterized curve $\psi(t)$ in $\mathbb{R}^n$ is called a *trajectory*, and the oriented but unparameterized curve $\psi(t)$ is called an *orbit*. An orbit is the set $\{\psi(t) : t \in (\alpha, \omega)\}$ with the orientation coming from the orientation of (α, ω) in $\mathbb{R}$, and a trajectory is the map $\psi : (\alpha, \omega) \to \mathbb{R}$. In dynamical systems, the geometry of the trajectories/orbits in $\mathbb{R}^m$ is the object of study.

If $\psi_i(t)$, $i = 1, 2$, are two solutions with $\psi_1(t_1) = \psi_2(t_2)$, then $\chi(t) = \psi_2(t - t_1 + t_2)$ is also a solution of (6.1) with $\chi(t_1) = \psi_2(t_2) = \psi_1(t_1)$, so, by the uniqueness theorem for differential equations, $\chi(t) = \psi_2(t - t_1 + t_2) \equiv \psi_1(t)$. So if two solutions meet, they are simply time translates of each other and have the same orbit in $\mathcal{O}$. Thus orbits never cross in $\mathcal{O}$.

Let $\phi(t, \xi)$ denote the general solution of (6.1) that is the maximal solution of (6.1) which satisfies $\phi(0, \xi) = \xi$ for $\xi \in \mathcal{O}$.

Lemma 6.1.1. *If t and τ are such that $\phi(\tau, \xi)$ and $\phi(t + \tau, \xi)$ are defined, then*

$$\phi(t, \phi(\tau, \xi)) = \phi(t + \tau, \xi). \tag{6.2}$$

Proof. Both sides of (6.2) are solutions of (6.1) and are equal to $\phi(\tau, \xi)$ when $t = 0$. Thus by the uniqueness theorem for differential equations, they are equal where defined.

Lemma 6.1.2. *Let $\xi_0 \in \mathcal{O}$ be an equilibrium point, i.e., $f(\xi_0) = 0$. If $\phi(t, \xi') \to \xi_0$ as $t \to t'+$ (respectively, $t \to t'-$) and $\xi' \neq \xi_0$, then $t' = +\infty$ (respectively, $t' = -\infty$.) It takes an infinite amount of time to come to rest!*

Proof. Assume not, so t' is finite. Then $\eta(t) \equiv \xi_0$ is one solution through ξ_0 at time t' and so is $\psi(t)$, where $\psi(t) = \phi(t, \xi')$ for $t < t'$ and $\psi(t) = \xi_0$ for $t \geq t'$. But this contradicts the uniqueness theorem for differential equations.

Let $g : \mathcal{O} \to \mathbb{R}$ be smooth and positive, then a reparameterization of the solutions of (6.1) is defined by

$$dt = g(x)d\tau, \tag{6.3}$$

and if $' = d/d\tau$, then Equation (6.1) becomes

$$x' = f(x)g(x). \tag{6.4}$$

The solution curves of (6.1) and (6.2) are the same, only their parameterizations are different.

Lemma 6.1.3. *There exists a reparameterization of* (6.1) *such that all solutions are defined for all t.*

Proof. We only prove this theorem when $\mathcal{O} = \mathbb{R}^m$. Let $g(x) = 1/(1+\|f(x)\|)$. The equation $x' = f(x)g(x) = h(x)$ satisfies $\|h(x)\| \leq 1$ for all x. A solution $\psi(t)$ is either defined for all time or tends to ∞ in finite time. But $\|\psi(t)\| \leq 1$ implies $\|\psi(t)\| \leq \|\psi(0)\| + t$, so, ψ must be defined for all t.

In the general case when $\mathcal{O} \neq \mathbb{R}^m$, one can construct a smooth function $g : \mathcal{O} \to \mathbb{R}$ such that $g(x) \to 0$ and $f(x)g(x) \to 0$ as $x \to \partial\mathcal{O}$ where $\partial\mathcal{O}$ is the boundary of $\mathcal{O}$. By the above argument, the solutions of $x' = h(x) = f(x)g(x)$ will be defined for all t. By defining $h(x) = 0$ for $x \notin \mathcal{O}$, the equations and solutions would be defined for all $x \in \mathbb{R}^m$ also.

Assume that the function $f(x)$ in (6.1) is defined and smooth on all of $\mathcal{O}$ and that all the solutions of (6.1) are defined for all $t \in \mathbb{R}$. By the discussion given above, these assumptions are always valid after a reparameterization, so these assumptions do not limit the discussion of the geometry of the orbits. Let $\phi(t,\xi)$ be the solution of (6.1) that satisfies $\phi(0,\xi) = \xi$ and define $\phi_t(\xi) = \phi(t,\xi)$. By this definition and Lemma 6.1.1, the family $\{\phi_t\}$ satisfies

$$\phi_0 = id = \text{the identity map on } O, \tag{6.5}$$
$$\phi_t \circ \phi_\tau = \phi_{t+\tau}.$$

This implies ϕ_t has an inverse ϕ_{-t} and so ϕ_t is a homeomorphism for all t. Any family of smooth mappings satisfying (6.5) defines a dynamical system or a *flow* on $\mathcal{O}$. If (6.1) is a Hamiltonian system of equations, then ϕ_t is symplectic for all t by Theorem 2.6.2. In this case, the family of smooth maps ϕ_t defines a symplectic or Hamiltonian dynamical system or a Hamiltonian flow. Sometimes the name dynamical system is used even if the solutions are not defined for all t.

A trajectory $\phi_t(\xi_0) = \phi(t,\xi_0)$ is periodic if there is a $T \neq 0$ such that $\phi(t+T,\xi_0) = \phi(t,\xi_0)$ for all $t \in \mathbb{R}$. The number T is called a period, and the least positive T is called the period.

Two dynamical systems $\phi_t : \mathcal{O} \to \mathcal{O}$ and $\psi_t : \mathcal{Q} \to \mathcal{Q}$ are (topologically) equivalent if there is a homeomorphism $h : \mathcal{O} \to \mathcal{Q}$ that carries orbits of ϕ_t onto orbits of ψ_t and vice versa. Usually it is required to preserve the sense or orientation of the orbits also. Thus two dynamical systems are equivalent if the geometry of their orbits is the same, but the timing may not be the same. The homeomorphism will take equilibrium points to equilibrium points and periodic orbits to periodic orbits. The dynamical systems defined by the two harmonic oscillators $\dot{x} = \omega_i y$, $\dot{y} = -\omega_i x$, $i = 1, 2$ $\omega_1 > \omega_2 > 0$ are equivalent, because the identity map takes orbits to orbits.

Lemma 6.1.4. $\phi_t(\xi_0)$ *is periodic with period T if and only if $\phi_T(\xi_0) = \xi_0$.*

Proof. If $\phi_t(\xi_0)$ is periodic, then set $t = 0$ in $\phi(t + T, \xi_0) = \phi(t, \xi_0)$ to get $\phi(T, \xi_0) = \phi(0, \xi_0) = \xi_0$. If $\phi_T(\xi_0) = \xi_0$, then apply ϕ_t to both sides and apply (6.5) to get $\phi(t + T, \xi_0) = \phi_t \circ \phi_T(\xi_0) = \phi_t(\xi_0) = \phi(t, \xi_0)$.

An *invariant set* is a subset $Q \subset \mathcal{O}$ such that if $\xi \in Q$, then $\phi(t, \xi) \in Q$ for all t. That is an invariant set is a union of orbits.

A linear equation $\dot{x} = Ax$, A a constant $m \times m$ matrix, defines a linear dynamical system $\psi_t(x) = e^{At}x$ on $\mathbb{R}^m$. If A is a Hamiltonian matrix, then the map is a symplectomorphism. The origin is an equilibrium point. If $x_0 = u_0 + iv_0$ is an eigenvector corresponding to a pure eigenvalue $\lambda = i\omega$, $\omega \neq 0$, then $e^{At}u$ is a $2\pi/\omega$ periodic solution. In fact, the two-dimensional real linear space span $\{u, v\}$ is filled with $2\pi/\omega$ periodic solutions.

If none of the eigenvalues of the matrix A are pure imaginary, then the matrix A is called *hyperbolic*, and the equilibrium point at the origin is called hyperbolic also. If all the eigenvalues of A have real parts less (greater) than zero, then the $e^{At}x \to 0$ as $t \to +\infty$ (respectively as $t \to -\infty$) for all x. Neither of these cases happens for a Hamiltonian matrix A because the eigenvalues of a Hamiltonian matrix are symmetric with respect to the imaginary axis. If A has k eigenvalues with negative real parts and $m - k$ eigenvalues with positive real parts, then by the Jordan canonical form theorem there is a nonsingular $m \times m$ matrix P such that $P^{-1}AP = A' = \text{diag}(B, C)$, where B is a $k \times k$ matrix with eigenvalues with negative real parts, and C is an $(m-k) \times (m-k)$ matrix with eigenvalues with positive real parts. The matrix P can be thought of as the matrix of a change of variables, so that in the new variables, A has the form $A' = \text{diag}(B, C)$. Thus in this case A preserves the splitting $\mathbb{R}^m = \mathbb{R}^k \times \mathbb{R}^{m-k}$, i.e., the coordinate planes $\mathbb{R}^k \times \{0\}$ and $\{0\} \times \mathbb{R}^{m-k}$ are invariant sets. If $x \in \mathbb{R}^k \times \{0\}$, then $e^{A't}x \to 0$ as $t \to +\infty$, and if $x \in \{0\} \times \mathbb{R}^{m-k}$, then $e^{A't}x \to 0$ as $t \to -\infty$. Figure 6.2a indicates the orbit structure for the hyperbolic, symplectic matrix $A' = \text{diag}(-1, +1)$.

If all the eigenvalues of the matrix A are pure imaginary and A is simple (diagonalizable), then A is called elliptic, and the equilibrium point at the origin is called elliptic. By the Jordan canonical form theorem, there is a nonsingular, $m \times m$ matrix P such that $P^{-1}AP = A' = \text{diag}(\omega_1 J, \ldots, \omega_k J, 0, \ldots, 0)$ where J is the 2×2 matrix

$$J = \begin{bmatrix} 0 & 1 \\ -1 & 0 \end{bmatrix}.$$

Then $e^{A't} = \text{diag}(R(\omega_1 t), \ldots, R(\omega_k t), 1, \ldots, 1)$, where $R(\theta)$ is the rotation matrix

$$R(\theta) = \begin{bmatrix} \cos\theta & \sin\theta \\ -\sin\theta & \cos\theta \end{bmatrix}.$$

Figure 6.2b indicates the orbit structure for the elliptic, symplectic matrix $R(\theta)$.

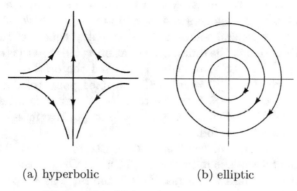

(a) hyperbolic   (b) elliptic

Figure 6.2. Linear dynamical systems.

If p is an equilibrium point for the nonlinear equation (6.1), then the equation $\dot{x} = Ax$, where $A = Df(p) = \partial f(p)/\partial x$, is called the linearization of (6.1) at p. The equilibrium point p is called hyperbolic or elliptic as matrix A is called.

A theorem of Hartman (1964) says that the flow near a hyperbolic equilibrium point is equivalent to a linear flow. That is, if p is a hyperbolic equilibrium for (6.1), then there are neighborhoods $\mathcal{O}$ of p and $\mathcal{Q}$ of the $0 \in \mathbb{R}^m$ and a homeomorphism $h : \mathcal{O} \to \mathcal{Q}$ such that h maps orbits of (6.1) onto orbits of $\dot{x} = Ax$. No such theorem is true for elliptic equilibrium points.

6.2 Discrete Dynamical Systems

Closely related to differential equations are diffeomorphisms that define discrete dynamical systems.

6.2.1 Diffeomorphisms and Symplectomorphisms

A map $\psi : \mathcal{O} \to \mathbb{R}^m, \mathcal{O}$ open in $\mathbb{R}^m$, is a diffeomorphism if ψ is differentiable and has a differentiable inverse. In particular a diffeomorphism is a homeomorphism of $\mathcal{O}$ onto $\psi(\mathcal{O})$. In many cases it is required that ψ take $\mathcal{O}$ into $\mathcal{O}, \psi(\mathcal{O}) = \mathcal{O}$, in which case ψ is said to be a diffeomorphism of $\mathcal{O}$. Let k be a positive integer, and let $\psi^k = \psi \circ \psi \circ \ldots \circ \psi$, k times, be the kth composition of ψ with itself. So $\psi^1 = \psi$. Define $\psi^0 = id$, the identity map $(id(x) = x)$ and $\psi^{-k} = \psi^{-1} \circ \psi^{-1} \circ \cdots \circ \psi^{-1}$, k times, be the kth composition of ψ^{-1}, the inverse of ψ. If ψ is a diffeomorphism of $\mathcal{O}$, then ψ^k is defined for all k and is a diffeomorphism of $\mathcal{O}$ for all k. In general ψ^k may be defined for some k and on only a part of $\mathcal{O}$. In either case, it is easy to verify that $\psi^{k+s} = \psi^k \circ \psi^s$ whenever the two sides are defined. If ψ is a symplectic diffeomorphism, then ψ is called a *symplectomorphism.*

A discrete dynamical system is simply a diffeomorphism ψ of a set $\mathcal{O}$. A discrete Hamiltonian dynamical system is simply a symplectomorphism of a set $\mathcal{O}$. If we let $\mathbb{Z}$ be the integers and $\Psi(k, \xi) = \psi^k(\xi)$, then $\Psi : \mathbb{Z} \times \mathbb{R}^m \to \mathbb{R}^m$ is analogous to the general solution of a differential equation. In fact $\Psi(k, \xi)$ is the general solution of the difference equation $x(k+1) = \psi(x(k)), x(0) = \xi$.

The set $\{\psi^n(p) : -\infty < n < +\infty\}$ is called the *orbit* of the point p. A point $p \in \mathcal{O}$ such that $\psi(p) = p$ is called a fixed point (of ψ), and a point $p \in \mathcal{O}$ such that $\psi^k(p) = p$, for some positive integer k, is called a periodic point (of ψ), and k is called a period. The least positive integer k such that $\psi^k(p) = p$ is called the period.

Two discrete dynamical systems $\phi : \mathcal{O} \to \mathcal{O}$ and $\psi : \mathcal{Q} \to \mathcal{Q}$ are *(topologically) equivalent* if there is a homeomorphism $h : \mathcal{O} \to \mathcal{Q}$ that carries orbits of ϕ onto orbits of ψ and vice versa. This is the same as $h \circ \phi = \psi \circ h$.

A nonsingular linear map $x \to Ax$, A a constant $m \times m$ matrix, defines a discrete dynamical system $\psi(x) = Ax$ on $\mathbb{R}^m$. If A is a symplectic matrix, then the map is a symplectomorphism. The origin is a fixed point. If x_0 is an eigenvector corresponding to an eigenvalue λ which is a kth root of unity, $\lambda^k = 1$, then x_0 is a period point of period k, because $A^k x_0 = \lambda^k x_0 = x_0$.

If none of the eigenvalues of the matrix A have modulus 1, then the matrix A is called *hyperbolic*, and the fixed point at the origin is called hyperbolic also. If all the eigenvalues of A have modulus less (respectively, greater) than one, then $A^n x \to 0$ as $n \to +\infty$ (respectively, as $n \to -\infty$) for all x. Neither of these cases happens for a symplectic matrix A. If A has k eigenvalues with modulus less than 1 and $m - k$ eigenvalues with modulus greater than 1, then by the Jordan canonical form theorem, there is a nonsingular, $m \times m$ matrix P such that $P^{-1}AP = A' = \mathrm{diag}(B, C)$, where B is a $k \times k$ matrix with eigenvalues of modulus less than 1, and C is an $(m - k) \times (m - k)$ matrix with eigenvalues of modulus greater than 1. The matrix P can be thought of as the matrix of a change of variables, so that in the new variables, A has the form $A' = \mathrm{diag}(B, C)$. Thus in this case, A' preserves the splitting $\mathbb{R}^m = \mathbb{R}^k \times \mathbb{R}^{m-k}$, i.e., A carries the coordinate plane $\mathbb{R}^k \times \{0\}$ into itself and the coordinate plane $\{0\} \times \mathbb{R}^{m-k}$ into itself. If $x \in \mathbb{R}^k \times \{0\}$, then $A'^n x \to 0$ as $n \to +\infty$ and if $x \in \{0\} \times \mathbb{R}^{m-k}$, then $A'^n x \to 0$ as $n \to -\infty$. Figure 6.2a indicates the orbit structure for the hyperbolic, symplectic matrix $A = \mathrm{diag}(1/2, 2)$, but now remember that an orbit is a sequence of discrete points not a continuous curve.

If all the eigenvalues of the matrix A have modulus 1 and A is diagonalizable, then A is called *elliptic* and the fixed point at the origin is called *elliptic*. By the Jordan canonical form theorem, there is a nonsingular, $m \times m$ matrix

P such that $P^{-1}AP = A' = \operatorname{diag}(R(\theta_1), \ldots, R(\theta_k), \pm 1, \ldots, \pm 1)$ where $R(\theta)$ is the rotation matrix

$$R(\theta) = \begin{bmatrix} \cos\theta & \sin\theta \\ -\sin\theta & \cos\theta \end{bmatrix}$$

or the reflection matrix

$$R(\theta) = \begin{bmatrix} \cos 2\theta & \sin 2\theta \\ \sin 2\theta & -\cos 2\theta \end{bmatrix}.$$

A' is the direct sum of rotations and reflections. Figure 6.2b indicates the orbit structure for the elliptic, symplectic matrix $R(\theta)$, but now remember that an orbit is a sequence of discrete points not a continuous curve.

If ψ is a general nonlinear diffeomorphism with fixed point p (respectively, periodic point p of period k), then p is called hyperbolic or elliptic as matrix $D\psi(p)$ (respectively, $D\psi^k(p)$) is called.

A theorem of Hartman (1964) says that near a hyperbolic fixed point, a diffeomorphism is equivalent to its linear part. That is, if p is a hyperbolic point for ψ, then there are neighborhoods $\mathcal{O}$ of p and $\mathcal{Q}$ of the $0 \in \mathbb{R}^m$ and a homeomorphism $h : \mathcal{O} \to \mathcal{Q}$ such that h maps orbits of ψ onto orbits of $x \to Ax$. No such theorem is true for elliptic fixed points.

6.2.2 The Time τ–Map

If $\phi(t, \xi)$ is the general solution of the autonomous differential equation $\dot{x} = f(x)$, then for a fixed τ the map $\psi : \xi \to \phi(\tau, \xi)$ is a diffeomorphism because its inverse is $\psi^{-1} : \xi \to \phi(-\tau, \xi)$. It is called the time τ-map. If the differential equation is Hamiltonian, then ψ is a symplectomorphism.

Let p be a fixed point of ψ, i.e., $\psi(p) = \phi(\tau, p) = p$. Then by Lemma 6.1.4 p is an initial condition for a periodic solution of period τ. In a like manner a periodic point of period k is an initial condition for a periodic solution of period $k\tau$. This example seems somewhat artificial because the choice of τ was arbitrary. There is no clock in an autonomous system.

The harmonic oscillator $\dot{q} = \partial H/\partial p = \omega p$, $\dot{p} = -\partial H/\partial q = -\omega q$, $H = (\omega/2)(q^2 + p^2)$ defines the discrete Hamiltonian system

$$\begin{bmatrix} q \\ p \end{bmatrix} \to \begin{bmatrix} \cos\omega\tau & \sin\omega\tau \\ -\sin\omega\tau & \cos\omega\tau \end{bmatrix} \begin{bmatrix} q \\ p \end{bmatrix}, \tag{6.6}$$

which is a rotation of the plane by an angle $\omega\tau$ and the origin is an elliptic fixed point for this system.

6.2.3 The Period Map

Consider a periodic differential equation

$$\dot{x} = f(t, x), \qquad f(t + \tau, x) \equiv f(t, x), \ \tau > 0. \tag{6.7}$$

(Now there is a clock.) Let $\phi(t, \xi)$ be the general solution, so, $\phi(0, \xi) = \xi$. The mapping $\psi : \xi \to \phi(\tau, \xi)$ is called the period map (sometimes the Poincaré map). If Equation (6.7) is defined for all $x \in \mathbb{R}^m$ and the solutions for all t, $0 \leq t \leq \tau$, then ψ defines a discrete dynamical system; and if the equation is Hamiltonian, then ψ defines a discrete Hamiltonian system. By the same argument as above, a fixed point of ψ is the initial condition of a periodic solution of period τ, and a periodic point of period k is the initial condition of a periodic solution of period $k\tau$.

A natural question to ask is whether all diffeomorphisms and symplectomorphisms are time τ-maps of autonomous equations or period maps of periodic systems. Later we show that time τ-maps of autonomous differential equations are much simpler than general diffeomorphisms but that period maps are essentially the same as diffeomorphisms.

A diffeomorphism $\psi : \mathbb{R}^m \to \mathbb{R}^m$ is isotopic to the identity through diffeomorphisms if there exists a smooth function $\Phi : [0, \tau] \times \mathbb{R}^m \to \mathbb{R}^m$ such that for each $t \in [0, \tau]$ the map $\Phi(t, \cdot) : \mathbb{R}^m \to \mathbb{R}^m$ is a diffeomorphism and $\Phi(0, \xi) \equiv \xi$ and $\Phi(\tau, \xi) \equiv \psi(\xi)$ for all $\xi \in \mathbb{R}^m$. There is a similar definition where diffeomorphism and m are replaced by symplectomorphism and $2n$ throughout. The period map $\psi(\xi) = \phi(\tau, \xi)$ of a periodic system is clearly isotopic to the identity through diffeomorphisms. In fact:

Theorem 6.2.1. *A necessary and sufficient condition for a diffeomorphism $\psi : \mathbb{R}^m \to \mathbb{R}^m$ to be isotopic to the identity through diffeomorphisms is that ψ is the period map of a periodic system of the form* (6.7). *Also if ψ is isotopic to the identity through symplectomorphisms, then Equation* (6.7) *is Hamiltonian.*

Proof. First we give a little trickery on smooth functions. The function α defined by $\alpha(t) = 0$ for $t \leq 0$ and $\alpha(t) = \exp(-1/t)$ for $t > 0$ is a smooth function. (It is an easy argument to show that all the right and left derivatives of α are zero at $t = 0$ by l'Hôpital's rule.) The function $\beta(t) = \alpha(t - \tau/3)/(\alpha(t - \tau/3) + \alpha(2\tau/3 - t))$ is smooth, and $\beta(t) \equiv 0$ for $t \leq \tau/3$, and $\beta(t) \equiv 1$ for $t \geq 2\tau/3$.

Let $\Phi : [0, \tau] \times \mathbb{R}^m \to \mathbb{R}^m$ be the isotopy. Define $\Xi(t, \xi) = \Phi(\beta(t), \xi)$, so, $\Xi(t, \cdot)$ is the identity map for $0 \leq t \leq \tau/3$ and is the diffeomorphism ψ for $2\tau/3 \leq t \leq \tau$. Let $X(t, \eta)$ be the inverse of $\Xi(t, \xi)$, so, $X(t, \Xi(t, \xi)) \equiv \xi$. Now

$$\frac{\partial \Xi}{\partial t}(t, \xi) = \frac{\partial \Xi}{\partial t}(t, X(t, \Xi(t, \xi))) = F(t, \Xi(t, \xi)),$$

where

$$F(t, x) = \frac{\partial \Xi}{\partial t}(t, X(t, x)).$$

So $\Xi(t, \xi)$ is the general solution of $\dot{x} = F(t, x)$. Because Ξ is constant in t for $0 \leq t \leq \tau/3$ and $2\tau/3 \leq t \leq \tau$, F is identically zero for $0 \leq t \leq \tau/3$ and $2\tau/3 \leq t \leq \tau$. Therefore, the τ−periodic extension of F is smooth. Thus Ξ is the general solution of a τ-periodic system, and ψ is a period map because $\psi(\xi) = \Xi(\tau, \xi)$. If Φ is symplectic, then F is Hamiltonian by Theorem 2.6.2.

For example, let $\psi(x) = Ax + g(x)$, where $g(0) = \partial g(0)/\partial x = 0$, so, the origin is a fixed point. If A has a logarithm, so that $A = \exp B$, then $\Phi(t, x) = \exp(Bt) + tg(x)$ is an isotopy through diffeomorphisms near the origin.

For a symplectic map you must be a little more careful. First, if $\psi'(x) = x + g'(x)$, where $g'(0) = \partial g'(0)/\partial x = 0$, then by Theorem 8.1.1, $\psi : (q, p) \to (Q, P)$, where $q = \partial S(p, Q)/\partial p$, $P = \partial S(p, Q)/\partial Q$, $S(p, Q) = p^T Q + s(p, Q)$, where s is second order at the origin. Then $S(t, p, Q) = p^T Q + ts(p, Q)$ generates an isotopy to the identity through symplectomorphisms for ψ, $\Phi'(t, x)$; (i.e., $\Phi(1, \cdot) - \psi$ and $\Phi(0, \cdot) = $ identity). Now if $\psi(x) = Ax + g(x)$ write $\psi(x) = A(x + g'(x))$, and an isotopy for ψ is $\Phi(t, x) = \exp(Bt)\Phi'(t, x)$, where B is a Hamiltonian logarithm of A.

6.3 Flow Box Theorems

This section investigates the local flow and local integrals near a nonequilibrium point. Consider first an ordinary differential equation

$$\dot{x} = f(x), \tag{6.8}$$

where f is smooth on $\mathcal{O}$, an open set in $\mathbb{R}^m$, and let $\phi(t, \xi)$ be the solution of (6.8) such that $\phi(0, \xi) = \xi$. The analogous results for diffeomorphism are developed in the Problem section. A point $r \in \mathcal{O}$ is an ordinary point for (6.8) if $f(r) \neq 0$, otherwise r is an equilibrium point or a rest point.

Theorem 6.3.1. *Let $r \in \mathcal{O}$ be an ordinary point for (6.8), then there exists a change of coordinates $y = \psi(x)$ defined near r such that in the new coordinates, Equations (6.8) define a parallel flow; in particular, the equations become*

$$\dot{y}_1 = 1, \quad \dot{y}_i = 0, \quad i = 2, \ldots, m. \tag{6.9}$$

This coordinate system near an ordinary point is often called a *flow box*.

Proof. Let r be the origin in $\mathbb{R}^m$. Because $f(0) = a \neq 0$, one component of a is nonzero, say $a_1 \neq 0$. The solutions cross the hyperplane $x_1 = 0$ transversely, and so the new coordinates will be the time from the crossing of this hyperplane and the $n - 1$ coordinates where the solution hits the hyperplane; see Figure 6.3. That is, define the change of variables by

$$x = \phi(y_1, 0, y_2, y_3, \ldots, y_n). \tag{6.10}$$

Inasmuch as $\phi(0, \xi) = \xi$, $\dfrac{\partial \phi}{\partial x}(0, 0) = I$, and so

$$\frac{\partial x}{\partial y}(0) = \begin{bmatrix} a_1 & 0 & \cdots & 0 \\ a_2 & 1 & \cdots & 0 \\ \vdots & \vdots & & \vdots \\ a_n & 0 & \cdots & 1 \end{bmatrix}, \tag{6.11}$$

which is nonsingular because $a_1 \neq 0$. Thus (6.10) defines a valid change of coordinates near 0. The first variable, y_1, is time, so, $\dot{y}_1 = 1$ and the variables $y_2, \ldots, y_n$ are initial conditions, so, $\dot{y}_2 = \dot{y}_3 = \cdots = \dot{y}_n = 0$.

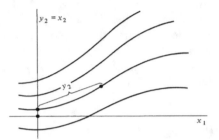

Figure 6.3. The flow box.

A set of smooth functions $F_1, \ldots, F_k$ defined on $\mathcal{O} \subset \mathbb{R}^m$ are independent at $r \in \mathcal{O}$ if the vectors $\nabla F_1(r), \ldots, \nabla F_k(r)$ are linearly independent.

Corollary 6.3.1. *Near an ordinary point, the system* (6.8) *has $m - 1$ independent integrals.*

Proof. In the y coordinates, $y_2, \ldots, y_m$ are constants of the motion and clearly independent, therefore they are independent integrals.

Consider a Hamiltonian system

$$\dot{z} = J\nabla H(z), \quad \text{or} \tag{6.12}$$
$$\dot{q} = H_p, \quad \dot{p} = -H_q,$$

where $z = (q, p)$ and H is a smooth function defined on the open set $\mathcal{O} \subset \mathbb{R}^{2n}$.

Theorem 6.3.2. *Let $r \in \mathcal{O} \subset \mathbb{R}^{2n}$ be an ordinary point for (6.12), then there exist symplectic coordinates $\{y\}$ defined near r such that the Hamiltonian becomes $H(y) = y_{n+1}$, and the equations of motion become*

$$\dot{y}_1 = \frac{\partial H}{\partial y_{n+1}} = 1, \quad \dot{y}_i = 0 \text{ for } i = 2, \ldots, 2n. \tag{6.13}$$

Proof. Again let r be the origin in $\mathbb{R}^{2n}$. Make a linear symplectic coordinates change so that $J\nabla H(0) = (1, 0, \ldots, 0)$ (see the Problems section). Let $q = \bar{q}(t, \xi, \eta), p = \bar{p}(t, \xi, \eta)$ be the general solution of *(6.12)* with $\bar{q}(0, \xi, \eta) = \xi$ and

$\bar{q}(0, \xi, \eta) = \eta$. For small values of t, these functions can be inverted to give $\xi = \xi(t, q, p)$, $\eta = \bar{\eta}(t, q, p)$. Because $J\nabla H(0) = (1, 0, \ldots, 0)$, we can solve the equation $\xi_1(t, q, p) = 0$ for t to give $t = \bar{t}(q, p)$.

Define the new coordinates by

$$y_1 = \bar{t}(q, p), \qquad\qquad y_{n+1} = H(q, p),$$

$$y_i = \xi_i(\bar{t}(q, p), q, p), \quad y_{n+i} = \bar{\eta}_i(\bar{t}(q, p), q, p), \quad i = 2, \ldots, n. \tag{6.14}$$

By Theorems (2.6.2) and (2.6.3), for fixed t, $\{\bar{\xi}_i, \bar{\xi}_j\} = \{\bar{\eta}_i, \bar{\eta}_j\} = 0$ and $\{\bar{\xi}_i, \bar{\eta}_j\} = \delta_{ij}$. Now check that (6.14) is symplectic. Let $2 \leq i, j \leq n$.

$$\{y_i, y_j\} = \sum_{k=1}^{n} \left(\frac{\partial y_i}{\partial q_k} \frac{\partial y_j}{\partial p_k} - \frac{\partial y_i}{\partial p_k} \frac{\partial y_j}{\partial q_k} \right) \tag{6.15}$$

$$= \sum_{k=1}^{n} \left[\left(\frac{\partial \bar{\xi}_i}{\partial t} \frac{\partial \bar{t}}{\partial q_k} + \frac{\partial \bar{\xi}_i}{\partial q_k} \right) \left(\frac{\partial \bar{\xi}_j}{\partial t} \frac{\partial \bar{t}}{\partial p_k} + \frac{\partial \bar{\xi}_j}{\partial p_k} \right) \right.$$

$$\left. - \left(\frac{\partial \bar{\xi}_i}{\partial t} \frac{\partial \bar{t}}{\partial p_k} + \frac{\partial \bar{\xi}_i}{\partial p_k} \right) \left(\frac{\partial \bar{\xi}_j}{\partial t} \frac{\partial \bar{t}}{\partial q_k} + \frac{\partial \bar{\xi}_j}{\partial q_k} \right) \right]$$

$$= \{\bar{\xi}_i, \bar{\xi}_j\} + \sum_{k=1}^{n} \left[\frac{\partial \bar{\xi}_i}{\partial t} \left(\frac{\partial \bar{t}}{\partial q_k} \frac{\partial \bar{\xi}_j}{\partial p_k} - \frac{\partial \bar{t}}{\partial p_k} \frac{\partial \bar{\xi}_j}{\partial q_k} \right) \right.$$

$$\left. - \frac{\partial \bar{\xi}_j}{\partial t} \left(\frac{\partial \bar{t} \partial \bar{\xi}_i}{\partial q_k \partial p_k} - \frac{\partial \bar{t}}{\partial p_k} \frac{\partial \bar{\xi}_i}{\partial q_k} \right) \right]$$

$$= \{\bar{\xi}_i, \bar{\xi}_j\} - \left[\frac{\partial \bar{\xi}_i}{\partial t} \{\bar{\xi}_1, \bar{\xi}_j\} - \frac{\partial \bar{\xi}_j}{\partial t} \{\bar{\xi}_1, \bar{\xi}_i\} \right] \left(\frac{\partial \bar{\xi}_1}{\partial t} \right)^{-1} = 0.$$

The simplification from the second to last line follows by the identities $\partial \bar{t}/\partial q_k = -(\partial \bar{\xi}_1/\partial q_k)/(\partial \bar{\xi}_1/\partial t)$ and $\partial \bar{t}/\partial p_k = -(\partial \bar{\xi}_1/\partial p_k)/(\partial \bar{\xi}_1/\partial t)$. In a similar way, $\{y_i, y_j\} = 0$ when $n + 2 \leq i, j \leq 2n$ and $\{y_i, y_{n+j}\} = \delta_{ij}$ for $i = 1, \ldots, n$.

$\{y_1, y_{1+n}\} = \{t, H\} = 1$ because the time rate of change of t along a solution is 1. Because $\bar{\xi}_i$ and $\bar{\eta}_i$ are integrals and $y_{1+n} = H$, it follows that $\{y_i, y_{1+n}\} = 0$ for $i = 2, \ldots, 2n$.

Let $2 \leq i \leq n$. Then

$$\{y_1, y_i\} = \sum_{k=1}^{n} \left(\frac{\partial t}{\partial q_k} \frac{\partial y_i}{\partial p_k} - \frac{\partial t}{\partial p_k} \frac{\partial y_i}{\partial q_k} \right)$$

$$= \sum_{k=1}^{n} \left[\frac{\partial \bar{t}}{\partial q_k} \left(\frac{\partial \bar{\xi}_i}{\partial t} \frac{\partial \bar{t}}{\partial p_k} + \frac{\partial \bar{\xi}_i}{\partial p_k} \right) - \frac{\partial \bar{t}}{\partial p_k} \left(\frac{\partial \bar{\xi}_i}{\partial t} \frac{\partial \bar{t}}{\partial q_k} + \frac{\partial \bar{\xi}_i}{\partial q_k} \right) \right] \tag{6.16}$$

$$= \sum_{k=1}^{n} \left(\frac{\partial \bar{t}}{\partial q_k} \frac{\partial \bar{\xi}_i}{\partial p_k} - \frac{\partial \bar{t}}{\partial p_k} \frac{\partial \bar{\xi}_i}{\partial q_k} \right) = \{\bar{\xi}_1, \bar{\xi}_i\} \left(\frac{\partial \bar{\xi}_1}{\partial t} \right)^{-1} = 0.$$

In a similar manner $\{y_1, y_i\} = 0$ for $i = n + 2, \ldots, 2n$.

This proof is due to Qiudong Wang, for an alternate proof see Moser and Zehnder (2005). A discussion of global integrable systems is given in Section 7.8.

A set of smooth functions $F_1, \ldots, F_k$ defined on $\mathcal{O} \subset \mathbb{R}^{2n}$ are in involution if $\{F_i, F_j\} \equiv 0$ for $1 \leq i, j \leq k$.

Corollary 6.3.2. *Near an ordinary point in the Hamiltonian system* (6.12) *has n independent integrals in involution.*

Proof. In the coordinates of Theorem 6.3.2, the coordinates $\eta_1, \ldots, \eta_n$ are independent integrals in involution.

If H is independent of one coordinate, say q_i, then $\dot{p}_i = \partial H / \partial q_i = 0$ or p_i is an integral. Similarly if H is independent of p_i, then q_i is an integral. If H is independent of one coordinate, then it is called an *ignorable coordinate*, and its conjugate is an integral.

Let q_1 be ignorable, so, p_1 is an integral, and $p_1 = \alpha$, a constant. When the variable p_1 is replaced by the parameter α in (6.12), the equations in (6.12) are independent of q_1 and p_1. The equations for $q_2, \ldots, q_n, p_2, \ldots, p_n$ are the equations of a Hamiltonian system of $n - 1$ degrees of freedom that depend on a parameter α. If these equations are solved explicitly in terms of t, these solutions can be substituted into the q_1 equation and q_1 can be determined by a single integration or quadrature. Thus an ignorable coordinate reduces the equations to a system of equations of $n - 1$ degrees of freedom containing a parameter and a quadrature.

In Hamiltonian systems, an integral gives rise to an ignorable coordinate and many integrals in involution give rise to many ignorable coordinates.

Theorem 6.3.3. *Let $F_1, \ldots, F_k$, $1 \leq k \leq n$, be smooth functions on $\mathcal{O}$, which are in involution and independent at a point $r \in \mathcal{O} \subset \mathbb{R}^{2n}$. Then there exist symplectic coordinates (ξ, η) at r such that in these coordinates $F_i = \eta_i$ for $i = 1, \ldots, k$.*

Proof. This theorem is left as an exercise. Use induction on k, the number of functions.

Theorem 6.3.4. *Assume that the Hamiltonian system* (6.12) *has k integrals $F_1, \ldots, F_k$, in involution which are independent at some point $r \in \mathcal{O}$. Then there exist symplectic coordinates ξ, η such that $\xi_1, \ldots, \xi_k$ are ignorable. So the system can be reduced locally to a Hamiltonian system with $n - k$ degrees of freedom depending on k parameters and k quadratures.*

For the N-body problem in Jacobi coordinates (see Section 7.2) the three components of the center of mass $g = (m_1 q_1 + \cdots + m_N q_N)/M$ are ignorable, and the conjugate momenta are the three components of total linear momentum, $G = p_1 + \cdots + p_N$. Jacobi coordinates effect a reduction in the number of degrees of freedom by 3.

The planar Kepler problem admits the z component of angular momentum as an integral. In polar coordinates r, θ, R, Θ of Section 7.2, θ is an ignorable coordinate, and its conjugate momentum, Θ, angular momentum, is an integral.

6.4 Periodic Solutions and Cross Sections

In view of the results of the previous sections, it would seem that the next question to be considered is the nature of the flow near an equilibrium point. This is one of the central and difficult questions in local geometric theory. It turns out that many of the questions about equilibrium points are very similar to questions about periodic solutions. This section introduces this similarity.

6.4.1 Equilibrium Points

Consider first a general system

$$\dot{x} = f(x), \tag{6.17}$$

where $f : \mathcal{O} \to \mathbb{R}^m$ is smooth, and $\mathcal{O}$ is open in $\mathbb{R}^m$. Let the general solution be $\phi(t, \xi)$. An equilibrium solution $\phi(t, \xi')$ is such that $\phi(t, \xi') \equiv \xi'$ for all t. Obviously $\phi(t, \xi')$ is an equilibrium solution if and only if $f(\xi') = 0$. So questions about the existence and uniqueness of equilibrium solutions are finite-dimensional questions. The eigenvalues of $\partial f(\xi')/\partial x$ are called the (characteristic) exponents of the equilibrium point. If $\partial f(\xi')/\partial x$ is nonsingular, or equivalently the exponents are all nonzero, then the equilibrium point is called elementary.

Proposition 6.4.1. *Elementary equilibrium points are isolated.*

Proof. $f(\xi') = 0$ and $\partial f(\xi')/\partial x$ is nonsingular, so, the implicit function theorem applies to f; thus, there is a neighborhood of ξ' with no other zeros of f.

The analysis of stability, bifurcations, etc. of equilibrium points starts with an analysis of the linearized equations. For this reason, one shifts the equilibrium point to the origin, and (6.17) is rewritten

$$\dot{x} = Ax + g(x), \tag{6.18}$$

where $A = \partial f(0)/\partial x$, $g(x) = f(x) - Ax$, so, $g(0) = 0$ and $\partial g(0)/\partial x = 0$. The eigenvalues of A are the exponents. The reason the eigenvalues of A are called exponents is that the linearized equations (e.g., $g(x) \equiv 0$ in (6.18)) have solutions which contain terms such as $\exp(\lambda t)$, where λ is an eigenvalue of A.

6.4.2 Periodic Solutions

A periodic solution is a solution $\phi(t, \xi')$, such that $\phi(t + T, \xi') \equiv \phi(t, \xi')$ for all t, where T is some nonzero constant. T is called a period, and the least positive T which satisfies that relation is called *the* period or the least period. In general, an equilibrium solution will not be considered as a periodic solution; however, some statements have a simpler statement if equilibrium solutions are considered as periodic solutions. It is easy to see that if the solution is periodic and not an equilibrium solution, then the least period exists, and all periods are integer multiples of it.

Lemma 6.4.1. *A necessary and sufficient condition for $\phi(t, \xi')$ to be periodic with a period T is*

$$\phi(T, \xi') = \xi', \tag{6.19}$$

where T is nonzero.

Proof. This is a restatement of Lemma 6.1.4.

This lemma shows that questions about the existence and uniqueness of periodic solutions are ultimately finite-dimensional questions. The analysis and topology of finite-dimensional spaces should be enough to answer all such questions.

Let $\phi(t, \xi')$ be periodic with least period T. The matrix $\partial \phi(T, \xi')/\partial \xi$ is called the monodromy matrix, and its eigenvalues are called the (characteristic) multipliers of the period solution. It is tempting to use the implicit function theorem on (6.19) to find a condition for local uniqueness of a periodic solution. To apply the implicit function theorem to (6.19) the matrix $\partial \phi(T, \xi')/\partial \xi - I$ would have to be nonsingular, or $+1$ would not be a multiplier. But this will never happen.

Lemma 6.4.2. *Periodic solutions of (6.17) are never isolated, and $+1$ is always a multiplier. In fact, $f(\xi')$ is an eigenvector of the monodromy matrix corresponding to the eigenvalue $+1$.*

Proof. Because (6.17) is autonomous, it defines a local dynamical system, so, a translate of a solution is a solution. Therefore, the periodic solution is not isolated. Differentiating the group relation $\phi(\tau, \phi(t, \xi')) = \phi(t + \tau, \xi')$ with respect to t and setting $t = 0$ and $\tau = T$ gives

$$\frac{\partial \phi}{\partial \xi}(T, \xi')\phi(0, \xi') = \phi(T, \xi'),$$

$$\frac{\partial \phi}{\partial \xi}(T, \xi')f(\xi') = f(\xi').$$

Inasmuch as the periodic solution is not an equilibrium point, $f(\xi') \neq 0$.

Because of this lemma, the correct concept is "isolated periodic orbit." In order to overcome the difficulties implicit in Lemma 6.4.2, one introduces a *cross section*. Let $\phi(t, \xi')$ be a periodic solution. A cross section to the periodic solution, or simply a section, is a hyperplane Σ of codimension one through ξ' and transverse to $f(\xi')$. For example, Σ would be the hyperplane $\{x : a^T(x - \xi') = 0\}$, where a is a constant vector with $a^T f(\xi') \neq 0$. The periodic solution starts on the section and, after a time T, returns to it. By the continuity of solutions with respect to initial conditions, nearby solutions do the same. See Figure 6.4. So if ξ is close to ξ' on Σ, there is a time $\mathcal{T}(\xi)$ close to T such that $\phi(\mathcal{T}(\xi), \xi)$ is on Σ. $\mathcal{T}(\xi)$ is called the first return time. The *section map*, or *Poincaré map*, is defined as the map $P : \xi \to \phi(\mathcal{T}(\xi), \xi)$ which is a map from a neighborhood N of ξ' in Σ into Σ.

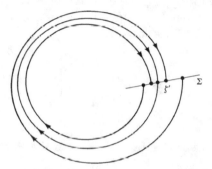

Figure 6.4. The cross section.

Lemma 6.4.3. *If the neighborhood N of ξ' in Σ is sufficiently small, then the first return time, $\mathcal{T} : N \to \mathbb{R}$, and the Poincaré map, $P : N \to \Sigma$, are smooth.*

Proof. Let $\Sigma = \{x : a^T(x - \xi') = 0\}$, where $a^T f(\xi') \neq 0$. Consider the function $g(t, \xi) = a^T(\phi(t, \xi) - \xi')$. Because $g(T, \xi') = 0$ and $\partial g(T, \xi')/\partial t = a^T \phi(T, \xi')/\partial t = a^T f(\xi') \neq 0$, the implicit function theorem gives a smooth function $\mathcal{T}(\xi)$ such that $g(\mathcal{T}(\xi), \xi) = 0$. g being zero defines Σ so the first return time, $\mathcal{T}$, is smooth. The Poincaré map is smooth because it is the composition of two smooth maps.

The periodic solution now appears as a fixed point of P; indeed, any fixed point, ξ'', of P is the initial condition for a periodic solution of period $\mathcal{T}(\xi'')$, because $(\mathcal{T}(\xi''), \xi'')$ would satisfy (6.19). A point $\xi'' \in N$ such that $P^k(\xi'') = \xi''$ for some integer $k > 0$ is called a periodic point of P of period k. The solution (6.17) through such a periodic point will be periodic with period nearly kT.

The analysis of stability, bifurcations, etc. of fixed points starts with an analysis of the linearized equations. For this reason, one shifts the fixed point to the origin and writes the Poincaré map

$$P(y) = Ay + g(y), \tag{6.20}$$

where $A = \partial P(0)/\partial y$, $g(y) = P(y) - Ay$, so $g(0) = 0$, and $\partial g(0)/\partial y = 0$. The eigenvalues of A are the multipliers of the fixed point. The reason the eigenvalues, λ_i, of A are called multipliers is that the linearized map (e.g., $g(x) \equiv 0$) in (6.20) takes an eigenvector to λ_i times itself. A fixed point is called *elementary* if none of its multipliers is equal to $+1$.

Lemma 6.4.4. *If the multipliers of the periodic solution are* $1, \lambda_2, \ldots, \lambda_m$, *then the multipliers of the corresponding fixed point of the Poincaré map are* $\lambda_2, \ldots, \lambda_m$.

Proof. Rotate and translate the coordinates so that $\xi' = 0$ and $f(\xi') = (1, 0, \ldots, 0)$, so, Σ is the hyperplane $x_1 = 0$. Let $B = \partial \phi(T, \xi')/\partial \xi$, the monodromy matrix. By Lemma 6.4.1, $f(\xi')$ is an eigenvector of B corresponding to the eigenvalue $+1$. In these coordinates,

$$B = \begin{bmatrix} 1 & x & x & x & x \\ 0 & & & & \\ \vdots & & A & & \\ 0 & & & & \end{bmatrix}. \tag{6.21}$$

Clearly the eigenvalues of B are $+1$ along with the eigenvalues of A.

We also call the eigenvalues $\lambda_2, \ldots, \lambda_n$ the *nontrivial multipliers* of the periodic orbit. Recall that an orbit is the solution considered as a curve in $\mathbb{R}^n$, and so is unaffected by reparameterization. A periodic orbit of period T is isolated if there is a neighborhood L of it with no other periodic orbits in L with period near to T. There may be periodic solutions of much higher period near an isolated periodic orbit. A periodic orbit is isolated if and only if the corresponding fixed point of the Poincaré map is an isolated fixed point. A periodic orbit is called *elementary* if none of its nontrivial multipliers is $+1$.

Proposition 6.4.2. *Elementary fixed points and elementary periodic orbits are isolated.*

Proof. Apply the implicit function theorem to the Poincaré map.

6.4.3 A Simple Example

Consider the system

$$\dot{x} = y + x(1 - x^2 - y^2), \tag{6.22}$$
$$\dot{y} = -x + y(1 - x^2 - y^2),$$

which in polar coordinates is

$$\dot{r} = r(1 - r^2), \tag{6.23}$$
$$\dot{\theta} = -1;$$

see Figure 6.5. The origin is an elementary equilibrium point, and the unit circle is an elementary periodic orbit. To see the latter claim, consider the cross section $\theta \equiv 0 \bmod 2\pi$. The first return time is 2π. The linearized equation about $r = 1$ is $r = -2r$, and so the linearized Poincaré map is $r \to r \exp(-4\pi)$. The multiplier of the fixed point is $\exp(-4\pi)$.

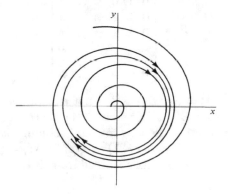

Figure 6.5. Phase portrait of the example.

6.4.4 Systems with Integrals

By Lemma 6.4.2, the monodromy matrix of a periodic solution has $+1$ as a multiplier. If Equation (6.17) were Hamiltonian, the monodromy matrix would be symplectic by Theorem 2.6.2, and so by Proposition 2.3.1, the algebraic multiplicity of the eigenvalue $+1$ would be even and so at least 2. Actually, this is simply due to the fact that an autonomous Hamiltonian system has the Hamiltonian as an integral.

Throughout this subsection assume that (6.17) admits an integral F, where F is a smooth map from $\mathcal{O}$ to $\mathbb{R}$, and assume that $\phi(t, \xi')$ is a periodic solution of period T. Furthermore assume that the integral F is nondegenerate on this periodic solution, i.e., $\nabla F(\xi')$ is nonzero. For a Hamiltonian system the Hamiltonian H is always nondegenerate on a nonequilibrium solution because $\nabla H(\xi') = 0$ would imply that ξ' is an equilibrium.

Lemma 6.4.5. *If F is nondegenerate on the periodic solution $\phi(t, \xi')$, then the multiplier $+1$ has algebraic multiplicity at least 2. Moreover, the row vector $\partial F(\xi')/\partial x$ is a left eigenvector of the monodromy matrix corresponding to the eigenvalue $+1$.*

Proof. Differentiating $F(\phi(t, \xi)) \equiv F(\xi)$ with respect to ξ and setting $\xi = \xi'$ and $t = T$ yields

$$\frac{\partial F(\zeta')}{\partial x} \frac{\partial \phi(T, \xi')}{\partial \zeta} = \frac{\partial F(\zeta')}{\partial x}, \tag{6.24}$$

which implies the second part of the lemma. Choose coordinates so that $f(\xi')$ is the column vector $(1, 0, \ldots, 0)^T$ and $\partial F(\xi')/\partial x$ is the row vector $(0, 1, 0, \ldots, 0)$. Because $f(\xi')$ is a right eigenvector and $\partial F(\xi')/\partial x$ is a left eigenvector, the monodromy matrix $B = \partial \phi(T, \xi')/\partial \xi$ has the form

$$B = \begin{bmatrix} 1 & x & x & x & \ldots & x \\ 0 & 1 & 0 & 0 & \ldots & 0 \\ 0 & x & x & x & \ldots & x \\ 0 & x & x & x & \ldots & x \\ \vdots & & & & & \vdots \\ 0 & x & x & x & \ldots & x \end{bmatrix}. \tag{6.25}$$

Expand by minors $p(\lambda) = \det(B - \lambda I)$. First expand along the first column to get $p(\lambda) = (1 - \lambda) \det(B' - \lambda I)$, where B' is the $(m - 1) \times (m - 1)$ matrix obtained from B by deleting the first row and column. Expand $\det(B' - \lambda I)$ along the first row to get $p(\lambda) = (1 - \lambda)^2 \det(B'' - \lambda I) = (1 - \lambda)^2 q(\lambda)$, where B'' is the $(m - 2) \times (m - 2)$ matrix obtained from B by deleting the first two rows and columns.

Again there is a good geometric reason for the degeneracy implied by this lemma. The periodic solution lies in an $(m - 1)$-dimensional level set of the integral, and typically in nearby level sets of the integral, there is a periodic orbit. So periodic orbits are not isolated.

Consider the Poincaré map $P : N \to \Sigma$, where N is a neighborhood of ξ' in Σ. Let u be flow box coordinates so that ξ' corresponds to $u = 0$. Equations (6.17) are $u_1 = 1$, $u_2 = 0, \ldots, u_m = 0$, and $F(u) = u_2$. In these coordinates, we may take Σ to be $u_1 = 0$. Because u_2 is the integral in these coordinates, P maps the level sets $u_2 = $ constant into themselves, so, we can ignore the u_2 component of P. Let $e = u_2$; let Σ_e be the intersection of Σ and the level set $F = e$; and let $y_1 = u_3, \ldots, y_{m-2} = u_m$ be coordinates in Σ_e. Here e is considered as a parameter (the value of the integral). In these coordinates, the Poincaré map P is a function of y and the parameter e. So $P(e, y) + (e, Q(e, y))$, where for fixed e, $Q(e, \cdot)$ is a mapping of a neighborhood N_e of the origin in Σ_e into Σ_e. Q is called the Poincaré map in an integral surface, see Figure 6.6. The eigenvalues of $\partial Q(0, 0)/\partial y$ are called the multipliers of the fixed point in the integral surface or the nontrivial multipliers. By the same argument as above, we have the following lemma.

Lemma 6.4.6. *If the multipliers of the periodic solution of a system with nondegenerate integral are $1, 1, \lambda_3, \ldots, \lambda_m$, then the multipliers of the fixed point in the integral surface are $\lambda_3, \ldots, \lambda_m$.*

Lemma 6.4.7. *If the system is Hamiltonian, then the Poincaré map in an integral surface is symplectic.*

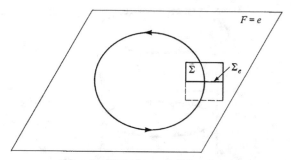

Figure 6.6. Poincaré map in an integral surface.

Proof. In this case, use the Hamiltonian flow box theorem (Theorem 6.3.2). In this case, $H = \eta_1$, and the equations are $\xi_1 = 1$, $\xi_i = 0$ for $i = 2, \ldots, n$ and $\eta_i = 0$ for $i = 1, \ldots, n$. The cross section is $\xi_1 = 0$, and the integral parameter is $\eta_1 = e$. The Poincaré map in an integral surface in these coordinates is in terms of the symplectic coordinates $\xi_2, \ldots, \xi_n, \eta_2, \ldots, \eta_n$ on Σ_e. Because the total map $x \rightarrow \phi(T, x)$ is symplectic (Theorem 2.6.2), the map $y \rightarrow Q(e, y)$ is symplectic.

If none of the trivial multipliers is 1, and the integral is nondegenerate on the periodic solution, then we say that the periodic solution (or fixed point) is elementary.

Theorem 6.4.1 (The cylinder theorem). *An elementary periodic orbit of a system with integral lies in a smooth cylinder of periodic solutions parameterized by the integral F. (See Figure 6.7.)*

Proof. Apply the implicit function theorem to $Q(e, y) - y - 0$ to get a one-parameter family of fixed points $y^*(e)$ in each integral surface $F = e$.

6.5 The Stable Manifold Theorem

In this section we discuss some important theorems about the local structure of differential equations near equilibrium points and diffeomorphisms near fixed points by introducing the concept of a hyperbolic point. Hamiltonian dynamics is more the study of elliptic points than hyperbolic points, therefore we concentrate on the elliptic case and refer the reader to the literature for some of the proofs for the hyperbolic theorems.

Let the equation

$$\dot{x} = f(x) \tag{6.26}$$

have an equilibrium point at $x = p$, so, $f(p) = 0$. Let $A = \partial f(p)/\partial x$, so, A is an $m \times m$ constant matrix. The eigenvalues of A are called the exponents

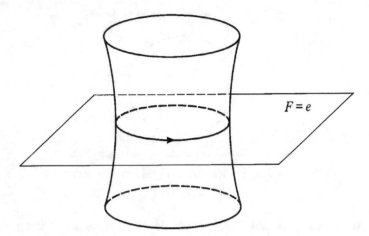

Figure 6.7. Cylinder of periodic solutions.

of equilibrium point p. The linearization of (6.26) about $x = p$ is $y = Ay$, where $y = x - p$. We say that p is a hyperbolic equilibrium point for (6.26), if A has no eigenvalues with zero real part, so, all the eigenvalues have either positive real parts or negative real parts. Thus the solutions of the linearized equation are sums of exponentially increasing and decreasing terms. The set of all solutions tending to zero is a linear subspace, as is the set of all solutions tending away from zero. The full nonlinear equations have similar sets, which is the subject of the following theorems.

At first the results are local, so, one can shift the equilibrium point to the origin and write (6.26) in the form

$$\dot{x} = Ax + g(x), \tag{6.27}$$

where $g(x) = f(x) - Ax$, so, $g(0) = Dg(0) = 0$, and A is an $m \times m$ real constant matrix with no eigenvalue with zero real part. Let $\phi(t, \xi)$ be the general solution of (6.27), so, $\phi(0, \xi) = \xi$. Let $\varepsilon > 0$ be given; then the local stable manifold is

$$W^s(\varepsilon) = \{\xi \in \mathbb{R}^m : |\phi(t, \xi)| < \varepsilon \text{ for all } t \leq 0\}. \tag{6.28}$$

and the local unstable manifold is

$$W^u(\varepsilon) = \{\xi \in \mathbb{R}^m : |\phi(t, \xi)| < \varepsilon \text{ for all } t \leq 0\}. \tag{6.29}$$

If the reader is not familiar with the definition of a manifold, simply read the remark following the statement of the theorem.

Theorem 6.5.1 (Local stable manifold for equations). *Let A have d eigenvalues with negative real parts and $m - d$ eigenvalues with positive real* *parts. Let g be as above. Then for ε sufficiently small, $\mathcal{W}^s(\varepsilon)$ and $\mathcal{W}^u(\varepsilon)$ are smooth manifolds of dimensions d and m − d, respectively. If $\xi \in \mathcal{W}^s(\varepsilon)$ (respectively, $\xi \in \mathcal{W}^u(\varepsilon)$), then $\phi(t, \xi) \to 0$ as $t \to +\infty$ (respectively, $\phi(t, \xi) \to 0$ as $t \to -\infty$). Actually, there is a smooth, near identity change of coordinates that takes the stable and unstable manifolds to (different) coordinate planes.*

Proof. See Hale (1972) or Chicone (1999).

Remarks. By a linear change of coordinates, if necessary, we may assume that

$$A = \begin{bmatrix} B & 0 \\ 0 & C \end{bmatrix},$$

where B is a $d \times d$ matrix with eigenvalues with negative real parts, and C is an $(m-d) \times (m-d)$ matrix with positive real parts. Writing $\mathbb{R}^m = \mathbb{R}^d \times \mathbb{R}^{m-d}$, $(w, z) \in \mathbb{R}^m = \mathbb{R}^d \times \mathbb{R}^{m-d}$, so, Equation (6.27) becomes

$$\dot{z} = Bz + h(z, w), \tag{6.30}$$
$$\dot{w} = Cw + k(z, w),$$

where h, g, and their first partials vanish at the origin. One proof of the existence of the stable manifold establishes the existence of a change of coordinates of the form $\xi = z$, $\eta = w - u(z)$, which makes the ξ coordinate hyperplane invariant or at least locally invariant. The function u is shown to be smooth and small with $u(0) = Du(0) = 0$. Thus in the new coordinates, the local stable manifold is a piece of the d-dimensional linear space $\eta = 0$. In the original coordinates, the local stable manifold is the graph of the function u. Because $u(0) = Du(0) = 0$, the stable manifold is tangent to the d-dimensional linear space $z = 0$. See Figure 6.8.

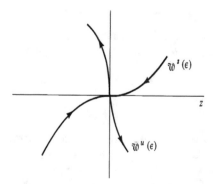

Figure 6.8. Local stable and unstable manifolds.

This is a local result, so, a natural question to ask is what happens to these manifolds. Now we show how to continue these manifolds. Assume f in (6.26) is globally defined and let the general solution $\phi(t, \xi)$ of (6.26) be globally defined, so, ϕ defines a dynamical system. The (global) stable manifold is

$$\mathcal{W}^s = \mathcal{W}^s(p) = \{\xi \in \mathbb{R}^m : \phi(t, \xi) \to p \text{ as } t \to +\infty\}, \tag{6.31}$$

and the (global) unstable manifold is

$$\mathcal{W}^u = \mathcal{W}^u(p) = \{\xi \in \mathbb{R}^m : \phi(t, \xi) \to p \text{ as } t \to -\infty\}. \tag{6.32}$$

Theorem 6.5.2 (Global stable manifold for equations). *Let p be a hyperbolic equilibrium point with d exponents with negative real parts and $m - d$ exponents with positive real parts. Then the stable manifold is an immersed d-dimensional submanifold. That is, there exists a smooth function $\Gamma : \mathbb{R}^d \to \mathbb{R}^n$ which is globally one-to-one, and $D\Gamma$ has rank d everywhere. Similarly, the unstable manifold is an immersed $m - d$ submanifold.*

Proof. See Palis and de Melo (1980).

There are similar theorems for diffeomorphisms. Consider a diffeomorphism

$$\psi : \mathbb{R}^m \to \mathbb{R}^m \tag{6.33}$$

with a fixed point p. Let $A = \partial\psi(p)/\partial x$, so, A is an $m \times m$ constant matrix. The eigenvalues of A are called the multipliers of p. The linearization of (6.33) about $x = p$ is $y \to Ay$, where $y = x - p$. We say that p is a hyperbolic fixed point if A has no eigenvalues with absolute value equal to 1. The set of all trajectories tending to zero is a linear subspace, as is the set of all solutions tending away from zero.

The first theorem is local, so, shift the fixed point to the origin and consider

$$x \to \Phi(x) = Ax + g(x), \tag{6.34}$$

where g is defined and smooth in a neighborhood of the origin in $\mathbb{R}^m$ with $g(0) = 0$, $Dg(0) = 0$. Define the stable manifold as

$$\mathcal{W}^s(\varepsilon) = \{x \in \mathbb{R}^m : |\Phi^k(x)| < \varepsilon \text{ for all } k \geq 0\} \tag{6.35}$$

and the unstable manifold similarly.

Theorem 6.5.3 (Local stable manifold for diffeomorphisms). *Let A have d eigenvalues with absolute value less than 1 and $m - d$ eigenvalues with absolute value greater than 1. Let g be as above. Then for ε sufficiently small, $\mathcal{W}^s(\varepsilon)$ and $\mathcal{W}^u(\varepsilon)$ are smooth manifolds of dimensions d and $m - d$, respectively. If $\xi \in \mathcal{W}^s(\varepsilon)$ (respectively, $\xi \in \mathcal{W}^u(\varepsilon)$), then $\Phi^k(\xi) \to 0$ as $k \to +\infty$ (respectively, $\Phi^k(\xi) \to 0$ as $k \to -\infty$). Actually there is a smooth, near identity change of coordinates that takes the stable and unstable manifolds to (different) coordinate planes.*

Assume ψ in (6.33) is a global diffeomorphism, so, it defines a dynamical system. The (global) stable manifold is

$$\mathcal{W}^s = \mathcal{W}^s(p) = \{\xi \in \mathbb{R}^m : \psi^k(\xi) \to p \text{ as } k \to +\infty\}, \qquad (6.36)$$

and the (global) unstable manifold is similarly defined.

Theorem 6.5.4 (Global stable manifold for diffeomorphisms). *Let p be a hyperbolic fixed point for ψ with d multipliers with absolute value less than 1 and $m - d$ multipliers with absolute value greater than 1. Then the stable manifold is an immersed d-dimensional submanifold. That is, there exists a smooth function $\Gamma : \mathbb{R}^d \to \mathbb{R}^n$ that is globally one-to-one, and $D\Gamma$ has rank d everywhere. Similarly, the unstable manifold is an immersed $m - d$ submanifold.*

For the rest of the section let ψ in (6.33) be a diffeomorphism of $\mathbb{R}^2$, p be a hyperbolic fixed point with one multiplier greater than one and one multiplier less than one so that the stable and unstable manifolds of p are smooth curves. A point $q \in \mathcal{W}^s(p) \cap \mathcal{W}^u(p)$, $q \neq p$, is called a *homoclinic point* (homoclinic to p). Because q is in both the stable and unstable manifolds of p, $\psi^k(q) \to p$ as $k \to \pm\infty$, the orbit of q is said to be doubly asymptotic to p. If the curves $\mathcal{W}^s(p)$ and $\mathcal{W}^u(p)$ are not tangent at q, the intersection is said to be *transversal*, and q is said to be a transversal homoclinic point. Henceforth, let q be a transversal homoclinic point, homoclinic to p; see Figure 6.9.

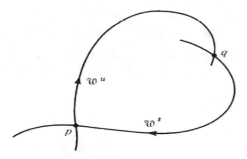

Figure 6.9. Transverse intersection of stable and unstable manifolds.

The stable and unstable manifolds are invariant, and so $\psi^k(q) \in \mathcal{W}^s(p) \cap \mathcal{W}^u(p)$ for all k, or the whole orbit of a homoclinic point consists of homoclinic points. In a neighborhood of the hyperbolic point p, the diffeomorphism ψ is well approximated by its linear part, and so it contracts in the stable manifold direction and expands in the unstable direction. This results in the following:

Theorem 6.5.5 (Palis' lambda lemma). *Let Λ be any interval in the unstable manifold with p in its interior. Let λ be a small segment transverse to the stable manifold. Then for any $\varepsilon > 0$, there is a K such that for $k \geq K$, $\psi^k(\lambda)$ is within the ε neighborhood of Λ. Moreover, if $a \in \Lambda$, $b \in \psi^k(\lambda)$, and $\mathrm{dist}\,(a, b) < \varepsilon$ then the tangents to Λ and $\psi(\lambda)^k$ are within ε.*

Proof. See Palis and de Melo (1980) and Figure 6.10.

This theorem says that the small segment λ of the unstable manifold is stretched out and that C^1 approximates the whole unstable manifold by iterates of ψ.

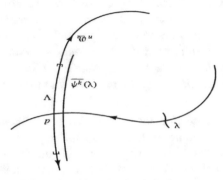

Figure 6.10. The lambda lemma.

Now let q be a transverse homoclinic point and λ a small segment of the unstable manifold at q. Images of this small segment λ are stretched out along the unstable manifold until the image again intersects the stable manifold as shown in Figure 6.11. So a homoclinic point begets another homoclinic point near the first. Repeating the argument you get the following.

Theorem 6.5.6 (Poincaré's homoclinic theorem). *A transversal homoclinic point is the limit of transversal homoclinic points.*

Proof. See Poincaré (1899) and Figure 6.11.

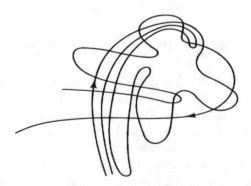

Figure 6.11. The Poincaré tangle.

6.6 Problems

1. Let $\{\phi_t\}$ be a smooth dynamical system, i.e., $\{\phi_t\}$ satisfies (6.5). Prove that $\phi(t,\xi) = \phi_t(\xi)$ is the general solution of an autonomous differential equation.

2. Let ψ be a diffeomorphism of $\mathbb{R}^m$, so, it defines a discrete dynamical system. A nonfixed point is called an ordinary point. So $p \in \mathbb{R}^m$ is an ordinary point if $\psi(p) \neq p$. Prove that there are local coordinates x at an ordinary point p and coordinates y at $q = \psi(p)$ such that in these local coordinates $y_1 = x_1, \ldots, y_m = x_n$. (This is the analog of the flow box theorem for discrete systems.)

3. Let ψ be as in Problem 2. Let p be a fixed point p of ψ. The eigenvalues of $\partial\psi(p)/\partial x$ are called the (characteristic) multipliers of p. If all the multipliers are different from $+1$, then p is called an elementary fixed point of ψ. Prove that elementary fixed points are isolated.

4. a) Let $0 < a < b$ and $\xi \in \mathbb{R}^m$ be given. Show that there is a smooth nonnegative function $\gamma : \mathbb{R}^m \to \mathbb{R}$ which is identically $+1$ on the ball $\|x - \xi\| \le a$ and identically zero for $\|x - \xi\| \ge b$.

    b) Let $\mathcal{O}$ be any closed set in $\mathbb{R}^m$. Show that there exists a smooth, nonnegative function $\delta : \mathbb{R}^m \to \mathbb{R}$ which is zero exactly on $\mathcal{O}$.

5. Let $H(q_1, \ldots, q_N, p_1, \ldots, p_N), q_i, p_i \in \mathbb{R}^3$ be invariant under translation, so, $H(q_1 + s, \ldots, q_N + s, p_1, \ldots, p_N) = H(q_1, \ldots, q_N, p_1, \ldots, p_N)$ for all $s \in \mathbb{R}^3$. Show that total linear momentum, $L = \sum p_i$, is an integral. This is another consequence of the Noether theorem.

6. An $m \times m$ nonsingular matrix T is such that $T^2 = I$ is a discrete symmetry of (or a reflection for) $\dot{x} = f(x)$ if and only if $f(Tx) = -Tf(x)$ for all $x \in \mathbb{R}^m$. This equation is also called reversible in this case.

    a) Prove: If T is a discrete symmetry of (1), then $\phi(t, T\xi) \equiv T\phi(-t, \xi)$ where $\phi(t, \xi)$ is the general solution of $\dot{x} = f(x)$.

    b) Consider the 2×2 case and let $T = \text{diag}(1, -1)$. What does $f(Tx) = -Tf(x)$ mean about the parity of f_1 and f_2? Show that Part (a) means that a reflection of a solution in the x_1 axis is a solution.

7. Let $\mathcal{G}$ be a matrix Lie group, i.e., $\mathcal{G}$ is a closed subgroup of the general group $GL(m, \mathbb{R})$. (See the Problem section at the end of Chapter 2.) $\mathcal{G}$ is a symmetry group for $\dot{x} = f(x)$ if $Tf(x) = f(Tx)$ for all $T \in \mathcal{G}$ and $x \in \mathbb{R}^m$.

    a) Prove: If $\mathcal{G}$ is a symmetry group, then $\phi(t, T\xi) \equiv T\phi(t, \xi)$, where ϕ is the general solution of $\dot{x} = f(x)$.

    b) Consider the 2×2 case where $\mathcal{G}$ is $SO(2, \mathbb{R})$ the group of rotations of the plane (i.e., orthogonal matrices with determinant $+1$.) In polar coordinates (r, θ), $\dot{x} = f(x)$ becomes $\dot{r} = R(r, \theta)$, $\dot{\theta} = \Theta(r, \theta)$. Prove that the symmetry condition implies R and Θ are independent of θ.

8. Now let $\dot{x} = f(x)$ be Hamiltonian with Hamiltonian $H : \mathbb{R}^{2n} \to \mathbb{R}$; so, $f(x) = J\nabla H(x)$. A matrix T is antisymplectic if $T^T J T = -J$.

An antisymplectic matrix T such that $T^2 = I$ is a discrete symplectic symmetry for H if $H(Tx) \equiv H(x)$.

a) Prove: A discrete symplectic symmetry of the Hamiltonian is a discrete symmetry of the equation $\dot{x} = f(x)$.

b) Consider a general Newtonian system as discussed in Section 1.1 of the form $H(x,p) = \frac{1}{2}p^T M^{-1} p + F(x)$ where $x, p \in \mathbb{R}^n$ and M is a nonsingular, symmetric matrix. De fine $T = \text{diag}(I, -I)$, so, $T^2 = I$, show that T is antisymplectic and $H(T(x,p)) = H(x,p)$.

c) Consider the restricted 3-body problem as discussed in Section 3.1. Let $T = \text{diag}(1, -1, -1, 1)$; show $H(T(x,y)) = H(x,y)$ where H is the Hamiltonian of the restricted 3-body problem (4.1).

d) What is FIX of Problem 7 for these two examples?

9. Use Problems 7 and 9.

a) Show that the solution of the restricted problem which crosses the x_1 axis (the line of syzygy) at a time t_1 and later at a time t_2 is a period of period $2(t_2 - t_1)$.

b) Show that the above criterion is $x_2 = y_1 = 0$ at times t_1 and t_2 in rectangular coordinates, and $\theta = n\pi$ (n an integer), $R = 0$ in polar coordinates.

10. Let $\mathcal{G}$ be a matrix Lie group of symplectic matrices, i.e., $\mathcal{G}$ is a closed subgroup of the symplectic group $Sp(2n, \mathbb{R})$. Let $x = f(x)$ be Hamiltonian with Hamiltonian H. $\mathcal{G}$ is a symmetry group for the Hamiltonian H if $H(Tx) = H(x)$ for all $T \in \mathcal{G}$. Prove: A symmetry group for the Hamiltonian H is a symmetry group for the equations of motion.

7. Symplectic Geometry

This chapter gives a survey of the general global questions in the theory of Hamiltonian systems. It tries to answer the questions: What is J? How to define a Hamiltonian system on a manifold? What is the global reduction theorem for symplectic group actions?

This chapter assumes a basic knowledge of differential manifolds by the reader. We give a quick introduction to the symplectic differential form, symplectic manifolds, Lie group actions, the general Noether's theorem, the Marsden-Wienstein-Meyer reduction theorem, and Arnold's theorem on completely integrable systems. A complete discussion with proofs requires massive tome like Abraham and Marsden (1978). The material in this chapter is used sparingly in later chapters, so it can be skipped on the first reading.

But first, what is a symplectic manifold? Quite simply it is just a space where Hamiltonian systems can be defined. That is, a *symplectic manifold* is a smooth differential manifold M of dimension $2n$ with a symplectic atlas $\mathcal{A}$ such that if ϕ, $\psi \in \mathcal{A}$ then $\phi \circ \psi^{-1}$ and $\psi \circ \phi^{-1}$ are symplectic maps of subsets of $\mathbb{R}^{2n}$ where defined. With such an atlas a system of Hamiltonian differential equations can be consistently defined on all of M. An alternate equivalent definition will be given later in Section 7.4.

This is a perfectly satisfactory and workable definition, but fails to shed light on the role of J. Since $J = J_{ij}$ is a doubly index array that is skew-symmetric it leads us to a discussion of multilinear algebra and differential forms next.

7.1 Exterior Algebra

Let $\mathbb{V}$ be a vector space of dimension m over the real numbers $\mathbb{R}$. Let $\mathbb{V}^k$ denote k copies of $\mathbb{V}$, i.e., $\mathbb{V}^k = \mathbb{V} \times \cdots \times \mathbb{V}$ (k times). A function $\phi : \mathbb{V}^k \longrightarrow \mathbb{R}$ is called em k-multilinear if it is linear in each argument; so,

© Springer International Publishing AG 2017
K.R. Meyer, D.C. Offin, *Introduction to Hamiltonian Dynamical Systems and the N-Body Problem*, Applied Mathematical Sciences 90, DOI 10.1007/978-3-319-53691-0_7

$$\phi(a_1, \ldots, a_{r-1}, \alpha u + \beta v, a_{r+1}, \ldots, a_k)$$
$$= \alpha\phi(a_1, \ldots, a_{r-1}, u, a_{r+1}, \ldots, a_k) + \beta\phi(a_1, \ldots, a_{r-1}, v, a_{r+1}, \ldots, a_k)$$

for all $a_1, \ldots, a_k, u, v \in \mathbb{V}$, all $\alpha, \beta \in \mathbb{R}$, and all arguments, $r = 1, \ldots, k$.

A 1-multilinear map is a linear functional called a em covector or *1-form*. In $\mathbb{R}^m$ the scalar product $(a, b) = a^T b$ is 2-multilinear, in $\mathbb{R}^{2n}$ the symplectic product $\{a, b\} = a^T J b$ is 2-multilinear, and the determinant of an $m \times m$ matrix is m-multilinear in its m rows (or columns).

A k-multilinear function ϕ is *skew-symmetric* or *alternating* if interchanging any two arguments changes its sign. For a skew-symmetric k-multilinear ϕ,

$$\phi(a_1, \ldots, a_r, \ldots, a_s, \ldots, a_k) = -\phi(a_1, \ldots, a_s, \ldots, a_r, \ldots, a_k)$$

for all $a_1, \ldots, a_k \in \mathbb{V}$ and all $r, s = 1, \ldots, k, r \neq s$. Thus ϕ is zero if two of its arguments are the same. We call an alternating k-multilinear function a k-linear form or k-form for short. The symplectic product $\{a, b\} = a^T J b$ and the determinant of an $m \times m$ matrix are alternating.

Let $\mathbb{A}^0 = \mathbb{R}$ and $\mathbb{A}^k = \mathbb{A}^k(\mathbb{V})$ be the space of all k-forms for $k \geq 1$. It is easy to verify that $\mathbb{A}^k$ is a vector space when using the usual definition of addition of functions and multiplication of functions by a scalar.

In $\mathbb{E}^3$ a linear functional (a 1-form) acting on a vector v can be thought of as the scalar project of v in a particular direction. A physical example is work. The work done by a uniform force is a linear functional on the displacement vector of a particle.

Two vectors in $\mathbb{E}^3$ determine a plane through the origin and a parallelogram in that plane. The oriented area of this parallelogram is a 2-form. Two vectors in $\mathbb{E}^3$ determine (i) a plane, (ii) an orientation in the plane, and (iii) a magnitude, the area of the parallelogram. Physical quantities that also determine a plane, an orientation, and a magnitude are torque, angular momentum, and magnetic field.

Three vectors in $\mathbb{E}^3$ determine a parallelepiped whose oriented volume is a 3-form. The flux of a uniform vector field, v, crossing a parallelogram determined by two vectors a and b is a 3-form.

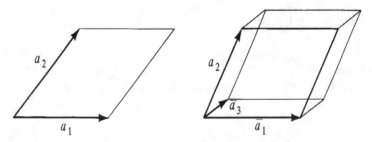

Figure 7.1. Multilinear functions.

If ψ is a 2-multilinear function, then ϕ defined by $\phi(a, b) = \{\psi(a, b) - \psi(b, a)\}/2$ is alternating and is called the *alternating part* of ψ. If ψ is already alternating, then $\phi = \psi$. If α and β are 1-forms, then $\phi(a, b) = \alpha(a)\beta(b) - \alpha(b)\beta(a)$ is a 2-form. This construction can be generalized. Let P_k be the set of all permutations of the k numbers $1, 2, \ldots, k$ and sign:$P_k \longrightarrow \{+1, -1\}$ the function that assigns $+1$ to an even permutation and -1 to an odd permutation. So if ϕ is alternating, $\phi(a_{\sigma(1)}, \ldots, a_{\sigma(k)}) = \mathrm{sign}(\sigma)\phi(a_1, \ldots, a_k)$. If ψ is a k-multilinear function, then ϕ defined by

$$\phi(a_1, \ldots, a_k) = \frac{1}{k!} \sum_{\sigma \in P} \mathrm{sign}(\sigma)\psi(a_{\sigma(1)}, \ldots, a_{\sigma(k)})$$

is alternating. We write $\phi = \mathrm{alt}\,(\psi)$. If ψ is already alternating, then $\psi = \mathrm{alt}\,(\psi)$. If $\alpha \in \mathbb{A}^k$ and $\beta \in \mathbb{A}^r$, then define $\alpha \wedge \beta \in \mathbb{A}^{k+r}$ by

$$\alpha \wedge \beta = \frac{(k+r)!}{k!r!}\,\mathrm{alt}\,(\alpha\beta)$$

or

$$\alpha \wedge \beta(a_1, \ldots, a_{k+r})$$
$$= \sum_{\sigma \in P} \mathrm{sign}(\sigma)\alpha(a_{\sigma(1)}, \ldots, a_{\sigma(k)})\beta(a_{\sigma(k+1)}, \ldots, a_{\sigma(k+r)}).$$

The operator $\wedge : \mathbb{A}^k \times \mathbb{A}^r \longrightarrow \mathbb{A}^{k+r}$ is called the *exterior product* or *wedge product*.

Lemma 7.1.1. *Let α be a k-form, β and δ be r-forms, and γ be a s-form then:*

1. $\alpha \wedge (\beta + \delta) = \alpha \wedge \beta + \alpha \wedge \delta$.
2. $\alpha \wedge (\beta \wedge \gamma) = (\alpha \wedge \beta) \wedge \gamma$.
3. $\alpha \wedge \beta = (-1)^{kr}\beta \wedge \alpha$.

Proof. The first two parts are fairly easy and are left as exercises. Let τ be the permutation $\tau : (1, \ldots, k, k+1, \ldots, k+r) \longrightarrow (k+1, \ldots, k+r, 1, \ldots, k)$; i.e., τ interchanges the first k entries and the last r entries. By thinking of τ as being the sequence

$$(1, \ldots, k, k+1, \ldots, k+r) \longrightarrow (k+1, 1, \ldots, k, k+2, \ldots, k+r)$$
$$\longrightarrow (k+1, k+2, 1, \ldots, k+3, \ldots, k+r) \longrightarrow \cdots \longrightarrow (k+1, \ldots, k+r, 1, \ldots, k),$$

it is easy to see that $\mathrm{sign}(\tau) = (-1)^{\mathrm{rk}}$. Now

$$\alpha \wedge \beta(a_1, \ldots, a_{k+r})$$

$$= \sum_{\sigma \in P} \mathrm{sign}(\sigma)\alpha(a_{\sigma(1)}, \ldots, a_{\sigma(k)})\beta(a_{\sigma(k+1)}, \ldots, a_{\sigma(k+r)})$$

$$= \sum_{\sigma \in P} \mathrm{sign}(\sigma \circ \tau)\alpha(a_{\sigma\circ\tau(1)}, \ldots, a_{\sigma\circ\tau(k)})\beta(a_{\sigma\circ\tau(k+1)}, \ldots, a_{\sigma\circ\tau(k+r)})$$

$$= \sum_{\sigma \in P} \mathrm{sign}(\sigma)\mathrm{sign}(\tau)\beta(a_{\sigma(1)}, \ldots, a_{\sigma(r)})\alpha(a_{\sigma(r+1)}, \ldots, a_{\sigma(k+r)})$$

$$= (-1)^{\mathrm{rk}}\beta \wedge \alpha.$$

Let $e_1, \ldots, e_m$ be a basis for $\mathbb{V}$ and $f^1, \ldots, f^m$ be the dual basis for the dual space $\mathbb{V}^*$, i.e., $f^i(e_j) = \delta^i_j$ where

$$\delta^i_j = \begin{cases} 1 & \text{if } i = j \\ 0 & \text{if } i \neq j. \end{cases}$$

This is our first introduction to the subscript-superscript convention of differential geometry and classical tensor analysis.

Lemma 7.1.2. $\dim \mathbb{A}^k = \binom{m}{k}$. *In particular a basis for $\mathbb{A}^k$ is*

$$\{f^{i_1} \wedge f^{i_2} \wedge \cdots \wedge f^{i_k} : 1 \leq i_1 < i_2 < \cdots < i_k \leq m\}.$$

Proof. Let I denote the set $\{(i_1, \ldots, i_k) : i_j \in \mathbb{Z}, 1 \leq i_1 < \cdots < i_k \leq m\}$ and $f^i = f^{i_1} \wedge \cdots \wedge f^{i_k}$ when $i \in I$. From the definition, $f^{i_1} \wedge f^{i_2} \wedge \cdots \wedge f^{i_k}(e_{j_1}, \ldots, e_{j_k})$ equals 1 if $i, j \in I$ and $i = j$ and equals 0 otherwise; in short, $f^i(e_j) = \delta^i_j$.

Let ϕ be a k-form and define

$$\psi = \sum_{i \in I} \phi(e_{i_1}, \ldots, e_{i_k}) f^{i_1} \wedge f^{i_2} \wedge \cdots \wedge f^{i_k} = \sum_{i \in I} \phi(e_i) f^i.$$

Let $v_i = \sum a^j_i e_j, i = 1, \ldots, k$, be k arbitrary vectors. By the multilinearity of ϕ and ψ, one sees that $\phi(v_1, \ldots, v_k) = \psi(v_1, \ldots, v_k)$; so, they agree on all vectors and, therefore, are equal. Thus the set $\{f^i : i \in I\}$ spans $\mathbb{A}^k$.

Assume that

$$\sum_{i \in I} a_{i_1 \ldots i_k} f^{i_1} \wedge f^{i_2} \wedge \cdots \wedge f^{i_k} = 0. \tag{7.1}$$

For a fixed set of indices $s_1, \ldots, s_k$, let $r_{k+1}, \ldots, r_m$ be a complementary set, i.e., $s_1, \ldots, s_k, r_{k+1}, \ldots, r_m$ is just a permutation of the integers $1, \ldots, m$. Take the wedge product of (7.1) with $f^{r_{k+1}} \wedge \cdots \wedge f^{r_m}$ to get

$$\sum_{i \in I} a_{i_1 \ldots i_k} f^{i_1} \wedge f^{i_2} \wedge \cdots \wedge f^{i_k} \wedge f^{r_{k+1}} \wedge \cdots \wedge f^{r_m} = 0. \tag{7.2}$$

The only term in the above sum without a repeated f in the wedge is the one with $i_1 = s_1, \ldots, i_k = s_k$, and so it is the only nonzero term. Because $s_1, \ldots, s_k, r_{k+1}, \ldots, r_m$ is just a permutation of the integers $1, \ldots, m$, $f^{s_1} \wedge f^{s_2} \wedge \cdots \wedge f^{s_k} \wedge f^{r_{k+1}} \wedge \cdots \wedge f^{r_m} = \pm f^1 \wedge \cdots \wedge f^m$. Thus applying the sum in (7.1) to $e_1, \ldots, e_m$ gives $\pm a_{s_1 \ldots s_k} = 0$. Thus the $f^i, i \in I$, are independent.

In particular, the dimension of $\mathbb{V}^m$ is 1, and the space has as a basis the single element $f^1 \wedge \cdots \wedge f^m$.

Lemma 7.1.3. *Let $g^1, \ldots, g^r \in \mathbb{V}^*$. Then $g^1, \cdots, g^r$ are linearly independent if and only if $g^1 \wedge \cdots \wedge g^r \neq 0$.*

Proof. If the gs are dependent, then one of them is a linear combination of the others, say $g^r = \sum_{s=1}^{r-1} \alpha_s g^s$. Then $g^1 \wedge \cdots \wedge g^r = \sum_{s=1}^{r-1} \alpha_s g^1 \wedge \cdots \wedge g^{r-1} \wedge g^s$. Each term in this last sum is a wedge product with a repeated entry, and so by the alternating property, each term is zero. Therefore $g^1 \wedge \cdots \wedge g^r = 0$.

Conversely, if $g^1, \ldots, g^r$ are linearly independent, then extend them to a basis $g^1, \ldots, g^r, \ldots, g^m$. By Lemma 7.1.2, $g^1 \wedge \cdots \wedge g^r \wedge \cdots \wedge g^m \neq 0$, so $g^1 \wedge \cdots \wedge g^r \neq 0$.

A linear map $L : \mathbb{V} \longrightarrow \mathbb{V}$ induces a linear map $L_k : \mathbb{A}^k \longrightarrow \mathbb{A}^k$ by the formula $L_k \phi(a_1, \ldots, a_k) = \phi(La_1, \ldots, La_k)$. If M is another linear map of $\mathbb{V}$ onto itself, then $(LM)_k = M_k L_k$, because $(LM)_k \phi(a_1, \ldots, a_k) = \phi(LMa_1, \ldots, LMa_k) = L_k \phi(Ma_1, \ldots, Ma_k) = M_k L_k \phi(a_1, \ldots, a_k)$. Recall that $\mathbb{A}^1 = \mathbb{V}^*$ is the dual space, and $L_1 = L^*$ is called the dual map.

If $\mathbb{V} = \mathbb{R}^m$ (column vectors), then we can identify the dual space $\mathbb{V}^* = \mathbb{A}^1$ with $\mathbb{R}^m$ by the convention $f \longleftrightarrow \hat{f}$, where $f \in \mathbb{V}^*$, $\hat{f} \in \mathbb{R}^m$, and $f(x) = \hat{f}^T x$. In this case, L is an $m \times m$ matrix, and Lx is the usual matrix product. $L_1 f$ is defined by $L_1 f(x) = f(Lx) = \hat{f}^T Lx = (L^T \hat{f})^T x$; so, the matrix representation of L_1 is the transpose of L, i.e., $L_1(f) = L^T \hat{f}$. The matrix representation of L_k is discussed in Flanders (1963).

By Lemma 7.1.2, $\dim \mathbb{A}^m = 1$, and so every element in $\mathbb{A}^m$ is a scalar multiple of a single element. L_m is a linear map; so, there is a constant ℓ such that $L_m f = \ell f$ for all $f \in \mathbb{A}^m$. Define the *determinant* of L to be this constant ℓ, and denote it by $\det(L)$; so, $L_m f = \det(L)f$ for all $f \in \mathbb{A}^m$.

Lemma 7.1.4. *Let L and $M : \mathbb{V} \longrightarrow \mathbb{V}$ be linear. Then*

1. $\det(LM) = \det(L)\det(M)$.
2. $\det(I) = 1$, *where $I : \mathbb{V} \longrightarrow \mathbb{V}$ is the identity map.*
3. *L is invertible if and only if $\det(L) \neq 0$, and, if L is invertible, $\det(L^{-1}) = \det(L)^{-1}$.*

Proof. Part (1) follows from $(LM)_m = M_m L_m$ which was established above. (2) follows from the definition. Let L be invertible; so, $LL^{-1} = I$, and by (1) and (2), $\det(L)\det(L^{-1}) = 1$; so, $\det(L) \neq 0$ and $\det(L^{-1}) = 1/\det(L)$. Conversely assume L is not invertible so there is an $e \in \mathbb{V}$ with $e \neq 0$ and $Le = 0$. Extend e to a basis, $e_1 = e, e_2, \ldots, e_m$. Then for any m-form ϕ, $L_m \phi(e_1, \ldots, e_m) = \phi(Le_1, \ldots, Le_m) = \phi(0, Le_2, \ldots, Le_m) = 0$. So $\det(L) = 0$.

Let $\mathbb{V} = \mathbb{R}^m$, $e_1, e_2, \ldots, e_m$ be the standard basis of $\mathbb{R}^m$, and let L be the matrix $L = (L_i^j)$; so, $Le_i = \sum_j L_i^j e_j$. Let ϕ be a nonzero element of $\mathbb{A}^m$.

$$\det(L)\phi(e_1, \ldots, e_m) = L_m \phi(e_1, \ldots, e_m) = \phi(Le_1, \ldots, Le_m)$$

$$= \sum_{j_1} \cdots \sum_{j_m} \phi(L_1^{j_1} e_{j_1}, \ldots, L_m^{j_m} e_{j_m})$$

$$= \sum_{j_1} \cdots \sum_{j_m} L_1^{j_1} \cdots L_m^{j_m} \phi(e_{j_1}, \ldots, e_{j_m})$$

$$= \sum_{\sigma \in P} \mathrm{sign}(\sigma) L_1^{\sigma(1)} \cdots L_m^{\sigma(m)} \phi(e_1, \ldots, e_m).$$

In the second to last sum above the only nonzero terms are the ones with distinct e's. Thus the sum over the nonzero terms is the sum over all permutations of the e's. From the above,

$$\det(L) = \sum_{\sigma \in P} \text{sign}(\sigma) L_1^{\sigma(1)} \cdots L_m^{\sigma(m)},$$

which is one of the classical formulas for the determinant of a matrix.

7.1.1 Linear Symplectic Form

Now let $(\mathbb{V}, \omega)$ be a symplectic space of dimension 2n. Recall that in Chapter 2 a symplectic form ω (on a vector space $\mathbb{V}$) was defined to be a nondegenerate, alternating bilinear form on $\mathbb{V}$, and the pair $(\mathbb{V}, \omega)$ was called a symplectic space.

Theorem 7.1.1. *There exists a basis* $f^1, \ldots, f^{2n}$ *for* $\mathbb{V}^*$ *such that*

$$\omega = \sum_{i=1}^{n} f^i \wedge f^{n+i}. \tag{7.3}$$

Proof. By Corollary 2.2.1, there is a symplectic basis $e_1, \ldots, e_{2n}$ so that the matrix of the form ω is the standard $J = (J_{ij})$ or $J_{ij} = \omega(e_i, e_j)$. Let $f^1, \ldots, f^{2n} \in \mathbb{V}^*$ be the basis dual to the symplectic basis $e_1, \ldots, e_{2n}$. The 2-form given on the right in (7.3) above agrees with ω on the basis $e_1, \ldots, e_{2n}$.

The basis $f^1, \ldots, f^{2n}$ is a symplectic basis for the dual space $\mathbb{V}^*$. By the above, $\omega^n = \omega \wedge \omega \wedge \cdots \wedge \omega$ (n times $) = \pm n! f^1 \wedge f^2 \wedge \cdots \wedge f^{2n}$, where the sign is plus if n is even and minus if n is odd. Thus ω^n is a nonzero element of $\mathbb{A}^{2n}$. Because a symplectic linear transformation preserves ω, it preserves ω^n, and therefore, its determinant is $+1$. (This is another of our four proofs of this fact.)

Corollary 7.1.1. *The determinant of a symplectic linear transformation (or matrix) is* $+1$.

Actually, using the above arguments and the full statement of Theorem 2.2.1, we can prove that a 2-form ν on a linear space of dimension $2n$ is nondegenerate if and only if ν^n is nonzero.

7.2 Tangent and Cotangent Spaces

Let M be a smooth differential manifold of dimension m and $(\mathcal{Q}, \phi)$ a chart, so $\phi : \mathcal{Q} \to \mathcal{O} \subset \mathbb{V}$ where $\mathcal{O}$ be an open set in a real m-dimensional vector space $\mathbb{V}$. First we work in a chart, i.e., in $\mathcal{O}$.

Take a basis $e_1, \ldots, e_m$ for $\mathbb{V}$, and $f^1, \ldots, f^m$ the dual basis. Let $x = (x^1, \ldots, x^m)$ be coordinates in $\mathbb{V}$ relative to $e_1, \ldots, e_m$ and also coordinates in V^* relative to the dual basis. Let $\mathbb{I} = (-1, 1) \subset \mathbb{R}^1$, and let t be a coordinate in $\mathbb{R}^1$. Think of $\mathbb{V}$ as $\mathbb{R}^m$. (We use the more general notation because it is helpful to keep a space and its dual distinct.) $\mathbb{R}^m$ and its dual are often identified with each other which can lead to confusion.

Much of analysis reduces to studying maps from an interval in $\mathbb{R}^1$ into $\mathcal{O}$ (curves, solutions of differential equations, etc.) and the study of maps from $\mathcal{O}$ into $\mathbb{R}^1$ (differentials of functions, potentials, etc.). The linear analysis of these two types of maps is, therefore, fundamental. The linearization of a curve at a point gives rise to a tangent vector, and the linearization of a function at a point gives rise to a cotangent vector. These are the concepts of this section.

A tangent vector at $p \in \mathcal{O}$ is to be considered as the tangent vector to a curve through p. Let $g, g^\dagger : \mathbb{I} \longrightarrow \mathcal{O} \subset \mathbb{V}$ be smooth curves with $g(0) = g^\dagger(0) = p$. We say g and $g^\dagger$ are equivalent at p if $Dg(0) = Dg^\dagger(0)$. Because $Dg(0) \in \mathcal{L}(\mathbb{R}, \mathbb{V})$, we can identify $\mathcal{L}(\mathbb{R}, \mathbb{V})$ with $\mathbb{V}$ by letting $Dg(0)(1) = dg(0)/dt \in \mathbb{V}$. Being equivalent at p is an equivalence relation on curves, and an equivalence class (a maximal set of curves equivalent to each other) is defined to be a tangent vector or a vector to $\mathcal{O}$ at p. That is, a tangent vector, $\{g\}$, is the set of all curves equivalent to g at p, i.e., $\{g\} = \{g^\dagger : \mathbb{I} \longrightarrow \mathcal{O} : g^\dagger(0) = p$ and $dg(0)/dt = dg^\dagger(0)/dt\}$. In the x coordinates, the derivative is $dg(0)/dt = (dg^1(0)/dt, \ldots, dg^m(0)/dt) = (\gamma^1, \ldots, \gamma^m)$; so, $(\gamma^1, \ldots, \gamma^m)$ are coordinates for the tangent vector $\{g\}$ relative to the x coordinates. The set of all tangent vectors to $\mathcal{O}$ at p is called the tangent space to $\mathcal{O}$ at p and is denoted by $T_p\mathcal{O}$. This space can be made into a vector space by using the coordinate representation given above. The curve $\xi_i : t \longrightarrow p + te_i$ has $d\xi_i(0)/dt = e_i$ which is $(0, \ldots, 0, 1, 0, \ldots, 0)$ (1 in the ith position) in the x coordinates. The tangent vector consisting of all curves equivalent to ξ_i at p is denoted by $\partial/\partial x^i$. The vectors $\partial/\partial x^1, \ldots, \partial/\partial x^m$ form a basis for $T_p\mathcal{O}$. A typical vector $v_p \in T_p\mathcal{O}$ can be written $v_p = \gamma^1 \partial/\partial x_1 + \cdots + \gamma^m \partial/\partial x_m$. In classical tensor notation, one writes $v_p = \gamma^i \partial/\partial x_i$; it was understood that a repeated index, one as a superscript and one as a subscript, was to be summed from 1 to m. This was called the *Einstein convention* or the *summation convention*.

A cotangent vector (or *covector* for short) at p is to be considered as the differential of a function at p. Let $h, h^\dagger : \mathcal{O} \longrightarrow \mathbb{R}^1$ be two smooth functions. We say h and $h^\dagger$ are equivalent at p if $Dh(p) = Dh^\dagger(p)$. ($Dh(p)$ is the same as the differential $dh(p)$.) This is an equivalence relation. A cotangent vector or a covector to $\mathcal{O}$ at p is by definition an equivalence class of functions. That is, a covector $\{h\}$ is the set of functions equivalent to h at p, i.e., $\{h\} = \{h^\dagger : \mathcal{O} \longrightarrow \mathbb{R}^1 : Dh^\dagger(p) = Dh(p)\}$. In the x coordinate, $Dh(p) = (\partial h(p)/\partial x^1, \ldots, \partial h(p)/\partial x^m) = (\eta_1, \ldots, \eta_m)$; so, $(\eta_1, \ldots, \eta_m)$ are coordinates for the covector $\{h\}$. The set of all covectors at p is called the *cotangent* *space* to $\mathcal{O}$ at p and is denoted by $T_p^*\mathcal{O}$. This space can be made into a vector space by using the coordinate representation given above. The function $x^i : \mathcal{O} \longrightarrow \mathbb{R}^1$ defines a cotangent vector at p, which is $(0,\ldots,1,\ldots 0)$ (1 in the ith position). The covector consisting of all functions equivalent to x^i at p is denoted by dx^i. The covectors $dx^1,\ldots,dx^m$ form a basis for $T_p^*\mathcal{O}$. A typical covector $v^p \in T_p^*\mathcal{O}$ can be written $\eta_1 dx^1 + \cdots + \eta_m dx^m$ or $\eta_i dx^i$ using the Einstein convention.

In the above two paragraphs there is clearly a parallel construction being carried out. If fact they are dual constructions. Let g and h be as above; so, $h \circ g : I \subset \mathbb{R}^1 \longrightarrow \mathbb{R}^1$. By the chain rule, $D(h \circ g)(0)(1) = Dh(p) \circ Dg(0)(1)$ which is a real number; so, $Dh(p)$ is a linear functional on tangents to curves. In coordinates, if

$$\{g\} = v_p = \frac{dg^1}{dt}(0)\frac{\partial}{\partial x_1} + \cdots + \frac{dg^m}{dt}(0)\frac{\partial}{\partial x_m} = \gamma^1 \frac{\partial}{\partial x_1} + \cdots + \gamma^m \frac{\partial}{\partial x_m}$$

and

$$\{h\} = v^p = \frac{\partial h}{\partial x_1}(p)dx^1 + \cdots + \frac{\partial h}{\partial x_m}(p)dx^m = \eta_1 dx^1 + \cdots + \eta_m dx^m,$$

then

$$v^p(v_p) = D(h \circ g)(0)(1)$$

$$= \frac{dg^1}{dt}(0)\frac{\partial h}{\partial x_1}(p) + \cdots + \frac{dg^m}{dt}(0)\frac{\partial h}{\partial x_m}(p)$$

$$= \gamma^1 \eta_1 + \cdots + \gamma^m \eta_m$$

$$= \gamma^i \eta_i \qquad \text{(Einstein convention)}.$$

Thus $T_p\mathcal{O}$ and $T_p^*\mathcal{O}$ are dual spaces.

At several points in the above discussion the coordinates $x^1,\ldots,x^m$ were used. The natural question to ask is to what extent do these definitions depend on the choice of coordinates. Let $y^1,\ldots,y^m$ be another coordinate system that may not be linearly related to the x's. Assume that we can change coordinates by $y = \phi(x)$ and back by $x = \psi(y)$, where ϕ and ψ are smooth functions with nonvanishing Jacobians, $D\phi$ and $D\psi$. In classical notation, one writes $x^i = x^i(y)$, $y^j = y^j(x)$, and $D\phi = \{\partial y^j/\partial x^i\}, D\psi = \{\partial x^i/\partial y^j\}$.

Let $g : \mathbb{I} \longrightarrow \mathcal{O}$ be a curve. In x coordinates let $g(t) = (a^1(t),\ldots,a^m(t))$ and in y coordinates let $g(t) = (b^1(t),\ldots,b^m(t))$. The x coordinate for the tangent vector $v_p = \{g\}$ is $\mathbf{a} = (da^1(0)/dt,\ldots,da^m(0)/dt) = (\alpha^1,\ldots,\alpha^m)$, and the y coordinate for $v_p = \{g\}$ is $\mathbf{b} = (db^1(0)/dt,\ldots,db^m(0)/dt) = (\beta^1,\ldots,\beta^m)$. Recall that we write vectors in the text as row vectors, but they are to be considered as column vectors. Thus $\mathbf{a}$ and $\mathbf{b}$ are column vectors. By the change of variables, $a(t) = \psi(b(t))$; so, differentiating gives $\mathbf{a} = D\psi(p)\mathbf{b}$. In classical notation $a^i(t) = x^i(b(t))$; so, $da^i/dt = \sum_i(\partial x^i/\partial y^j)db^j/dt$ or

$$\alpha^i = \sum_{j=1}^{m} \frac{\partial x^i}{\partial y^j}\beta^j \quad (= \frac{\partial x^i}{\partial y^j}\beta^j \text{ Einstein convention}). \tag{7.4}$$

This formula tells how the coordinates of a tangent vector are transformed. In classical tensor jargon, this is the transformation rule for a contravariant vector.

Let $h : \mathcal{O} \longrightarrow \mathbb{R}^1$ be a smooth function. Let h be $a(x)$ in x coordinates and $b(y)$ in y coordinates. The cotangent vector $v^p = \{h\}$ in x coordinates is $\mathbf{a} = (\partial a(p)/\partial x^1, \dots, \partial a(p)/\partial x^m) = (\alpha_1, \dots, \alpha_m)$ and in y coordinates it is $\mathbf{b} = (\partial b(p)/\partial y^1, \dots, \partial b(p)/\partial y^m) = (\beta_1, \dots, \beta_m)$. By the change of variables $a(x) = b(\phi(x))$; so, differentiating gives $\mathbf{a} = D\phi(p)^T\mathbf{b}$. In classical notation $a(x) = b(y(x))$; so, $\alpha_i = \partial a/\partial x^i = \sum_j (\partial b/\partial y^j)(\partial y^j/\partial x^i) = \sum_j \beta_j(\partial y^j/\partial x^i)$ or

$$\alpha_i = \sum_{j=1}^{m} \frac{\partial y^j}{\partial x^i}\beta_j \quad (= \frac{\partial y^j}{\partial x^i}\beta_j \text{ Einstein convention}). \tag{7.5}$$

This formula tells how the coordinates of a cotangent vector are transformed. In classical tensor jargon this is the transformation rule for a covariant vector.

The *tangent bundle* of M is defined as $TM = \cup_{p \in M} T_p M$ and the *cotangent bundle* is $T^*M = \cup_{p \in M} T_p^* M$. Using the coordinate systems given above it is clear the these spaces are differential manifolds of dimension $2m$.

Careful! the tangent bundle is not in general a product. If $TS^2 = S^2 \times R^2$, then $S^2 \times (1,0) \subset S^2 \times R^2$ would be a smooth choice of a nonzero vector at each point of the sphere. So we could define 'east' at each point of the surface of the earth which geographers know is impossible.

7.3 Vector Fields and Differential Forms

Continue the notation of the last section. A tangent vector field on M is a smooth map

$$V : M \to TM : p \to v_p \in T_p M,$$

and a cotangent field is a smooth map

$$U : M \to T^*M : p \to u_p \in T_p^* M.$$

In $\mathcal{O}$ it is a smooth choice of a tangent (cotangent) vector at each point of $\mathcal{O}$.

In coordinates, a tangent vector field, V, can be written in the form

$$V = V(x) = \sum_{i=1}^{m} v^i(x)\frac{\partial}{\partial x^i} \quad (= v^i(x)\frac{\partial}{\partial x^i}), \tag{7.6}$$

where the $v^i : \mathcal{O} \longrightarrow \mathbb{R}^1$, $i = 1, \dots, m$, are smooth functions, and a cotangent vector field U can be written in the form

$$U = U(x) = \sum_{i=1}^{m} u_i(x)dx^i \quad (= u_i(x)dx^i), \tag{7.7}$$

where $u_i : \mathcal{O} \longrightarrow \mathbb{R}^1, i = 1, \ldots, m$, are smooth functions.

A tangent vector field V gives a tangent vector $V(p) \in T_p\mathcal{O}$ which was defined as the tangent vector of some curve. A different curve might be used for each point of $\mathcal{O}$; so, a natural question to ask is whether there exist a curve $g : \mathbb{I} \subset \mathbb{R} \longrightarrow \mathcal{O}$ such that $dg(t)/dt = V(g(t))$. In coordinates this is

$$\frac{dg^i(t)}{dt} = v^i(g(t)).$$

This is the same as asking for a solution of the differential equation $\dot{x} = V(x)$. Thus a tangent vector field is an ordinary differential equation on a manifold. Of course, the standard existence theorem for ordinary differential equations gives such a curve. In classical tensor jargon a tangent vector field is called a *contravariant vector field*.

A vector field acts on a function giving the directional derivative of the function. Let $f : M \to \mathbb{R}^1$ and V the vector field given above then

$$Vf = \sum_{i=1}^{m} v^i \frac{\partial f}{\partial x^i} = v^i \frac{\partial f}{\partial x^i}$$

which is again a smooth function from M to $\mathbb{R}^1$. Let $W = \sum w_j \partial/\partial x^j$ be another smooth vector field. VW does not behave well under changes of variables due to the existence of mixed partials. However $VW - WV$ does, so we define the *Lie bracket* as the vector field

$$[V, W] = VW - WV = \sum_{j=1}^{m} \left(v^i \frac{\partial u^j}{\partial x^i} - u^i \frac{\partial v^j}{\partial x^i} \right) \frac{\partial}{\partial x^j}.$$

It is left as an exercise to verify that $[V, W]$ transforms like a vector field. Clearly the Lie bracket is linear in each argument and skew-symmetric. A tedious computation yields the Jacobi identity

$$[V, [W, Z]] + [W, [Z, V]] + [Z, [V, W]] = 0. \tag{7.8}$$

The above shows that $\mathfrak{X}(M)$ the set of all smooth vector fields on M equipped with this Lie bracket is a Lie algebra. (See the definition given in Problem 2 of Chapter 1.)

A cotangent vector field U gives a cotangent vector $U(p) \in T_p^*\mathcal{O}$ which was defined as the differential of a function at p. A different function might be used for each point of $\mathcal{O}$; so, a natural question to ask is whether there exists a function $h : \mathcal{O} \longrightarrow \mathbb{R}^1$ such that $dh(x) = U(x)$. The answer to this question is no in general. Certain integrability conditions discussed below must be satisfied before a cotangent vector field is a differential of a function. If this cotangent vector field is a field of work elements, i.e., a field of forces, then if $dh = -U$, the h would be a potential and the force field would be conservative. But, as we show, not all forces are conservative.

Let $p \in \mathcal{O}$, and denote by $\mathbb{A}_p^k \mathcal{O}$ the linear space of k-forms on the tangent space $T_p \mathcal{O}$. A k-differential form or k-form on $\mathcal{O}$ is a smooth choice of a k-linear form in $\mathbb{A}_p^k \mathcal{O}$ for all $p \in \mathcal{O}$. That is, a k-form, F, can be written

$$F = \sum_{1 \leq i_1 < \cdots < i_k \leq m} f_{i_1 i_2 \ldots i_k}(x_1, x_2, \ldots, x_m) dx^{i_1} \wedge \cdots \wedge dx^{i_k}$$

$$= \sum_{i \in I} f_i(x) dx^i,$$

(7.9)

where the functions $f_{i_1 \ldots i_k} : \mathcal{O} \longrightarrow \mathbb{R}$ are smooth. In the last expression in (7.9), I denotes the set $\{(i_1, \ldots, i_k) : i_j \in \mathbb{Z}, 1 \leq i_1 < \cdots < i_k \leq m\}$, and $dx^i = dx^{i_1} \wedge \cdots \wedge dx^{i_k}$. Because $\mathbb{A}_p^0 \mathcal{O} = \mathbb{R}$, 0-forms are simply smooth functions, and because $\mathbb{A}_p^1 \mathcal{O} = T_p^* \mathcal{O}$, 1-forms are covector fields.

The global space of k forms is $\mathbb{A}^k M = \cup_{p \in M} \mathbb{A}_p^k M$. A k differential form on M is a smooth map

$$F : M \to \mathbb{A}^k M : p \to f_p \in \mathbb{A}_p^k M.$$

In classical analysis, everything was a vector. In $\mathbb{R}^3$, 1-forms are often identified with (or confused with) vector fields. For example, the differential of a function, $df = f_x dx + f_y dy + f_z dz$, is treated as a vector field by writing $\nabla f = \text{grad } f = f_x \mathbf{i} + f_y \mathbf{j} + f_z \mathbf{k}$. That is why one calls a force a vector and not a covector even when it is the gradient of a potential function.

Also, because the dimension of the space of 2-linear forms in a 3 dimensional space is

$$\binom{3}{2} = 3$$

classically 2-forms in $\mathbb{R}^3$ were identified with (or confused with) vector fields. Usually one identifies $a(\mathbf{j} \wedge \mathbf{k}) + b(\mathbf{k} \wedge \mathbf{i}) + c(\mathbf{i} \wedge \mathbf{j})$ with the vector $a\mathbf{i} + b\mathbf{j} + c\mathbf{k}$. Think about the cross product of vectors. This is why angular momentum and magnetic fields are sometimes misrepresented as vectors.

Given a 0-form F, (i.e., a function) dF is a 1-form. The natural generalization is the exterior derivative operator d which converts a k-form F as given in (7.9) into a $(k+1)$-form dF by the formula

$$dF = \sum_{j=1}^{m} \sum_{1 \leq i_1 < \cdots < i_k \leq m} \frac{\partial f_{i_1 \cdots i_k}}{\partial x^j} dx^j \wedge dx^{i_1} \wedge \cdots \wedge dx^{i_k}$$

(7.10)

$$= \sum_{i \in I} df_i \wedge dx^i.$$

Lemma 7.3.1. *Let F and G be smooth forms defined on an open set $\mathcal{O}$. Then*

1. $d(F + G) = dF + dG$.
2. $d(F \wedge G) = dF \wedge G + (-1)^{\deg(F)} F \wedge dG$.

3. $d(dF) = 0$ *for all F.*
4. *If F is a function, then dF agrees with the standard definition of the differential of F,*
5. *The operator d is uniquely defined by the properties given above.*

Proof. Part (4) is obvious, and parts (1), (2), and (5) are left as exercises. Only part (3) is proved here. Let i be a multiple index, and so the summations on i range over I. Let $F = \sum_i f_i dx^i$. Then

$$d(dF) = \sum_i \sum_{j=1}^{m} \sum_{k=1}^{m} \left(\frac{\partial^2 f_i}{\partial x_j \partial x_k} \right) dx^k \wedge dx^j \wedge dx^i$$

$$= \sum_i \sum_{1 \le j < k \le m} \left(\frac{\partial^2 f_i}{\partial x_j \partial x_k} - \frac{\partial^2 f_i}{\partial x_k \partial x_j} \right) dx^k \wedge dx^j \wedge dx^i$$

$$= 0.$$

The last sum is zero by the equality of mixed partial derivatives.

Remark: The first four parts of this lemma can be used as a coordinate-free definition of the operator d. Formula (7.10) shows its existence, and part (v) shows its uniqueness.

Let (x, y, z) be the standard coordinates in $\mathbb{R}^3$ and $\mathbf{i}, \mathbf{j}, \mathbf{k}$ the usual unit vectors. If $F(x, y, z)$ is a function, then

$$dF = \frac{\partial F}{\partial x} dx + \frac{\partial F}{\partial y} dy + \frac{\partial F}{\partial z} dz$$

is the usual differential. The classical approach is to make the differential a vector field by defining

$$\nabla F = \operatorname{grad} F = \frac{\partial F}{\partial x} \mathbf{i} + \frac{\partial F}{\partial y} \mathbf{j} + \frac{\partial F}{\partial z} \mathbf{k}.$$

Next consider a 1-form $F = a(x, y, z)dx + b(x, y, z)dy + c(x, y, z)dz$; then

$$dF = \left(\frac{\partial c}{\partial y} - \frac{\partial b}{\partial z} \right) dy \wedge dz + \left(\frac{\partial a}{\partial z} - \frac{\partial c}{\partial x} \right) dz \wedge dx + \left(\frac{\partial b}{\partial x} - \frac{\partial a}{\partial y} \right) dx \wedge dy.$$

The classical approach is to make this F a vector field $F = a\mathbf{i} + b\mathbf{j} + c\mathbf{k}$ and to define a new vector field by

$$\nabla \times F = \operatorname{curl} F = \left(\frac{\partial c}{\partial y} - \frac{\partial b}{\partial z} \right) \mathbf{i} + \left(\frac{\partial a}{\partial z} - \frac{\partial c}{\partial x} \right) \mathbf{j} + \left(\frac{\partial b}{\partial x} - \frac{\partial a}{\partial y} \right) \mathbf{k}.$$

Now let F be a 2-form so $F = a(x, y, z)dy \wedge dz + b(x, y, z)dz \wedge dx + c(x, y, z)dx \wedge dy$ and

$$dF = \left(\frac{\partial a}{\partial x} + \frac{\partial b}{\partial y} + \frac{\partial c}{\partial z} \right) dx \wedge dy \wedge dz.$$

The classical approach would have considered F as a vector field $F = a\mathbf{i} + b\mathbf{j} + c\mathbf{k}$ and defined a scalar function

$$\nabla \cdot F = \operatorname{div} F = \left(\frac{\partial a}{\partial x} + \frac{\partial b}{\partial y} + \frac{\partial c}{\partial z} \right).$$

7.3.1 Poincaré's Lemma

The statement that $d(dF) = 0$, or $d^2 = 0$, contains the two classical statements $\operatorname{curl}(\operatorname{grad} F) = 0$ and $\operatorname{div}(\operatorname{curl} F) = 0$.

A k-form F is *closed* if $dF = 0$. A k-form F is *exact* if there is a $(k-1)$-form G such that $F = dG$. Part (3) of Lemma 7.3.1 says that an exact form is closed. A partial converse is also true.

Theorem 7.3.1 (Poincaré's lemma). *Let $\mathcal{O}$ be a ball in $\mathbb{R}^m$ and F a k-form such that $dF = 0$. Then there is a $(k-1)$-form g on $\mathcal{O}$ such that $F = dg$.*

Remark: This is a partial converse to $d(dg) = 0$. Note that the domain of definition, $\mathcal{O}$, of the form F is required to be a ball. The theorem says that in a ball, a closed form is exact. The 1-form, $F = (x\,dy - y\,dx)/(x^2 + y^2)$, satisfies $dF = 0$, but there does not exist a function, g, defined on all of $\mathbb{R}^2 \setminus (0,0)$ such that $dg = F$. The form F is the differential of the polar angle $\theta = \arctan(y/x)$ that is not a single-valued function defined on all of $\mathbb{R}^2 \setminus (0,0)$. However, it can be made single valued in a neighborhood of any point in $\mathbb{R}^2 \setminus (0,0)$, e.g., for any point not on the negative x-axis, one can take $-\pi < \theta < \pi$, and for points on the negative x-axis, one can take $0 < \theta < 2\pi$. Because F locally defines a function we have $dF = 0$.

Poincaré's lemma contains the classical theorems: (i) if F is a vector field defined on a ball in $\mathbb{R}^3$ with $\operatorname{curl} F = 0$, then there is a smooth function g such that $F = \operatorname{grad}(g)$, and (ii) if F is a smooth vector field defined on a ball such that $\operatorname{div} F = 0$, then there is a smooth vector field g such that $F = \operatorname{curl} g$.

Proof. The full statement of the Poincaré lemma is not needed here; only the case when $k = 1$ is used in subsequent chapters. Therefore, only that case is proved here. The proof of the full theorem can be found in Flanders (1963), or Spivak (1965) or Abraham and Marsden (1978).

Let $F = \sum_i f_i(x)dx^i$ be a given 1-form.

$$dF = \sum_i \sum_j \left(\frac{\partial f_i}{\partial x^j} \right) dx^j \wedge dx^i = \sum_{i<j} \left(\frac{\partial f_i}{\partial x^j} - \frac{\partial f_j}{\partial x^i} \right) dx^j \wedge dx^i.$$

So $dF = 0$ if and only if $\partial f_i / \partial x^j = \partial f_j / \partial x^i$. Define

$$g(x) = \int_0^1 \sum_i f_i(tx)x^i \, dt.$$

So

$$\frac{\partial g(x)}{\partial x^j} = \int_0^1 \left\{ \sum_i \frac{\partial f_i(tx)}{\partial x^j} tx^i + f_j(tx) \right\} dt$$

$$= \int_0^1 \left\{ \frac{t d f_j(tx)}{dt} + f_j(tx) \right\} dt$$

$$= \int_0^1 \frac{d}{dt} \{ t f_j(tx) \} \, dt$$

$$= t f_j(tx) \Big|_0^1 = f_j(x).$$

Thus $dg = F$.

Note that the function g defined in the proof given above is a line integral and the condition $dF = 0$ is the condition that a line integral be independent of the path.

Corollary 7.3.1. *Let $F = (F^1, \ldots, F^m)$ be a vector valued function defined in a ball $\mathcal{O}$ in $\mathbb{R}^m$. Then a necessary and sufficient condition for F to be the gradient of a function $g : \mathcal{O} \longrightarrow \mathbb{R}$ is that the Jacobian matrix $(\partial F^i / \partial x^j)$ be symmetric.*

Proof. First, to see that it is a corollary, consider F as the differential form $F = F^1 dx^1 + \cdots + F^m dx^m$. Then by the above,

$$dF = \sum_{i<j} \left(\frac{\partial F^i}{\partial x^j} - \frac{\partial F^j}{\partial x^i} \right) dx^i \wedge dx^j.$$

So $dF = 0$ if and only if the Jacobian $(\partial F^i / \partial x^j)$ is symmetric. The corollary follows from part 3 of Lemma 7.3.1 and Theorem 7.3.1.

7.3.2 Changing Variables

To change coordinates for vector fields or differential forms, simply transform the coordinates as was done in Section 7.2 using the Jacobian of the transformation. In particular, let x and y be coordinates on $\mathcal{O}$, and assume that the change of coordinates is given by $x = \psi(y)$ and the change back by $y = \phi(x)$, or in classical notation $x = x(y)$ and $y = y(x)$. Assume the Jacobians, $D\phi = (\partial y^j / \partial x^i)$ and $D\psi = (\partial x^i / \partial y^j)$, are nonsingular.

If a vector field V is given by

$$V = \sum_{i=1}^{m} \alpha^i(x) \frac{\partial}{\partial x^i} = \sum_{i=1}^{m} \beta^i(x) \frac{\partial}{\partial y^i},$$

and we set $\mathbf{a}(x) = (\alpha^1(x), \ldots, \alpha^m(x))$, $\mathbf{b}(y) = (\beta^1(y), \ldots, \beta^m(y))$, then

$$\mathbf{a} = D\psi(\mathbf{b}) \quad \text{or} \quad \alpha^i = \sum_{j=1}^{m} \frac{\partial x^i}{\partial y^j} \beta^j. \tag{7.11}$$

If a differential 1-form is given by

$$F = \sum_{i=1}^{m} \alpha_i(x) dx^i - \sum_{i=1}^{m} \beta_i(y) dy^i,$$

and we set $\mathbf{a}(x) = (\alpha_1(x), \ldots, \alpha_m(x))$, $\mathbf{b}(y) = (\beta_1(y), \ldots, \beta_m(y))$, then

$$\mathbf{a} = \mathbf{b} D\phi \quad \text{or} \quad \alpha_i = \sum_{j=1}^{m} \frac{\partial y^j}{\partial x^i} \beta_j. \tag{7.12}$$

If a differentiable 2-form F is given by

$$F = \sum_{i=1}^{m} \sum_{j=1}^{m} \alpha_{ij}(x) dx^i \wedge dx^j = \sum_{i=1}^{m} \sum_{j=1}^{m} \beta_{ij}(y) dy^i \wedge dy^j, \tag{7.13}$$

and we set $\mathbf{A} = (\alpha_{ij})$, $\mathbf{B} = (\beta_{ij})$ ($\mathbf{A}$ and $\mathbf{B}$ are skew-symmetric matrices), then

$$\mathbf{A} = D\psi^T \mathbf{B} D\psi \quad \text{or} \quad \alpha_{ij} = \sum_{s=1}^{m} \sum_{r=1}^{m} \frac{\partial y^s}{\partial x^i} \frac{\partial y^r}{\partial x^j} \beta_{sr}. \tag{7.14}$$

Let $\mathcal{O}$ be an open set in $\mathbb{R}^{2n}$. A 2-form F on $\mathcal{O}$ is *nondegenerate* if $F^n = F \wedge F \wedge \cdots \wedge F$ (n times) is nonzero. As we saw above, the coefficients in a coordinate system of a 2-form can be represented as a skew-symmetric matrix. As we saw in Section 7.1.1, a linear 2-form is nondegenerate if and only if the coefficient matrix is nonsingular. Thus the 2-form F in (7.13) is nondegenerate if and only if $\mathbf{A}$ (or $\mathbf{B}$) is nonsingular on all of $\mathcal{O}$.

7.4 Symplectic Manifold

A *symplectic manifold* is a pair (M, Ω) where M is a $2n$-dimensional differentiable manifold and Ω is a smooth closed nondegenerate 2-form on M. Ω is called the symplectic structure or symplectic form.

The standard example is $\mathbb{R}^{2n}$ with

$$\Omega = \sum_{i=1}^{n} dq^i \wedge dp^i = \sum_{i=1}^{n} dz^i \wedge dz^{n+i} = \frac{1}{2} J_{ij} dz^i dz^j. \tag{7.15}$$

where $z = (z^1, \ldots, z^{2n}) = (q^1, \ldots, q^n, p^1, \ldots, p^n)$ are coordinates in $\mathbb{R}^{2n}$. The coefficient matrix of Ω is just J.

In a natural way a cotangent bundle of a manifold N is a symplectic manifold. Let $q = (q^1, \ldots, q^n)$ be coordinates in a chart on N. Then $\sigma = p_1 dq^1 + \cdots + p_n dq^n$ is a one-form on N and $p = (p_1, \ldots, p_n)$ are coordinates in the tangent space to N at the point q. In other words (q, p) is a chart in T^*N. Then $\Omega = d\sigma = \sum_{i=1}^{n} dq^i \wedge dp_i$ is a globally defined 2-form on T^*N and it is closed since $d\Omega = dd\sigma = 0$. There is a slight problem with the subscript-superscript convention used here. In σ the p's are functions on M and so are subscripted, but in Ω they are coordinates in T^*M and so should be superscripted. Introducing extra notation to distinguish the two would only add more confusion.

Classically, if a mechanical system in $\mathbb{R}^{n+d}$ is subject to d constraints of the form $f_i(x_1, \ldots, x_{n+d}) = 0$, $i = 1, \ldots, d$ which defines an n dimensional submanifold N of $\mathbb{R}^{n+d}$ then the system is said to be subject to *holonomic constraints*. Of course, this system defines a Hamiltonian system on the cotangent bundle of N by the above. See Whittaker (1937) for this classic approach.

In general

$$\Omega = \frac{1}{2} \sum_{i=1}^{2n} \sum_{j=1}^{2n} \omega_{ij}(z) dz^i \wedge dz^j = \frac{1}{2} \omega_{ij} dz^i \wedge dz^j. \tag{7.16}$$

A symplectic structure creates a way to convert a one-form like dH to a vector field like $\dot{z} = J\nabla H$ which is at the heart of the matter. At each point $p \in M$ the symplectic form defines a nondegenerate bilinear map $\Omega_p : T^*M_p \times T^*M_p \to \mathbb{R}^1$, so the map $\flat : T_pM \to T_p^*M : v \mapsto v^\flat = \Omega_p(v, \cdot)$ is a well defined invertible linear map with inverse $\sharp : T_p^*M \to T_pM$.

Let $v \in T_pM$ so

$$v = \sum_{i=1}^{2n} v^i \frac{\partial}{\partial z^i} = v^i \frac{\partial}{\partial z^i}$$

then

$$v^\flat = \sum_{j=1}^{2n} \omega_{ij} v^i dx^j = \omega_{ij} v^i dx^j.$$

The form Ω is nondegenerate, so the coefficient matrix has a inverse; let it be ω^{ij} so that $\omega_{is}\omega^{sj} = \delta_i^j$ the Kronecker delta. Let $w = \sum w_i dx^i$ then

$$w^\sharp = \sum_{i=1}^{2n} \omega^{ij} w_i \frac{\partial}{\partial x^j} = \omega^{ij} w_i \frac{\partial}{\partial x^j}.$$

Thus, if the Hamiltonian is H, then the Hamiltonian vector field is $dH^\sharp$.

A *symplectic submanifold* is simply a submanifold such that Ω is nondegenerate on it and so it is a symplectic manifold in its own right.

A *Lagrangian submanifold* $L \subset M$ is a submanifold of dimension n such that $\Omega|L = 0$, that is, it is a submanifold of maximal dimension where Ω is zero. Or a Lagrangian submanifold L is a submanifold of M such that T_pL is a Lagrangian subspace of T_pM for all $p \in L$ in the sense of Section 2.2.

For example, each $T_pN \subset TN$ is Lagrangian. The zero section $\mathcal{Z} : N \to TN : p \mapsto 0_p$ takes N into a Lagrangian submanifold of TN. In coordinates (q, p) where $\Omega = \sum dq^i \wedge dp^i$ the sets where $q = 0$ or where $p = 0$ are Lagrangian submanifolds.

Recall from the discussion in Section 5.3 that if A is real Hamiltonian matrix all of whose eigenvalues have nonzero real parts, then there exists Lagrangian linear subspaces L, L^* such that $\mathbb{R}^{2n} = L \oplus L^*$ and all the solutions of $\dot{x} = Ax$ in L (respectively L^*) tend to 0 as $t \to +\infty$ (resp. $t \to -\infty$). These are the stable and unstable sets. The nonlinear version of this gives rise to the stable and unstable manifold of a hyperbolic critical point discussed in Section 6.5. They are Lagrangian submanifolds.

7.4.1 Darboux's Theorem

By Corollary 2.2.1, there is a linear change of coordinates so that the coefficient matrix of a nondegenerate 2-form is J at one point. Nature generalization is:

Theorem 7.4.1 (Darboux's theorem). *If F is a symplectic structure on an open ball in $\mathbb{R}^{2n}$, then there exists a coordinate system z such that F in this coordinate system is the standard symplectic structure Ω.*

Proof. See Abraham and Marsden (1978).

A coordinate system for which a symplectic structure is Ω is called a *symplectic coordinate system* (for this form). A *symplectic transformation*, ϕ, is one that preserves the form Ω or preserves the coefficient matrix J, i.e., $D\phi^T J D\phi = J$.

Darboux's Theorem shows that the definition of symplectic manifold given in the introduction to this chapter is the same as the definition given above in Section 7.4.

7.5 Lie Groups

A *Lie group* is a smooth manifold $\mathcal{G}$ with a group structure that is smooth, that is, the group multiplication

$$\mu : \mathcal{G} \times \mathcal{G} \to \mathcal{G} : (g, h) \mapsto \mu(g, h) = gh$$

is a C^∞ map. Let e denote the identity element of $\mathcal{G}$. A simple application of the implicit function theorem shows that the map $g \to g^{-1}$ is a smooth diffeomorphism of $\mathcal{G}$ (solve $\mu(g, h) = e$ for h).

We have already seen examples of Lie groups namely the matrix groups: $Gl(n, \mathbb{R})$ the general linear group, $So(n, \mathbb{R})$ the special orthogonal group, $Sp(2n, \mathbb{R})$ the symplectic group, etc. More mundane examples are $\mathbb{R}$ and $\mathbb{R}^n$ with the group operation just addition, $\mathbb{S}$ the unit circle in $\mathbb{C}$ with complex multiplication, $T^n = \mathbb{S} \times \cdots \times \mathbb{S}$ the torus and $\mathbb{S}^3$ the unit quaterions with quaterions multiplications. Except for $Gl(n, \mathbb{R})$ all these examples are connected, but $Gl(n, \mathbb{R})$ has two components one with $\det A > 0$ and one with $\det A < 0$.

The theory of Lie groups is vast. Here we just skim over the top. Some basic structure theorems are:

A closed subgroup of a Lie group is a Lie group.

A Lie group is isomorphic to a closed subgroup of $Gl(n, \mathbb{R})$.

An Abelian Lie group is isomorphic to $T^n \times \mathbb{R}^d$.

The group action transforms various structures all over the manifold $\mathcal{G}$. The left translation map $L_g : \mathcal{G} \to \mathcal{G} : h \mapsto gh$ is global diffeomorphism of $\mathcal{G}$ for each $g \in \mathcal{G}$.

In particular they carry the tangent space of the identity, T_0, to the tangent space at g, so the map

$$\mathcal{G} \times T_0\mathcal{G} \to T\mathcal{G} : (g, v) \mapsto DL_g(v)$$

is a global diffeomorphism, showing that the tangent bundle of $\mathcal{G}$ is a product, i.e., $T\mathcal{G} = \mathcal{G} \times T_0\mathcal{G}$.

Also to each $v \in T_0\mathcal{G}$ we can define a vector field X_v on $\mathcal{G}$ by $X_v(g) = DL_g(v)$, which is left invariant since $DL_g X = X$ for all $g \in \mathcal{G}$. Recall that $\mathfrak{X}(\mathcal{G})$ is the Lie algebra of all smooth vector fields on the manifold $\mathcal{G}$ equipped with the bracket $[\cdot, \cdot]$. Now the set $\mathfrak{X}_L(\mathcal{G})$ of all left invariant vector fields is a subalgebra of $\mathfrak{X}(\mathcal{G})$ since the bracket of two left invariant vector fields is left invariant. This subalgebra is called the Lie algebra of $\mathcal{G}$ and denoted by $\mathcal{A}$. One can consider this bracket as defined on the tangent space of the identity by

$$[\xi, \eta] = [X_\xi, X_\eta] \quad \text{for all} \ \ \xi, \eta \in T_0\mathcal{G}.$$

Either interpretation defines the algebra $\mathcal{A}$. When the group is one of the matrix groups already introduced its algebra is a subalgebra of $gl(n, \mathbb{R})$ with the bracket on matrices give by $[A, B] = AB - BA$ with the usual matrix product.

A Lie algebra is called *Abelian* if the bracket is trivial, i.e., $[X, Y] = 0$. If the group is Abelian then so is its algebra and conversely for connected groups.

If X_ξ is a left invariant vector field for $\xi \in \mathcal{A}$ then, since such a vector field defines an autonomous differential differential equation there is a unique solution curve $\gamma_\xi : \mathbb{R} \to \mathcal{G}$ starting at e when $t = 0$. By the autonomous nature of the equation

$$\gamma_\xi(s + t) = \gamma_\xi(s)\gamma_\xi(t),$$

or in other words γ_ξ is a one parameter subgroup of $\mathcal{G}$. One can show that the solution is defined for all t and indeed all one parameter subgroups arise in this manner.

The *exponential map* $\exp : \mathcal{A} \to \mathcal{G}$ is defined by

$$\exp(\xi) = \gamma_\xi(1),$$

and by the autonomous nature of the equation

$$\exp(s\xi) = \gamma_\xi(s),$$

and that exp is smoothly invertible. So exp defines a diffeomorphism of a neighborhood of 0 in $\mathcal{A}$ to a neighborhood of e in $\mathcal{G}$.

7.6 Group Actions

An *action* of a Lie group $\mathcal{G}$ on a manifold M is a smooth mapping $\Phi : \mathcal{G} \times M \to M$ such that
  (i) $\Phi(e, x) = x$ for all $x \in M$,
  (ii) $\Phi(g, \Phi(h, x)) = \Phi(gh, x)$ for all $g, h \in \mathcal{G}$ and all $x \in M$.
  Sometimes we just write $\Phi(g, x) = gx$. Let $\Phi_g : M \to M : x \mapsto \Phi(g, x)$, so that the two basic axioms become (i) $\Phi_e = id$ and (ii) $\Phi_{gh} = \Phi_g \circ \Phi_h$.

The standard examples are the action of a matrix group on $\mathbb{R}^m$ by matrix multiplication and the action of $\mathbb{R}$ on a manifold given as solutions of a differential equation. In Section 1.4, where two harmonic oscillators are discussed, the case when the two frequencies are the same gives a pictorial representation of the group S^1 acting on the sphere S^3.

The *orbit* of x is $\mathbb{O}(x) = \{\Phi_g(x) : g \in \mathcal{G}\} \subset M$. The *orbit space*, denoted by M/G, is the set of all orbits with the quotient topology. Sometimes the orbit space is nice as in the example of two harmonic oscillators with the same frequencies where all the orbits on the 3-sphere are circles and the orbit space is a 2-sphere. But sometimes the orbit space is not even Hausdorff as for example for the irrational flow on the torus discussed in Section 1.5.

The action is *free* if it has no fixed points, i.e., $\Phi_g(x) = x$ implies $g = e$. The action is called *proper* if the mapping

$$\tilde{\Phi} : \mathcal{G} \times M \to M \times M : (g, x) \mapsto (x, \Phi(g, x))$$

is proper, i.e., inverse images of compact sets are compact. If $\mathcal{G}$ is compact the action is proper.

Proposition 7.6.1. *If the action is free and proper then M/G is a smooth manifold and the projection $\pi : M \to M/G$ is a smooth submersion.*

See Proposition 4.2.23 of Abraham and Marsden (1978).

7.7 Moment Maps and Reduction

Now let M be a symplectic manifold and $\Phi : \mathcal{G} \times M \to M$ a *symplectic action*, i.e., $\Phi_g : M \to M$ is symplectic for all $g \in \mathcal{G}$.

Previously we defined a linear map which takes $\xi \in \mathcal{A}$ to a left invariant vector field $X_\xi \in \mathfrak{X}_L$ and since the action is symplectic the vector field is locally Hamiltonian.

Assume that to each such X_ξ there is a globally defined Hamiltonian H_ξ and the map is linear. That is, assume there is a linear map $S : \mathcal{A} \to \mathcal{F}(M) :$ $\xi \mapsto H_\xi$ where $\mathcal{F}(M)$ is the space of all smooth functions from M to $\mathbb{R}^1$. So $S(\xi)$ is function and $S(\xi)(z) \in \mathbb{R}^1$. By assumption $S(\xi)(z)$ is linear in ξ and so it is a linear functional on $\mathcal{A}$, so we define $\mathbf{S} : M \to \mathcal{A}^*$ by

$$< \mathbf{S}(z), \xi >= S(\xi)(z),$$

where $< \cdot, \cdot >$ is the linear pairing of a functional on a vector. The function $\mathbf{S}$ is called the *moment map*[1].

Translational Symmetry: Let $\mathcal{G} = \mathbb{R}^3$ with the group action being vector addition act on $M = \mathbb{R}^3 \times \mathbb{R}^3$ by the symplectic action $\Phi : (u, (g, p)) \mapsto (q + u, p)$. The algebra and its dual are $\mathcal{A} = \mathcal{A}^* = \mathbb{R}^3$ where the pairing is by the inner product. For $\xi \in \mathcal{A}$ the flow is $(q + t\xi, p)$ which is generated by the Hamiltonian $H_\xi = \xi^T p$. So $S : \xi \to \xi^T p$ and $< \mathbb{S}, \xi >= \xi^T p$ or $\mathbb{S}(q, p) = p$.

Rotational Symmetry: Let $\mathcal{A} = So(3, \mathbb{R})$ act on $M = \mathbb{R}^3 \times \mathbb{R}^3$ by the symplectic action $\Phi : (A, (q, p)) \mapsto (Aq, Ap)$. The algebra is $\mathcal{A} = so(3, \mathbb{R})$ the algebra of 3×3 skew-symmetric matrices.

We can identify $\xi \in \mathbb{R}^3$ and $R_\xi \in so(3, \mathbb{R})$ by

$$\xi = \begin{bmatrix} \xi_1 \\ \xi_2 \\ \xi_3 \end{bmatrix} \iff R_\xi = \begin{bmatrix} 0 & \xi_1 & -\xi_2 \\ -\xi_1 & 0 & \xi_3 \\ \xi_2 & -\xi_3 & 0 \end{bmatrix}.$$

A direct computation gives

$$\mathrm{trace} R_\xi^T R_\eta = 2\xi \cdot \eta, \quad [R_\xi, R_\eta] = R_{x \times y}.$$

The first identity shows that the standard inner product on $\mathbb{R}^3$ induces an invariant inner product on $so(3, \mathbb{R})$ and so the dual $so^*(3, \mathbb{R})$ can be identified with $so(3, \mathbb{R})$ using this inner product.

Let $\xi \in \mathcal{A}$ so the flow generated by ξ is $(e^{\xi t} q, e^{\xi t} p)$ which is generated by the Hamiltonian $-q^T \xi p$, so let $S : \xi \to -q^T \xi p$. The moment map is

$$\mathbf{S} : \mathbb{R}^3 \times \mathbb{R}^3 \to so^*(3, \mathbb{R}) : (q, p) \mapsto R_{q \times p}$$

where

---

[1]M has already been used too much, so we use $\mathbf{S}$ after J. M. Souriau.

$$R_{q \times p} = \begin{bmatrix} 0 & q_2 p_3 - q_3 p_2 & q_1 p_3 - q_3 p_1 \\ -q_2 p_3 + q_3 p_2 & 0 & q_1 p_2 - q_2 p_1 \\ -q_1 p_3 + q_3 p_1 & -q_1 p_2 + q_2 p_1 & 0 \end{bmatrix}.$$

Thus holding $R_{q \times p}$ gives the same result as holding $q \times p$ fixed in the classical case.

A Hamiltonian H admits $\mathcal{G}$ as a symmetry, or H is symmetric with respect to $\mathcal{G}$, if $H(\Phi(g, x)) = H(gx) = H(x)$ for all $g \in \mathcal{G}$, $x \in M$. This discussion leads to

Theorem 7.7.1 (Noether-Souriau). *Let the action $\Phi : \mathcal{G} \times M \to M$ be symplectic, the moment map $\mathbf{S} : M \to \mathcal{A}^*$ be defined and $H : M \to \mathbb{R}^1$ a smooth Hamiltonian which admits the action Φ as a symmetry. Then $\mathbf{S}$ is constant along the solutions of $dH^\sharp$.*

Proof. By the more earthy version of Noether's theorem, Theorem 3.4.1, S is an integral so then $\mathbf{S}$ is also.

Let $\eta \in \mathcal{A}^*$ then by the above $M_\eta = \mathbf{S}^{-1}(\eta)$ is an invariant set for the Hamiltonian flow defined by H and if η is a regular value then M_η is a smooth submanifold of M. Let $\mathcal{G}_\eta \subset \mathcal{G}$ be the subgroup that leaves M_η fixed. The reduced space is $\mathbf{B}_\eta = M_\eta / \mathcal{G}_\eta$.

Theorem 7.7.2 (Reduction Theorem). *Let $\mathcal{A}, M, \mathbf{S}, H$ be as in Theorem 7.7.1. If the action of $\mathcal{G}_\eta$ on M_η is free and proper then the reduced space $\mathbf{B}_\eta$ is a symplectic manifold and the flow of $H|M_\eta$ drops down in a natural way to a Hamiltonian flow on $\mathbf{B}_\eta$.*

Marsden and Weinstein (1974) is usually cited for this theorem, but a more prosaic form of the same theorem is found in Meyer (1973).

Proof. By all the assumptions $\mathbf{B}_\eta$ is a smooth manifold, so all that needs to be done is show that the symplectic form is nondegenerate on it. To that end simply apply Lemma 2.2.2 by letting $p \in M_\eta$, $\mathbb{V} = T_p^* M$, $\mathbb{U} = T_p^* M_\eta$.

While the notation is defined it is easy to state another interesting result.

Proposition 7.7.1. *Use the notation and assumptions of Theorem 7.7.2. Let $\phi(t)$ be a periodic solution of $dH^\#$ that does not become an equilibrium solution on $\mathbf{B}_\eta$, i.e., it is not a relative equilibrium. Then the geometric multiplicity of $+1$ is at least $1 + \dim \mathcal{G}$ and the algebraic multiplicity of multiplier $+1$ is at least $2 + \dim \mathcal{G} + \dim \mathcal{G}_\eta$.*

A proof can be found in Meyer (1973).

7.8 Integrable Systems

We have seen several examples of integrable systems, e.g., a pair of harmonic oscillators in Section 1.5 and the Kepler problem in $\mathbb{R}^2$ in Section 3.3. Indeed the classical text Whittaker (1937) is replete with such examples. Why are they so special? As we shall see the answer is action-angle variables.

Let M be a symplectic manifold of dimension $2n$ and $H : M \to \mathbb{R}$ a Hamiltonian. H is said to be integrable if there exists n independent integrals $F_i : M \to \mathbb{R}$, $i = 1, \ldots, n$ that are in involution (of course H may be one of them.). That is

(1) $dF_1, \ldots, dF_n$ are independent at some point in M,
(2) $\{H, F_i\} = 0$ for $i = 1, \ldots, n$,
(3) $\{F_i, F_j\} = 0$ for all i, j.

Actually we need a bit more in the noncompact case. Assume there is an open ball $B \subset \mathbb{R}^n$ filled with regular values of $F = (F_1, \ldots, F_n) : M \to \mathbb{R}^n$. Let $\mathbb{B} = F^{-1}(B) \subset M$ and assume the flow defined by each F_i is complete on $\mathbb{B}$, i.e., the solutions of $dF_i^\sharp$ exist for all t.

With the above notation:

Theorem 7.8.1. *For each $b \in B$, $F^{-1}(b) \subset M$ is diffeomorphic to $T^k \times \mathbb{R}^{n-k}$ where $T^k = S^1 \times S^1 \times \cdots S^1$ is the k torus. If M is compact then $k = n$.*

There exist symplectic action-angle coordinates, $(I_1, \ldots, I_n, \theta_1, \ldots, \theta_n)$ on $\mathbb{B}$ where $\theta_1, \ldots, \theta_k$ are angular coordinates defined mod 2π on T^k and $\theta_{k+1}, \ldots, \theta_n$ are rectangular coordinates on $\mathbb{R}^{n-k}$.

In these coordinates $H = H(I_1, \ldots, I_n)$ and so the equations of motion are

$$\dot{I}_i = \frac{\partial H}{\partial \theta_i} = 0, \ \dot{\theta}_i = -\frac{\partial H}{\partial I_i} = -\omega(I_1, \ldots, I_n), \ i = 1, \ldots, n.$$

This theorem is usually attributed to Arnold, but a proof can be found in Markus and Meyer (1974). In this same reference you will see that integrable systems are rare and completely integrable systems are rarer still.

In the planar Kepler problem there are two integrals in involution, namely, the Hamiltonian itself and the z-component of angular momentum, so it is integrable. But in the spatial Kepler problem at first sight there seems to be a problem because the three components of angular momentum are not in involution. (The Poisson bracket of two of the components gives $\pm$ the third component.) But we have seen in Section 8.9 in the construction of the spatial Delaunay variables the correct choice is the Hamiltonian, the z-component of angular momentum and the magnitude of total angular momentum.

Consider the central force problem in $\mathbb{R}^n$, i.e.,

$$H = \frac{1}{2}|p|^2 - K(|q|), \quad K(\beta) = \int^\beta k(\tau)d\tau.$$

where $k : \mathbb{R}^1 \to \mathbb{R}^1$ and $q, p \in \mathbb{R}^n$. For the Kepler problem take $k(\tau) = 1/\tau$. The symmetry group is now $So(n, \mathbb{R})$ and so by Noether's theorem it has the $n(n-1)/2$ integrals

$$A_{ij} = q_i p_j - q_j p_i, \quad 1 \le i < j \le n.$$

The central force problem is integrable by the following proposition.[2]

Proposition 7.8.1. *The n integrals of the central force problem*

$$\mathcal{A}_k = \sum_{1 \le i < j \le k} A_{ij}^2$$

$k = 1, 2, \ldots, n$ are independent and in involution.

Proof. The $\mathcal{A}_k$ are functions of integrals, so are integrals. Since $\mathcal{A}_{k-1}$ is independent of q_k, p_k and $\mathcal{A}_k = \mathcal{A}_{k-1} +$ new terms in q_k, p_k the functions are independent.

$$\{\mathcal{A}_k, \mathcal{A}_{ab}\} = \sum_{1 \le i < j \le k} \{A_{ij}^2, \mathcal{A}_{ab}\},$$

$$= \sum_{1 \le i < j \le k} 2A_{ij}\{A_{ij}, \mathcal{A}_{ab}\},$$

$$= \sum_{i,j=1}^{k} A_{ij}(A_{ia}\delta_{jb} - A_{ja}\delta_{ib} - A_{ib}\delta_{ja} + A_{jb}\delta_{ia}),$$

$$- 2 \sum_{i,j=1}^{k} A_{ij}(A_{ia}\delta_{jb} - A_{ib}\delta_{ja}).$$

In the above we used $A_{ij} = -A_{ji}$ and $A_{ii} = 0$. Therefore, for $a, b \le k$ we have

$$\{\mathcal{A}_k, \mathcal{A}_{ab}\} = 2 \sum_{i=1}^{k} (A_{ib}A_{ia} - A_{ia}A_{ib}) = 0,$$

and hence for $\ell < k$

$$\{\mathcal{A}_k, \mathcal{A}_\ell\} = 2 \sum_{1 \le a < b \le \ell} \{\mathcal{A}_k, \mathcal{A}_{ab}\}\mathcal{A}_{ab} = 0,$$

which proves the proposition.

This proof was adapted from Moser and Zehnder (2005).

---

[2] Actually the n-dimensional Kepler problem is completely integrable. See Moser (1970).

7.9 Problems

1. Show that if $f^1, \ldots, f^k$ are 1-forms, then

$$
f^1 \wedge \cdots \wedge f^k (a_1, \ldots, a_k) = \det
\begin{bmatrix}
f^1(a_1) & \cdots & f^k(a_1) \\
\vdots & & \vdots \\
f^1(a_k) & \cdots & f^k(a_k)
\end{bmatrix}.
$$

2. Show that the mapping $(f^1, f^2) \longrightarrow f^1 \wedge f^2$ is a skew-symmetric bilinear map from $V^* \times V^* \longrightarrow \mathbb{A}^2$.

3. Let F and G be 0-, 1-, or 2-forms in $\mathbb{R}^3$. Verify Lemma 7.3.1 in this case.

4.   a) Let $F = a\,dx + b\,dy + c\,dz$ be a 1-form in $\mathbb{R}^3$ such that $dF = 0$. Verify that $\partial a / \partial y = \partial b / \partial x$, $\partial a / \partial z = \partial c / \partial x$, $\partial c / \partial y = \partial b / \partial z$. Also verify that if

$$
f(x, y, z) = \int_0^1 (a(tx, ty, tz)x + b(tx, ty, tz)y + c(tx, ty, tz)z)\,dt,
$$

then $F = df$.

   b) Let F be a 2-form in $\mathbb{R}^3$ such that $dF = 0$. Verify that if $F = a\,dy \wedge dz + b\,dz \wedge dx + c\,dx \wedge dy$, then $\partial a / \partial x + \partial b / \partial y + \partial c / \partial z = 0$. Also verify that $F = df$ where

$$
f = \left(\int_0^1 a(tx, ty, tz)t\,dt \right) (y\,dz - z\,dy)
$$

$$
+ \left(\int_0^1 b(tx, ty, tz)t\,dt \right) (z\,dx - x\,dz)
$$

$$
+ \left(\int_0^1 c(tx, ty, tz)t\,dt \right) (x\,dy - y\,dx).
$$

5. Prove that the $\wedge$ operator is bilinear and associative. (See Lemma 7.1.1.)

6.   a) Show that the operator d which operates on smooth forms is linear, i.e., $d(F + G) = dF + dG$.

   b) Show that d satisfies a product rule, $d(F \wedge G) = dF \wedge G + (-1)^{\deg(F)} F \wedge dG$.

   c) Show that if δ is a mapping which takes smooth k-forms to $(k+1)$-forms and satisfies

   i. $\delta(F + G) = \delta F + \delta G$,
   ii. $\delta(F \wedge G) = \delta F \wedge G + (-1)^{\deg(F)} F \wedge \delta G$,
   iii. $\delta(\delta F) = 0$ for all F,
   iv. If F is a function, then δF agrees the standard definition of the differential of F, then δ is the same as the operator d given by the formula in (7.10).

7. Let $Q(q, p)$ and $P(q, p)$ be smooth functions defined on an open set in $\mathbb{R}^2$. Consider the four differential forms $\Omega_1 = PdQ - pdq$, $\Omega_2 = PdQ + qdp$, $\Omega_3 = QdP + pdq$, $\Omega_4 = QdP - qdp$.

   a) Show that Ω_i is exact if and only if Ω_j is exact for $i \neq j$.

   b) Show that Ω_i is closed if and only if Ω_j is closed for $i \neq j$.

   c) Show that if Ω_i is exact (or closed) then so is $\Theta = (Q - q)d(P + p) - (P - p)d(Q + q)$. (Hint: $d(qp) = qdp + pdq$ is exact.)

8. Special Coordinates

Celestial mechanics is replete with special coordinate systems some of which bear the names of the greatest mathematicians of all times. There is an old saying in celestial mechanics: "No set of coordinates is good enough."

We consider some of these special coordinates in this chapter. Because the topics of this chapter are special in nature, the reader is advised to glance at the sections first and then refer back to this chapter when the need arises.

8.1 Differential Forms and Generating Functions

Definition (2.33) is easy to check a posteriori, but it is difficult to use this definition to generate a symplectic transformation with desired properties. This section contains only a local analysis, so, we assume that everything is defined in some ball about the origin in $\mathbb{R}^{2n}$.

8.1.1 The Symplectic Form

Recall that in Chapter 7 we defined the (standard) symplectic form to be

$$\Omega = \frac{1}{2}\sum_{i=1}^{n}\sum_{j=1}^{n} J_{ij}dz^i \wedge dz^j = \sum_{i=1}^{n} dz^i \wedge dz^{i+n} = \sum_{i=1}^{n} dq^i \wedge dp^i = dq \wedge dp. \quad (8.1)$$

Here we have used the notation of differential geometry by using superscripts for components instead of subscripts. Also we have $z = (z^1, \ldots, z^{2n}) = (q^1, \ldots, q^n, p^1, \ldots, p^n)$ as usual. Ω is closed, $d\Omega = 0$, but, in fact, it is exact because

$$\Omega = d\alpha, \qquad \alpha = \sum_{i=1}^{n} q^i dp^i = qdp. \quad (8.2)$$

In short, Ω is a closed nondegenerate (the coefficient matrix is nonsingular) 2-form. By Darboux's theorem for any closed, nondegenerate 2-form, there are local coordinates such that in these coordinates the 2-form is given

© Springer International Publishing AG 2017
K.R. Meyer, D.C. Offin, *Introduction to Hamiltonian Dynamical Systems and the N-Body Problem*, Applied Mathematical Sciences 90, DOI 10.1007/978-3-319-53691-0_8

by (8.1). This says that J is simply the coefficient matrix of a closed, non-degenerate 2-form in Darboux coordinates. The left-hand side of (2.33) is just the transformation law for a 2-form with coefficient matrix J, so, a symplectic transformation is a transformation that preserves the special form of the differential form Ω. In two-dimensions $\Omega = dq \wedge dp$, the area form in $\mathbb{R}^2$, and so we see again that a two-dimensional symplectic transformation is area-preserving. In higher dimensions, being symplectic is far more restrictive than simply being volume-preserving.

8.1.2 Generating Functions

Let $Q = Q(q, p), P = P(q, p)$ be a change of variables, and assume the functions Q and P are defined in a ball in $\mathbb{R}^{2n}$. This change of variables is symplectic if and only if

$$dq \wedge dp = dQ \wedge dP.$$

This is equivalent to $d(qdp - QdP) = 0$ or that $\sigma_1 = qdp - QdP$ is exact. σ_1 is exact if and only if $\sigma_2 = \sigma_1 + d(QP) = qdp + PdQ$ is exact. In a similar manner the change of variables $Q = Q(q, p), P = P(q, p)$ is symplectic if and only if any one of the following forms is exact:

$$\sigma_1 = qdp - QdP, \qquad \sigma_2 = qdp + PdQ,$$
$$\sigma_3 = pdq - PdQ, \qquad \sigma_4 = pdq + QdP. \tag{8.3}$$

Because the functions Q and P are defined in a ball, closed forms are exact by Poincaré's lemma, so, the change of variables is symplectic if and only if one of the functions S_1, S_2, S_3, S_4 exists and satisfies

$$dS_1(p, P) = \sigma_1, \qquad dS_2(p, Q) = \sigma_2,$$
$$dS_3(q, Q) = \sigma_3, \qquad dS_4(q, P) = \sigma_4.$$

In the above formulas, there is an implied summation over the components.

These statements give an easy way to construct a symplectic change of variables. Assume that there exists a function $S_1(p, P)$ such that $dS_1 = \sigma_1$; so,

$$dS_1 = \frac{\partial S_1}{\partial p} dp + \frac{\partial S_1}{\partial P} dP = qdp - QdP.$$

So if

$$q = \frac{\partial S_1}{\partial p}(p, P), \qquad Q = -\frac{\partial S_1}{\partial P}(p, P) \tag{8.4}$$

defines a change of variables from (q, p) to (Q, P), then it is symplectic. By the implicit function theorem, the equations in (8.4) are solvable for P as a function of q and p and for p as a function of Q and P when the Hessian of S_1 is nonsingular. Thus in a similar manner we have the following.

Theorem 8.1.1. *The following define a local symplectic change of variables:*

$$q = \frac{\partial S_1}{\partial p}(p, P), \quad Q = -\frac{\partial S_1}{\partial P}(p, P) \quad \text{when } \frac{\partial^2 S_1}{\partial p \partial P} \text{ is nonsingular;}$$

$$q = \frac{\partial S_2}{\partial p}(p, Q), \quad P = \frac{\partial S_2}{\partial Q}(p, Q) \quad \text{when } \frac{\partial^2 S_2}{\partial p \partial Q} \text{ is nonsingular;}$$

$$p = \frac{\partial S_3}{\partial q}(q, Q), \quad P = -\frac{\partial S_3}{\partial Q}(q, Q) \quad \text{when } \frac{\partial^2 S_3}{\partial q \partial Q} \text{ is nonsingular;} \qquad (8.5)$$

$$p = \frac{\partial S_4}{\partial q}(q, P), \quad Q = \frac{\partial S_4}{\partial P}(q, P) \quad \text{when } \frac{\partial^2 S_4}{\partial q \partial P} \text{ is nonsingular.}$$

The functions S_i are called *generating functions*. For example, if $S_2(p, Q) = pQ$, then the identity transformation $Q = q, P = p$ is symplectic, or if $S_1(p, P) = pP$, then the switching of variables $Q = -p, P = q$ is symplectic.

8.1.3 Mathieu Transformations

If you are given a point transformation $Q = f(q)$, with $\partial f/\partial q$ invertible, then the transformation can be extended to a symplectic transformation by defining $S_4(q, P) = f(q)^T P$ and

$$Q = f(q), \qquad p = \frac{\partial f}{\partial q}(q)P.$$

One can also add a function $F(q)$ to S_4 to get $S_4(q, P) = f(q)^T P + F(q)$ and

$$Q = f(q), \qquad p = \frac{\partial f}{\partial q}(q)P + \frac{\partial F}{\partial q}(q).$$

These transformations were studied by Mathieu (1874).

8.2 Jacobi Coordinates

Jacobi coordinates are ideal coordinates for investigations of the N-body problem. First, one coordinate locates the center of mass of the system, and so it can be set to zero and ignored in subsequent considerations. Second, one coordinate is the vector from one particle to another, and this is useful when studying the case when two particles are close together. Third, another coordinate is the vector from the center of mass of the first $N - 1$ particles to the Nth, and this is useful when studying the case when one particle is far from the others.

Let $q_i, p_i \in \mathbb{R}^3$ for $i = 1, \ldots, N$, be the coordinates of the N-body problem as discussed in Section 3.1. Define a sequence of transformations starting with $g_1 = q_1$ and $\mu_1 = m_1$ and proceed inductively by

$$T_k : \begin{cases} u_k = q_k - g_{k-1}, \\ g_k = (1/\mu_k)(m_k q_k + \mu_{k-1} g_{k-1}), \\ \mu_k = \mu_{k-1} + m_k \end{cases} \tag{8.6}$$

for $k = 2, \ldots, N$. μ_k is the total mass, and g_k is the position vector of the center of mass of the system of particles with indices $1, 2, \ldots, k$. The vector u_k is the position of the kth particle relative to the center of mass of the previous $k - 1$ particles (see Figure 8.1). Consider T_k as a change of coordinates from $g_{k-1}, u_2, \ldots, u_{k-1}, q_k, \ldots, q_N$ to $g_k, u_2, \ldots, u_k, q_{k+1}, \ldots, q_N$ or simply from g_{k-1}, q_k to g_k, u_{k+1}. The inverse of T_k is

$$T_k^{-1} : \begin{cases} q_k = (\mu_{k-1}/\mu_k)u_k + g_k, \\ g_{k-1} = (-m_k/\mu_k)u_k + g_k. \end{cases} \tag{8.7}$$

This is a linear transformation on the q variables only, i.e., on a Lagrangian subspace, so, Lemma 2.2.5 forces the transformation on the p variables in order to have a symplectic transformation. To make the symplectic completion of T_k, define $G_1 = p_1$ and

$$Q_k : \begin{cases} v_k = (\mu_{k-1}/\mu_k)p_k - (m_k/\mu_k)G_{k-1}, \\ G_k = p_k + G_{k-1}, \end{cases} \tag{8.8}$$

$$Q_k^{-1} : \begin{cases} p_k = v_k + (m_k/\mu_k)G_k, \\ G_{k-1} = -v_k + (\mu_{k-1}/\mu_k)G_k. \end{cases} \tag{8.9}$$

If we denote the coefficient matrix in (8.6) by A, then the coefficient matrices in (8.7), (8.8), and (8.9) are A^{-1}, A^{-T}, and A^T, respectively, so, the pair T_k, Q_k is a symplectic change of variables. Thus, the composition of all these changes is symplectic and the total set $g_N, u_2, \ldots, u_N, G_N, v_2, \ldots, v_N$ forms a symplectic coordinate system known as the Jacobi coordinates.

These variables satisfy the identities

$$g_{k-1} \times G_{k-1} + q_k \times p_k = g_k \times G_k + u_k \times v_k$$

and

$$\frac{\|G_{k-1}\|^2}{2\mu_{k-1}} + \frac{\|p_k\|^2}{2m_k} = \frac{\|G_k\|^2}{2\mu_k} + \frac{\|v_k\|^2}{2M_k},$$

where $M_k = m_k \mu_{k-1}/\mu_k$. Thus kinetic energy is

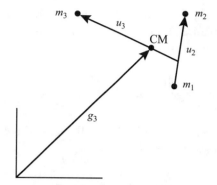

Figure 8.1. Jacobi coordinates for the 3-body problem.

$$KE = \sum_{k=1}^{N} \frac{\|p_k\|^2}{2m_k} = \frac{\|G_N\|^2}{2\mu_N} + \sum_{k=2}^{N} \frac{\|v_k\|^2}{2M_k},$$

and total angular momentum is

$$A = \sum_{1}^{N} q_k \times p_k = g_N \times G_N + \sum_{2}^{N} u_k \times v_k. \tag{8.10}$$

Also, g_N is the center of mass of the system, and G_N is the total linear momentum.

Unfortunately, the formulas for the variables u_k and v_k are not simply expressed in terms of the variables q_k and p_k. Note that

$$u_2 = q_2 - q_1.$$

Let $d_{ij} = q_i - q_j$ and so the Hamiltonian of the N-body problem in Jacobi coordinates is

$$H = \frac{\|G_N\|^2}{2\mu_N} + \sum_{k=2}^{N} \frac{\|v_k\|^2}{2M_k} - \sum_{1 \le i < j \le N} \frac{m_i m_j}{\|d_{ij}\|}. \tag{8.11}$$

Note that the Hamiltonian is independent of g_N, it's ignorable, so $\dot{G}_N = 0$ or G_N is an integral of the system. Because $\dot{g}_N = G_N/\mu_N$, the center of gravity moves with uniform rectilinear motion. In general, one may assume that the center of mass is fixed at the origin of the system and so sets $g_N = G_N = 0$, which reduces the problem by three degrees of freedom in the spatial problem.

In the planar problem, one verifies

$$\sum_{1}^{N} q_k^T K p_k = g_N^T K G_N + \sum_{2}^{N} u_k^T K v_k,$$

which is the same as the formula for angular momentum given above. So the Hamiltonian of the planar N-body problem in rotating coordinates with the center of mass fixed at the origin is

$$H = \sum_{k=2}^{N} \frac{\|v_k\|^2}{2M_k} + \sum_{2}^{N} u_k^T K v_k - \sum_{1 \le i < j \le N} \frac{m_i m_j}{\|d_{ij}\|}. \tag{8.12}$$

The 2-Body Problem in Jacobi Coordinates When $N = 2$, then (8.11) with $g_2 = G_2 = 0$ takes the simple form

$$H = \frac{\|v\|^2}{2M} - \frac{m_1 m_2}{\|u\|}, \tag{8.13}$$

where $v = v_2, u = u_2 = q_2 - q_1, M = m_1 m_2/(m_1 + m_2)$. This is just the Kepler problem, and so in Jacobi coordinates the 2-body problem is just the Kepler problem. This says that the motion of one body, say the moon, when viewed from another, say the earth, is as if the earth were a fixed body and the moon were attracted to the earth by a central force. See the discussion of the 2-body problem in Section 3.2.

The 3-Body Problem in Jacobi Coordinates In the 3-body problem the distances between the bodies, and hence the potential, are not too complicated in Jacobi coordinates. Let

$$M_2 = \frac{m_1 m_2}{m_1 + m_2}, \qquad M_3 = \frac{m_3(m_1 + m_2)}{m_1 + m_2 + m_3},$$

$$\alpha_0 = \frac{m_2}{m_1 + m_2}, \qquad \alpha_1 = \frac{m_1}{m_1 + m_2}; \tag{8.14}$$

then the Hamiltonian of the 3-body problem with center of mass fixed at the origin and zero linear momentum in Jacobi coordinates is

$$H = \frac{\|v_2\|^2}{2M_2} + \frac{\|v_3\|^2}{2M_3} - \frac{m_1 m_2}{\|u_2\|} - \frac{m_1 m_3}{\|u_3 + \alpha_0 u_2\|} - \frac{m_2 m_3}{\|u_3 - \alpha_1 u_2\|}. \tag{8.15}$$

See Figure 8.1. Sometimes one numbers the N-bodies from 0 to $N-1$. In this case all the subscripts in (8.15) except the subscripts of the $\alpha's$ are reduced by 1, which looks nicer to some people.

8.3 Action–Angle Variables

The change from rectangular coordinates q, p to polar coordinates r, ϕ is not symplectic, but because

$$dq \wedge dp = r dr \wedge d\phi = d(r^2/2) \wedge d\phi,$$

the following define a symplectic change of variables.

$$I = \tfrac{1}{2}(q^2 + p^2) = r^2/2, \qquad \phi = \arctan(p/q),$$

$$q = \sqrt{2I}\cos\phi, \qquad\qquad p = \sqrt{2I}\sin\phi. \qquad (8.16)$$

Therefore, I, ϕ are symplectic (or canonical) coordinates called *action–angle coordinates*. In Chapter 2, we saw that the harmonic oscillator could be written as a Hamiltonian system with Hamiltonian

$$H = \frac{\omega}{2}(q^2 + p^2) = \omega I,$$

and in action–angle coordinates, the equations of motion are

$$\dot{I} - \frac{\partial H}{\partial \phi} - 0, \qquad \dot{\phi} = -\frac{\partial H}{\partial I} = -\omega.$$

So the solutions move on the circles $I = constant$ with uniform angular frequency ω in a counterclockwise direction.

Action–angle variables are used quite often in perturbation theory, for example, Duffing's equation has a Hamiltonian

$$\dot{H} = \frac{1}{2}(q^2 + p^2) + \frac{\gamma}{4}q^4,$$

where γ is a constant. Writing Duffing's Hamiltonian in action–angle variables gives

$$H = I + \gamma I^2 \cos^4\phi = I + \frac{\gamma}{8}I^2\{3 + 4\cos 2\phi + \cos 4\phi\}.$$

The last term in the braces in the above formula is an example of a Poisson. A *Poisson series* in $r = \sqrt{2I}$ and ϕ is a Fourier series in ϕ with coefficients that are polynomials in r. Such series arise from substituting action–angle variables into a power series expansion in q and p, but not all Poisson series come about in this manner. Action–angle variables are used and misused so often in celestial mechanics that we investigate this point next.

8.3.1 d'Alembert Character

Consider a Poisson series

$$g(r, \phi) = \sum_i a_{i0}r^i + \sum\sum(a_{ij}r^i \cos j\phi + b_{ij}r^i \sin j\phi).$$

The Poisson series $g(r, \phi)$ comes from a power series $f(q, p) = \sum f_{ij}q^i p^j$ if $g(r, \theta) = f(r\cos\phi, r\sin\phi)$.

The Poisson series g has the *d'Alembert character* if $a_{ij} = 0, b_{ij} = 0$ unless $i \geq |j|$, and $i \equiv j \bmod 2$ (i.e., i and j have the same parity).

Theorem 8.3.1. *The Poisson series g comes from a power series if and only if it has the d'Alembert character.*

Proof. $x^i y^j = r^{i+j} \cos^i \phi \sin^j \phi$. Claim: $\cos^i \phi \sin^j \phi$ has a Fourier polynomial of the form $a_0 + \Sigma\{a_k \cos k\phi + b_k \sin k\phi\}$, where $a_k = b_k = 0$ unless $k \leq i+j$ and $k \equiv i+j \bmod 2$. The claim is clearly true for $i+j = 1$; so, assume it's true for $i+j < N$, and let $i+j = N$. Let $i \neq 0$; then

$$\cos^i \phi \sin^j \phi = \cos \phi [\cos^{i-1} \phi \sin^j \phi]$$

$$= \cos \phi [\alpha_0 + \sum\{\alpha_k \cos k\phi + \beta_k \sin k\phi\}]$$

$$= \alpha_0 \cos \phi + \sum (\alpha_k/2)[\cos(k+1)\phi + \cos(k-1)\phi]$$

$$+ \sum (\beta_k/2)[\sin(k+1)\phi + \sin(k-1)\phi].$$

The induction hypothesis gives $\alpha_k = 0, \beta_k = 0$ unless $k \leq i+j-1$, and $k \equiv i+j-1 \bmod 2$. The last polynomial above shows that the induction hypothesis is true for $i+j = N$. Similar formulas hold when $j \neq 0$. So a power series gives rise to a Poisson series with the d'Alembert character.

Conversely, if $r^a \cos b\phi$ satisfies $a \geq b$ and $a \equiv b \bmod 2$, then $\cos b\phi = \Re(\exp i\phi)^b =$ a sum of terms like $\cos(b - 2s)\phi \sin 2s\phi$. The d'Alembert character gives $a = b + 2p$, so $r^a \cos b\phi =$ a sum of terms like $r^{2p}\{r^{b-2p} \cos(b - 2p\phi)\}\{r^{2p} \sin(2p\phi)\} = (x^2 + y^2)^p x^{b-2p} y^2$.

Perturbation analysis is often done in action–angle variables. Keeping track of the d'Alembert character of the change of variables in action–angle variables is important in order to keep track of the fact that the change of variables is analytic in rectangular variables.

Say you want an analytic Hamiltonian with a fivefold symmetry. In polar coordinates the functions $r^k, \cos 5\theta, \sin 5\theta$ are all invariant under the rotation $\theta \to \theta + 2\pi/5$ but are not analytic in rectangular coordinates. The functions $r^{2k}, r^{2k+5} \cos 5\theta$, and $r^{2k+5} \sin 5\theta$ all are invariant under the rotation $\theta \to \theta + 2\pi/5$ and have the d'Alembert character; therefore, any linear combination (finite or uniformly convergent for $r < \rho, \rho > 0$) gives an analytic function in rectangular coordinates with a fivefold symmetry.

8.4 General Action–Angle Coordinates

Consider a Hamiltonian $H(x, y)$ defined in a neighborhood of the origin in $\mathbb{R}^2$, such that the origin is a center for the Hamiltonian flow. Thus the origin is encircled by periodic orbits. Assume that the origin is a local minimum of H, and $H(0, 0) = 0$. We seek symplectic action–angle variables (L, ℓ) where ℓ is an angle defined mod 2π and the Hamiltonian is to be of the form $H(L, \ell) = \Omega(L)$.

Let $R(h)$ = the component of $\{(x, y) \in \mathbb{R}^2 : H(x, y) \leq h\}$ that contains the origin. Because $dx \wedge dy = dL \wedge d\ell$, we must have

$$\int\int_{R(h)} dx \wedge dy = \int\int_{R(h)} dL \wedge d\ell = 2\pi L,$$

or

$$L = \frac{1}{2\pi} \int\int_{R(h)} dx \wedge dy = \frac{1}{2\pi} \oint_{\partial R(h)} x \, dy.$$

Thus the variable L is just the area of the region $R(h)$. The last integral in the formula for L is classically called the *action*. The equations of motion in the new coordinates are

$$\dot{L} = 0, \qquad \dot{\ell} = -\Omega'(L); \tag{8.17}$$

therefore L is a constant and $\ell = \ell_0 - \Omega'(L)t$. Thus ℓ is a scaled time, scaled so that it is 2π-periodic.

Let the period of the orbit be $p(h)$. We wish to find symplectic variables (L, ℓ) defined in a neighborhood of the origin with ℓ an angle defined mod 2π and L is its conjugate momentum so that the Hamiltonian becomes a function of L alone. With $H = \Omega(L)$ the equations become (8.17). Thus the period of the orbit is $2\pi/|\Omega'(L)|$. Therefore, we must have

$$\frac{2\pi}{\Omega'(L)} = \pm p(\Omega(L)),$$

or

$$\Omega' = \pm \frac{2\pi}{p(\Omega)}.$$

The differential equation above defines the function $\Omega(L)$ in terms of $p(h)$. The sign is chosen so that $\pm 2\pi/\Omega'(L)$ is positive (the period).

In the Kepler problem $H = -1/2a$, and $P = 2\pi a^{3/2}$ where a is the semi-major axis. So $p(h) = \pi 2^{-1/2}(-h)^{-3/2}$. Thus the equation to solve is

$$\Omega' = 2^{3/2}(-\Omega)^{3/2}.$$

Separating variables gives

$$(-\Omega)^{-3/2} d\Omega = 2^{3/2} dL$$

$$(-\Omega)^{-1/2} = 2^{1/2} L.$$

Thus the Hamiltonian must be

$$\Omega = -\frac{1}{2L^2}. \tag{8.18}$$

Now consider the Hamiltonian system

$$H = \frac{1}{2}y^2 + F(x), \quad F(x) = \int_0^x f(\tau)d\tau,$$

where $xf(x) > 0$ for $0 < x < x_0$, so the origin is a center. So for small positive h, the set $H = h$ is a closed orbit. Because $\dot{\ell}$ is constant, ℓ is a constant multiple of time t, in particular $\ell = \ell_0 - \Omega'(L_0)t$. In order to fix initial conditions, let t and ℓ be measured from the positive x-axis. A one degree of freedom Hamiltonian system is integrable up to quadrature. To this end, let a denote the point on the positive x-intercept of $H = h$, so $F(a) = \Omega(L)$. From the equation

$$\frac{1}{2}y^2 + F(x) = h,$$

solve for y,

$$y = \frac{dx}{dt} = \pm\{2h - 2F(x)\}^{1/2},$$

separate variables

$$dt = \pm\{2h - 2F(x)\}^{-1/2}dx,$$

and so

$$t = \pm\int_a^x \frac{d\xi}{\{2h - 2F(\xi)\}^{1/2}}. \tag{8.19}$$

The angle ℓ is $-\Omega'(L_0)t$. The orbit is swept out in a clockwise direction, so if the angle ℓ is to be measured in the usual counterclockwise direction for small ℓ and t, the minus sign should be used in (8.19).

To construct the symplectic change of variables, consider the generating function

$$W(x, L) = \int_a^x \{2\Omega(L) - 2F(\xi)\}^{1/2}d\xi$$

with

$$y = \frac{\partial W}{\partial x} = \{2\Omega(L) - 2F(x)\}^{1/2},$$

so

$$\frac{1}{2}y^2 + F(x) = \Omega(L)$$

and

$$\ell = \frac{-\partial W}{\partial L} = -\{2\Omega(L) - 2F(a)\}^{1/2}\frac{da}{dL} - \int_a^x \frac{\Omega'(L)d\xi}{\{2\Omega(L_0) - 2F(\xi)\}^{1/2}}$$

$$= -t\Omega'(L).$$

(The first term in the formula for ℓ is zero by the definition of a.) Thus the change of variables is symplectic. We call these variables action–angle variables for the Hamiltonian H. If $H = \frac{1}{2}(x^2 + y^2)$, the usual harmonic oscillator, then $\ell = \tan^{-1}(y/x)$ and $L = \frac{1}{2}(x^2 + y^2)$. A more interesting example is given in Section 8.9 on the Delaunay elements.

8.5 Lemniscate Coordinates

Recall from Section 1.6 the sine lemniscate function, $sl\,(t)$, which is the solution of
$$\ddot{x} + 2x^3 = 0, \quad x(0) = 0, \quad \dot{x}(0) = 1,$$
and the cosine lemniscate function $cl\,(t) = \dot{sl}\,(t)$, both of which are τ periodic. This is a Hamiltonian system with Hamiltonian

$$L = \frac{1}{2}(y^2 + x^4).$$

Just as we use the solutions of the harmonic oscillator to define action-angle coordinates, we will use these new functions to define coordinates (K, κ) which we will call *lemniscate coordinates*. They will be used in several stability proofs in Chapter 12.

Define these coordinates by

$$x = K^{1/3}sl\,(\kappa), \qquad y = -K^{2/3}cl\,(\kappa),$$

and check:

$$\begin{aligned}
dx \wedge dy &= (\tfrac{1}{3}K^{-2/3}sl\,(\kappa)dK + K^{1/3}cl\,(\kappa)d\kappa)\wedge \\
&\quad (-\tfrac{2}{3}K^{-1/3}cl\,(\kappa)dK - K^{2/3}(-2sl^3(\kappa))d\kappa) \\
&= \tfrac{2}{3}sl^4(\kappa)dK \wedge d\kappa - \tfrac{2}{3}cl^2(\kappa)d\kappa \wedge dK \\
&= \tfrac{2}{3}dK \wedge d\kappa
\end{aligned}$$

which is symplectic with multiplier $\dfrac{3}{2}$.

The Hamiltonian for the lemniscate function becomes

$$L = \frac{1}{2}(y^2 + x^4) = \frac{3}{2}\frac{1}{2}\left(K^{4/3}cl^2(\kappa) + K^{4/3}sl^4(\kappa)\right) = \frac{3}{4}K^{4/3},$$

and the equations of motion become

$$\dot{K} = \frac{\partial L}{\partial \kappa} = 0, \quad \dot{\kappa} = -\frac{\partial L}{\partial K} = -K^{1/3}.$$

So the period T of the a solution with starting at (K_0, κ_0) is

$$T = \frac{4\tau}{K_0^{1/3}}.$$

This shows the dependence of the period on the amplitude K_0.

8.6 Polar Coordinates

Let x, y be the usual coordinates in the plane and X, Y their conjugate momenta. Suppose we wish to change to polar coordinates, r, θ in the x, y-plane and to extend this point transformation to a symplectic change of variables. Let R, Θ be conjugate to r, θ. Use the generating function $S = S_2 = Xr\cos\theta + Yr\sin\theta$, and so

$$x = \frac{\partial S}{\partial X} = r\cos\theta, \qquad y = \frac{\partial S}{\partial Y} = r\sin\theta,$$

$$R = \frac{\partial S}{\partial r} = X\cos\theta + Y\sin\theta = \frac{xX + yY}{r},$$

$$\Theta = \frac{\partial S}{\partial \theta} = -Xr\sin\theta + Yr\cos\theta = xY - yX.$$

If we think of a particle of mass m moving in the plane, then $X = m\dot{x}$ and $Y = m\dot{y}$ are linear momenta in the x and y directions; so, $R = m\dot{r}$ is the linear momentum in the r direction, and $\Theta = mx\dot{y} - my\dot{x} = mr^2\dot{\theta}$ is the angular momentum. The inverse transformation is

$$X = R\cos\theta - \left(\frac{\Theta}{r}\right)\sin\theta,$$

$$Y = R\sin\theta + \left(\frac{\Theta}{r}\right)\cos\theta.$$

8.6.1 Kepler's Problem in Polar Coordinates

The Hamiltonian of the planar Kepler's problem in polar coordinates is

$$H = \frac{1}{2}(X^2 + Y^2) - \frac{\mu}{\sqrt{x^2 + y^2}} = \frac{1}{2}\left(R^2 + \frac{\Theta^2}{r^2}\right) - \frac{\mu}{r}. \tag{8.20}$$

Because H is independent of θ, it is an ignorable coordinate, and Θ is an integral. The equations of motion are

$$\dot{r} = R, \qquad \dot{\theta} = \frac{\Theta}{r^2},$$

$$\dot{R} = \frac{\Theta^2}{r^3} - \frac{\mu}{r^2}, \qquad \dot{\Theta} = 0. \tag{8.21}$$

These equations imply that Θ, angular momentum, is constant, say c; so,

$$\ddot{r} = \dot{R} = \frac{c^2}{r^3} - \frac{\mu}{r^2}. \tag{8.22}$$

This is a one degree of freedom equation for r, so, it is solvable by the method discussed in Section 1.7. Actually, this equation for r can be solved explicitly.

Assume $c \neq 0$, so, the motion is not collinear. In (8.22) make the changes of variables $u = 1/r$ and $dt = (r^2/c)d\theta$ so

$$\ddot{r} = \frac{c}{r^2}\frac{d}{d\theta}\left\{\frac{c}{r^2}\frac{dr}{d\theta}\right\} = c^2 u^2 \frac{d}{d\theta}\left\{u^2 \frac{du^{-1}}{d\theta}\right\} = -c^2 u^2 u''$$

$$= \frac{c^2}{r^3} - \frac{\mu}{r^2} = c^2 u^3 - \mu u^2,$$

or

$$u'' + u = \mu/c^2, \tag{8.23}$$

where $' = d/d\theta$. Equation (8.23) is just the nonhomogeneous harmonic oscillator which has the general solution $u = (\mu/c^2)(1 + e\cos(\theta - g))$, where e and g are integration constants. Let $f = \theta - g$, so,

$$r = \frac{(c^2/\mu)}{1 + e\cos f}. \tag{8.24}$$

Equation (8.24) is the equation of a conic section in polar coordinates. Consider a line d in Figure 8.2 that is perpendicular to the ray at angle g through the origin and at a distance c^2/μ. Rewrite (8.24) as

$$r = e\left(\frac{c^2}{e\mu} - r\cos f\right),$$

which says that the distance of the particle to the origin r is equal to e times the distance of the particle to the line d,

$$\rho = \frac{c^2}{e\mu} - r\cos f.$$

This is one of the many definitions of a conic section. One focus is at the origin, e is the eccentricity, and the locus is a circle if $e = 0$, an ellipse if $0 < e < 1$, a parabola if $e = 1$, and a hyperbola if $e > 1$.

The point of closest approach in Figure 8.2 is called the perihelion if the sun is the attractor at the origin or the perigee if the earth is. The angle f is called the true anomaly and g the argument of the perihelion (perigee).

For a complete discussion of how to find the position of a particle on the conic as a function of time see Pollard (1966).

8.6.2 The 3-Body Problem in Jacobi–Polar Coordinates

Consider the 3-body problem in Jacobi coordinates with center of mass at the origin and linear momentum zero; i.e., the Hamiltonian (8.15). Introduce polar coordinates for u_2 and u_3 as discussed above. That is, let

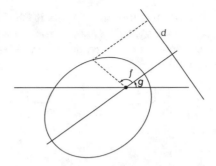

Figure 8.2. The elements of a Kepler motion.

$$u_2 = (r_1 \cos\theta_1, r_1 \sin\theta_1), \quad u_3 = (r_2 \cos\theta_2, r_2 \sin\theta_2),$$

$$v_2 = \left(R_1 \cos\theta_1 - \left(\frac{\Theta_1}{r_1}\right) \sin\theta_1, R_1 \sin\theta_1 + \left(\frac{\Theta_1}{r_1}\right) \cos\theta_1 \right),$$

$$v_3 = \left(R_2 \cos\theta_2 - \left(\frac{\Theta_2}{r_2}\right) \sin\theta_2, R_2 \sin\theta_2 + \left(\frac{\Theta_2}{r_2}\right) \cos\theta_2 \right),$$

so, the Hamiltonian (8.15) becomes

$$H = \frac{1}{2M_2}\left\{ r_1^2 + \left(\frac{\Theta_1^2}{r_1^2}\right) \right\} + \frac{1}{2M_3}\left\{ r_2^2 + \left(\frac{\Theta_2^2}{r_2^2}\right) \right\} - \frac{m_0 m_1}{r_1}$$

$$- \frac{m_0 m_2}{\sqrt{r_2^2 + \alpha_0^2 r_1^2 - 2\alpha_0 r_1 r_2 \cos(\theta_2 - \theta_1)}} \tag{8.25}$$

$$- \frac{m_1 m_2}{\sqrt{r_2^2 + \alpha_1^2 r_1^2 - 2\alpha_1 r_1 r_2 \cos(\theta_2 - \theta_1)}}.$$

The constants are the same as in (8.14), but here we number the masses m_0, m_1, m_2. Note that the Hamiltonian only depends on the difference of the polar angles, $\theta_2 - \theta_1$.

8.7 Spherical Coordinates

Sometimes the discussion of a spatial problem is easier in spherical coordinates which were briefly introduced in Section 1.10. This section is used to define the three-dimensional Delaunay and Poincaré elements which in turn are used to establish three-dimensional periodic solutions of the restricted problem.

Let x, y, z be the usual coordinates in space and X, Y, Z their conjugate momenta. We wish to change to spherical coordinates (ρ, θ, ϕ), the radius, longitude, and colatitude and their conjugate momenta P, Θ, Φ. The standard definition of spherical coordinates is

$$x = \rho \sin \phi \cos \theta, \qquad y = \rho \sin \phi \sin \theta, \qquad z = \rho \cos \phi. \qquad (8.26)$$

To extend this point transformation use the Mathieu generating function

$$S = S_2 = X\rho \sin \phi \cos \theta + Y\rho \sin \phi \sin \theta + Z\rho \cos \phi,$$

so

$$P = \frac{\partial S}{\partial \rho} = X \sin \phi \cos \theta + Y \sin \phi \sin \theta + Z \cos \phi$$
$$= (xX + yY + zZ)/\rho = \dot{\rho},$$

$$\Theta = \frac{\partial S}{\partial \theta} = -X\rho \sin \phi \sin \theta + Y\rho \sin \phi \cos \theta = -Xy + Yx = \rho^2 \dot{\theta} \qquad (8.27)$$

$$\Phi = \frac{\partial S}{\partial \phi} = X\rho \cos \phi \cos \theta + Y\rho \cos \phi \sin \theta - Z\rho \sin \phi = \rho^2 \cos^2 \phi \dot{\phi}.$$

Thus R is the radial momentum, and Θ is the z-component of angular momentum. From these expressions compute

$$Z - P \cos \phi - (\Phi/\rho) \sin \phi,$$

$$P \sin \phi + (\Phi/\rho) \cos \phi = X \cos \theta + Y \sin \theta,$$

$$\Theta/(\rho \sin \phi) = -X \sin \theta + Y \cos \theta.$$

From the last two formulas, compute $X^2 + Y^2$ without computing X and Y. You will find that the Hamiltonian of the Kepler problem in spherical coordinates is

$$H = \frac{1}{2}\left\{ P^2 + \frac{\Phi^2}{\rho^2} + \frac{\Theta^2}{\rho^2 \sin^2 \phi} \right\} - \frac{1}{\rho}, \qquad (8.28)$$

and the equations of motion are

$$\dot{\rho} = H_P = P, \qquad \dot{P} = -H_\rho = \frac{\Phi^2}{\rho^3} + \frac{\Theta^2}{\rho^3 \sin^2 \phi} - \frac{1}{\rho^2},$$

$$\dot{\theta} = H_\Theta = \frac{\Theta}{\rho^2 \sin^2 \phi}, \qquad \dot{\Theta} = -H_\theta = 0, \qquad (8.29)$$

$$\dot{\phi} = H_\Phi = \frac{\Phi}{\rho^2}, \qquad \dot{\Phi} = -H_\phi = \left(\frac{\Theta^2}{\rho^2}\right) \frac{\cos \phi}{\sin^3 \phi}.$$

Clearly, Θ, the z-component of angular momentum, is an integral, but so is G defined by

$$G^2 = \left(\frac{\Theta^2}{\sin^2 \phi} + \Phi^2 \right). \tag{8.30}$$

We show that G is the magnitude of total angular momentum after we investigate the invariant plane in spherical coordinates.

The equation of a plane through the origin is of the form $\alpha x + \beta y + \gamma z = 0$, or in spherical coordinates

$$\alpha \sin \phi \cos \theta + \beta \sin \phi \sin \theta + \gamma \cos \phi = 0,$$

or

$$a \sin(\theta - \theta_0) = b \cot \phi. \tag{8.31}$$

Let the plane meet the x, y-plane in a line through the origin with polar angle $\theta = \Omega$ (the longitude of the node) and be inclined to the x, y-plane by an angle i (the inclination).

When $\theta = \Omega, \phi = \pi/2$ so we may take $\theta_0 = \Omega$. Let ϕ_m be the minimum ϕ takes on the plane, so $\phi_m + i = \pi/2$. ϕ_m gives the maximum value of $\cot \phi$ and $\sin$ has its maximum value of $+1$. Thus from (8.31) $a = b \cot \phi_m$ or $a \sin \phi_m = b \cos \phi_m$. Take $a = \cos \phi_m = \sin i$ and $b = \sin \phi_m = \cos i$. Therefore, the equation of a plane in spherical coordinates with the longitude of the node Ω and inclination i is

$$\sin i \sin(\theta - \Omega) = \cos i \cot \phi. \tag{8.32}$$

Use (8.30) to solve for Φ and substitute it into the equation for $\dot{\phi}$, then eliminate ρ^2 from the equations for $\dot{\phi}$ and $\dot{\theta}$, to obtain

$$\dot{\phi} = \frac{\Phi}{\rho^2} = \left\{ G^2 - \frac{\Theta^2}{\sin^2 \phi} \right\}^{1/2} \frac{1}{\rho^2} = \left\{ G^2 - \frac{\Theta^2}{\sin^2 \phi} \right\}^{1/2} \left\{ \frac{\sin^2 \phi \dot{\theta}}{\Theta} \right\}.$$

Separate variables and let $\theta = \Omega$ when $\phi = \pi/2$, so that Ω is the longitude of the node. Thus

$$\int_0^\phi \left\{ G^2 - \frac{\Theta^2}{\sin^2 \phi} \right\}^{-1/2} \sin^{-2} \phi d\phi = \int_\Omega^\theta \Theta^{-1} d\theta = (\theta - \Omega)/\Theta$$

$$= - \int_0^u \{ G^2 - \Theta^2 (1 + u^2) \}^{-1/2} du \tag{8.33}$$

$$= -\Theta^{-1} \int_0^u \{ \beta^2 - u^2 \}^{-1/2} du$$

$$= \Theta^{-1} \sin^{-1}(u/\beta).$$

The first substitution is $u = \cot\phi$ and β is defined by $\beta^2 = (G^2 - \Theta^2)/\Theta^2$. Therefore,

$$- \cot\phi = \pm\beta\sin(\theta - \Omega).$$

Finally

$$\cos i \cot\phi = \sin i \sin(\theta - \Omega), \tag{8.34}$$

where

$$\beta^2 = \frac{G^2 - \Theta^2}{\Theta^2} = \tan^2 i = \frac{\sin^2 i}{\cos^2 i}. \tag{8.35}$$

Equation (8.34) is the equation of the invariant plane. The above gives $\Theta = \pm G\cos i$. Because i is the inclination and Θ is the z-component of angular momentum this means that G is the magnitude of total angular momentum. In the above take θ_0 to be Ω, the longitude of the node.

8.8 Complex Coordinates

There are many different conventions used with complex coordinates in Hamiltonian mechanics. We present two of them here.

First, we look at a one degree of freedom problem defined in the plane $\mathbb{R}^2$ with coordinates x, y. Here x and y are conjugate variables in the sense of Hamiltonian mechanics.

Because the real numbers are a subset of the complex numbers we now think of x, y as coordinates in $\mathbb{C}^2$. Consider the change of coordinates from x, y to z, w given by

$$z = x - iy, \qquad w = x + iy,$$

$$x = \frac{z + w}{2}, \qquad y = \frac{w - z}{2i}. \tag{8.36}$$

If x, y are real numbers then z and w are conjugate complex numbers and conversely. That is, the real plane $\{(x, y) : x, y \in \mathbb{R}\} \subset \mathbb{C}^2$ is mapped linearly by (8.36) onto the plane $\{(z, w) : z, w \in \mathbb{C}^2,\ w = \bar{z}\}$ and conversely.

Consider a formal or convergent series

$$f(x, y) = \sum_i \sum_j a_{ij} x^i y^j,$$

and

$$g(z, w) = f\left(\frac{z + w}{2}, \frac{w - z}{2i}\right) = \sum_m \sum_n A_{mn} z^m w^n;$$

i.e., g is obtained from f by the change of variables (8.36). It is easy to see that f is a real series (i.e., all a_{ij} are real) if and only if

$$A_{mn} = \bar{A}_{nm}.$$

In other words $\overline{f(x,y)} \equiv f(\bar{x}, \bar{y})$ if and only if $\overline{g(z,w)} \equiv \bar{g}(w,z)$. These restrictions on g are called *reality conditions*.

The transformation (8.36) is symplectic with multiplier $2i$ and so the Hamiltonian $H(x,y)$ is replaced by $2iH((z+w)/2, (w-z)/(2i))$. (To eliminate the multiplier some people divide some of the terms in (8.36) by $2i$ or even $\sqrt{2i}$.)

As an example, consider the Hamiltonian of the harmonic oscillator

$$H = \frac{1}{2}(x^2 + y^2),$$

which is transformed by (8.36) to

$$H = izw$$

and the equations of motion become

$$\dot{z} = \frac{\partial H}{\partial w} = iz, \qquad \dot{w} = -\frac{\partial H}{\partial z} = -iw.$$

Because we are interested in real problems with real x and y we are only interested in z and w with $w = \bar{z}$. Thus we eliminate the proliferation of symbols by using z and $\bar{z}$.

For example, the Hamiltonian of the harmonic oscillator is written $H = iz\bar{z}$ and the equation of motion is $\dot{z} = iz$.

Second, consider a two degree of freedom problem defined in $\mathbb{R}^4$ with coordinates x, y where $x = (x_1, x_2)$ and $y = (y_1, y_2)$ are real 2-vectors and are conjugate in the sense of Hamiltonian mechanics. Consider the change of coordinates from (x_1, x_2, y_1, y_2) to $(z, \bar{z}, w, \bar{w})$ given by

$$z = x_1 + ix_2, \qquad w = \tfrac{1}{2}(y_1 - iy_2),$$

$$\bar{z} = x_1 - ix_2, \qquad \bar{w} = \tfrac{1}{2}(y_1 + iy_2). \tag{8.37}$$

This is a symplectic change of variables in as much as

$$dx_1 \wedge dy_1 + dx_2 \wedge dy_2 = dz \wedge dw + d\bar{z} \wedge d\bar{w}.$$

A prime example of the use of these coordinates is the regularization of collisions in the Kepler problem.

8.8.1 Levi–Civita Regularization

The Kepler problem has a removable singularity, namely the collision. The process of removing the singularity has become known as Levi–Civita regularization. Consider the planar Kepler problem with Hamiltonian

$$H = \frac{1}{2}|y|^2 - \frac{1}{|x|} + \frac{d^2}{2},$$

where now x, y are to be considered as complex numbers. The additive constant $d^2/2$ does not affect the equations of motion. Note that $H = 0$ corresponds to the traditional Hamiltonian equal to $-d^2/2$; i.e., $|y|^2/2 - 1/|x| = -d^2/2$.

First make a symplectic change of variables from x, y to z, w using the generating function

$$S(y, z) = yz^2,$$

$$x = \frac{\partial S}{\partial y} = z^2, \quad w = \frac{\partial S}{\partial z} = 2yz.$$

The Hamiltonian becomes

$$H = \frac{1}{8|z|^2}\left\{|w|^2 + 4d^2|z|^2 - 8\right\}.$$

The expression in the braces above is the Hamiltonian of two harmonic oscillators with equal frequencies. To remove the multiplicative factor we use a trick of Poincaré to change the time scale and yet still keep the Hamiltonian character of the problem.

The Poincaré trick applies to a general Hamiltonian of the form $H(x) = \phi(x)L(x)$ where $\phi(x)$ is a positive function. The equations of motion are $\dot{x} = J\nabla H = \phi J \nabla L + L J \nabla \phi$. On the level set $L = 0$, the equations of motion become $\dot{x} = \phi J \nabla L$ or $x' = J \nabla L$ where we reparameterize the time by $d\tau = \phi(x)dt$ and $' = d/d\tau$. Thus the flow defined by L on $L = 0$ is a reparameterization of the flow defined by H on $H = 0$.

Returning to the Kepler problem we change time by

$$4|z|^2 d\tau = dt,$$

and consider the Hamiltonian

$$L = \frac{1}{2}(|w|^2 + 4d^2|z|^2).$$

By the above remarks, the flow defined by H on the set $H = 0$ (energy equal to $-d^2$) and the flow defined by L on the set $L = 4$ are simply reparameterizations of each other.

Notice that the change has $x = z^2$ so z and $-z$ are mapped to the same x. The change becomes one-to-one if one identifies the antipodal points (z, w) and $(-z, -w)$. The set where $L = 4$ is a 3-sphere, S^3, and the 3-sphere with antipodal points identified is real projective three space RP^3. The flow defined by L on $L = 4$ does not have a singularity and so a collision orbit is carried to a regular orbit of the system defined by L.

The flow defined by L is two harmonic oscillators. One can think of it as the Hamiltonian of a particle of mass 1 that moves in the w-plane which is subjected to the force of a linear spring attached to the origin with spring constant $4d^2$. This is a super-integrable system that admits the three independent integrals

$$E_1 = \frac{1}{2}(w_1^2 + 4d^2 z_1^2), \quad E_2 = \frac{1}{2}(w_2^2 + 4d^2 z_2^2), \quad A = z_1 w_2 - z_2 w_1,$$

which are energy in the z_1 direction plane, energy in the z_2 direction, and angular momentum.

8.9 Delaunay and Poincaré Elements

There are two more sets of coordinates that make the 2-body problem particularly simple and thus simplify perturbation arguments. The first set of variables, the Delaunay elements, are valid for the elliptic orbits, and the second set, the Poincaré elements, are valid near the circular orbits of the Kepler problem

8.9.1 Planar Delaunay Elements

Here the ideas discussed in Section 8.4 are used to create action–angle variables for the Kepler problem. These variables are called Delaunay elements, named after the French astronomer of the 19th century who used these coordinates to develop his theory of the moon. Delaunay elements are valid only in the domain in phase space where there are elliptic orbits for the Kepler problem.

The Hamiltonian of the Kepler problem in symplectic polar coordinates (r, θ, R, Θ) is

$$H = \frac{1}{2}\left\{ R^2 + \frac{\Theta^2}{r^2} \right\} - \frac{1}{r}. \tag{8.38}$$

Angular momentum, Θ, is an integral, so for fixed $\Theta \neq 0$ this is a one degree of freedom system of the form discussed in Section 8.4, except the origin of the center is at $r = \Theta$ (the circular orbit). Set $H = -1/2L^2$ and solve for the r value of perigee (when $R = 0$) to get that $a = L[L - (L^2 - \Theta^2)^{1/2}]$. Note that $L^2 \geq \Theta^2$ and $L = \pm\Theta$ correspond to the circular orbits of the Kepler problem.

To change to Delaunay variables (ℓ, g, L, G) we use the generating function

$$W(r, \theta, L, G) = \theta G + \int_a^r \left\{ -\frac{G^2}{\xi^2} + \frac{2}{\xi} - \frac{1}{L^2} \right\}^{1/2} d\xi,$$

where $a = a(L, G) = L[L - (L^2 - G^2)^{1/2}]$. Thus

$$R = \frac{\partial W}{\partial r} = \left\{ -\frac{G^2}{r^2} + \frac{2}{r} - \frac{1}{L^2} \right\}^{1/2},$$

$$\Theta = \frac{\partial W}{\partial \theta} = G,$$

$$\ell = \frac{\partial W}{\partial L} = -\left\{ -\frac{G^2}{a^2} + \frac{2}{a} - \frac{1}{L^2} \right\}^{1/2} \frac{\partial a}{\partial L} + \int_a^r \left\{ -\frac{G^2}{\xi^2} + \frac{2}{\xi} - \frac{1}{L^2} \right\}^{-1/2} d\xi L^{-3}$$

$$= t/L^3,$$

$$g = \frac{\partial W}{\partial G} = \theta + \int_a^r \left\{ -\frac{G^2}{\xi^2} + \frac{2}{\xi} - \frac{1}{L^2} \right\}^{-1/2} \left(-\frac{G}{\xi^2} \right) d\xi = \theta - f$$

$$(8.39)$$

$G = \Theta$, so G is angular momentum. Solving for $-1/2L^2$ in the expression for R and using (8.38) yields $H = -1/2L^2$ as is expected.

The first quantity in the definition of ℓ is zero by the definition of a and the integral in the second quantity is just time, t. So $\ell = -t/L^3$ where t is measured from perigee, so ℓ, is measured from perigee also. Recall that to change independent variable from t (time) to f (true anomaly) in the solution of the Kepler problem we set $df = (\Theta/r^2)dt = (G/r^2)dt$. Thus because $d\ell = (-G^2/\xi^2 + 2/\xi - 1/L^2)^{-1/2}d\xi$ the integrand in the definition of g is just df, and the integral gives the true anomaly, f, measured from perigee. Thus $g = \theta - f$ is the argument of the perigee.

ℓ is known as the mean anomaly and it is an angular variable in that it is defined modulo 2π, but it is not measured in radians as is true anomaly f. Radian measure of an angle is the ratio of the arc length subtended by the angle to the circumference of the circle normalized by 2π. Because the time derivative of ℓ is constant it measures the area swept out by Kepler's second law. Thus a definition of the measure of ℓ is the ratio of the area sector swept to the total area normalized by 2π. That is,

$$\ell = 2\pi \frac{\text{area of sector swept out from perigee}}{\text{area of ellipse}}.$$

See Figure 8.3.

One says that the Kepler problem is solved, but that is true only up to a certain point. From the initial conditions one can easily compute the orbit of the solution, but not where the particle is on that orbit as a function of time. Refer to any classical text on celestial mechanics for a discussion of the relations between true anomaly f and mean anomaly ℓ.

8.9.2 Planar Poincaré Elements

The argument of the perihelion is clearly undefined for circular orbits, so, Delaunay elements are not valid coordinates in a neighborhood of the circular orbits. To overcome this problem, Poincaré introduced what he called Kepler variables but which have become known as Poincaré elements. Make the symplectic change of variables from the Delaunay variables (ℓ, g, L, G) to the Poincaré variables (Q_1, Q_2, P_1, P_2) by

$$Q_1 = \ell + g, \qquad Q_2 = [2(L-G)]^{1/2} \cos \ell,$$

$$P_1 = L, \qquad P_2 = [2(L-G)]^{1/2} \sin \ell.$$

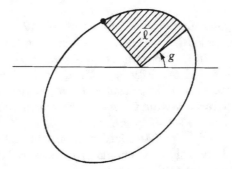

Figure 8.3. Delaunay angles.

The Hamiltonian of the Kepler problem (3.8) becomes

$$H = \frac{1}{2}(X^2 + Y^2) - \frac{1}{(x^2 + y^2)} = -\frac{1}{2P_1^2},$$

and the Hamiltonian of the Kepler problem in rotating coordinates becomes

$$H = \frac{1}{2}(X^2 + Y^2) - (xY - yX) - \frac{1}{(x^2 + y^2)} = -\frac{1}{2P_1^2} - P_1 + \frac{1}{2}(Q_2^2 + P_2^2).$$

Q_1 is an angular coordinate defined modulo 2π, and the remaining coordinates $Q_2, P_1,$ and P_2 are rectangular variables. $Q_2 = P_2 = 0$ correspond to the circular orbits of the 2-body problem. Even though these new coordinates are defined from the Delaunay elements, which are not defined on the circular orbits, it can be shown that these are valid coordinates in a neighborhood of the direct circular orbits. There is a similar set for the retrograde orbits.

8.9.3 Spatial Delaunay Elements

We change from spherical coordinates $(\rho, \theta, \phi, P, \Theta, \Phi)$ to Delaunay elements $(\ell, g, k, \ L, G, K)$ where the first three variables are angles defined $\mod 2\pi$. Consider the generating function

$$W(\rho, \theta, \phi, L, G, K) = \theta K + \int_{\pi/2}^{\phi} \left\{ G^2 - \frac{K^2}{\sin^2 \zeta} \right\}^{1/2} d\zeta +$$

$$\int_a^\rho \left\{ -\frac{G^2}{\xi^2} + \frac{2}{\xi} - \frac{1}{L^2} \right\}^{1/2} d\xi, \qquad (8.40)$$

where $a = a(L, G) = L[1 - (L^2 - G^2)]$. The change of coordinates is

$$P = \frac{\partial W}{\partial \rho} = \left\{ -\frac{G^2}{\rho^2} + \frac{2}{\rho} - \frac{1}{L^2} \right\}^{1/2}$$

$$\Theta = \frac{\partial W}{\partial \theta} = K$$

$$\Phi = \frac{\partial W}{\partial \phi} = \left\{ G^2 - \frac{K^2}{\sin^2 \phi} \right\}^{1/2}$$

$$\ell = \frac{\partial W}{\partial L} = \int_a^r \left\{ -\frac{G^2}{\xi^2} + \frac{2}{\xi} - \frac{1}{L^2} \right\}^{-1/2} d\xi L^{-3} = -t/L^3 \qquad (8.41)$$

$$g = \frac{\partial W}{\partial G} = -\int_{\pi/2}^{\phi} \left\{ G^2 - \frac{K^2}{\sin^2 \zeta} \right\}^{-1/2} G d\zeta$$

$$\qquad - \int_a^r \left\{ -\frac{G^2}{\xi^2} + \frac{2}{\xi} - \frac{1}{L^2} \right\}^{-1/2} \left(\frac{G}{\xi^2} \right) d\xi$$

$$\qquad = \sigma - f$$

$$k = \frac{\partial W}{\partial K} = \theta - \int_{\pi/2}^{\phi} \left\{ G^2 - \frac{K^2}{\sin^2 \zeta} \right\}^{-1/2} \left(\frac{K}{\sin^2 \zeta} \right) d\zeta = \Omega.$$

Because $\Theta = K$, K is the z-component of angular momentum, the expression for Φ gives that G is the magnitude of total angular momentum, and the expression for P ensures that $H = -1/2L^2$. $\ell = -t/L^3$, where t is measured from perigee, and so ℓ is the mean anomaly. The integral in the definition of k is the first integral in (8.33), so $k = \theta - (\theta - \Omega) = \Omega$, the longitude of the node.

The first integral in the formula for g is integrated as follows.

$$\int_{\pi/2}^{\phi} \left\{ G^2 - \frac{K^2}{\sin^2 \zeta} \right\}^{-1/2} G d\zeta = \int_{\pi/2}^{\phi} \left\{ 1 - \frac{\cos^2 i}{\sin^2 \zeta} \right\}^{-1/2} d\zeta$$

$$= \int_{\pi/2}^{\phi} \frac{\sin \zeta d\zeta}{\sqrt{\sin^2 \zeta - \cos^2 i}}$$

$$= -\int_{0}^{\cos \phi} \frac{du}{\sqrt{1 - u^2 - \cos^2 i}}, \qquad (u = \cos \zeta)$$

$$= -\int_{0}^{\cos \phi} \frac{du}{\sqrt{\sin^2 i - u^2}}$$

$$= -\sin^{-1} \left(\frac{\cos \phi}{\sin i} \right)$$

$$= -\sigma.$$

Therefore,

$$\sin \sigma = \frac{\sin(\pi/2 - \phi)}{\sin i} = \frac{\sin \psi}{\sin i}.$$

The angle σ is defined by the spherical triangle with sides (arcs measured in radians) θ, $\psi = \pi/2 - \phi$, σ, and spherical angle i. Recall the law of sines for spherical triangles and see Figure 8.4.

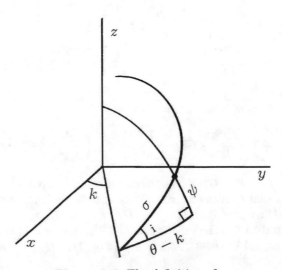

Figure 8.4. The definition of σ.

Thus σ measures the position of the particle in the invariant plane. Because f is the true anomaly measured from the perigee in the invariant plane, $g = \sigma - f$ is the argument of the perigee measured in the invariant plane.

8.10 Pulsating Coordinates

A generalization of rotating coordinates are pulsating coordinates. These are coordinates that rotate and scale with a solution of the Kepler problem. These coordinates are the natural coordinates to use when discussing the elliptic restricted 3-body problem.

Let us recall some basic formulas from the Kepler problem and its solution. Let $\phi = (\phi_1, \phi_2)$ be any solution of the planar Kepler problem, r the length of ϕ, and c its angular momentum, so that

$$\ddot{\phi} = -\frac{\phi}{\|\phi\|^3}, \qquad r = \sqrt{\phi_1^2 + \phi_2^2}, \qquad c = \phi_1\dot{\phi}_2 - \phi_2\dot{\phi}_1, \tag{8.42}$$

where the independent variable is t, time, and $\dot{} = d/dt$, $\ddot{} = d^2/dt^2$. Rule out collinear solutions by assuming that $c \neq 0$ and then scale time so that $c = 1$. The units of distance and mass are chosen so that all other constants are 1. In polar coordinates (r, θ), the equations become

$$\ddot{r} - r\dot{\theta}^2 = -1/r^2, \qquad d(r^2\dot{\theta})/dt = dc/dt = r\ddot{\theta} + 2\dot{r}\dot{\theta} = 0. \tag{8.43}$$

The fact that $c = r^2\dot{\theta} = 1$ is a constant of motion yields

$$\ddot{r} - 1/r^3 = -1/r^2. \tag{8.44}$$

Equation (8.44) is reduced to a harmonic oscillator $u'' + u = 1$ by letting $u = 1/r$ and changing from time t to f the true anomaly of the Kepler problem, by $dt = r^2 df$ and $' = d/df$. The general solution is then

$$r = r(f) = 1/(1 + e\cos(f - \omega)), \tag{8.45}$$

where e and ω are integration constants, e being the eccentricity and ω the argument of the pericenter. When $e = 0$, the orbit is a circle, when $0 < e < 1$, an ellipse, when $e = 1$, a parabola, and when $e > 1$, a hyperbola. There is no harm in assuming that the argument of the pericenter is zero, so henceforth $\omega = 0$.

Define a 2×2 matrix by

$$A = \begin{bmatrix} \phi_1 & -\phi_2 \\ \phi_2 & \phi_1 \end{bmatrix}, \tag{8.46}$$

so $A^{-1} = (1/r^2)A^T$ and $A^{-T} = (A^T)^{-1} = (1/r^2)A$, where A^T denotes the transpose of A.

Consider the planar 3-body problem in fixed rectangular coordinates (q, p) given by the Hamiltonian

$$H = H_3 = \sum_{i=1}^{3} \frac{\|p_i\|^2}{2m_i} - U(q), \quad U(q) = \sum_{1 \le i < j \le 3} \frac{m_i m_j}{\|q_i - q_j\|}. \tag{8.47}$$

Pulsating coordinates are the symplectic coordinates defined below by two symplectic coordinate changes. First, make the symplectic change of coordinates

$$q_i = AX_i, \quad p_i = A^{-T} Y_i = (1/r^2) AY_i \quad \text{for } i = 1, 2, 3. \tag{8.48}$$

Recall that if $H(z)$ is a Hamiltonian and $z = T(t)u$ is a linear symplectic change of coordinates, then the Hamiltonian becomes $H(u) + (1/2)u^T W(t)u$, where W is the symmetric matrix $W = JT^{-1}\dot{T}$. Compute

$$W = \begin{bmatrix} 0 & I \\ -I & 0 \end{bmatrix} \begin{bmatrix} r^{-2}A^T & 0 \\ 0 & A^T \end{bmatrix} \begin{bmatrix} \dot{A} & 0 \\ 0 & (r^{-2}\dot{A} - 2r^{-3}\dot{r}A) \end{bmatrix}$$

$$= \begin{bmatrix} 0 & -r^{-2}(A^T \dot{A})^T \\ -r^{-2}A^T \dot{A} & 0 \end{bmatrix}. \tag{8.49}$$

Recall that W is symmetric, or use $A^T A = r^2 I$, to get the 1,2 position. Now

$$-r^{-2}A^T \dot{A} = r^{-2} \begin{bmatrix} -r\dot{r} & 1 \\ -1 & -r\dot{r} \end{bmatrix}. \tag{8.50}$$

Note that $\|AX\| = r\|X\|$, so the Hamiltonian becomes

$$H = \frac{1}{r^2} \sum_{i=1}^{N} \frac{\|Y_i\|^2}{2m_i} - \frac{1}{r} U(X) - \frac{\dot{r}}{r} \sum_{i=1}^{N} X_i^T Y_i - \frac{1}{r^2} \sum_{i=1}^{N} X_i^T J Y_i. \tag{8.51}$$

Change the independent variable from time t to f, the true anomaly of the Kepler problem, by $dt = r^2 df, ' = d/df, H \to r^2 H$ so that

$$H = \sum_{i=1}^{N} \frac{\|Y_i\|^2}{2m_i} - rU(X) - \frac{r'}{r} \sum_{i=1}^{N} X_i^T Y_i - \sum_{i=1}^{N} X_i^T J Y_i. \tag{8.52}$$

The second symplectic change of variables changes only the momentum by letting

$$X_i = Q_i, \quad Y_i = P_i + \alpha_i Q_i, \tag{8.53}$$

where the $\alpha_i = \alpha_i(f)$ are to be determined. This defines the pulsating coordinates (Q_i, P_i) for $i = 1, \dots, N$. To compute the remainder term, consider

$$R_i = \begin{bmatrix} 0 & I \\ -I & 0 \end{bmatrix} \begin{bmatrix} I & 0 \\ -\alpha_i I & I \end{bmatrix} \begin{bmatrix} 0 & 0 \\ \alpha_i' & 0 \end{bmatrix} = \begin{bmatrix} \alpha_i' & 0 \\ 0 & 0 \end{bmatrix}. \tag{8.54}$$

Thus the remainder term is $(1/2) \sum \alpha_i'(f) Q_i^T Q_i$ and the Hamiltonian becomes

$$H = \sum_{i=1}^{N} \frac{\|P_i\|^2}{2m_i} - rU(Q) + \left(\frac{\alpha_i}{m_i} - \frac{r'}{r}\right) \sum_{i=1}^{N} Q_i^T P_i$$

$$- \sum_{i=1}^{N} Q_i^T J P_i + \sum_{i=1}^{N} \left(\frac{1}{2}\alpha_i' + \frac{1}{2}\frac{\alpha_i^2}{m_i} - \frac{r'}{r}\alpha_i\right) Q_i^T Q_i. \tag{8.55}$$

Choose α_i so that the third term on the right in (8.55) vanishes; i.e., take $\alpha_i = m_i r'/r$. To compute the coefficient of $Q_i^T Q_i$ in the last sum in (8.55), note that

$$\left(\frac{r'}{r}\right)' - \left(\frac{r'}{r}\right)^2 = \frac{rr'' - 2r(r')^2}{r^2} = r\frac{d}{df}\left(\frac{r'}{r^2}\right) = r\frac{d\dot{r}}{df} = r^3\ddot{r} = 1 - r, \tag{8.56}$$

where the last equality comes from the formula (8.44). Thus the Hamiltonian of the N-body problem in pulsating coordinates is

$$H = \sum_{i=1}^{N} \frac{\|P_i\|^2}{2m_i} - rU(Q) - \sum_{i=1}^{N} Q_i^T J P_i + \frac{(1-r)}{2} \sum_{i=1}^{N} m_i Q_i^T Q_i, \tag{8.57}$$

and the equations of motion are

$$Q'_i = \frac{P_i}{m_i} - JQ_i,$$

$$P'_i = r\frac{\partial U}{\partial Q_i} - JP_i - (1-r)m_iQ_i. \tag{8.58}$$

These are particularly simple equations considering the complexity of the coordinate change.

8.10.1 The Elliptic Restricted 3-Body Problem

Consider the 3-body problem in pulsating coordinates but index from 0 to 2. Recall that a central configuration of the 3-body problem is a solution (Q_0, Q_1, Q_2) of the system of nonlinear algebraic equations

$$\frac{\partial U}{\partial Q_i} + \lambda m_i Q_i = 0 \quad \text{for } i = 0, 1, 2 \tag{8.59}$$

for some scalar λ. By scaling the distance, λ may be taken as 1. Thus a central configuration is a geometric configuration of the three particles so that the force on the ith particle is proportional to m_i times the position. This is the usual definition of a central configuration. Define a relative equilibrium as a critical point of the Hamiltonian of the N-body problem in pulsating coordinates. This is slightly different from the usual definition of a relative equilibrium.

Lemma 8.10.1. *The relative equilibria are central configurations.*

Proof. The critical points of (8.57) satisfy

$$\frac{\partial H}{\partial Q_i} = -r\frac{\partial U}{\partial Q_i} + JP_i + (1-r)m_iQ_i = 0,$$

$$\frac{\partial H}{\partial P_i} = \frac{P_i}{m_i} - JQ_i = 0.$$

From the second equation $P_i = m_iJQ_i$. Plugging this into the first equation gives

$$-r\partial U/\partial Q_i - m_iQ_i + (1-r)m_iQ_i = -r\{\partial U/\partial Q_i + m_iQ_i\} = 0.$$

Because r is positive, this equation is satisfied if and only if $\partial U/\partial Q_i + m_iQ_i = 0$.

Let H_3 and U_3 be the Hamiltonian and self-potential of the 3-body problem written in pulsating coordinates. Consider also the 2-body problem with particles indexed from 1 to 2 with H_2 and U_2 the Hamiltonian and self-potential of the 2-body problem written in pulsating coordinates. We have

$$H_3 = \sum_{i=0}^{2} \frac{\|P_i\|^2}{m_i} - rU_2(Q) - \sum_{i=0}^{2} Q_i^T JP_i + \frac{(1-r)}{2}\sum_{i=0}^{N} m_iQ_i^T Q_i$$

$$\tag{8.60}$$

$$= \frac{\|P_0\|^2}{2m_0} - r\sum_{j=1}^{N} \frac{m_0m_j}{\|Q_0 - Q_j\|} - Q_0^T JP_0 + \frac{(1-r)}{2}m_0Q_0^T Q_0 + H_2.$$

Assume that one mass is small by setting $m_0 = \varepsilon^2$. The zeroth body is known as the infinitesimal and the other two bodies are known as the primaries. Let Z be the coordinate vector for the 2-body problem, so $Z = (Q_1, Q_2, P_1, P_2)$, and let $Z^* = (a_1, a_2, b_1, b_2)$ be any central configuration for the 2-body problem. By Lemma 8.10.1, $\nabla H_2(Z^*) = 0$. The Taylor expansion for H_2 is

$$H_2(Z) = H_2(Z^*) + \frac{1}{2}(Z - Z^*)^T S(f)(Z - Z^*) + \cdots,$$

where $S(f)$ is the Hessian of H_N at Z^*. Forget the constant term $H_2(Z^*)$. Change coordinates by

$$Q_0 = \xi, \quad P_0 = \varepsilon^2\eta, \quad Z - Z^* = \varepsilon V. \tag{8.61}$$

This is a symplectic transformation with multiplier ε^{-2}. Making this change of coordinates in (8.60) yields

$$H_3 = R + \frac{1}{2}V^T S(f)V + O(\varepsilon), \tag{8.62}$$

where R is the Hamiltonian of the conic (i.e., circular, elliptic, etc.) restricted 3-body problem given by

$$R = \frac{1}{2}\|\eta\|^2 - r\sum_{i=1}^{N}\frac{m_i}{\|\xi - a_i\|} - \xi^T J\eta + \frac{(1-r)}{2}\xi^T\xi. \qquad (8.63)$$

To the zeroth order, the equations of motion are

$$\xi' = \eta + J\xi,$$

$$\eta' = r\sum_{i=1}^{N}\frac{m_i(\xi - a_i)}{\|\xi - a_i\|^3} + J\eta - (1-r)\xi, \qquad (8.64)$$

$$V' = D(f)V, \qquad D(f) = JS(f). \qquad (8.65)$$

The equations in (8.64) are the equations of the conic restricted problem and those in (8.65) are the linearized equations of motion about the relative equilibrium.

When $e = 0$, Equations (8.64) and (8.65) are time-independent and (8.63) is the Hamiltonian of the circular restricted 3-body problem. When $0 < e < 1$, Equations (8.64) and (8.65) are 2π-periodic in f and (8.63) is the Hamiltonian of the elliptic restricted 3-body problem.

In the classical elliptic restricted 3-body problem the masses of the primaries are $m_1 = 1 - \mu > 0, m_2 = \mu > 0$, and they are located at $a_1 = (-\mu, 0), a_2 = (1 - \mu, 0)$. The parameter μ is called the mass ratio parameter. Thus the Hamiltonian of the classical elliptic 3-body problem is

$$R = \frac{1}{2}\|\eta\|^2 - r\left(\frac{1-\mu}{d_1} + \frac{\mu}{d_2}\right) - \xi^T J\eta + \frac{(1-r)}{2}\xi^T\xi, \qquad (8.66)$$

where

$$d_1 = \{(\xi_1 + \mu)^2 + \xi_2^2\}^{1/2}, \qquad d_2 = \{(\xi_1 - 1 + \mu)^2 + \xi_2^2\}^{1/2},$$

$$r = r(f) = 1/(1 + e\cos f), \qquad 0 < e < 1. \qquad (8.67)$$

8.11 Problems

1. Write the functions $r^{2k}, r^{2k+5}\cos 5\theta$, and $r^{2k+5}\sin 5\theta$ in rectangular coordinates. Sketch the level curves of $r^2 + r^5\cos 5\theta$.
2. In Section 8.6.1, the equations for the Kepler problem were written in polar coordinates. Because angular momentum, Θ, is a constant set $\Theta = c$, investigate the equation for r, $\ddot{r} = \dot{R} = -c^2/r^3 + \mu/r^2$, using geometric methods.

3. In Section 8.8.1 the three integrals E_1, E_2, A of the regularized Kepler problem were given. Compute the total algebra of integrals of the regularized Kepler problem.

9. Poincaré's Continuation Method

This chapter shows how to introduce and exploit a small parameter in order to obtain periodic solutions in some simple Hamiltonian systems. When the small parameter is zero, a periodic solution is obvious, and then by using the implicit function theorem it is shown that the periodic solution is in a family of periodic solutions. Poincaré used these ideas extensively, and they have become known as the *Poincaré's continuation method*; see Poincaré (1899).

After giving some elementary general results, a variety of families of periodic solutions in the restricted problem and the 3-body problem are given. The first result is a simple proof of the Lyapunov center theorem with applications to the five libration points in the restricted problem. Then in the next three sections, the circular orbits of Kepler's problem are continued into the restricted problem when one mass is small (Poincaré's solutions), when the infinitesimal is near a primary (Hill's solutions), and when the infinitesimal is near infinity (comet solutions). Lastly, a general theorem on the continuation of periodic solutions from the restricted problem to the full 3-body problem is given. A more complete discussion with more general results can be found in Moulton (1920) and Meyer (1999).

9.1 Continuation of Solutions

Assume that the differential equations depend on some parameters, so consider

$$\dot{x} = f(x, \nu), \tag{9.1}$$

where $f : \mathcal{O} \times \mathcal{Q} \longrightarrow \mathbb{R}^m$ is smooth, $\mathcal{O}$ (the domain) is open in $\mathbb{R}^m$, and $\mathcal{Q}$ (the parameter space) is open in $\mathbb{R}^k$. The general solution $\phi(t, \xi, \nu)$ is smooth in the parameter ν also.

First let ξ' be an equilibrium point when $\nu = \nu'$, i.e., $f(\xi', \nu') = 0$. A continuation of this equilibrium point is a smooth function $u(\nu)$ defined for ν near ν' such that $u(\nu') = \xi'$, and $u(\nu)$ is an equilibrium point for all ν, i.e., $f(u(\nu), \nu) = 0$.

© Springer International Publishing AG 2017
K.R. Meyer, D.C. Offin, *Introduction to Hamiltonian Dynamical Systems and the N-Body Problem*, Applied Mathematical Sciences 90,
DOI 10.1007/978-3-319-53691-0_9

Recall (see Section 6.4) that an equilibrium point ξ' for (9.1) when $\nu = \nu'$, satisfies $f(\xi', \nu') = 0$, and it is elementary if $\partial f(\xi', \nu')/\partial \xi$ is nonsingular, i.e., if zero is not an exponent.

Now let the solution $\phi(t, \xi', \nu')$ be T-periodic. A continuation of this periodic solution is a pair of smooth functions, $u(\nu), \tau(\nu)$, defined for ν near ν' such that $u(\nu') = \xi', \tau(\nu') = T$, and $\phi(t, u(\nu), \nu)$ is $\tau(\nu)$-periodic. One also says that the periodic solution can be continued. This means that the solution persists when the parameters are varied, and the periodic solution does not change very much with the parameters.

The solution $\phi(t, \xi', \nu')$ is T-periodic if and only if $\phi(T, \xi', \nu') = \xi'$. This periodic solution is elementary if $+1$ is an eigenvalue of the monodromy matrix $\partial \phi(T, \xi', \nu')/\partial \xi$ with multiplicity one for a general autonomous differential equation and of multiplicity two for a system with a nondegenerate integral (e.g., a Hamiltonian system). Recall that the eigenvalues of $\partial \phi(T, \xi', \nu')/\partial \xi$ are called the multipliers (of the periodic solution). Drop one $+1$ multiplier for a general autonomous equation, and drop two $+1$ multipliers from the list of multipliers for an autonomous system with an integral to get the nontrivial multipliers. If the nontrivial multipliers are not equal to one then the periodic solution is called elementary.

Proposition 9.1.1. *An elementary equilibrium point, or an elementary periodic solution, or an elementary periodic solution in a system with a nondegenerate integral can be continued.*

Proof. For equilibrium points, apply the implicit function theorem to the function $f(x, \nu) = 0$. By assumption, $f(\xi', \nu') = 0$, and $\partial f(\xi', \nu')/\partial x$ is nonsingular, so the implicit function theorem asserts the existence of the function $u(\nu)$ such that $u(\nu') = \xi'$. and $f(u(\nu), \nu) \equiv 0$.

Now consider the periodic case. Because the existence of the first return time and the Poincaré map depended on the implicit function theorem, these functions depend smoothly on the parameter ν. For the rest of the proposition, apply the implicit function theorem to $P(x, \nu) - x = 0$, where $P(x, \nu)$ is the Poincaré map of the cross-section to the periodic solution when $\nu = \nu'$

Similarly, if the system has an integral $I(x, \nu)$, then the construction of the Poincaré map in an integral surface depends smoothly on ν. Again apply the implicit function theorem to the map $Q(x, \nu) - x = 0$, where $Q(x, \nu)$ is the Poincaré map in the integral surface of the cross-section to the periodic solution when $\nu = \nu'$.

There is a similar definition of continuation and a similar lemma for fixed points.

Corollary 9.1.1. *The exponents of an elementary equilibrium point, and the multipliers of an elementary periodic solution (with or without nondegenerate integral) vary continuously with the parameter ν.*

Proof. For equilibrium points, the implicit function theorem was applied to $f(x, \nu) = 0$ to get a function $u(\nu)$ such that $u(\nu') = \xi'$ and $f(u(\nu), \nu) \equiv 0$. The exponents of the equilibrium $u(\nu)$ are the eigenvalues of $\partial f(u(\nu), \nu)/\partial x$. This matrix varies smoothly with the parameter ν, and so its eigenvalues vary continuously with the parameter ν. (See the Problem section for an example where the eigenvalues are not smooth in a parameter.) The other parts of the theorem are proved using the same idea applied to the Poincaré map.

Corollary 9.1.2. *A small perturbation of an elliptic (respectively, a hyperbolic) periodic orbit of a Hamiltonian system of two degrees of freedom is elliptic (respectively, hyperbolic).*

Proof. If the system has two degrees of freedom, then a periodic solution has as multipliers $+1, +1, \lambda, \bar{\lambda} = \lambda^{-1}$, and so the multipliers lie either on the real axis or on the unit circle. If the periodic solution is hyperbolic, then λ and λ^{-1} lie on the real axis and are not 0 or ± 1. A small change cannot make these eigenvalues lie on the unit circle or take the value 0 or ± 1. Thus a small change in a hyperbolic periodic solution is hyperbolic. A similar argument holds for elliptic periodic solutions.

9.2 Lyapunov Center Theorem

An immediate consequence of the discussion in the previous section is the following celebrated theorem.

Theorem 9.2.1 (Lyapunov center theorem). *Assume that the system*

$$\dot{x} = f(x)$$

admits a nondegenerate integral and has an equilibrium point with exponents $\pm \omega i, \lambda_3, \ldots, \lambda_m$, where $i\omega \neq 0$ is pure imaginary. If $\lambda_j/i\omega$ is never an integer for $j = 3, \ldots, m$, then there exists a one-parameter family of periodic orbits emanating from the equilibrium point. Moreover, when approaching the equilibrium point along the family, the periods tend to $2\pi/\omega$ and the nontrivial multipliers tend to $\exp(2\pi\lambda_j/\omega), j = 3, \ldots, m$.

Remark. The Hamiltonian is always a nondegenerate integral.

Proof. Say that $x = 0$ is the equilibrium point, and the equation is

$$\dot{x} = Ax + g(x),$$

where $g(0) = \partial g(0)/\partial x = 0$. Because we seek periodic solutions near the origin, scale by $x \to \epsilon x$ where ϵ is to be considered as a small parameter. The equation becomes

$$\dot{x} = Ax + O(\epsilon),$$

and when $\epsilon = 0$, the system is linear. Because this linear system has exponents $\pm\omega i$, it has a periodic solution of period $2\pi/\omega$ of the form $\exp(At)a$, where a is a fixed nonzero vector. The multipliers of this periodic solution are the eigenvalues of $\exp(A2\pi/\omega)$, or $1, 1, \exp(2\pi\lambda_j/\omega)$. By assumption, the nontrivial multipliers are not $+1$, and so this periodic solution is elementary. From Proposition 9.1.1, there is a periodic solution of the form $\exp(At)a + O(\epsilon)$. In the unscaled coordinates, the solution is of the form $\epsilon \exp(At)a + O(\epsilon^2)$, and the result follows.

A more careful analysis shows that this family of periodic solutions lies on a smooth surface which is even analytic if the equation is analytic – see Siegel and Moser (1971).

9.2.1 Lyapunov Families at the Euler and Lagrange points

In Section 4.5, the linearized equations at the five libration (equilibrium) points of the restricted 3-body problem were analyzed. The eigenvalues at the three collinear libration points L_1, L_2, L_3 of Euler were shown to be a pair of real eigenvalues and a pair of pure imaginary eigenvalues. Thus the Lyapunov center theorem implies that there is a one-parameter family of periodic solutions emanating from each of these libration points.

By symmetry we may assume that $0 < \mu \le 1/2$. At the equilateral triangle libration points L_4, L_5 of Lagrange, the characteristic equation of the linearized system was found to be

$$\lambda^4 + \lambda^2 + \frac{27}{4}\mu(1 - \mu),$$

where μ is the mass ratio parameter. The roots of this polynomial satisfy

$$\lambda^2 = \frac{1}{2}(-1 \pm \sqrt{1 - 27\mu(1 - \mu)}),$$

which implies that for $0 < \mu < \mu_1 = (1 - \sqrt{69}/9)/2 \simeq 0.0385$, the eigenvalues are distinct pure imaginary numbers $\pm\omega_1 i, \pm\omega_2 i$, with $0 < \omega_2 < \omega_1$. Because $i\omega_2/i\omega_1$ is less than 1 in modulus, the Lyapunov center theorem implies that there is a family of periodic orbits emanating from $\mathcal{L}_4$ with period approaching $2\pi/\omega_1$ for all $\mu, 0 < \mu < \mu_1$. This family is called the *short period family*.

Define μ_r to be the value of μ for which $\omega_1/\omega_2 = r$. If $0 < \mu < \mu_1$ and $\mu \ne \mu_r, r = 1, 2, \ldots$, then the Lyapunov center theorem implies that there is a family of periodic orbits emanating from $\mathcal{L}_4$ with period approaching $2\pi/\omega_2$. This family is called the *long period family*.

The mass ratio μ_r satisfies

$$\mu_r = \frac{1}{2} - \frac{1}{2}\left\{1 - \frac{16r^2}{27(r^2+1)^2}\right\}^{1/2},$$

so, $0\cdots < \mu_3 < \mu_2 < \mu_1$.

What else happens at and near μ_r has generated a lot of papers, see Meyer et al. (2015) and the references therein.

9.3 Poincaré's Orbits

The essence of the continuation method is that the equations contain a parameter, and for one value of the parameter, there is a periodic solution whose multipliers can be computed. The restricted 3-body problem has a parameter μ, the mass ratio parameter, and when $\mu = 0$, the equation is just the Kepler problem in rotating coordinates.

The Kepler problem has many periodic solutions, but they all have their multipliers equal to $+1$ in fixed coordinates, but as we shall see the circular orbits have nontrivial multipliers in rotating coordinates. Thus the circular solutions of the Kepler problem can be continued into the restricted problem for small values of μ.

The reason that all the multipliers are $+1$ for the Kepler problem in fixed coordinates is that all the periodic solutions in an energy level have the same period and so are not isolated in an energy level (see Proposition 6.4.2).

Recall that the Hamiltonian of the restricted problem is

$$H = \frac{\|y\|^2}{2} - x^T K y - \frac{\mu}{d_1} - \frac{1-\mu}{d_2}, \tag{9.2}$$

where $d_1^2 = (x_1 - 1 + \mu)^2 + x_2^2$, $d_2^2 = (x_1 + \mu)^2 + x_2^2$, and

$$K = \begin{bmatrix} 0 & 1 \\ -1 & 0 \end{bmatrix}.$$

The term $x^T K y$ comes from that fact that the Hamiltonian is written in rotating coordinates. It is the Coriolis term. Consider μ as a small parameter, so the Hamiltonian is of the form

$$H = \frac{\|y\|^2}{2} - x^T K y - \frac{1}{\|x\|} + O(\mu). \tag{9.3}$$

Be careful of the $O(\mu)$ term because it has terms that go to infinity near the primaries; therefore, a neighborhood of the primaries must be excluded. When $\mu = 0$, this is the Kepler problem in rotating coordinates. Put this problem in polar coordinates (see Section 8.1) to get (when $\mu = 0$)

$$H = \frac{1}{2}\left(R^2 + \frac{\Theta^2}{r^2}\right) - \Theta - \frac{1}{r}, \tag{9.4}$$

$$\dot{r} = R, \qquad \dot{R} = \frac{\Theta^2}{r^3} - \frac{1}{r^2},$$

$$\dot{\theta} = \frac{\Theta}{r^2} - 1, \quad \dot{\Theta} = 0. \tag{9.5}$$

Angular momentum Θ is an integral, so let $\Theta = c$ be a fixed constant. For $c \neq 1$, the circular orbit $R = 0$, $r = c^2$ is a periodic solution with period $|\, 2\pi c^3/(1 - c^3)\,|$ (this is the time for θ to increase by 2π). Linearizing the r and R equations about this solution gives

$$\dot{r} = R, \qquad \dot{R} = -c^{-6}r,$$

which has solutions of the form $\exp(\pm it/c^3)$, and so the nontrivial multipliers of the circular orbits are $\exp(\pm i2\pi/(1 - c^3))$ which are not $+1$, provided $1/(1 - c^3)$ is not an integer. Thus we have proved the following theorem.

Theorem 9.3.1 (Poincaré). *If $c \neq 1$ and $1/(1 - c^3)$ is not an integer, then the circular orbits of the Kepler problem in rotating coordinates with angular momentum c can be continued into the restricted problem for small values of μ. These orbits are elliptic.*

The rotating coordinates used here rotate counterclockwise, and so in fixed coordinates the primaries rotate clockwise. If $c < 0$, then $\dot{\theta} < 0$, and $1/(1-c^3)$ is never an integer. Orbits with negative angular momentum rotate clockwise in either coordinate system and so are called retrograde orbits.

If $c > 0$, $c \neq 1$, and $1/(1 - c^3)$ is not an integer, then in the fixed coordinates, these orbits rotate counterclockwise and so are called direct orbits. The circular orbits of the Kepler problem when $1/(1 - c^3)$ is an integer, say k, undergo a bifurcation when $\mu \neq 0$, but this is too lengthy a problem to be discussed here. For a start see Schmidt (1972).

9.4 Hill's Orbits

Another way to introduce a small parameter is to consider the infinitesimal particle to be very near one of the primaries. This is usually referred to as Hill's problem because he extensively investigated the motion of the moon, which to a first approximation is like this problem.

Consider the restricted problem where one primary is at the origin, i.e., replace x_1 by $x_1 - \mu$ and y_2 by $y_2 - \mu$, so the Hamiltonian (9.2) becomes

$$H = \frac{\|y\|^2}{2} - x^T K y - \frac{\mu}{d_1} - \frac{1 - \mu}{d_2} + \mu x_1 - \frac{1}{2}\mu^2,$$

where $d_1^2 = (x_1 - 1)^2 + x_2^2$, $d_2^2 = x_1^2 + x_2^2$. Introduce a scale parameter ϵ by changing coordinates by $x = \epsilon^2 \xi$, $y = \epsilon^{-1} \eta$, which is a symplectic change of coordinates with multiplier ϵ^{-1}. In the scaling, all constant terms are dropped from the Hamiltonian because they do not affect the equations of motion. Note that if $\|\xi\|$ is approximately 1, then $\|x\|$ is about ϵ^2, or $\|x\|$ is very small when ϵ is small. Thus ϵ is a measure of the distance of the infinitesimal particle from the primary at the origin and so is considered as the small parameter. We fix the mass ratio parameter, μ, as arbitrary (i.e., not small), and for simplicity we set $c^2 = 1 - \mu, c > 0$. The Hamiltonian becomes

$$H = \epsilon^{-3} \left\{ \frac{\|\eta\|^2}{2} - \frac{c^2}{\|\xi\|} \right\} - \xi^T K \eta + O(\epsilon).$$

The dominant term is the Hamiltonian of the Kepler problem, and the next most important term is the rotational term, so this formula says that when the infinitesimal is close to the primary that has mass $c^2 = 1 - \mu$ the main force on it is the gravitational force of that primary. The next most important term is the Coriolis term.

Kepler's third law says that the period of a circular orbit varies with the radius to the 3/2 power, so time should be scaled by $t \longrightarrow \epsilon^{-3} t$ and $H \longrightarrow \epsilon^3 H$, and the Hamiltonian is

$$H = \left\{ \frac{\|\eta\|^2}{2} - \frac{c^2}{\|\xi\|} \right\} - \epsilon^3 \xi^T K \eta + O(\epsilon^4).$$

Introduce polar coordinates as before, so,

$$H = \frac{1}{2} \left(R^2 + \frac{\Theta^2}{r^2} \right) - \frac{c^2}{r} - \epsilon^3 \Theta + O(\epsilon^4),$$

$$\dot{r} = R, \qquad \dot{R} = \frac{\Theta^2}{r^3} - \frac{c^2}{r^2},$$

$$\dot{\theta} = \frac{\Theta}{r^2} - \epsilon^3, \qquad \dot{\Theta} = 0.$$

(9.6)

In (9.6) the terms of order ϵ^4 have been omitted. Omitting these terms makes Θ an integral. The two solutions, $\Theta = \pm c, R = 0, r = 1$, are periodic solutions of (9.6) of period $2\pi/(c \mp \epsilon^3)$. Linearizing the r and R equations about this solution gives

$$\dot{r} = R, \qquad \dot{R} = -c^2 r.$$

These linear equations have solutions of the form $\exp(\pm ict)$, and so the nontrivial multipliers of the circular orbits of (9.6) are $\exp(\pm ic2\pi/(c \mp \epsilon^3)) = +1 \pm \epsilon^3 2\pi i/c + O(\epsilon^6)$.

Consider the period map in a level surface of the Hamiltonian about this circular orbit. Let u be the coordinate in this surface, with $u = 0$, corresponding to the circular orbit when $\epsilon = 0$. From the above, the period map has a fixed point at the origin up to terms of order ϵ^3 and is the identity up to terms of order ϵ^2, and at ϵ^3 there is a term whose Jacobian has eigenvalues $\pm 2\pi i/c$. That is, the period map is of the form $P(u) = u + \epsilon^3 p(u) + O(\epsilon^4)$, where $p(0) = 0$, and $\partial p(0)/\partial u$ has eigenvalues $\pm 2\pi i/c$, so, in particular, $\partial p(0)/\partial u$ is nonsingular. Apply the implicit function theorem to $G(u, \epsilon) = (P(u) - u)/\epsilon^3 = p(u) + O(\epsilon)$. Because $G(0, 0) = 0$ and $\partial G(0, 0)/\partial u = \partial p(0)/\partial u$, there is a smooth function $\bar{u}(\epsilon)$ such that $G(\bar{u}(\epsilon), \epsilon) = 0$ for all ϵ sufficiently small. Thus the two solutions can be continued from the equations in (9.6) to the full equations, where the $O(\epsilon^4)$ terms are included. These solutions are elliptic also.

In the scaled variables, these solutions have $r, \simeq 1$ and period $T \simeq 2\pi$. In the original unscaled variables, the periodic solution has $\|x\| \simeq \epsilon^2$ with period $T \simeq 2\pi \epsilon^{-3}$.

Theorem 9.4.1. *There exist two one-parameter families of nearly circular elliptic periodic solutions of the restricted 3-body problem that encircle a primary for all values of the mass ratio parameter.*

9.5 Comets

Another way to introduce a small parameter is to consider orbits that are close to infinity. In the Hamiltonian of the restricted problem (9.2), scale the variables by $x \longrightarrow \epsilon^{-2}x$, $y \longrightarrow \epsilon y$; this is symplectic with multiplier ϵ. The Hamiltonian becomes

$$H = -x^T K y + \epsilon^3 \left\{ \frac{\|y\|^2}{2} - \frac{1}{\|x\|} \right\} + O(\epsilon^5). \tag{9.7}$$

Now ϵ small means that the infinitesimal is near infinity, and (9.7) says that near infinity the Coriolis force dominates, and the next most important force looks like a Kepler problem with both primaries at the origin. Again change to polar coordinates to get

$$H = -\Theta + \epsilon^3 \left[\frac{1}{2} \left(R^2 + \frac{\Theta^2}{r^2} \right) - \frac{1}{r} \right] + O(\epsilon^5),$$

$$\dot{r} = \epsilon^3 R, \qquad\qquad \dot{R} = \epsilon^3 \left(\frac{\Theta}{r^3} - \frac{1}{r^2} \right),$$

$$\dot{\theta} = -1 + \epsilon^3 \frac{\Theta}{r^2}, \qquad \dot{\Theta} = 0. \tag{9.8}$$

As before, the terms of order ϵ^5 have been dropped from the equations in (9.8), and to this order of approximation, Θ is a integral. A pair of circular periodic solutions of (9.8) are $\Theta = \pm 1$, $R = 0$, $r = 1$, which are periodic of period $2\pi/(1 \mp \epsilon^3)$. Linearizing the r and R equations about these solutions gives

$$\dot{r} = \epsilon^3 R, \qquad \dot{R} = -\epsilon^3 r.$$

These linear equations have solutions of the form $\exp(\pm i\epsilon^3 t)$, and so the nontrivial multipliers of the circular orbits of (9.8) are $\exp(\pm i\epsilon^3 2\pi/(1 \mp \epsilon^3)) = +1 \pm \epsilon^3 2\pi i + O(\epsilon^6)$. Repeat the argument given in the last section to continue these solutions into the restricted problem.

Theorem 9.5.1. *There exist two one-parameter families of nearly circular large elliptic periodic solutions of the restricted 3-body problem for all values of the mass ratio parameter. These family of orbits tend to infinity.*

9.6 From the Restricted to the Full Problem

In this chapter, four classes of periodic solutions of the restricted problem have been established: Lyapunov centers at the libration points, Poincaré's orbits of the first kind, Hill's lunar orbits, and comet orbits. All of these families are elementary, and most are elliptic. In this section, these solutions and more are continued into the full 3-body problem, where one of the three particles has small mass.

Periodic solutions of the N-body problem are never elementary because the N-body problem has many symmetries and integrals. As was shown in Section 6.4, an integral for the system implies $+1$ as a multiplier of a periodic solution. In fact, the multiplicity of $+1$ as a multiplier of a periodic solution is at least 8 in the planar N-body problem and at least 12 in space. The only way around this problem is to exploit the symmetries and integrals themselves and to go directly to the reduced space as discussed in Section 7.7.

A solution of the N-body problem is called a (relative) periodic solution if its projection on the reduced space is periodic. Note that it need not be periodic in phase space; in fact, it is not usually. A periodic solution of the N-body problem is called an elementary periodic solution if its projection on the reduced space is periodic and the multiplicity of the multiplier of the periodic solution on the reduced space is exactly 2, i.e., it is elementary on the reduced space.

The main result of this section is the following general theorem.

Theorem 9.6.1. *Any elementary periodic solution of the planar restricted 3-body problem whose period is not a multiple of 2π can be continued into the full 3-body problem with one small mass.*

See Hadjidemetriou (1975). The proof is an easy consequence of two procedures that have previously been discussed: the scaling of Section 2.7 and the reduction in Section 7.7. These facts are recalled now before the formal proof of this theorem is given.

Recall the scaling given in Section 2.7, i.e., consider the planar 3-body problem in rotating coordinates with one small particle, $m_3 = \epsilon^2$. The Hamiltonian is then of the form

$$H_3 = \frac{\|v_3\|^2}{2\epsilon^2} - u_3^T K v_3 - \sum_{i=1}^{2} \frac{\epsilon^2 m_i}{\|u_i - u_3\|} + H_2, \tag{9.9}$$

where H_2 is the Hamiltonian of the 2-body problem in rotating coordinates, ϵ is a small parameter that measures the smallness of one mass. A small mass should make a small perturbation on the other particles, thus, ϵ should measure the smallness of the mass and how close the two finite particles' orbits are to circular. To accomplish this, use one variable that represents the deviation from a circular orbit.

Let $Z = (u_1, u_2, v_1, v_2)$, so H_2 is a function of the 8-vector Z. A circular solution of the 2-body problem is a critical point of the Hamiltonian of the 2-body problem in rotating coordinates, i.e., H_2. Let $Z^* = (a_1, a_2, b_1, b_2)$ be such a critical point. By Taylor's theorem

$$H_2(Z) = H_2(Z^*) + \frac{1}{2}(Z - Z^*)^T S(Z - Z^*) + O(\|Z - Z^*\|^3),$$

where S is the Hessian of H_2 at Z^*. Because the equations of motion do not depend on constants, drop the constant term in the above. Change variables by $Z - Z^* = \epsilon \Psi$, $u_3 = \xi$, $v_3 = \epsilon^2 \eta$, which is a symplectic change of variables with multiplier ϵ^{-2}. The Hamiltonian becomes

$$H_3 = G + \tfrac{1}{2}\Psi^T S\Psi + O(\epsilon),$$

$$G = \frac{1}{2}\|\eta\|^2 - \xi^T K\eta - \sum_{i=1}^{2} \frac{m_i}{\|\xi - a_i\|}. \tag{9.10}$$

G in (9.10) is the Hamiltonian of the restricted 3-body problem if we take $m_1 = \mu$, $m_2 = 1 - \mu$, $a_1 = (1 - \mu, 0)$, $a_2 = (-\mu, 0)$. (Because it is necessary to discuss several different Hamiltonians in the same section, our usual convention of naming all Hamiltonians H would lead to mass confusion.) The quadratic term above simply gives the linearized equations about the circular solutions of the 2-body problem in rotating coordinates. Thus to first order in ϵ, the Hamiltonian of the full 3-body problem decouples into the sum of the Hamiltonian for the restricted problem and the Hamiltonian of the linearized equations about the circular solution.

Now look at this problem on the reduced space. Let $\Psi = (q_1, q_2, p_1, p_2)$ and $M = \epsilon^2 + m_1 + m_2 = \epsilon^2 + 1$ (total mass), so $u_i = a_i - \epsilon q_i$ and $v_i = -m_i K a_i - \epsilon p_i$. The center of mass C, linear momentum L, and angular momentum A in these coordinates are

$$C = \frac{1}{M}\{\epsilon^2 \xi - \epsilon(m_1 q_1 + m_2 q_2)\},$$

$$L = \epsilon^2 \eta - \epsilon(p_1 + p_2),$$

$$A = \epsilon^2 \xi^T K \eta - (a_1 - \epsilon q_1)^T K (m_1 K a_1 + \epsilon p_1) - (a_2 - \epsilon q_2)^T K (m_2 K a_2 + \epsilon p_2).$$

Note that when $\epsilon = 0$, these three quantities depend only on the variables of the 2-body problem, $\Psi - (q_1, q_2, p_1, p_2)$, and are independent of the variables of the restricted problem, ξ, η. So when $\epsilon = 0$, the reduction is on the 2-body problem alone.

Look at the reduction of the 2-body problem in rotating coordinates with masses μ and $1 - \mu$, and let $\nu = \mu(1 - \mu)$. Fixing the center of mass at the origin and ignoring linear momentum is done by moving to Jacobi coordinates which are denoted by (α, β); see Section 8.2. Replace (u, v) with (α, β) to get

$$T = \frac{\|\beta\|^2}{2\nu} - \alpha^T K \beta - \frac{\nu}{\|\alpha\|}.$$

Put this problem in polar coordinates (see Section 8.1) to get

$$T = \frac{1}{2\nu}\left(R^2 + \frac{\Theta^2}{r^2}\right) - \Theta - \frac{\nu}{r},$$

$$\dot{r} = \frac{R}{\nu}, \qquad \dot{R} = \frac{\Theta^2}{\nu r^3} - \frac{\nu}{r^2},$$

$$\dot{\theta} = \frac{\Theta}{\nu r^2} - 1, \qquad \dot{\Theta} = 0.$$

The reduction to the reduced space is done by holding the angular momentum, Θ, fixed and ignoring the angle θ (mod out the rotational symmetry). The distance between the primaries has been chosen as 1, so the relative equilibrium must have $r = 1$; therefore, $\Theta = \nu$. The linearization about this critical point is

$$\dot{r} = R/\nu, \qquad \dot{R} = -\nu r.$$

This linear equation is a harmonic oscillator with frequency 1 and comes from the Hamiltonian $S = \frac{1}{2}(R^2/\nu + \nu r^2)$.

In summary, the Hamiltonian of the 3-body problem on the reduced space is

$$H_R = G + S = G + \frac{1}{2}(R^2/\nu + \nu r^2) + O(\varepsilon), \tag{9.11}$$

where G is the Hamiltonian of the restricted three body problem. We call this the *reduced problem*. Sometimes we will scale r, R so that $\nu = 1$.

Proof. Let $\xi = \phi(t)$, $\eta = \psi(t)$ be a T-periodic solution of the restricted problem with multipliers $1, 1, \tau, \tau^{-1}$. By assumption $\tau \neq 1$ and $T \neq k2\pi$, where k is an integer. Now $\xi = \phi(t)$, $\eta = \psi(t)$, $r = 0$, $R = 0$ is a T-periodic solution of the system whose Hamiltonian is H_R, i.e., the Hamiltonian of the 3-body problem on the reduced space with $\epsilon = 0$. The multipliers of this periodic solution are $1, 1, \tau, \tau^{-1}, \exp(iT), \exp(-iT)$. Because T is not a multiple of 2π, $\exp(\pm iT) \neq 1$, and so this solution is elementary. By Proposition 9.1.1 this solution can be continued into the full problem with $\epsilon \neq 0$, but small.

9.7 Some Elliptic Orbits

All of the multipliers of the elliptic solutions of the Kepler problem in either fixed or rotating coordinates are $+1$, because they are not isolated in an energy level; see Proposition 6.4.2. Thus there is no hope of using the methods used previously, however, the restricted problem has a symmetry that when exploited properly proves that some elliptic orbits can be continued from the Kepler problem into the restricted problem. The main idea is given in the following lemma.

Lemma 9.7.1. *A solution of the restricted 3-body problem* (9.2) *that crosses the line of syzygy (the x_1-axis) orthogonally at a time $t = 0$ and later at a time $t = T/2 > 0$ is T-periodic and symmetric with respect to the line of syzygy.*

Proof. This is an easy consequence of the exercises following Chapter 6.

That is, if $x = \phi(t)$, $y = \psi(t)$ is a solution of the restricted problem such that $x_2(0) = y_1(0) = x_2(T/2) = y_1(T/2) = 0$, where $T > 0$, then this solution is T-periodic and symmetric in the x_1-axis.

In Delaunay coordinates (ℓ, g, L, G) (see Section 8.9), an orthogonal crossing of the line of syzygy at a time t_0 is

$$\ell(t_0) = n\pi, \qquad g(t_0) = m\pi, \qquad n, m \text{ integers.} \qquad (9.12)$$

These equations are solved using the implicit function theorem to yield the following theorem.

Theorem 9.7.1. *Let m, k be relatively prime integers and $T = 2\pi m$. Then the elliptic T-periodic solution of the Kepler problem in rotating coordinates that satisfies*

$$\ell(0) = \pi, \qquad g(0) = \pi, \qquad L^3(0) = m/k$$

and does not go through $x = (1, 0)$ can be continued into the restricted problem for μ small. This periodic solution is symmetric with respect to the line of syzygy.

Proof. (See Barrar (1965).) The Hamiltonian of the restricted 3-body problem in Delaunay elements for small μ is

$$H = -\frac{1}{2L^2} - G + O(\mu),$$

and the equations of motion are

$$\dot{\ell} = 1/L^3 + O(\mu), \qquad \dot{L} = 0 + O(\epsilon),$$

$$\dot{g} = -1 + O(\mu), \qquad \dot{G} = 0 + O(\epsilon). \tag{9.13}$$

Let $L_0^3 = m/k$, and let $\ell(t, \Lambda, \mu)$, $g(t, \Lambda, \mu)$, $L(t, \Lambda, \mu)$, $G(t, \Lambda, \mu)$ be the solution that goes through $\ell = \pi$, $g = \pi$, $L = \Lambda$, $G = anything$ at $t = 0$, so it is a solution with an orthogonal crossing of the line of syzygy at $t = 0$.

From (9.13) $\ell(t, \Lambda, 0) = t/\Lambda^3 + \pi$, $g(t, \Lambda, 0) = -t + \pi$. Thus $\ell(T/2, L_0, 0) = (1+k)\pi$ and $g(T/2, L_0, 0) = (1-m)\pi$, and so when $\mu = 0$, this solution has another orthogonal crossing at time $T/2 = m\pi$. Also

$$\det \begin{bmatrix} \partial\ell/\partial t & \partial\ell/\partial\Lambda \\ \partial g/\partial t & \partial g/\partial\Lambda \end{bmatrix}_{t=T/2, L=L_0, \mu=0} = \det \begin{bmatrix} k/m & -3\pi(k^4/m)^{1/3} \\ -1 & 0 \end{bmatrix} \neq 0.$$

Thus the theorem follows by the implicit function theorem.

It is not too hard to show that for a fixed m and k, only a finite number of such elliptic orbits pass through the singularity at the other primary, $x = (1, 0)$. This rules out a finite number of collision orbits and a finite number of Gs.

It is only a little more difficult to establish the existence of symmetric elliptic periodic solutions near a primary as in Section 9.4 (see Arenstorf (1968)). It is also easy to show the existence of symmetric elliptic periodic solutions near infinity as in Section 9.5 (see Meyer (1981a)).

9.8 Problems

1. Consider a periodic system of equations of the form $\dot{x} = f(t, x, \nu)$ where ν is a parameter, and f is T-periodic in t, $f(t + T, x, \nu) = f(t, x, \nu)$. Let $\phi(t, \xi, \nu)$ be the general solution, $\phi(0, \xi, \nu) = \xi$.
    a) Show that $\phi(t, \xi', \nu')$ is T-periodic if and only if $\phi(T, \xi', \nu') = \xi'$.
    b) A T-periodic solution $\phi(t, \xi', \nu')$ can be continued if there is a smooth function $\bar{xi}(\nu)$ such that $\bar{\xi}(\nu') = \xi'$, and $\phi(T, \bar{\xi}(\nu'), \nu')$ is T-periodic. The multipliers of the T-periodic solution $\phi(t, \xi', \nu')$ are the eigenvalues of $\partial\phi(T, \xi', \nu')/\partial\xi$. Show that a T-periodic solution can be continued if all of its multipliers are different from $+1$.

2. Consider the classical Duffing's equation $\ddot{x}+x+\gamma x^3 = A\cos\omega t$ or $\dot{x} = y = \partial H/\partial y$, $\dot{y} = -x-\gamma x^3 + A\cos\omega t = -\partial H/\partial x$, where $H = (1/2)(y^2+x^2) + \gamma x^4/4 - Ax\cos\omega t$. Show that if $1/\omega \neq 0, \pm1, \pm2, \pm3, \ldots$, then for small forcing A and small nonlinearity γ there is a small periodic solution of the forced Duffing's equation with the same period as the external forcing, $T = 2\pi/\omega$. In the classical literature this solution is sometimes referred to as the harmonic. (Hint: Set the parameters γ and A to zero; then the equation is linear and solvable. Note that zero is a T-periodic solution).

3. Show that the eigenvalues of

$$\begin{bmatrix} 0 & -1 \\ \mu & 0 \end{bmatrix}$$

are continuous in μ but not smooth in μ.

4. Hill's lunar problem is defined by the Hamiltonian

$$H = \frac{\|y\|^2}{2} - x^T K y - \frac{1}{\|x\|} - \frac{1}{2}(3x_1^2 - \|x\|^2),$$

where $x, y \in \mathbb{R}^2$. Show that it has two equilibrium points on the x_1 axis. Linearize the equations of motion about these equilibrium points, and discuss how Lyapunov's center and the stable manifold theorem apply.

5. Show that the scaling used in Section 9.4 to obtain Hill's orbits for the restricted problem works for Hill's lunar problem (Problem 4) also. Why doesn't the scaling for comets work?

6. Prove Lemma 9.7.1, and verify that (9.12) is the condition for an orthogonal crossing of the line of syzygy in Delaunay elements.

10. Normal Forms

Perturbation theory is one of the few ways that one can bridge the gap between the behavior of a real nonlinear system and its linear approximation. Because the theory of linear systems is so much simpler, investigators are tempted to fit the problem at hand to a linear model without proper justification. Such a linear model may lead to quantitative as well as qualitative errors. On the other hand, so little is known about the general behavior of a nonlinear system that some sort of approximation has to be made.

Many interesting problems can be formulated as a system of equations that depend on a small parameter ε with the property that when $\varepsilon = 0$ the system is linear or at least integrable. This chapter develops a very powerful and general method for handling the formal aspects of perturbations of linear and integrable systems, and Chapters 11 and 12 contain rigorous results that depend on these formal considerations.

10.1 Normal Form Theorems

In this section the main theorems about the normal form at an equilibrium and at a fixed point developed in this chapter are summaries without proof. Upon a first sitting a reader may want to read this section, skip the details in the rest of the chapter, and go on to other topics.

10.1.1 Normal Form at an Equilibrium Point

Consider a Hamiltonian system of the form

$$H_{\#}(x) = \sum_{i=0}^{\infty} H_i(x). \tag{10.1}$$

In order to study this system we change coordinates so that the system in the new coordinates is simpler. The definition of simpler depends on the problem at hand. In this chapter we construct formal, symplectic, near-identity

© Springer International Publishing AG 2017   239
K.R. Meyer, D.C. Offin, *Introduction to Hamiltonian Dynamical*
Systems and the N-Body Problem, Applied Mathematical Sciences 90,
DOI 10.1007/978-3-319-53691-0_10

changes of variables $x = X(y) = y + \cdots$, such that in the new coordinates the Hamiltonian becomes

$$H^{\#}(y) = \sum_{i=0}^{\infty} H^i(y). \tag{10.2}$$

If the Hamiltonian $H^{\#}$ meets the criteria for being simple then the system is said to be in normal form. It is important to understand the implications of a formal transformation. Even though the original system (10.1) is a convergent series for x in some domain, the series expansion for the change of variables $X(y)$ will not converge in general. Thus the series (10.2) does not necessarily converge. The only way to obtain rigorous results based on this theory is to truncate the series expansion for X at some finite order to obtain a finite (hence convergent) series for X. In this case only the first few terms of $H^{\#}$ are in normal form. In general, if the series for X is truncated after the Nth term then the series for $H^{\#}$ will be convergent, but only the terms up to and including the Nth will be in normal form.

Various methods for transforming a system into normal form have been given since the middle of the nineteenth century, but we present the method of Lie transforms because of its great generality and simplicity. The simplicity of this method is the result of its recursive algorithmic definition which lends itself to easy computer implementation.

Our first example is the classical theorem on the normal form for a Hamiltonian system at a simple equilibrium point. Consider an analytic Hamiltonian, $H_{\#}$, which has an equilibrium point at the origin in $\mathbb{R}^{2n}$, and assume that the Hamiltonian is zero at the origin. Then $H_{\#}$ has a Taylor series expansion of the form (10.1) where H_i is a homogeneous polynomial in x of degree $i + 2$, so, $H_0(x) = \frac{1}{2} x^T S x$, where S is a $2n \times 2n$ real symmetric matrix, and $A = JS$ is a Hamiltonian matrix. The linearized equation about the critical point $x = 0$ is

$$\dot{x} = Ax = JSx = J\nabla H_0(x), \tag{10.3}$$

and the general solution of (10.3) is $\phi(t, \xi) = \exp(At)\xi$. A traditional analysis is to solve (10.3) by linear algebra techniques and then hope that the solutions of the nonlinear problem are not too dissimilar from the solutions of the linear equation. In many cases this hope is unjustifiable. The next best thing is to put the equations in normal form and to study the solutions of the normal form equations. This too has its pitfalls.

Theorem 10.1.1. *Let A be diagonalizable. Then there exists a formal, symplectic change of variables, $x = X(y) = y + \cdots$, which transforms the Hamiltonian (10.1) to (10.2) where H^i is a homogeneous polynomial of degree $i + 2$ such that*

$$H^i(e^{At}y) \equiv H^i(y), \tag{10.4}$$

for all $i = 0, 1, \ldots$, all $y \in \mathbb{R}^{2n}$, and all $t \in \mathbb{R}$.

For example consider a two degree of freedom system in the case when the matrix A is diagonalizable and has distinct pure imaginary eigenvalues $\pm i\omega_1, \pm i\omega_2$. In this case we may assume that after a symplectic change of variables the quadratic terms are

$$H_0(x) = \frac{\omega_1}{2}(x_1^2 + x_3^2) + \frac{\omega_2}{2}(x_2^2 + x_4^2) = \omega_1 I_1 + \omega_2 I_2, \qquad (10.5)$$

where in the second form we use the action–angle coordinates

$$I_1 = \frac{1}{2}(x_1^2 + x_3^2), \quad I_2 = \frac{1}{2}(x_2^2 + x_4^2), \quad \phi_1 = \tan^{-1}\frac{x_3}{x_1}, \quad \phi_2 = \tan^{-1}\frac{x_4}{x_2}.$$

The linear equations (10.3) in action–angle coordinates become

$$\dot{I}_1 = 0, \quad \dot{I}_2 = 0, \quad \dot{\phi}_1 = -\omega_1, \quad \dot{\phi}_2 = -\omega_2.$$

The condition (10.4) requires the terms in the normal form to be constant on the solutions of the above equations. These equations have as solutions $I_1 = I_1^0$ and $I_2 = I_2^0$ where I_1^0 and I_2^0 are constants. $I_1 = I_1^0$ and $I_2 = I_2^0$ where $I_1^0 > 0$ and $I_2^0 > 0$ defines a 2-torus with angular coordinates ϕ_1 and ϕ_2. This type of flow on a torus was discussed in detail in Section 1.5.

There are two cases depending on whether the ratio ω_1/ω_2 is rational or irrational. In the case when the ratio is irrational the flow on the torus defined by the equations above is dense on the torus and so the only continuous functions defined on the torus are constants, therefore, the terms in the normal form will depend only on the action variables I_1, I_2. On the other hand, if the ratio is rational, say $\omega_1/\omega_2 = p/q$, then the terms in the normal form may contain a dependence on the single angle $\psi = q\phi_1 - p\phi_2$.

Thus: *If H_0 in (10.1) is of the form (10.5) then the normal form for the system is*

$$H^{\#} = \sum_{i=0}^{\infty} H^i(I_1, I_2)$$

when the ratio ω_1/ω_2 is irrational, and

$$H^{\#} = \sum_{i=0}^{\infty} H^i(I_1, I_2, q\phi_1 - p\phi_2)$$

when $\omega_1/\omega_2 = p/q$.

This covers the normal form at the equilibrium point $\mathcal{L}_4$ of the restricted 3-body problem when $0 < \mu < \mu_1$. A multitude of interesting stability and bifurcation results follow from simple inequalities on a finite number of terms in this normal form.

In the case where the matrix A is not diagonalizable the only change in the statement of Theorem 10.1.1 is that the condition (10.4) is replaced by

$$H^i(e^{A^T t} y) \equiv H^i(y),$$

where A^T is the transpose of A.

Consider a two degree of freedom Hamiltonian system at an equilibrium point when the exponents are $\pm i\omega$ with multiplicity two and the linearized system is not diagonalizable. The normal form for the quadratic part of such a Hamiltonian was given as

$$H_0 = \omega(x_2 y_1 - x_1 y_2) + \frac{\delta}{2}(x_1^2 + x_2^2),$$

where $\omega \neq 0$ and $\delta = \pm 1$. In this case

$$A = \begin{bmatrix} 0 & \omega & 0 & 0 \\ -\omega & 0 & 0 & 0 \\ -\delta & 0 & 0 & \omega \\ 0 & -\delta & -\omega & 0 \end{bmatrix}.$$

The normal form in this case depends on the four quantities

$$\Gamma_1 = x_2 y_1 - x_1 y_2, \qquad \Gamma_2 = \tfrac{1}{2}(x_1^2 + x_2^2),$$
$$\Gamma_3 = \tfrac{1}{2}(y_1^2 + y_2^2), \qquad \Gamma_4 = x_1 y_1 + x_2 y_2.$$

Note that $\{\Gamma_1, \Gamma_2\} = 0$ and $\{\Gamma_1, \Gamma_3\} = 0$. The system is in Sokol'skii normal form if the higher-order terms depend on the two quantities Γ_1 and Γ_3, that is, the Hamiltonian is of the form

$$H^{\#} = \omega(x_2 y_1 - x_1 y_2) + \frac{\delta}{2}(x_1^2 + x_2^2) + \sum_{k=1}^{\infty} H_{2k}(x_2 y_1 - x_1 y_2, y_1^2 + y_2^2),$$

where H_{2k} is a polynomial of degree k in two variables. The first few terms of this normal form determine the nature of the stability and bifurcations at the equilibrium point $\mathcal{L}_4$ of the restricted problem when $\mu = \mu_1$.

10.1.2 Normal Form at a Fixed Point

The study of the stability and bifurcation of a periodic solution of a Hamiltonian system of two degrees of freedom can be reduced to the study of the Poincaré map in an energy level (i.e., level surface of the Hamiltonian). Sometimes the value of the Hamiltonian must be treated as a parameter.

Consider a diffeomorphism of the form

$$F_{\#} : N \subset \mathbb{R}^2 \to \mathbb{R}^2 : x \to f(x), \tag{10.6}$$

where N is a neighborhood of the origin in $\mathbb{R}^2$, and f is a smooth function such that

$$f(0) = 0, \qquad \det \frac{\partial f}{\partial x}(x) \equiv 1.$$

The origin is a fixed point for the diffeomorphism because $f(0) = 0$, and it is orientation-preserving and area-preserving because $\det \partial f / \partial x \equiv 1$. This map should be considered as the Poincaré map associated with a periodic solution of a two degree of freedom Hamiltonian system.

The linearization of this map about the origin is $x \to Ax$ where A is the 2×2 matrix $(\partial f/\partial x)(0)$. Because the determinant of A is 1 the product of its eigenvalues must be 1. The eigenvalues λ, λ^{-1} of A are called the multipliers of the fixed point. There are basically four cases:

1. Hyperbolic fixed point: multipliers real and $\lambda \neq \pm 1$
2. Elliptic fixed point: multipliers complex conjugates and $\lambda \neq \pm 1$
3. Shear fixed point: $\lambda = +1$, A not diagonalizable
4. Flip fixed point: $\lambda = -1$, A not diagonalizable

As before in order to study an area-preserving map we can change coordinates so that the map in the new coordinates is simpler. Here we consider a formal symplectic, near-identity change of variables $x = X(y) = y + \cdots$, such that in the new coordinates the map (10.6) becomes

$$F^{\#} : y \to g(y). \tag{10.7}$$

If the map $F^{\#}$ meets the criteria for being simple then the map is said to be in normal form. It is important to understand the implications of a formal transformation. Even though the original system (10.6) is a convergent series for x in some domain, the series expansion for the change of variables $X(y)$ will not converge in general. Thus the series (10.7) does not converge in general. The only way to obtain rigorous results based on this theory is to truncate the series expansion for X at some finite order to obtain a finite (hence convergent) series for X. In this case only the first few terms of $F^{\#}$ will be in normal form. In general, if the series for X is truncated after the Nth term then the series for $F^{\#}$ will be convergent, but only the terms up to and including the Nth will be in normal form.

Hyperbolic fixed point. In the hyperbolic case after a change of variables we may assume that

$$A = \begin{bmatrix} \lambda & 0 \\ 0 & \lambda^{-1} \end{bmatrix},$$

with $\lambda \neq \pm 1$ and real. The mapping (10.7) is in normal form with $F^{\#}$: $(u, v) \to (u', v')$ where $y = (u, v)$ with

$$u' = u\ell(uv)$$

$$v' = v\ell(uv)^{-1},$$

where ℓ is a formal series in one variable, $\ell(uv) = \lambda + \cdots$.

The map takes the hyperbolas $uv = constant$ into themselves. The transformation to normal form actually converges by a classical theorem of Moser (1956).

Elliptic fixed point. In the elliptic case when λ is a complex number of unit modulus certain reality conditions must be met. Consider the case when

A has eigenvalues $\lambda^{\pm 1} = \exp(\pm \omega i) \neq \pm 1$; i.e., the origin is an elliptic fixed point. First assume that λ is not a root of unity. Change to action–angle variables (I, ϕ); The normal form in action–angle variables in this case is $F^{\#} : (I, \phi) \to (I', \phi')$ where

$$I' = I, \qquad \phi' = \phi + \ell(I),$$

where ℓ has a formal expansion $\ell(I) = -\omega + \beta I \cdots$. If a diffeomorphism is in this form with $\beta \neq 0$, then the origin is called a general elliptic point, or $F^{\#}$ is called a twist map. This map takes circles, $I = const$, into themselves and rotates each circle by an amount $\ell(I)$.

Now consider the case when the diffeomorphism has an elliptic fixed point whose multiplier is a root of unity. Let λ be a kth root of unity, so, $\lambda^k = 1$, $k > 2$, and $\lambda = \exp(h2\pi i/k)$, where h is an integer. The origin is called a k-resonance elliptic point in this case. The normal form in action–angle variables in this case is $F^{\#} : (I, \phi) \to (I', \phi')$ where

$$I' = I + 2\alpha I^{k/2} \sin(k\phi) + \cdots,$$

$$\phi' = \phi + (2\pi h/k) + \alpha I^{(k-2)/2} \cos(k\phi) + \beta I + \cdots. \tag{10.8}$$

Shear fixed point. Consider the cases where the multiplier is $+1$. If $A = I$, the identity matrix, the system is so degenerate that there is no normal form in general. Otherwise, by a coordinate change we have

$$A = \begin{bmatrix} 1 & \pm 1 \\ 0 & 1 \end{bmatrix}. \tag{10.9}$$

The important terms of the normal form $F^{\#} : (u, v) \to (u', v')$ are

$$u' = u \pm v - \cdots,$$

$$v' = v - \beta u^2 + \cdots. \tag{10.10}$$

The ellipsis may contain other quadratic terms and higher-order terms.

Flip periodic point. Now consider the case when A has eigenvalue -1. In this case the generic form for A is

$$A = \begin{bmatrix} -1 & \pm 1 \\ 0 & -1 \end{bmatrix}.$$

The quadratic terms can be eliminated and the important terms of the normal form $F^{\#} : (u, v) \to (u', v')$ are

$$u' = -u - v + \cdots,$$

$$v' = -v + \beta u^3 + \cdots. \tag{10.11}$$

The ellipsis may contain other cubic terms and higher-order terms.

10.2 Forward Transformations

One of the most general methods of mathematics is to simplify a problem by a change of variables. The method of Lie transforms developed by Deprit (1969) and extended by Kamel (1970) and Henrard (1970) is a general procedure to change variables in a system of equations that depend on a small parameter. Deprit's original method was for Hamiltonian systems only, but the extensions by Kamel and Henrard handle non-Hamiltonian equations. Only the Hamiltonian case is treated here.

10.2.1 Near-Identity Symplectic Change of Variables

The general idea of this method is to generate a symplectic change of variables depending on a small parameter as the general solution of a Hamiltonian system of differential equations; see Theorem 2.6.2. $X(\varepsilon, y)$ is said to be a near-identity symplectic change of variables (or transformation), if X is symplectic for each fixed ε and is of the form $X(\varepsilon, y) = y + O(\varepsilon)$, i.e., $X(0, y) = y$. Because $X(0, y) = y$, $\partial X(\varepsilon, y)/\partial y$ is nonsingular for small ε so by the inverse function theorem, the map $y \to X(\varepsilon, y)$ has a differentiable inverse for small ε. Both X and its inverse are symplectic for fixed ε.

Consider the nonautonomous Hamiltonian system

$$\frac{dx}{d\varepsilon} = J\nabla W(\varepsilon, x) \tag{10.12}$$

and the initial condition

$$x(0) = y, \tag{10.13}$$

where W is smooth. The basic theory of differential equations asserts that the general solution of this problem is a smooth function $X(\varepsilon, y)$ such that $X(0, y) \equiv y$, and by Theorem 2.6.2, the function X is symplectic for fixed ε. That is, the differential equation (10.12) and the initial condition (10.13) define a near-identity symplectic change of variables.

Conversely, let $X(\varepsilon, y)$ be a near-identity symplectic change of variables with inverse function $Y(\varepsilon, x)$ such that $X(\varepsilon, Y(\varepsilon, x)) \equiv x$ and $Y(\varepsilon, X(\varepsilon, y)) \equiv y$ where defined. Y is symplectic too. Differentiating $Y(\varepsilon, X(\varepsilon, y)) \equiv y$ with respect to ε yields

$$\frac{\partial Y}{\partial x}(\varepsilon, X(\varepsilon, y))\frac{\partial X}{\partial \varepsilon}(\varepsilon, y) + \frac{\partial Y}{\partial \varepsilon}(\varepsilon, X(\varepsilon, y)) \equiv 0$$

or

$$\frac{\partial X}{\partial \varepsilon}(\varepsilon, y) \equiv -\left[\frac{\partial Y}{\partial x}(\varepsilon, X(\varepsilon, y))\right]^{-1}\frac{\partial Y}{\partial \varepsilon}(\varepsilon, X(\varepsilon, y)).$$

This means that $X(\varepsilon, y)$ is the general solution of

$$\frac{dx}{d\varepsilon} = -U(\varepsilon, x), \quad \text{where } U(\varepsilon, x) = \left[\frac{\partial Y}{\partial x}(\varepsilon, x)\right]^{-1}\frac{\partial Y}{\partial \varepsilon}(\varepsilon, x).$$

This equation is Hamiltonian, so, there is a function $W(\varepsilon, x)$ such that $U(\varepsilon, x) = J\nabla W(\varepsilon, x)$. This proves the following.

Proposition 10.2.1. $X(\varepsilon, y)$ *is a near-identity symplectic change of variables if and only if it is the general solution of a Hamiltonian differential equation of the form* (10.12) *satisfying initial condition* (10.13).

A Hamiltonian system of equations generates symplectic transformations directly, which is in contrast to the symplectic transformations given by the generating functions in Theorem 8.1.1, where the new and old variables are mixed.

10.2.2 The Forward Algorithm

Let $X(\varepsilon, y)$, $Y(\varepsilon, x)$, and $W(\varepsilon, x)$ be as above, so, $X(\varepsilon, y)$ is the solution of (10.12) satisfying (10.13). Think of $x = X(\varepsilon, y)$ as a change of variables $x \to y$ that depends on a parameter. Throughout this chapter, when we change variables, we do not change the parameter ε.

Let $H(\varepsilon, x)$ be a Hamiltonian and $G(\varepsilon, y) \equiv H(\varepsilon, X(\varepsilon, y))$, so, G is the Hamiltonian in the new coordinates. We call G the Lie transform of H (generated by W). Sometimes H is denoted by H_* and G by H^*, and sometimes G is denoted by $\mathcal{L}(W)H$ to show that G is the Lie transform of H generated by W. Let the function $H = H_*$, $G = H^*$, and W all have series expansions in the small parameter ε. The forward algorithm of the method of Lie transforms is a recursive set of formulas that relate the terms in these various series expansions.

In particular let

$$H(\varepsilon, x) = H_*(\varepsilon, x) = \sum_{i=0}^{\infty} \left(\frac{\varepsilon^i}{i!} \right) H_i^0(x), \qquad (10.14)$$

$$G(\varepsilon, y) = H^*(\varepsilon, y) = \sum_{i=0}^{\infty} \left(\frac{\varepsilon^i}{i!} \right) H_0^i(y), \qquad (10.15)$$

$$W(\varepsilon, x) = \sum_{i=0}^{\infty} \left(\frac{\varepsilon^i}{i!} \right) W_{i+1}(x). \qquad (10.16)$$

The method of Lie transforms introduces a double indexed array $\{H_j^i\}$, $i, j = 0, 1, \ldots$ which agrees with the definitions given in (10.14) and (10.15) when either i or j is zero. The other terms are intermediary terms introduced to facilitate the computation.

Theorem 10.2.1. *Using the notation given above, the functions* $\{H_j^i\}$, $i = 1, 2, \ldots$, $j = 0, 1, \ldots$ *satisfy the recursive identities*

$$H_j^i = H_{j+1}^{i-1} + \sum_{k=0}^{j} \binom{j}{k} \{H_{j-k}^{i-1}, W_{k+1}\}. \qquad (10.17)$$

Remarks. The above formula contains the standard binomial coefficient

$$\binom{j}{k} = \frac{j!}{k!(j-k)!}.$$

Note that because the transformation generated by W is a near identity transformation, the first term in H_* and H^* is the same, namely H_0^0. Also note that the first term in the expansion for W starts with W_1. This convention imparts some nice properties to the formulas in (10.17). Each term in 10.17 has indices summing to $i+j$, and each term on the right-hand side has upper index $i-1$.

In order to construct the change of variables $X(\varepsilon, y)$, note that X is the transform of the identity function or $X(\varepsilon, y) = \mathcal{L}(W)(id)$, where $id(x) = x$.

The interdependence of the functions $\{H_j^i\}$ can easily be understood by considering the Lie triangle

$$
\begin{array}{ccc}
H_0^0 & & \\
\downarrow & & \\
H_1^0 \to H_0^1 & & \\
\downarrow \quad\quad \downarrow & & \\
H_2^0 \to H_1^1 \to H_0^2. & & \\
\downarrow \quad\quad \downarrow \quad\quad \downarrow & &
\end{array}
$$

The coefficients of the expansion of the old function H_* are in the left column, and those of the new function H^* are on the diagonal. Formula (10.17) states that to calculate any element in the Lie triangle, you need the entries in the column one step to the left and up.

For example, to compute the series expansion for H^* through terms of order ε^2, you first compute H_0^1 by the formula

$$H_0^1 = H_1^0 + \{H_0^0, W_1\}, \tag{10.18}$$

which gives the term of order ε, and then you compute

$$H_1^1 = H_2^0 + \{H_1^0, W_1\} + \{H_0^0, W_2\},$$

$$H_0^2 = H_1^1 + \{H_0^1, W_1\}.$$

Then $H^*(\varepsilon, x) = H_0^0(x) + H_0^1(x)\varepsilon + H_0^2(x)(\varepsilon^2/2) + \cdots.$

Proof. (Theorem 10.2.1) Recall that $H^*(\varepsilon, y) = G(\varepsilon, y) = H(\varepsilon, X(\varepsilon, y))$, where $X(\varepsilon, y)$ is the general solution of (10.12). Define the differential operator $\mathcal{D} = \mathcal{D}_W$ by

$$\mathcal{D}F(\varepsilon, x) = \frac{\partial F}{\partial \varepsilon}(\varepsilon, x) + \{F, W\}(\varepsilon, x),$$

so that

$$\frac{d}{d\varepsilon}\left(F(\varepsilon,x)\Big|_{x=X(\varepsilon,y)}\right) = DF(\varepsilon,x)\Big|_{x=X(\varepsilon,y)}.$$

Define new functions by $H^0 = H$, $H^i = DH^{i-1}$, $i \geq 1$. Let these functions have series expansions

$$H^i(\varepsilon,x) = \sum_{k=0}^{\infty}\left(\frac{\varepsilon^k}{k!}\right)H^i_k(x)$$

so,

$$H^i(\varepsilon,x) = D\sum_{k=0}^{\infty}\left(\frac{\varepsilon^k}{k!}\right)H^{i-1}_k(x)$$

$$= \sum_{k=1}^{\infty}\left(\frac{\varepsilon^{k-1}}{(k-1)!}\right)H^{i-1}_k(x) + \left\{\sum_{k=0}^{\infty}\left(\frac{\varepsilon^k}{k!}\right)H^{i-1}_k(x), \sum_{s=0}^{\infty}W_{s+1}\right\}$$

$$= \sum_{j=0}^{\infty}\left(\frac{\varepsilon^j}{j!}\right)\left(H^{i-1}_{j+1} + \sum_{k=0}^{j}\binom{j}{k}\{H^{i-1}_{j-k},W_{k+1}\}\right).$$

So the functions H^i_j are related by (10.17). It remains to show that $H^* = G$ has the expansion (10.15). By the above and Taylor's theorem

$$G(\varepsilon,y) = \sum_{n=0}^{\infty}\left(\frac{\varepsilon^n}{n!}\right)\frac{d^n}{d\varepsilon^n}G(\varepsilon,y)\Big|_{\varepsilon=0}$$

$$= \sum_{n=0}^{\infty}\left(\frac{\varepsilon^n}{n!}\right)\frac{d^n}{d\varepsilon^n}\left(H(\varepsilon,x)\Big|_{x=X(\varepsilon,y)}\right)_{\varepsilon=0}$$

$$= \sum_{n=0}^{\infty}\left(\frac{\varepsilon^n}{n!}\right)\left(D^n H(\varepsilon,x)\Big|_{x=X(\varepsilon,y)}\right)_{\varepsilon=0}$$

$$= \sum_{n=0}^{\infty}\left(\frac{\varepsilon^n}{n!}\right)H^n_0(y).$$

10.2.3 The Remainder Function

Assume now that the Hamiltonian and hence the equations are time dependent, i.e., consider

$$\dot{x} = J\nabla H(\varepsilon,t,x), \tag{10.19}$$

where H has an expansion

$$H(\varepsilon, t, x) = H_*(\varepsilon, t, x) = \sum_{i=0}^{\infty} \left(\frac{\varepsilon^i}{i!}\right) H_i^0(t, x). \tag{10.20}$$

Make a symplectic change of coordinates, $x = X(\varepsilon, t, y)$, which transforms (10.19) to the Hamiltonian differential equation

$$\dot{y} = J\nabla G(\varepsilon, t, y) + J\nabla R(\varepsilon, t, y) = J\nabla K(\varepsilon, t, y),$$

where $G(\varepsilon, t, y) = H^*(\varepsilon, t, y) = H(\varepsilon, t, X(\varepsilon, t, y))$ is the Lie transform of H, R is the remainder function, and $K = G + R$ is the new Hamiltonian. Let G, R, and K have series expansions of the form

$$G(\varepsilon, t, y) - \sum_{i=0}^{\infty} \left(\frac{\varepsilon^i}{i!}\right) H_0^i(t, y), \qquad R(\varepsilon, t, y) - \sum_{i=0}^{\infty} \left(\frac{\varepsilon^i}{i!}\right) R_0^i(t, y)$$

$$K(\varepsilon, t, y) = \sum_{i=0}^{\infty} \left(\frac{\varepsilon^i}{i!}\right) K_0^i(t, y).$$

Let the symplectic change of variables $X(\varepsilon, t, y)$ be the general solution of the Hamiltonian system of equations

$$\frac{dx}{d\varepsilon} = J\nabla W(\varepsilon, t, x), \qquad x(0) = y,$$

where $W(\varepsilon, t, x)$ is a Hamiltonian function with a series expansion of the form

$$W(\varepsilon, t, x) = \sum_{i=0}^{\infty} \left(\frac{\varepsilon^i}{i!}\right) W_{i+1}(t, x).$$

The variable t is simply a parameter, and so the function $G = H^*$ can be computed by formulas (10.17) in Theorem 10.2.1 using the Lie triangle as a guide. The remainder term R needs further consideration.

Theorem 10.2.2. *The remainder function is given by*

$$R(\varepsilon, t, y) = - \int_0^{\varepsilon} \mathcal{L}(W) \left(\frac{\partial W}{\partial t}\right) (s, t, y) ds. \tag{10.21}$$

Proof. Making the symplectic change of variable $x = X(\varepsilon, t, y)$ in (10.19) directly gives

$$\dot{y} = \left(\frac{\partial X}{\partial y}\right)^{-1} (\varepsilon, t, y) J\nabla_x H(\varepsilon, t, X(\varepsilon, t, y)) - \left(\frac{\partial X}{\partial y}\right)^{-1} (\varepsilon, t, y) \frac{\partial X}{\partial t} (\varepsilon, t, y).$$

By the discussion in Section 2.6 the first term on the right-hand side is $J\nabla G$, and so,

$$J\nabla R(\varepsilon, t, y) = -\left(\frac{\partial X}{\partial y}(\varepsilon, t, y)\right)^{-1} \frac{\partial X}{\partial t}(\varepsilon, t, y).$$

$A(\varepsilon) = \partial X(\varepsilon, t, y)/\partial y$ is the fundamental matrix solution of the variational equation, i.e., it is the matrix solution of

$$\frac{dA}{d\varepsilon} = \left(J\frac{\partial^2 W}{\partial x^2}(\varepsilon, t, X(\varepsilon, t, y))\right) A, \qquad A(0) = I.$$

Differentiating $\partial X(\varepsilon, t, y)/\partial \varepsilon = J\nabla W(\varepsilon, t, X(\varepsilon, t, y))$ with respect to t shows that $B(\varepsilon) = \partial X(\varepsilon, t, y)/\partial t$ satisfies

$$\frac{dB}{d\varepsilon} = \left(J\frac{\partial^2 W}{\partial x^2}(\varepsilon, t, X(\varepsilon, t, y))\right) B + J\frac{\partial^2 W}{\partial x \partial t}(\varepsilon, t, X(\varepsilon, t, y)).$$

Because $X(0, t, y) \equiv y$, $B(0) = 0$, and so, by the variation of constants formula,

$$B(\varepsilon) = \int_0^\varepsilon A(\varepsilon) A(s)^{-1} J \frac{\partial^2 W}{\partial x \partial t}(s, t, X(s, t, y)) ds;$$

therefore,

$$J\nabla R(\varepsilon, t, y) = -\left(\frac{\partial X}{\partial y}(\varepsilon, t, y)\right)^{-1} \frac{\partial X}{\partial t}(\varepsilon, t, y) = -A(\varepsilon)^{-1} B(\varepsilon)$$

$$= -\int_0^\varepsilon A(s)^{-1} J \frac{\partial^2 W}{\partial x \partial t}(s, t, X(s, t, y)) ds$$

$$= -\int_0^\varepsilon J A(s)^T \frac{\partial^2 W}{\partial x \partial t}(s, t, X(s, t, y)) ds$$

$$= -J\frac{\partial}{\partial y} \int_0^\varepsilon \frac{\partial W}{\partial t}(s, t, X(s, t, y)) ds$$

$$= -J\frac{\partial}{\partial y} \int_0^\varepsilon \mathcal{L}(W)\left(\frac{\partial W}{\partial t}\right)(s, t, y) ds.$$

In the above, the fact that A is symplectic is used to make the substitution $A^{-1}J = JA^T$.

Thus, to compute the remainder function, first compute the transform of $-\partial W/\partial t$, and then integrate it. That is, let $S_*(\varepsilon, t, x) = \sum(\varepsilon^i/i!)S_i^0(t, x)$, where $S_i^0(t, x) = -\partial W_{i-1}(t, x)/\partial t$. Compute the Lie transform of S_* by the previous algorithms to get $\mathcal{L}(W)(S) = S^*(\varepsilon, t, x) = \sum(\varepsilon^i/i!)S_0^i(t, x)$. Then $R_0^i = S_0^{i-1}$.

For example, to compute the series expansion for $K = G + R$, the new Hamiltonian, through terms of order ε^2, set $K_0^0 = H_0^0$, then compute K_0^1 by the formulas

$$H_0^1 = H_1^0 + \{H_0^0, W_1\}, \quad R_0^1 = -\frac{\partial W_1}{\partial t}, \quad K_0^1 = H_0^1 + R_0^1,$$

which gives the term of order ε, and then compute

$$H_1^1 = H_1^0 + \{H_1^0, W_1\} + \{H_0^0, W_2\}, \quad H_0^2 = H_1^1 + \{H_0^1, W_1\},$$

$$R_0^2 - -\frac{\partial W_2}{\partial t} - \left\{\frac{\partial W_1}{\partial t}, W_1\right\}, \qquad K_0^2 = H_0^2 + R_0^2.$$

Then $K^*(\varepsilon, x) = K_0^0(x) + \varepsilon K_0^1(x) + \frac{\varepsilon^2}{2} K_0^2(x) + \cdots.$

10.3 The Lie Transform Perturbation Algorithm

In many of the cases of interest, the Hamiltonian is given, and the change of variables is sought to simplify it. When the Hamiltonian, and hence the equations, are in sufficiently simple form, they are said to be in "normal form," an expression whose meaning is discussed in detail later.

10.3.1 Example: Duffing's Equation

In (8.2) the Hamiltonian of Duffing's equation was given as

$$II - \frac{1}{2}(q^2 + p^2) + \frac{\gamma}{4}q^4 \tag{10.22}$$

in rectangular coordinates, (q, p), and in action–angle variables, (I, ϕ), it was given as

$$H = I + \frac{\gamma}{8}I^2(3 + 4\cos 2\phi + \cos 4\phi). \tag{10.23}$$

The Hamiltonian is analytic in rectangular coordinates, and so has the d'Alembert character. Consider γ as a small parameter by setting $\varepsilon = \gamma/8$, so, $H(\varepsilon, I, \phi) = H^*(\varepsilon, I, \phi) = H_0^0(I, \phi) + \varepsilon H_1^0(I, \phi)$, where

$$H_0^0 = I, \qquad H_1^0 = I^2(3 + 4\cos 2\phi + \cos 4\phi).$$

By formula (10.18),

$$H_0^1 = H_1^0 + \{H_0^0, W_1\};$$

so,

$$H_0^1 = I^2(3 + 4\cos 2\phi + \cos 4\phi) - \frac{\partial W_1}{\partial \phi}.$$

Choose W_1 so that H_0^1 contains as few terms as possible (one definition of normal form). For the transformation generated by W_1 to be analytic in rectangular coordinates, W must be a Poisson series with the d'Alembert character. Thus the simplest form for H_0^1 is

$$H_0^1 = 3I^2,$$

which is accomplished taking

$$W_1 = I^2 \left(2\sin 2\phi + \frac{1}{4}\sin 4\phi\right).$$

With this W_1, the Hamiltonian in the new coordinates, (J, θ), would be

$$H_*(\varepsilon, J, \theta) = J + \frac{3\gamma}{8}J^2 + O(\gamma^2),$$

and the equations of motion would be

$$\dot{J} = O(\gamma^2), \qquad \dot{\theta} = -1 - \frac{3\gamma}{4}J + O(\gamma^2).$$

In these coordinates, up to terms $O(\gamma^2)$, the solutions move on circles $J =$ constant with uniform angular frequency $-1 - (3\gamma/4)J$.

Let us do this simple example again, but this time in complex coordinates $z = q + ip$, $\bar{z} = q - ip$. This change of variables is symplectic with multiplier $2i$, so, the Hamiltonian becomes

$$H(z, \bar{z}) = iz\bar{z} + \frac{\gamma i}{32}(z^4 + 4z^3\bar{z} + 6z^2\bar{z}^2 + 4z\bar{z}^3 + \bar{z}^4).$$

H is real in the rectangular coordinates (q, p), so H is conjugated by interchanging z and $\bar{z}$, i.e., $H(z, \bar{z}) = H(\bar{z}, z)$. This is the reality condition in these variables. Let $\varepsilon = \gamma/32$ and

$$H_0^0 = iz\bar{z}, \quad H_1^0 = i(z^4 + 4z^3\bar{z} + 6z^2\bar{z}^2 + 4z\bar{z}^3 + \bar{z}^4);$$

so Equation (10.18) becomes

$$H_0^1 = i(z^4 + 4z^3\bar{z} + 6z^2\bar{z}^2 + 4z\bar{z}^3 + \bar{z}^4) + \frac{1}{2}\left(z\frac{\partial W}{\partial z} - \bar{z}\frac{\partial W}{\partial \bar{z}}\right).$$

Try $W = az^\alpha \bar{z}^\beta$; then $(z\partial W/\partial z + \bar{z}\partial W/\partial \bar{z})/2 = a(\alpha - \beta)z^\alpha \bar{z}^\beta/2$, so, all the terms in H_1^0 can be eliminated except those with $\alpha = \beta$. That is, if we take

$$W = -i(z^4/2 + 4z^3\bar{z} - 4z\bar{z}^3 - \bar{z}^4/2),$$

then

$$H_* = H_0^0 + H_1^0 = iz\bar{z} + (3\gamma i/16)(z\bar{z})^2 + O(\varepsilon^2).$$

Notice that both W and H_* satisfy the reality condition and so are real functions in the original coordinates (q, p). The two methods of solving the problem (action–angle variables and complex variables) give the same results when written in rectangular coordinates.

10.3.2 The General Algorithm

The main Lie transform algorithm starts with a given Hamiltonian that depends on a small parameter ε, and constructs a change of variables so that the Hamiltonian in the new variables is simple. The algorithm is built around the following observation.

Consider the Hamiltonian $H_*(\varepsilon, x)$ with series expansion as given in Equation (10.14), so, all the H_i^0 are known. Assume that all the entries in the Lie triangle are known down to the Nth row, so, the H_j^i are known for $i + j \leq N$, and assume that the W_i are known for $i \leq N$. Let $L_j^i, i + j \leq N$, be computed from the same initial Hamiltonian, but with $U_1, \ldots, U_N$ where $U_i = W_i$ for $i = 1, 2, \ldots, N - 1$ and $U_N = 0$. Then

$$H_j^i = L_j^i \text{ for } i + j < N$$

$$H_j^i = L_j^i + \{H_0^0, W_N\} \quad \text{for } i + j = N. \tag{10.24}$$

This is easily seen from the recursive formulas in Theorem 10.2.1. Recall the remark that the sum of all the indices must add to the row number, so, W_N does not affect the terms in the first $N - 1$ rows. The second equation in (10.24) follows from a simple induction across the Nth row.

From this observation, the algorithm is as follows. Assume all the rows in the Lie triangle have been computed down to the $(N - 1)^{st}$ row, that $W_1, \ldots, W_{N-1}$ have been determined, and that the terms $H_0^1, \ldots, H_0^{N-1}$ are in normal form; i.e., simple in some sense. Now it is time to compute W_N so that H_0^N is in normal form. To compute the Nth row do the following.

Step 1: Compute the Nth row from the formulas in Theorem 10.2.1 assuming that $W_N = 0$, and call these terms $L_j^i, i + j = N$.

Step 2: Solve the equation $H_0^N = L_0^N + \{H_0^0, W_N\}$ for W_N and H_0^N, so that H_0^N is in normal form or simple.

Step 3: Add $\{H_0^0, W_N\}$ to each term in the Nth row, i.e., calculate $H_j^i = L_j^i + \{H_0^0, W_N\}$ for all $i + j = N$.

Step 4: Repeat for the next row.

Of course the definition of normal form and simple depends on the equation $H_0^N = L_0^N + \{H_0^0, W_N\}$, which in turn depends on H_0^0. This equation is called the Lie equation or the homology equation.

10.3.3 The General Perturbation Theorem

The algorithm can be used to prove a general theorem that includes almost all applications. Use the notation of Section 10.2.

Theorem 10.3.1. *Let $\{\mathcal{P}_i\}_{i=0}^\infty$, $\{\mathcal{Q}_i\}_{i=1}^\infty$, and $\{\mathcal{R}_i\}_{i=1}^\infty$ be sequences of linear spaces of smooth functions defined on a common domain O in $\mathbb{R}^{2n}$ with the following properties.*

1. $Q_i \subset P_i$, $i = 1, 2, \ldots$.
2. $H_i^0 \in P_i$, $i = 0, 1, 2, \ldots$.
3. $\{P_i, R_j\} \subset P_{i+j}$, $i + j = 1, 2, \ldots$.
4. *for any $D \in P_i$, $i = 1, 2, \ldots$, there exist $B \in Q_i$ and $C \in R_i$ such that*

$$B = D + \{H_0^0, C\}. \tag{10.25}$$

Then there exists a W with a formal Hamiltonian of the form (10.16) with $W_i \in R_i, i = 1, 2, \ldots$, which generates a near-identity symplectic change of variables $x \to y$ such that the Hamiltonian in the new variables has a series expansion given by (10.15) with $H_0^i \in Q_i, i = 1, 2, \ldots$.

Remarks. The Lie equation (10.25) is the heart of a perturbation problem. H_0^0 defines the unperturbed system when $\varepsilon = 0$, so it is supposed to be well understood. For example, it might be the harmonic oscillator or the 2-body problem. Equation (10.25) can be rewritten

$$B = D + \mathcal{F}(C)$$

where $\mathcal{F} = \{H_0^0, \cdot\}$ is a linear operator on functions. One must analyze this operator to determine in what linear spaces the equation (10.25) is solvable. Roughly speaking the Hamiltonian (10.14) starts with terms in the P-spaces ($H_i^0 \in P_i$), and the equation in normal form has terms in the Q-space ($H_0^i \in Q_i$). The Q-spaces are smaller than the P-spaces ($Q_i \subset P_i$). So the normal form is "simpler." The transformation is generated by a Hamiltonian differential equation with Hamiltonian W in the R-spaces ($W_i \in R_i$). D is an old term, B is a new term, and C is a generator.

Proof. Use induction on the rows of the Lie triangle.

Induction hypothesis I_n: Let $H_j^i \in P_{i+j}$ for $0 \leq i + j \leq n$ and $W_i \in R_i, H_0^i \in Q_i$ for $1 \leq i \leq n$.

I_0 is true by assumption, and so assume I_{n-1}. By Equation (10.17)

$$H_{n-1}^1 = H_n^0 + \sum_{k=0}^{n-2} \binom{n-1}{k} \{H_{n-1-k}^0, W_{k+1}\} + \{H_0^0, W_n\}.$$

The last term is singled out because it is the only term that contains an element, W_n, which is not covered by the induction hypothesis or the hypothesis of the theorem. All the other terms are in P_n by I_{n-1} and (3). Thus

$$H_{n-1}^1 = E^1 + \{H_0^0, W_n\},$$

where $E^1 \in P_n$ is known. A simple induction on the columns of the Lie triangle using (10.17) shows that

$$H_{n-s}^s = E^s + \{H_0^0, W_n\},$$

where $E^s \in \mathcal{P}_n$ for $s = 1, 2, \ldots, n$, and so

$$H_n^0 = E^n + \{H_0^0, W_n\}.$$

By (4), solve for $W_n \in \mathcal{R}_n$ and $H_0^n \in \mathcal{Q}_i$. Thus I_n is true.

The theorem given above is formal in the sense that the convergence of the various series is not discussed. In interesting cases the series diverge, but useful information can be obtained in the first few terms of the normal form. One can stop the process at any order N to obtain a W that is a polynomial in ε and so converges. From the proof given above, it is clear that the terms in the series for H^* up to order N are unaffected by the termination. Thus the more useful form of Theorem 10.3.1 is the following.

Corollary 10.3.1. *Let $N \geq 1$ be given, and let $\{\mathcal{P}_i\}_{i=0}^N, \{\mathcal{Q}_i\}_{i=1}^N$, and $\{\mathcal{R}_i\}_{i=1}^N$ be sequences of linear spaces of smooth functions defined on a common domain O in $\mathbb{R}^{2n}$ with the following properties.*

1. *$\mathcal{Q}_i \subset \mathcal{P}_i$, $i = 1, 2, \ldots, N$.*
2. *$H_i^0 \in \mathcal{P}_i$, $i = 0, 1, 2, \ldots, N$.*
3. *$\{\mathcal{P}_i, \mathcal{R}_j\} \subset \mathcal{P}_{i+j}$, $i + j = 1, 2, \ldots, N$.*
4. *For any $D \in \mathcal{P}_i$, $i = 1, 2, \ldots, N$, there exist$B \in \mathcal{Q}_i$ and $C \in \mathcal{R}_i$ such that*

$$B = D + \{H_0^0, C\}. \tag{10.26}$$

Then there exists a polynomial W,

$$W(\varepsilon, x) = \sum_{i=0}^{N-1} \left(\frac{\varepsilon^i}{i!} \right) W_{i+1}(x), \tag{10.27}$$

with $W_i \in \mathcal{R}_i, i = 1, 2, \ldots, N$, such that the change of variables $x = X(\varepsilon, y)$ where $X(\varepsilon, y)$ is the general solution of $dx/d\varepsilon = J\nabla W(\varepsilon, x), x(0) = y$, transforms the convergent Hamiltonian

$$H(\varepsilon, x) = H_*(\varepsilon, x) = \sum_{i=0}^{\infty} \left(\frac{\varepsilon^i}{i!} \right) H_i^0(x) \tag{10.28}$$

to the convergent Hamiltonian

$$G(\varepsilon, x) = H^*(\varepsilon, y) = \sum_{i=0}^{N} \left(\frac{\varepsilon^i}{i!} \right) H_0^i(y) + O(\varepsilon^{N+1}), \tag{10.29}$$

with $H_0^i \in \mathcal{Q}_i, i = 1, 2, \ldots, N$.

The nonautonomous case. In the nonautonomous case, the algorithm is slightly different. The remainder function, $\mathcal{R}(\varepsilon, t, y)$, is the indefinite integral of $S^*(\varepsilon, t, y)$, where $S^*(\varepsilon, t, y) = -\mathcal{L}(W)(\partial W/\partial t)(s, t, y)$, the Lie transform of $S_* = -\partial W/\partial t$. One constructs two Lie triangles, one for the Hamiltonian H and one for the function S. Because $\mathcal{R}$ is the indefinite integral of S^*, if you want the new Hamiltonian up to terms of order ε^N, then you need all the Lie triangle for H_* down to the Nth row, but only down to the $(N-1)^{st}$ for S. One simply works down the two triangles together, but with the S triangle one row behind.

Assume that all the entries in the Lie triangle for H are known down to the Nth row $(H^i_j, i + j \leq N)$ and that all the entries in the Lie triangle for S_* are known down to the $(N-1)^{st}$ row $(S^i_j, i + j \leq N - 1)$ using the W_i for $i \leq N$. Let $G^i_j, i + j \leq N$, be computed from the same Hamiltonian, so, $G^0_i = H^0_i$ for all i, but with $U_1, \ldots, U_N$, where $U_i = W_i$ for $i = 1, 2, \ldots, N-1$ and $U_N = 0$. Let Q^i_j be the terms in the Lie triangle for the remainder using the U'_is. Then

$$H^i_j = G^i_j \text{ for } i + j < N, \qquad\qquad S^i_j = Q^i_j \text{ for } i + j < N - 1,$$

$$H^i_j = G^i_j + \{H^0_0, W_N\} \text{ for } i + j = N, \quad S^i_j = Q^i_j - \frac{\partial W_N}{\partial t} \text{ for } i + j = N - 1.$$
$$(10.30)$$

This is easily seen from the recursive formulas in Theorem 10.2.1.

From this observation, the algorithm is as follows. Assume that all the rows in the Lie triangle for H have been computed down to the $(N-1)^{st}$ row, that all the rows in the Lie triangle for S_* have been computed down to the $(N-2)$nd row and that $W_1, \ldots, W_{N-1}$ have been determined, and that the $H^1_0, \ldots, H^{N-1}_0$ are in normal form. Now it is time to compute W_N so that H^N_0 is in normal form.

Step 1: Compute the Nth row for H and the $(N-1)^{st}$ row for the remainder assuming that $W_N = 0$, and call these terms $G^i_j, i + j = N$, and $Q^i_j, i + j = N - 1$, respectively.

Step 2: Solve the equation $H^N_0 = G^N_0 + Q^{N-1}_0 + \{H^0_0, W_N\} - \partial W_N/\partial t$ for W_N and H^N_0 so that H^N_0 is in normal form or simple.

Step 3: Add $\{H^0_0, W_N\}$ to each term in the Nth row for H, and add $\partial W_N/\partial t$ to each term in the $(N-1)^{st}$ row for S.

Step 4: Repeat.

The nonautonomous version of Theorem 10.3.1 is as follows.

Theorem 10.3.2. *Let $\{\mathcal{P}_i\}_{i=0}^\infty$, $\{\mathcal{Q}_i\}_{i=1}^\infty$, and $\{\mathcal{R}_i\}_{i=1}^\infty$ be sequences of linear spaces of smooth functions defined on a common domain O in $\mathbb{R}^1 \times \mathbb{R}^{2n}$. Let $\dot{\mathcal{R}}_i$ be the space of all derivatives of functions in $\mathcal{R}_i$. Assume the following:*

1. *$\mathcal{Q}_i \subset \mathcal{P}_i, i = 1, 2, \ldots$.*
2. *$H^0_i \in \mathcal{P}_i, i = 0, 1, 2, \ldots$.*
3. *$\{\mathcal{P}_i, \mathcal{R}_j\} \subset \mathcal{P}_{i+j}$ and $\{\mathcal{P}_i, \dot{\mathcal{R}}_j\} \subset \mathcal{P}_{i+j}$. for $i + j = 1, 2, \ldots$.*
4. *For any $D \in \mathcal{P}_i, i = 1, 2, \ldots$, there exists $B \in \mathcal{Q}_i$ and $C \in \mathcal{R}_i$ such that*

$$B = D + \{H_0^0, C\} - \frac{\partial C}{\partial t}. \tag{10.31}$$

Then there exists a W with a formal Hamiltonian of the form (10.16) with $W_i \in \mathcal{R}_i, i = 1, 2, \ldots$, that generates a near-identity symplectic change of variables $x \to y$ such that the Hamiltonian in the new variables has a series expansion given by (10.15) with $H_0^i \in \mathcal{Q}_i, i = 1, 2, \ldots$.

Duffing's equation revisited. Consider the Hamiltonian (10.23) of Duffing's equation as written in action–angle variables. The operator $\{H_0^0, C\} = \partial C/\partial \phi$ is very simple to understand. Equation (10.31) becomes

$$B = D + \frac{\partial C}{\partial \phi}.$$

If D is a finite Poisson series with d'Alembert character, then by taking B to be the term of D that is independent of the angle ϕ and $C = \int (B - D) d\phi$, B and C satisfy this equation. This leads us to the following definitions of the spaces.

Let $\mathcal{P}_i$ be the space of all finite Poisson series with d'Alembert character corresponding to homogeneous polynomials of degree $2i + 2$ in rectangular coordinates. So an element in $\mathcal{P}_i$ is of the form I^{i+1} times a finite Fourier series in ϕ. Let $\mathcal{Q}_i$ be the space of all polynomials of the form AI^{i+1}, where A is a constant. Let $\mathcal{R}_i$ be the subspace of $\mathcal{P}_i$ of Poisson series without a term independent of ϕ. So $\mathcal{P}_i = \mathcal{Q}_i \oplus \mathcal{R}_i$. Because the Poisson bracket of homogeneous polynomials of degree $2i + 2$ and degree $2j + 2$ is a polynomial of degree $2(i+j) + 2$, and because symplectic changes of coordinates preserve Poisson brackets, we have $\{\mathcal{P}_i, \mathcal{R}_j\} \subset \mathcal{P}_{i+j}$. Thus by Corollary 10.3.1, there exists a formal, symplectic transformation that transforms the Hamiltonian of Duffing's equation into the form

$$H^*(\varepsilon, J) = \sum_{i=0}^{\infty} \left(\frac{\varepsilon^i}{i!} \right) H_0^i(J)$$

and the equations of motion become

$$\dot{J} = 0, \qquad \dot{\phi} = -\frac{\partial H}{\partial \phi}(\varepsilon, J) = -\omega(\varepsilon, J).$$

Thus formally, the solutions move on circles with a uniform frequency $\omega(\varepsilon, J)$, which depends on ε and J. By the theorems of Poincaré (1885) and Rüssman (1959) the series converges in this simple case.

Uniqueness of normal forms: One of the important special cases where Theorem 10.3.1 applies is when the operator $\mathcal{F}_i = \{H_0^0, \cdot\} : \mathcal{P}_i \to \mathcal{P}_i$ is simple, i.e., when $\mathcal{P}_i = \mathcal{Q}_i \oplus \mathcal{R}_i$, $\mathcal{Q}_i = \text{kernel}(\mathcal{F}_i)$, and $\mathcal{R}_i = \text{range}(\mathcal{F}_i)$. In this case, the Lie equation (10.25) has a unique solution. This is not enough to assure uniqueness of the normal form. One needs one extra condition.

Theorem 10.3.3. *Let $\{\mathcal{P}_i\}_{i=0}^{\infty}$ be sequences of linear spaces of smooth functions defined on a common domain O in $\mathbb{R}^{2n}$. Let $\mathcal{F}_i = \{H_0^0, \cdot\} : \mathcal{P}_i \to \mathcal{P}_i$ be simple, so, $\mathcal{P}_i = \mathcal{Q}_i \oplus \mathcal{R}_i$, $\mathcal{Q}_i = kernel(\mathcal{F}_i)$, $\mathcal{R}_i = range(\mathcal{F}_i)$. Assume*

1. *$H_i^0 \in \mathcal{P}_i$, $i = 0, 1, 2, \ldots$.*
2. *$\{\mathcal{P}_i, \mathcal{R}_j\} \subset \mathcal{P}_{i+j}$, $i + j = 1, 2, \ldots$.*

Then there exists a W with a formal expansion of the form (10.16) with $W_i \in \mathcal{R}_i, i = 1, 2, \ldots$, such that W generates a near-identity symplectic change of variables $x \to y$ which transforms the Hamiltonian $H_*(\varepsilon, x)$ with the formal series expansion given in Equation (10.14) to the Hamiltonian $H^*(\varepsilon, y)$ with the formal series expansion given by Equation (10.15) with $H_0^i \in \mathcal{Q}_i, i = 1, 2, \ldots$.

Moreover, if

$$\{\mathcal{Q}_i, \mathcal{Q}_j\} = 0, \quad i, j = 1, 2, \ldots,$$

then the terms in the normal form are unique.

Remark. All the obvious remarks about the time-dependent cases hold here also. The normal form is unique, but the transformation taking the equation need not be unique. Clearly this theorem applies to the Duffing example. We do not need this theorem in our development. See Liu (1985) for a proof or see the Problems section.

10.4 Normal Form at an Equilibrium

Consider an analytic Hamiltonian H that has an equilibrium point at the origin in $\mathbb{R}^{2n}$, and assume that the Hamiltonian is zero at the origin. Then H has a Taylor series expansion of the form

$$H(x) = H_{\#}(x) = \sum_{i=0}^{\infty} H_i(x), \tag{10.32}$$

where H_i is a homogeneous polynomial in x of degree $i + 2$, so, $H_0(x) = \frac{1}{2}x^T S x$, where S is a $2n \times 2n$ real symmetric matrix, and $A = JS$ is a Hamiltonian matrix. The linearized equations about the critical point $x = 0$ are

$$\dot{x} = Ax = JSx = J\nabla H_0(x), \tag{10.33}$$

and the general solution of (10.33) is $\phi(t, \xi) = \exp(At)\xi$.

The classical case. The matrix A is simple if it has $2n$ linearly independent eigenvectors that may be real or complex. The matrix A being simple is equivalent to A being similar to a diagonal matrix by a real or complex similarity transformation. This is why A is sometimes said to be diagonalizable. The classical theorem on normal forms is as follows.

Theorem 10.4.1. *Let A be simple. Then there exists a formal symplectic change of variables,*

$$x = X(y) = y + \cdots,\qquad(10.34)$$

that transforms the Hamiltonian (10.32) to

$$H^{\#}(y) = \sum_{i=0}^{\infty} H^i(y),\qquad(10.35)$$

where H^i is a homogeneous polynomial of degree $i+2$ such that

$$H^i(e^{At}y) \equiv H^i(y),\qquad(10.36)$$

for all $i = 0, 1, \ldots$, all $y \in \mathbb{R}^{2n}$, and all $t \in \mathbb{R}$.

Remark. Formula (10.36) is the classical definition of normal form for a Hamiltonian near an equilibrium point with a simple linear part. Formula (10.36) says that H^i is an integral for the linear system (10.33), so, by Theorem 1.2.1, (10.36) is equivalent to

$$\{H^i, H^0\} = 0\qquad(10.37)$$

for all i.

Proof. In order to study the solutions near the origin, scale the variables by $x \to \varepsilon x$. This is a symplectic transformation with multiplier ε^{-2}; so, the Hamiltonian becomes

$$H(\varepsilon, x) = H_*(\varepsilon, x) = \sum_{i=0}^{\infty} \left(\frac{\varepsilon^i}{i!}\right) H_i^0(x),\qquad(10.38)$$

where $H_i^0 = i! H_i$. Because we are working formally, we can set $\varepsilon = 1$ at the end, or we can rescale by $x \to \varepsilon^{-1}x$.

Let $\mathcal{P}_i$ be the linear space of all real homogeneous polynomials of degree $i + 2$, so, $H_i^0 \in \mathcal{P}_i$. Because A is simple, A has $2n$ linearly independent eigenvectors $s_1, \ldots, s_{2n}$ corresponding to the eigenvalues $\lambda_1, \ldots, \lambda_{2n}$. The s_i are row eigenvectors, so, $s_i A = \lambda_i s_i$. Let $2r$ of the eigenvalues be complex, and number them so that $\lambda_i = \bar{\lambda}_{n+i}$ for $i = 1, \ldots, r$. Choose the eigenvectors so that $s_i = \bar{s}_{n+i}$ for $i = 1, \ldots, r$. The other eigenvalues and eigenvectors are real. Let $K \in \mathcal{P}_i$, so, K is a homogeneous polynomial of degree $i + 2$. Because the s_i are independent, K may be written in the form

$$K = \sum \kappa_{m_1 m_2 \ldots m_{2n}} (s_1 x)^{m_1} (s_2 x)^{m_2} \cdots (s_{2n} x)^{m_{2n}},\qquad(10.39)$$

where the sum is over all $m_1 + \cdots + m_{2n} = i + 2$. So the monomials in

$$B = \{(s_1 x)^{m_1} (s_2 x)^{m_2} \cdots (s_{2n} x)^{m_{2n}} : m_1 + \cdots + m_{2n} = i + 2\}\qquad(10.40)$$

span $\mathcal{P}_i$. It is also clear that they are independent, so, form a basis for $\mathcal{P}_i$. The coefficients in (10.39) may be complex but must satisfy the reality condition that interchanging the subscripts m_i and m_{n+i} for $i = 1, \ldots, r$ in the κ coefficients is the same as conjugation.

Now let $\mathcal{F} = \mathcal{F}_i : \mathcal{P}_i \to \mathcal{P}_i$ be the linear operator of Theorem 10.3.3 as it applies to Hamiltonian systems, that is, define $\mathcal{F}$ by $\mathcal{F}(G) = \{H_0^0, G\} = -(\partial G/\partial x)Ax$, so,

$$\mathcal{F}((s_1 x)^{m_1}(s_2 x)^{m_2} \cdots (s_{2n} x)^{m_{2n}})$$

$$= -(m_1 \lambda_1 + \cdots + m_{2n} \lambda_{2n})(s_1 x)^{m_1}(s_2 x)^{m_2} \cdots (s_{2n} x)^{m_{2n}}.$$

So the elements of B are eigenvectors of $\mathcal{F}$ and the eigenvalues are $-(m_1 \lambda_1 + \cdots + m_{2n} \lambda_{2n})$, $m_1 + \cdots + m_{2n} = i + 2$. Thus we can define $\mathcal{F}$-invariant subspaces

$$\mathcal{K}_i = \mathrm{span}\{(s_1 x)^{m_1}(s_2 x)^{m_2} \cdots (s_{2n} x)^{m_{2n}} : m_1 + \cdots + m_{2n} = i+2,$$
$$m_1 \lambda_1 + \cdots + m_{2n} \lambda_{2n} = 0\},$$

$$\mathcal{R}_i = \{(s_1 x)^{m_1}(s_2 x)^{m_2} \cdots (s_{2n} x)^{m_{2n}} : m_1 + \cdots + m_{2n} = i+2,$$
$$m_1 \lambda_1 + \cdots + m_{2n} \lambda_{2n} \neq 0\}.$$

In summary, $\mathcal{K}_i = \mathrm{kernel}\,(\mathcal{F})$, $\mathcal{R}_i = \mathrm{range}\,(\mathcal{F})$, and $\mathcal{P}_i = \mathcal{K}_i \oplus \mathcal{R}_i$. Thus this classical theorem follows from the first part of Theorem 10.3.3 because we have shown that the operators $\mathcal{F}_i : \mathcal{P}_i \to \mathcal{P}_i$ are simple. However, the extra condition in Theorem 10.3.3 is not satisfied in general, so, the normal form may not be unique.

Birkhoff (1927) considered a special case of the above.

Corollary 10.4.1. *Assume that the quadratic part of* (10.32) *is of the form*

$$H_0(x) = \sum_{j=1}^{n} \lambda_j x_j x_{n+j}, \tag{10.41}$$

where the λ_js are independent over the integers, i.e., there is no nontrivial relation of the form

$$\sum_{i=1}^{n} k_j \lambda_j = 0, \tag{10.42}$$

where the k_j are integers. Then there exists a formal symplectic change of variables $x = X(y) = y + \cdots$ that transforms the Hamiltonian (10.32) *to the Hamiltonian* (10.35), *where $H^j(y)$ is a homogeneous polynomial of degree $j + 1$ in the n products $y_1 y_{n+1}, \ldots, y_n y_{2n}$. So, $H^\#(y_1, \ldots, y_{2n}) = H^\#(y_1 y_{n+1}, \ldots, y_n y_{2n})$ where $H^\#$ is a function of n variables. Moreover, in this case, the normal form is unique.*

Remark. Formally the equations of motion for the system in normal form are

$$\dot{y}_j = y_j D_j H^\#(y_1 y_{n+1}, \ldots, y_n y_{2n}),$$

$$\dot{y}_{j+n} = -y_{j+n} D_j H^\#(y_1 y_{n+1}, \ldots, y_n y_{2n}).$$

Here D_j stands for the partial derivative with respect to the jth variable. In this form, the system of equations has n formal integrals in involution, $I_1 = y_1 y_{n+1}, \ldots, I_n = y_n y_{2n}$.

In the case when the $\lambda_j = i\omega_j$ are pure imaginary and the y_j are the complex coordinates discussed in Lemma 2.3.4, then we can switch to action–angle variables by $y_j = \sqrt{I_j/2}e^{i\phi_j}$, $y_{n+j} = \sqrt{I_j/2}e^{i\phi_j}$. The Hamiltonian in normal form is a function of the action variables only, so, the Hamiltonian is $H^\dagger(I_1, \ldots, I_n)$, and the equations of motion are

$$\dot{I}_j = \frac{\partial H^\dagger}{\partial \phi_j} = 0, \qquad \dot{\phi}_j = -\frac{\partial H^\dagger}{\partial I_j} = \omega_j(I_1, \ldots, I_n).$$

Here $\omega_i(I_1, \ldots, I_n) = \pm\omega_i + \cdots$, and the sign is determined by the cases in Lemma 2.3.2. Setting the action variables equal to nonzero constants, $I_1 = c_1, \ldots, I_n = c_n$, defines an invariant set which is an n-torus with n angular coordinates $\phi_1, \ldots, \phi_n$. On each torus the angular frequencies $\omega_j(I_1, \ldots, I_n)$, are constant, and so, define a linear flow on the torus as discussed in Section 1.1. The frequencies vary from torus to torus in general.

Notation. For this proof, and subsequent discussions, some notation is useful. Let $\mathcal{Z} = \mathbb{Z}_+^{2n}$ denote the set of all $2n$-tuples of nonnegative integers, so, $k \in \mathcal{Z}$ means $k = (k_1, \ldots, k_{2n})$, $k_i \geq 0$, k_i an integer. Let $|k| = k_1 + \cdots + k_{2n}$. If $x \in \mathbb{R}^{2n}$ and $k \in \mathcal{Z}$, then define $x^k = x_1^{k_1} x_2^{k_2} \cdots x_{2n}^{k_{2n}}$.

Proof. The linear part is clearly simple. Let $H^i(y) = \sum h_k y^k$, where the sum is over $k \in \mathcal{Z}, |k| = i + 2$. The general solution of the linear system is $y_i = y_{i0}\exp(\lambda_i t), y_{i+n} = y_{i+n,0}\exp(-\lambda_i t)$ for $i = 1, \ldots, n$. Formula (10.36) implies that $\sum h_k \exp t\{(k_1 - k_{n+1})\lambda_1 + \cdots + (k_n - k_{2n})\lambda_n\}y^k$ is constant in t, and this implies that $\{(k_1 - k_{n+1})\lambda_1 + \cdots + (k_n - k_{2n})\lambda_n = 0$. But because the $\lambda_i's$ are independent over the integers, this implies $k_1 = k_{n+1}, \ldots, k_n = k_{2n}$. That is, H^i is a function of the products $y_1 y_{n+1}, \ldots, y_n y_{2n}$ only.

By the remark above, the kernel consists of those functions that depend only on $I_1, \ldots, I_n$ and not on the angles in action–angle variables. Therefore, the extra condition of Theorem 10.3.3 holds, and the normal form is unique.

Remark. If the condition (10.42) only holds for $|k_1| + \cdots + |k_n| \leq N$, then the terms in the Hamiltonian up to the terms of order N can be put in normal form, and these terms are unique.

The general equilibria. In the 1970s, the question of the stability of the Lagrange triangular point $\mathcal{L}_4$ was studied intensely. For Hamiltonian systems, it is not enough to look at the linearized system alone, because the higher-order terms in the normalized equations can change the stability (see the discussion in Chapter 11). The matrix of the linearization of the equations at $\mathcal{L}_4$ when $\mu = \mu_1$ is not simple as was seen in Section 4.5. The normal form for this case, and other similar cases were carried out by the Russian school; see Sololskii (1978). First Kummer (1976, 1978) and then Cushman et al. (1983), used group representation theory. Representation theory is very helpful in understanding the general case, but there are simpler ways to understand the basic ideas and examples. In Meyer (1984b) a theorem like Theorem 10.4.1 above was given for non-Hamiltonian systems but A was replaced by A^T in (10.36), so, the terms in the normal form are invariant under the flow $\exp(A^T t)$. A far better proof can be found in Elphick et al. (1987), which is what we present here.

The proof of Theorem 10.4.1 rested on the fact that for a simple matrix, A, the vector space $\mathbb{R}^{2n}$ is the direct sum of the range and kernel of A, and this held true for the operator $\mathcal{F} = \{H_0^0, \cdot\}$ defined on homogeneous polynomials as well. The method of Elphick et al. is based on the following simple lemma in linear algebra known as the Fredholm alternative and an inner product defined on homogeneous polynomials given after the lemma.

Lemma 10.4.1. *Let $\mathbb{V}$ be a finite-dimensional inner product space with inner product $(\cdot, \cdot)$. Let $A : \mathbb{V} \to \mathbb{V}$ be a linear transformation, and A^* its adjoint (so $(Ax, y) = (x, A^* y)$ for all $x, y \in \mathbb{V}$). Then $\mathbb{V} = R \oplus K^*$ where R is the range of A and K^* is the kernel of A^*.*

Proof. Let $x \in R$, so, there is a $u \in \mathbb{V}$ such that $Au = x$. Let $y \in K^*$, so, $A^* y = 0$. Because $0 = (u, 0) = (u, A^* y) = (Au, y) = (y, x)$, it follows that R and K^* are orthogonal subspaces. Let K be the kernel of A. In a finite dimensional space, $\dim \mathbb{V} = \dim R + \dim K$ and $\dim K = \dim K^*$. Because R and K^* are orthogonal, $\dim(R + K^*) = \dim R + \dim K^* = \dim \mathbb{V}$, so, $\mathbb{V} = R \oplus K^*$.

Let $\mathcal{P} = \mathcal{P}_j$ be the linear space of all homogeneous polynomials of degree j in $2n$ variables $x \in \mathbb{R}^{2n}$. So if $P \in \mathcal{P}$, then

$$P(x) = \sum_{|k|=j} p_k x^k = \sum_{|k|=j} p_{k_1 k_2 \ldots k_{2n}} x_1^{k_1} x_2^{k_2} \cdots x_{2n}^{k_{2n}}.$$

Define $P(\partial)$ to be the differential operator

$$P(\partial) = \sum_{|k|=j} p_k \frac{\partial^k}{\partial x^k},$$

where we have introduced the notation

$$\frac{\partial^k}{\partial x^k} = \frac{\partial^{k_1}}{\partial x_1^{k_1}} \frac{\partial^{k_2}}{\partial x_2^{k_2}} \cdots \frac{\partial^{k_{2n}}}{\partial x_{2n}^{k_{2n}}}.$$

Let $Q \in \mathcal{P}$, $Q(x) = \sum q_h x^h$ be another homogeneous polynomial, and define an inner product $< \cdot, \cdot >$ on $\mathcal{P}$ by

$$< P, Q >= P(\partial)Q(x).$$

To see that this is indeed an inner product, note that $\partial^k x^h / \partial x^k = 0$ if $k \neq h$ and $\partial^k x^h / \partial x^k = k! = k_1! k_2! \cdots k_{2n}!$ if $k = h$, so,

$$< P, Q >= \sum_{|k|=j} k! p_k q_k.$$

Let $A = JS$ be a Hamiltonian matrix where S is a symmetric matrix of the quadratic Hamiltonian H^0, so, $H^0(x) - \frac{1}{2} x^T S x$. From Theorem 10.3.1 and the proof of Theorem 10.4.1, the operator of importance is $\mathcal{F}(A) : \mathcal{P} \to \mathcal{P}$, where

$$\mathcal{F}(A)P = \{H_0^0, P\} = -\frac{\partial P}{\partial x} Ax = \frac{d}{dt} P(e^{At} x)\big|_{t=0}. \tag{10.43}$$

Lemma 10.4.2. *Let $A : \mathbb{R}^{2n} \to \mathbb{R}^{2n}$ be as above and A^T its transpose (so A^T is the adjoint of A with respect to the standard inner product in $\mathbb{R}^{2n}$). Then for all $P, Q \in \mathcal{P}$,*

$$< P(x), Q(Ax) >=< P(A^T x), Q(x) > \tag{10.44}$$

and

$$< P, \mathcal{F}(A)Q >=< \mathcal{F}(A^T)P, Q > . \tag{10.45}$$

That is, the adjoint of $\mathcal{F}(A)$ with respect to $< \cdot, \cdot >$ is $\mathcal{F}(A^T)$.

Proof. Equation (10.44) follows from (10.43) because (10.43) implies

$$< P(x), Q(e^{At} x) >=< P(e^{A^T t} x), Q(x) > .$$

Differentiating this last expression with respect to t and setting $t = 0$ gives (10.45).

Let $y = Ax$ (i.e., $y^i = \sum_j A_j^i x^j$) and $F(y) = F(Ax)$. Inasmuch as

$$\frac{\partial F(y)}{\partial x^j} = \sum_i \frac{\partial F(y)}{\partial y^i} \frac{\partial y^i}{\partial x^j} = \sum_i \frac{\partial F(y)}{\partial y^i} A_j^i,$$

it follows that $\partial / \partial x = A^T \partial / \partial y$.

$$< P(x), Q(Ax) >= P(\partial_x)Q(Ax) = P(A^T \partial_y)Q(y) =< P(A^T y), Q(y) > .$$

Theorem 10.4.2. *Let A be a Hamiltonian matrix. Then there exists a formal symplectic change of variables, $x = X(y) = y + \cdots$, that transforms the Hamiltonian (10.32) to*

$$H^{\#}(y) = \sum_{j=0}^{\infty} H^j(y), \tag{10.46}$$

where H^j is a homogeneous polynomial of degree $j + 2$ such that

$$H^j(e^{A^T t}y) \equiv H^j(y), \tag{10.47}$$

for all $j = 0, 1, \ldots$, all $y \in \mathbb{R}^{2n}$, and all $t \in \mathbb{R}$.

Remark. Let $H_0^T(x) = H_T^0(x) = \frac{1}{2}x^T R x$ be the quadratic Hamiltonian for the adjoint linear equation, so, $A^T = JR$. Then (10.47) is equivalent to

$$\{H^i, H_T^0\} = 0$$

for $j = 1, 2, \ldots$.

Proof. By Theorem 10.3.1, we must solve Equation (10.25) or $\mathcal{F}(A)C + D = B$, where $D \in P_j = P$ is given, and $C \in Q_j = P$, and $D \in Q_j =$ kernel $(\mathcal{F}(A^T))$. By Lemma 10.4.2, we can write $D = B - G$, where $B \in$ kernel$(\mathcal{F}(A^T))$, so, $\{B, H_T^0\} = 0$, and $G \in$ range $(\mathcal{F}(A))$, so, $G = \mathcal{F}(A)C$, $C \in P$. With these choices, (10.30) is solved. Verification of the rest of the hypothesis in Theorem 10.3.1 is just as in the proof of Theorem 10.4.1.

Theorem 10.4.1 is a corollary of this theorem because when A is simple, it is diagonalizable, and so, its own adjoint. We proved Theorem 10.4.1 separately, because the proof is constructive.

Examples of normal forms in the nonsimple case. Consider the Hamiltonian system (10.32), where $n = 1$ and $x = (q, p)$. Let

$$H_0(q, p) = p^2/2, \qquad H_0^T(q, p) = -q^2/2,$$

$$A = \begin{bmatrix} 0 & 1 \\ 0 & 0 \end{bmatrix} \qquad A^T = \begin{bmatrix} 0 & 0 \\ 1 & 0 \end{bmatrix}.$$

Because

$$\exp(A^T t) = \begin{bmatrix} 1 & 0 \\ 1+t & 1 \end{bmatrix},$$

(10.47) implies that the higher-order terms in the normal form are independent of p, or $H^i = H^i(p, \cdot)$. Thus the Hamiltonian in normal form is $p^2/2 + G(q)$, which is the Hamiltonian for the second-order equation $\ddot{q} + g(q) = 0$, where $g(q) = \partial G(q)/\partial q$.

Now consider a Hamiltonian system with two degrees of freedom with a linearized system with repeated pure imaginary roots that are nonsimple. In Section 5.3, the normal form for the quadratic part of such a Hamiltonian was given as

$$H_0 = \omega(\xi_2 \eta_1 - \xi_1 \eta_2) + \frac{\delta}{2}(\xi_1^2 + \xi_2^2),$$

where $\omega \neq 0$ and $\delta = \pm 1$. The linearized equations are

$$\begin{bmatrix} \dot{\xi}_1 \\ \dot{\xi}_2 \\ \dot{\eta}_1 \\ \dot{\eta}_2 \end{bmatrix} = \begin{bmatrix} 0 & \omega & 0 & 0 \\ -\omega & 0 & 0 & 0 \\ -\delta & 0 & 0 & \omega \\ 0 & -\delta & -\omega & 0 \end{bmatrix} \begin{bmatrix} \xi_1 \\ \xi_2 \\ \eta_1 \\ \eta_2 \end{bmatrix}.$$

The transpose is defined by the Hamiltonian

$$H_0^T = \omega(\xi_2\eta_1 - \xi_1\eta_2) - \frac{\delta}{2}(\eta_1^2 + \eta_2^2),$$

and the transposed equations are

$$\begin{bmatrix} \dot{\xi}_1 \\ \dot{\xi}_2 \\ \dot{\eta}_1 \\ \dot{\eta}_2 \end{bmatrix} = \begin{bmatrix} 0 & -\omega & -\delta & 0 \\ \omega & 0 & 0 & -\delta \\ 0 & 0 & 0 & -\omega \\ 0 & 0 & \omega & 0 \end{bmatrix} \begin{bmatrix} \xi_1 \\ \xi_2 \\ \eta_1 \\ \eta_2 \end{bmatrix}.$$

Sololskii (1978) suggested changing to polar coordinates (see Section 8.1) to make the transposed equations simple. That is, he changed coordinates by

$$\eta_1 = r\cos\theta, \qquad R = (\xi_1\eta_1 + \xi_2\eta_2)/r,$$

$$\eta_2 = r\sin\theta, \qquad \Theta = \eta_1\xi_2 - \eta_2\xi_1.$$

In these coordinates,

$$H_0^T = -\omega\Theta + \frac{\delta}{2}r^2, \qquad H_0 = \omega\Theta + \frac{\delta}{2}\left(R^2 + \frac{\Theta^2}{r^2}\right),$$

and the transposed equations are

$$\dot{r} = 0, \quad \dot{\theta} = \omega, \quad \dot{R} = \delta r, \quad \dot{\Theta} = 0.$$

Thus the higher order terms in the normal form are independent of θ and R and so depend only on $r^2 = \eta_1^2 + \eta_2^2$ and $\Theta = \eta_1\xi_2 - \eta_2\xi_1$.

Thus the theory of the normal form in this case depends on three qualities

$$\Gamma_1 = \xi_2\eta_1 - \xi_1\eta_2, \qquad \Gamma_2 = \frac{1}{2}(\xi_1^2 + \xi_2^2), \qquad \Gamma_3 = \frac{1}{2}(\eta_1^2 + \eta_2^2).$$

The Hamiltonian $H_0 = \omega\Gamma_1 + \Gamma_2$ and the higher-order terms in the normal form are functions of Γ_1 and Γ_3 only. This is known as Sokol'skij's normal form.

10.5 Normal Form at $\mathcal{L}_4$

Recall that in Section 4.5, we showed that the linearization of the restricted 3-body problem at the Lagrange triangular point $\mathcal{L}_4$ had two pairs of pure imaginary eigenvalues, $\pm i\omega_1, \pm i\omega_2$ when $0 < \mu < \mu_1 = \frac{1}{2}(1 - \sqrt{69}/9)$, and that there are symplectic coordinates so that the quadratic part of the Hamiltonian is

$$H_2 = \omega_1 I_1 - \omega_2 I_2,$$

where I_1, I_2, ϕ_1, ϕ_2 are action–angle variables.

Recall that in Section 6.4, we defined μ_r to be the value of μ for which $\omega_1/\omega_2 = r$, and that $0 \cdots < \mu_3 < \mu_2 < \mu_1$. When $0 < \mu < \mu_1$, and $\mu \neq \mu_2, \mu_3$ then by Corollary 10.4.1, the Hamiltonian of the restricted 3-body problem can be normalized through the fourth-order terms, so, the Hamiltonian becomes

$$H = \omega_1 I_1 - \omega_2 I_2 + \frac{1}{2}(AI_1^2 + 2BI_1I_2 + CI_2^2) + \cdots.$$

After six months of hand calculations, Deprit and Deprit-Bartholome computed:

$$A = \frac{1}{72}\omega_2^2 \frac{(81 - 696\omega_1^2 + 124\omega_1^4)}{(1 - 2\omega_1^2)^2(1 - 5\omega_1^2)},$$

$$B = -\frac{1}{6}\frac{\omega_1\omega_2(43 + 64\omega_1^2\omega_2^2)}{(1 - 2\omega_1^2)^2(1 - 5\omega_1^2)},$$

$$C(\omega_1, \omega_2) = A(\omega_2, \omega_1).$$

Meyer and Schmidt (1986) computed the normal form through terms of sixth-order by computer. The results are too lengthy to reproduce here. It did serve as an independent check of the calculations of Deprit and Deprit-Bartholome. In Section 4.5, the quadratic part of the Hamiltonian of the restricted 3-body problem at $\mathcal{L}_4$ for $\mu = \mu_1$ was brought into normal form by a linear symplectic change of coordinates. In these coordinates, the quadratic part of the Hamiltonian is of the form

$$H_0 = \omega(\xi_2\eta_1 - \xi_1\eta_2) + \frac{1}{2}(\xi_2^2 + \xi_1^2) = \omega\Gamma_1 + \Gamma_2,$$

where $\omega = \sqrt{2}/2$ and $\delta = +1$.

The normal form for the Hamiltonian of the restricted 3-body problem at $\mathcal{L}_4$ for $\mu = \mu_1$ is of the form

$$H = \omega\Gamma_1 + \Gamma_2 + c\Gamma_1^2 + 2d\Gamma_1\Gamma_3 + 4e\Gamma_3^2 + \cdots$$

$$= \omega(\xi_2\eta_1 - \xi_1\eta_2) + \frac{1}{2}(\xi_2^2 + \xi_1^2)$$

$$+ c(\xi_2\eta_1 - \xi_1\eta_2)^2 + d(\eta_1^2 + \eta_2^2)(\xi_2\eta_1 - \xi_1\eta_2) + e(\eta_1^2 + \eta_2^2)^2 + \cdots$$

where c, d, e are constants.

The quadratic part of the Hamiltonian of the restricted 3-body problem at the Lagrange triangular point, $\mathcal{L}_4$, for values of the mass ratio parameter $\mu = \mu_1 + \varepsilon$ can be brought into normal form by a linear symplectic change of coordinates. The normal form up to order 4 looks like

$$Q = \omega\Gamma_1 + \Gamma_2 + \varepsilon\{a\Gamma_1 + b\Gamma_3\} + \cdots$$
$$= \omega(\xi_2\eta_1 - \xi_1\eta_2) + \tfrac{1}{2}(\xi_1^2 + \xi_1^2)$$
$$+\varepsilon\{a(\xi_2\eta_1 - \xi_1\eta_2) + \tfrac{1}{2}b(\eta_1 + \eta_2) + \cdots.$$

Schmidt (1990) calculated that

$$a = 3\sqrt{69}/16, \qquad b - 3\sqrt{69}/8.$$

10.6 Normal Forms for Periodic Systems

This section reduces the study of the normal forms for symplectomorphisms to the study of normal forms of periodic systems. Then as examples, the normal forms for symplectomorphisms of the plane are given in preparation for the study of generic bifurcations of fixed points given in Chapter 11.

The reduction. The study of a neighborhood of a periodic solution of an autonomous Hamiltonian system was reduced to the study of the Poincaré map in an energy surface by the discussion in Section 6.4. This Poincaré map is a symplectomorphism with a fixed point corresponding to the periodic orbit.

Let the origin be a fixed point for the symplectomorphism

$$\Psi(x) = \Gamma x + \psi(x), \qquad (10.48)$$

where Γ is a $2n \times 2n$ symplectic matrix, and ψ is higher-order, i.e., $\psi(0) = \partial\psi(0)/\partial x = 0$. By Theorem 6.2.1 and the discussion following that theorem, if Γ has a logarithm, then (10.48) is the period map of a periodic Hamiltonian system. Because $\Psi^2(x) = \Gamma^2 x + \cdots$, and Γ^2 always has a logarithm, if Ψ is not a period map, then Ψ^2 is. Except for one example given at the end of this chapter, only the case when Γ has a real logarithm is treated here.

Given a periodic system, by the Floquet–Lyapunov theorem (see Theorem 2.5.2 and the discussion following it), there is a linear, symplectic, periodic change of variables that makes the linear part of Hamiltonian equations constant in t. Thus the study of symplectomorphisms near a fixed point is equivalent to studying a 2π-periodic Hamiltonian system of the form

$$H_\#(t, x) = \sum_{i=0}^{\infty} H_i(t, x), \qquad (10.49)$$

where H_i is a homogeneous polynomial in x of degree $i + 2$ with 2π-periodic coefficients, and $H_0(t, x) = \frac{1}{2} x^T S x$ where S is a $2n \times 2n$ real constant symmetric matrix, and $A = JS$ is a constant, real, Hamiltonian matrix. The linearized equations about the critical point $x = 0$ are

$$\dot{x} = Ax = JSx = J\nabla H_0(x), \tag{10.50}$$

and the general solution of (10.50) is $\phi(t, \xi) = \exp(At)\xi$.

The general periodic case. Here, the generalization of the general normal form given in Section 10.4 is extended to periodic systems. As before, we consider the periodic system (10.49) but no longer assume that the linear system is simple. First let us consider the generalization of Theorem 10.4.2.

Consider the 2π-periodic equations

$$\dot{x} = A(t)x + f(t), \tag{10.51}$$

$$\dot{x} = A(t)x, \tag{10.52}$$

$$\dot{y} = -A(t)^T y. \tag{10.53}$$

Equation (10.52) is the homogeneous equation corresponding to the nonhomogeneous equation (10.51), and (10.53) is the adjoint equation of (10.52).

Lemma 10.6.1. *The nonhomogeneous equation* (10.51) *has a 2π-periodic solution $\phi(t)$ if and only if*

$$\int_0^{2\pi} y^T(s) f(s) ds = 0,$$

for all 2π-periodic solutions $y(t)$ of the adjoint equation (10.53).

Proof. Let $x(t, x_0)$ be the solution of (10.51) with $x(0, x_0) = x_0$. Then

$$x(t, x_0) = X(t)x_0 + \int_0^t X(t)Y^T(s)f(s)ds,$$

where $X(t)$ and $Y(t)$ are the fundamental matrix solutions of (10.52) and (10.53), respectively, so, $X^{-1} = Y^T$. The solution is 2π-periodic if and only if $x(t, x_0) = x$, or

$$Bx_0^0 = g,$$

where

$$B = I - X(2\pi), \qquad g = \int_0^{2\pi} X(2\pi)Y^T(s)f(s)ds.$$

By Lemma 10.4.1, the linear equation $Bx_0 = g$ has a solution if and only if $v^T g = 0$ for all v with $B^T v = 0$. That is, there is a 2π-periodic solution if and only if

$$\int_0^{2\pi} v^T X(2\pi) Y^T(s) f(s) ds = 0 \quad \text{for all } v \text{ with } X(2\pi)^T v = v.$$

But if $X(2\pi)^T v = v$, then the integral above is $\int_0^{2\pi} v^T Y^T(s) f(s) ds = 0$. But $X(2\pi)^T v = v$ if and only if $Y(2\pi)v = v$ and if and only if $Y(s)v$ is a 2π-periodic solution of (10.53).

Consider the periodic Hamiltonian system (10.49). Scale by $x \to \varepsilon x$ as in the proof of Theorem 10.4.1, and use the same notation for the scaled Hamiltonian. By Theorem 10.3.2 we must define spaces $\mathcal{P}_i$, $\mathcal{Q}_i$, and $\mathcal{R}_i$ with $\mathcal{Q}_i \subset \mathcal{P}_i$, $H_i^0 \in \mathcal{P}_i$, $H_0^i \in \mathcal{Q}_i$, $W_i \in \mathcal{R}_i$. The Lie equation to be solved in this case is

$$E = D + \{H_0^0, C\} - \frac{\partial C}{\partial t},$$

where D is given in $\mathcal{P}_i$, and we are to find $E \in \mathcal{Q}_i$ and $C \in \mathcal{R}_i$.

Let B be the adjoint of A, i.e., the transpose in the real case. Define $K(x) = (1/2)x^T Rx$, where $B = JR$, so, K is the Hamiltonian of the adjoint linear system. Let $\mathcal{P}_i$ be the space of polynomials in x with coefficients that are smooth 2π-periodic functions of t. Let $\mathcal{F} = \{H_0^0, \cdot\} : \mathcal{P}_i \to \mathcal{P}_i$, and let $\mathcal{T} = \{K, \cdot\} : \mathcal{P}_i \to \mathcal{P}_i$. $\mathcal{T}$ is the adjoint of $\mathcal{F}$ if we use the metric defined by Elphick et al. that was used in Section 10.4. Therefore, given D, the Lie equation has a unique 2π-periodic solution, C, where E is a 2π-periodic solution of the homogeneous adjoint equation

$$0 = \{K, E\} + \frac{\partial E}{\partial t}. \tag{10.54}$$

Characterizing the 2π-periodic solutions of (10.54) defines the normal form. Expand the elements of $\mathcal{P}_i$ in Fourier series. Let $E = d(x)e^{imt}$, and substitute into (10.54) to get

$$0 = \{K, d\} + imd.$$

Thus one characterization of the normal form is in terms of the eigenvectors of $\mathcal{T} = \{K, \cdot\} : \mathcal{P}_i \to \mathcal{P}_i$. That is, $\mathcal{Q}_i$ has a basis of the form $\{d(x)e^{imt} : d$ is an eigenvector of $\mathcal{T}$ corresponding to the eigenvalue im. $\}$

Theorem 10.6.1. *Let $H^0(x) = H_0(x) = \frac{1}{2}x^T Sx$, where $A = JS$ is an arbitrary, constant Hamiltonian matrix, and let B be the adjoint of A. Then there exists a formal, symplectic, 2π-periodic change of variables $x = X(t, y) = y + \cdots$ which transforms the Hamiltonian (10.49) to the Hamiltonian system*

$$\dot{y} = J\nabla H^\#(t, y), \qquad H^\#(t, y) = \sum_{i=0}^{\infty} H^i(t, y), \tag{10.55}$$

where

$$\{H^i, K\} + \frac{\partial H^i}{\partial t} = 0 \quad \text{for } i = 1, 2, 3, \dots, \tag{10.56}$$

or equivalently,

$$H^i(t, e^{Bt}x) \equiv H^i(0, x) \quad for \ i = 1, 2, 3, \ldots. \tag{10.57}$$

Corollary 10.6.1. *Let A be simple and have eigenvalues $\pm\lambda_1, \ldots, \pm\lambda_n$. Assume that $\lambda_1, \ldots, \lambda_n$ and i are independent over the integers; i.e., there is no relation of the form $k_1\lambda_1 + \cdots + k_n\lambda_n = mi$, where $k_1, \ldots, k_n$ and m are integers. Then there exists a formal, symplectic, 2π-periodic change of variables $x = X(t, y) = y + \cdots$ which transforms the Hamiltonian (10.49) to an autonomous Hamiltonian system*

$$\dot{y} = J\nabla H^{\#}(y), \quad H^{\#}(y) = \sum_{i=0}^{\infty} H^i(y), \tag{10.58}$$

where $H^0 = H_0$, and

$$\{H^i, H^0\} = 0, \tag{10.59}$$

or equivalently,

$$H^i(e^{At}y) \equiv H^i(y) \tag{10.60}$$

for all $i = 0, 1, 2, \ldots, y \in \mathbb{R}^{2n}, t \in \mathbb{R}$.

Proof. Let $A = B = \text{diag}(\lambda_1, \ldots, \lambda_n, -\lambda_1, \ldots, -\lambda_n)$. A typical term in the normal form given by Theorem 10.6.1 is of the form $h(t, x) = h_k e^{imt} x^k$. Applying (10.57) to this term gives

$$h_k \exp\{im + (k_1 - k_{n+1})\lambda_1 + \cdots + (k_n - k_{2n})\lambda_n\}t = 0.$$

By the assumption on the independence, this can only hold if $m = 0, k_1 = k_{n+1}, \ldots, k_n = k_{2n}$. Thus the Hamiltonian is in the normal form of Birkhoff as described in Corollary 10.4.1.

Corollary 10.6.2. *Let Γ be simple and have a real logarithm. Then there exists a formal, near-identity, symplectic change of variables $x \to y$ such that in the new coordinates the symplectomorphism in (10.48) is of the form*

$$\Phi(y) = \Gamma y + \phi(y), \tag{10.61}$$

where

$$\phi(\Gamma y) \equiv \Gamma\phi(y) \quad or \quad \Phi(\Gamma y) \equiv \Gamma\Phi(y). \tag{10.62}$$

Proof. Let $\Gamma = \exp(2\pi A)$. Because Γ is simple, so is A, and therefore it can be taken as its own adjoint. Then by the reduction given above, the map (10.48) is the period map of a system of Hamiltonian differential equations. Assume that the symplectic change of coordinates has been made so that the Hamiltonian is in normal form, and let the equations in these coordinates be $\dot{y} = Ay + f(t, y)$. Condition (10.57) implies $f(t, e^{At}x) = e^{At}f(0, x)$, and this implies $f(t, \Gamma x) = \Gamma f(t, x)$. Let $\xi(t, \eta)$ be a solution of this equation with $\xi(0,\eta) = \eta$. Define $\zeta(t,\eta) = \Gamma\xi(t,\Gamma^{-1}\eta)$, so $\xi(0,\eta) = \zeta(0,\eta) = \eta$. $\dot{\xi} = \Gamma\{A\xi + f(t,\xi)\} = A\Gamma\xi + \Gamma f(t,\xi) = A\Gamma\xi + f(t,\Gamma\xi) = A\zeta + f(t,\zeta)$. By the uniqueness theorem for ordinary differential equations $\xi(t,\eta) = \zeta(t,\eta) = \Gamma\xi(t,\Gamma\eta)$, so, the period map satisfies (10.61).

General hyperbolic and elliptic points. Consider as examples the case when $n = 1$, so, Ψ in (10.48) is a symplectomorphism of the plane with a fixed point at the origin.

First, consider the case when Γ has eigenvalues μ, μ^{-1}, where $0 < \mu < 1$; i.e., the origin is a hyperbolic fixed point. By Lemma 2.4.1, there are symplectic coordinates, say x, so that

$$\Gamma = \begin{bmatrix} \mu & 0 \\ 0 & \mu^{-1} \end{bmatrix}.$$

Let $2\pi\alpha = \ln\mu$, so, $\Gamma = \exp(2\pi\Lambda)$ where

$$A = \begin{bmatrix} \alpha & 0 \\ 0 & -\alpha \end{bmatrix}.$$

By the discussion given above, the symplectomorphism Ψ is the period map of the 2π-periodic system (10.49) with $H_0(x) = \alpha x_1 x_2$. By Corollary 10.6.1, there is a formal, 2π-periodic, symplectic change of variables $x \to y$ that transforms (10.49) to the autonomous system (10.58) with (10.60) holding. The solution of the linear system is $y_1(t) = y_{10}e^{\alpha t}$, $y_2(t) = y_{20}e^{-\alpha t}$, therefore the condition (10.60) implies that the Hamiltonian (10.58) is a function of the product $y_1 y_2$ only. Let $H^{\#}(y) = K^{\#}(y_1 y_2) = \alpha y_1 y_2 + K(y_1 y_2)$. By the above discussion, the normal form for (10.48) is the time 2π-map of the autonomous system whose Hamiltonian is $K^{\#}$. The equations defined by $K^{\#}$ are

$$\dot{y}_1 = y_1(\alpha + k(y_1 y_2)),$$

$$\dot{y}_2 = -y_2(\alpha + k(y_1 y_2)),$$

where k is the derivative of K. These equations have $y_1 y_2$ as an integral, and so the equations are solvable, and the solution is

$$y_1(t) = y_{10} \exp(t(\alpha + k(y_1 y_2))),$$

$$y_2(t) = y_{20} \exp(-t(\alpha + k(y_1 y_2))).$$

Thus the normal form for (10.48) in this case is

$$\Psi(y) = \begin{bmatrix} y_1 g(y_1 y_2) \\ y_2 g(y_1 y_2)^{-1} \end{bmatrix},$$

where g has a formal expansion $g(u) = \mu u + \cdots$. If a symplectomorphism is in this form, then the origin is called a general hyperbolic point. This map takes the hyperbolas $y_1 y_2 = $ constant into themselves. In this case, the transformation to normal form converges by a classical theorem of Moser (1956).

Next consider the case when A has eigenvalues $\lambda = \alpha + \beta i$, $\bar{\lambda} = \alpha - \beta i$, where $\alpha^2 + \beta^2 = 1$, $\beta \neq 0$, i.e., the origin is an elliptic fixed point. There are symplectic coordinates, say x, so that

$$\Gamma = \begin{bmatrix} \lambda & 0 \\ 0 & \bar{\lambda} \end{bmatrix}.$$

Let $\Gamma = \exp(2\pi A)$, where

$$A = \begin{bmatrix} \omega i & 0 \\ 0 & -\omega i \end{bmatrix}.$$

Assume that ω is not an integer; that is, λ is not a root of unity. By the discussion given above, the symplectomorphism Ψ is the period map of the 2π-periodic system (10.49) with $H_0(x) = i\omega x_1 x_2$. By Corollary 10.6.1, there is a formal, 2π-periodic, symplectic change of variables, $x \to y$, which transforms (10.49) to the autonomous system (10.58) satisfying (10.60). Equation (10.60) implies that the Hamiltonian is a function of $y_1 y_2$ only. Let $H^{\#}(y) = K^{\#}(y_1 y_2) = i\omega y_1 y_2 + iK(y_1 y_2)$. By the above discussion, the normal form for (10.48) is the time 2π-map of the autonomous system whose Hamiltonian is $K^{\#}$. Change to action–angle variables (I, ϕ), so, the Hamiltonian becomes $H^{\#}(I, \phi) = K^{\#}(I) = \omega I + K(I)$. The equations defined by $K^{\#}$ are

$$\dot{I} = 0, \qquad \dot{\phi} = \omega - k(I)$$

where k is the derivative of K. These equations have I as an integral, and so the equations are solvable, and the solution is

$$I(t) = I_0, \qquad \phi(t) = \phi_0 + (-\omega + k(I_0))t.$$

Thus the normal form for (10.48) in action–angle variables in this case is

$$\Psi(I, \phi) = \begin{bmatrix} I \\ \phi + g(I) \end{bmatrix},$$

where g has a formal expansion $g(I) = -\omega + \beta I \cdots$. If a symplectomorphism is in this form with $\beta \neq 0$, then the origin is called a general elliptic point, or Ψ is called a twist map. This map takes circles into circles and rotates each circle by an amount $g(I)$.

Higher resonance in the planar case. Let us consider the case when $n = 1$, and the symplectomorphism Ψ has an elliptic fixed point whose multiplier is a root of unity. Theorem 10.6.1 and Corollary 10.6.2 apply as well.

Let Γ have eigenvalues $\lambda = \alpha + \beta i$, $\bar{\lambda} = \alpha - \beta i$, where λ is a kth root of unity, so, $\lambda^k = 1$, $k > 2$, and $\lambda = \exp(h 2\pi i / k)$, where h is an integer. The origin is called a k-resonance elliptic point in this case. There are symplectic coordinates, say x, so that

$$\Gamma = \begin{bmatrix} \lambda & 0 \\ 0 & \bar{\lambda} \end{bmatrix}.$$

Let $\Gamma = \exp(2\pi A)$, where

$$A = \begin{bmatrix} (h/k)i & 0 \\ 0 & -(h/k)i \end{bmatrix}.$$

Because A is diagonal, it is its own adjoint. By the discussion given above the symplectomorphism Ψ is the period map of the 2π-periodic system (10.49) with $H_0(x) = (hi/k)(x_1 x_2)$, where the reality condition is $\bar{x}_1 - x_2$. The normal form for the Hamiltonian is given by Theorem 10.6.1 above.

Let $h(t, x)$ be a typical term in the normal form expansion, so

$$h = e^{ist} x_1^{m_1} x_2^{m_2}.$$

The term h satisfies (10.57) if and only if

$$(h/k)(m_1 - m_2)i + si = 0;$$

so it is in the normal form if h is

$$(x_1 x_2)^m \quad \text{or} \quad x_1^{m_1} x_2^{m_2} e^{-rit},$$

where $r = (m_1 - m_2)h/k$, and m, m_1, m_2, r are integers.

In action–angle coordinates (I, ϕ), $H^0(I, \phi) = (h/k)I$, and the solution of the linear system is $I = I_0$, $\phi = \phi_0 - (h/k)t$. Thus $H^{\#}(t, I, \phi)$ is a function of I and $(k\phi + ht)$, so, let $H^{\#}(t, I, \phi) = K^{\#}(I, k\phi + ht) = (h/k)I + K(I, k\phi + ht)$.

The lowest-order terms that contain t, the new terms, are $x_1^k e^{-hit}$ and $x_2^k e^{hit}$. In action–angle coordinates these terms are like $I^{k/2} \cos(k\phi + ht)$ and $I^{k/2} \sin(k\phi + ht)$. Thus the normalized Hamiltonian is a function of I and $(k\phi + ht)$ only, and it is of the form

$$\begin{aligned} H^{\#}(t, I, \phi) = (h/k)I + aI^2 + bI^3 + \cdots + \\ + I^{k/2}\{\alpha \cos(k\phi + ht) + \beta \sin(k\phi + ht)\} + \cdots. \end{aligned} \tag{10.63}$$

The equations of motion are

$$\dot{I} = I^{k/2}\{-\alpha \sin(k\phi + ht) + \beta \cos(k\phi + ht)\} + \cdots,$$

$$\dot{\phi} = -\frac{h}{k} - 2aI - \frac{k}{2}I^{(k-2)/2}\{\alpha \cos(k\phi + ht) + \beta \sin(k\phi + ht)\} + \cdots.$$

$$\tag{10.64}$$

By a rotation, $\phi \to \phi + \delta$; the first sin term can be absorbed into the cos term, so there is no loss in generality in assuming that $\beta = 0$ in (10.63) and (10.64). Henceforth, we assume this rotation has been made, and so, $\beta = 0$.

Note that in the $\dot{\phi}$ equation in (10.64), there are two nonlinear terms. When $k > 4$, the term that contains the angle is of higher-order in I, whereas

$k = 3$ it is lower-order. When $k = 4$, the two terms are both of order I^1. We show in later chapters on applications that the cases when $k = 3$ or 4 must be treated separately.

The 2π-map is then of the form

$$I = I_0 - \alpha I_0^{k/2} \sin(k\phi_0) + \cdots,$$

$$\phi = \phi_0 - (2\pi h/k) - 4\pi a I_0 + \alpha \pi k I^{(k-2)/2} \cos(k\phi_0) + \cdots. \tag{10.65}$$

Normal forms when multipliers are ± 1. Consider the cases where the multiplier is $+1$ first. For this problem no trigonometric functions are used, therefore assume that the periodic systems are periodic with period 1. If Γ has the eigenvalue $+1$, then either Γ is the identity, and A is the zero matrix, or there are symplectic coordinates such that

$$\Gamma = \exp A = \begin{bmatrix} 1 & \pm 1 \\ 0 & 1 \end{bmatrix}, \quad \text{where } A = \begin{bmatrix} 0 & \pm 1 \\ 0 & 0 \end{bmatrix}. \tag{10.66}$$

In the first case, when $\Gamma = I$ and $A = 0$, Theorem 10.6.1 gives no information, and this is because the situation is highly degenerate and nongeneric.

Therefore, consider the case when Γ and A are as in (10.66) with the plus sign, so, the adjoint of A is B where

$$B = \begin{bmatrix} 0 & 0 \\ 1 & 0 \end{bmatrix}, \quad \exp(Bt) = \begin{bmatrix} 1 & 0 \\ t & 1 \end{bmatrix}.$$

Let $x = (u, v)$. Condition (10.57) of Theorem 10.6.1 is $H^i(u, v + ut, t) \equiv H^i(u, v, 0)$. This condition and the fact that H^i must be periodic in t implies that $H^i(u, v, t) = K^i(u)$. Thus the normal form is

$$H^\#(t, u, v) = v^2/2 + K(u) = v^2/2 + \beta u^3/3 + \cdots \tag{10.67}$$

and the equations of motion are

$$\dot{u} = v + \cdots,$$

$$\dot{v} = -\frac{\partial K}{\partial u}(u) = -\beta u^2 + \cdots. \tag{10.68}$$

The period map is not so easy to compute and is not so simple. Fortunately, in applications, the critical information occurs at a very low order. By using the Lie transform methods discussed in the Problem section one finds that the period map is $(u, v) \to (u', v')$ where

$$u' = u + v - \frac{\beta}{12}(6u^2 + 4uv + v^2) + \cdots,$$

$$v' = v - \frac{\beta}{3}(3u^2 + 3uv + v^2) + \cdots. \tag{10.69}$$

Now consider the case when Γ has eigenvalue -1. Consider the case when

$$\Gamma = \begin{bmatrix} -1 & 0 \\ 0 & -1 \end{bmatrix}$$

first because it has a real logarithm,

$$\Gamma = \exp 2\pi A, \quad A = \begin{bmatrix} 0 & 1/2 \\ -1/2 & 0 \end{bmatrix}.$$

This is almost the same as the higher-order resonance considered in the previous subsection. Corollary 10.6.2 implies that the normal form in this case is simply an odd function. That is, $\Phi(y) = -y + \phi(y)$ is in normal form when $\phi(-y) = -\phi(y)$.

Now consider the case when

$$\Gamma = \begin{bmatrix} -1 & -1 \\ 0 & -1 \end{bmatrix}.$$

We make two changes of coordinates to bring this case to normal form. First, instead of the usual uniform scaling, scale by $x_1 \to \varepsilon x_1$, $x_2 \to \varepsilon^2 x_2$ so that the map (10.48) becomes $\Psi(x) = -x + O(\varepsilon)$. This nonuniform scaling moves the off-diagonal term to the higher-order terms, and now the lead term is the same as discussed in the last paragraph. Thus there is a symplectic change of coordinates $z = R(x)$ such that in the new coordinates z, the map (10.48) is odd, i.e., $R \circ \Psi \circ R^{-1}(z) = \Xi(z) = \Gamma z + \cdots$ is odd.

Write

$$\Xi(z) = -\Lambda(z) = -\{\Omega z + \zeta(z)\}, \quad \text{where} \quad \Omega = \begin{bmatrix} 1 & 1 \\ 0 & 1 \end{bmatrix}.$$

Now Ω is of the form discussed above, and so, there is a symplectic change of coordinates $y = S(z)$ which puts Λ in the normal form given by the time 1-map of a Hamiltonian system of the form (10.67), where now $K(u)$ is even. Because Λ is odd, the transformation S can be made odd also; see problems. Thus $S \circ \Lambda \circ S^{-1} = \Theta$ is in the normal form given by the time 1-map of a Hamiltonian system of the form

$$H^{\#}(t, u, v) = v^2/2 + K(u) = v^2/2 + \beta u^4/4 + \cdots. \tag{10.70}$$

Using the method discussed in the problems gives $\Theta : (u, v) \to (u', v')$, where

$$u' = u + v - \frac{\beta}{20}(10u^3 + 10u^2v + 5uv^2 + v^3) + \cdots,$$

$$v' = v - \frac{\beta}{3}(4u^3 + 6u^2v + 4uv^2 + v^3) + \cdots. \tag{10.71}$$

Combining these changes of coordinates and using the fact that S is odd, it follows that $(S \circ R) \circ \Psi \circ (S \circ R)^{-1} = -\Theta$. That is, in the new coordinates, the map is just the negative of (10.71), or the normal form for the map is

$$u' = -u - v + \frac{\beta}{20}(10u^3 + 10u^2v + 5uv^2 + v^3) + \cdots,$$

$$(10.72)$$

$$v' = -v + \frac{\beta}{3}(4u^3 + 6u^2v + 4uv^2 + v^3) + \cdots.$$

10.7 Problems

1. a) The normal form for a Hamiltonian system with $H_0^0(q,p) = p^2/2$ is
       $H^*(q,p) = p^2/2 + Q(q)$. This normal form also appears in Section 10.6
       when the case of multipliers equal to $+1$ is discussed. Carefully draw
       the phase portrait for the system with Hamiltonian $H(q,p) = p^2/2 +
       \beta q^3$ when $\beta = +1$ and -1.
    b) In Section 10.6 when the multiplier -1 is discussed the normal form
       is $H^*(q,p) = p^2/2 + Q(q)$ with Q even. Carefully draw the phase
       portrait for the system with Hamiltonian $H(q,p) = p^2/2 + \beta q^4$ when
       $\beta = +1$ and -1.
2. a) Compute the next term in the normal form of the unforced Duffing
       equation (10.22) by hand. Recall that H_0^0, H_1^0, H_0^1 and W_1 are given
       in Section (10.3). (Hint: To get the next term you do not have to
       compute all of H_1^1, H_0^2 and W_2. H_0^2 is the term which is independent
       of ϕ in $H_1^1 + \{H_1^0, W_1\}$. Show that $\{H_1^0, W_1\}$ has no term independent
       of ϕ. Now $H_1^1 = H_2^0 + \{H_1^0, W_1\} + \{H_0^0, W_2\}, H_0^2 = 0$, so you need to
       compute the term independent of ϕ in $\{H_1^0, W_1\}$.)
    b) Using Maple, Mathematica, etc., find the first four terms in the nor-
       mal form for the unforced Duffing equation.
3. The Hamiltonian for Duffing's equation is of the form $(q^2 + p^2)/2 + P(p)$
    where P is an even polynomial.
    a) Show that such a Hamiltonian in action–angle variables is a Poisson
       series with only cosine terms.
    b) Show that the Poisson bracket of two Poisson series, one of which
       is a cosine series and the other of which is a sine series, is always a
       cosine series.
    c) Let H_j^i and W_i be from the normalization of such a Hamiltonian with
       an even potential. Show that H_j^i can always be taken as a cosine series
       and W_i as a sine series. (Hint: Define the spaces $\mathcal{P}_i, \mathcal{Q}_i$, and $\mathcal{R}_i$ of
       Theorem 10.3.1.)
4. Consider a Hamiltonian differential equation of the form

$$\dot{x} = \varepsilon F_\#(\varepsilon, t, x) = \varepsilon F_1(t, x) + \varepsilon^2 F_2(t, x) + \cdots,$$

where F is T-periodic in t. Show that there is a formal symplectic series
expansion $x = X(\varepsilon, t, y) = y + \cdots$ which is T-periodic in t and transforms

the equation to the autonomous Hamiltonian system $\dot{y} = \varepsilon F^{\#}(y) = \varepsilon F^1(y) + \varepsilon^2 F^2(y) + \cdots$. Show that $F^1(y) = (1/T)\int_0^T F_1(\tau, y)d\tau$, i.e., F^1 is the average of F_1 over a period. This is called the method of averaging. (Hint: Use Theorem 10.6.1 and remember $F_0^0 = 0$.)

5. Use the notation of the previous problem. Show that if $F^1(\xi) = 0$ and $\partial F^1(\xi)/\partial x$ is nonsingular, then the equation $\dot{x} = \varepsilon F_{\#}(\varepsilon, t, x)$ has a T-periodic solution $\phi(t) = \xi + O(\varepsilon)$.

6. Analyze the forced Duffing's equation,

$$\ddot{x} + x = \varepsilon\{\delta x + \gamma x^3 + A\cos t\} = 0$$

in three different ways, and show that the seemingly different methods give the same intrinsic results. The parameter δ is called the detuning and is a measure of the difference between the natural frequency and the external forcing frequency. Remember that a one degree of freedom autonomous system has a phase portrait given by the level lines of the Hamiltonian.

a) Write the system in action–angle coordinates, and compute the first term in the normal form, F_0^1, as was done for Duffing's equation. Analyze the truncated equation by drawing the level lines of the Hamiltonian. (See Section 9.2.)

b) Write the system in complex coordinates and compute the first term in the normal form, F_0^1, as was done for Duffing's equation in Section 10.3. Analyze the equation.

c) Make the "van der Pol" change of coordinates

$$\begin{bmatrix} x \\ y \end{bmatrix} = \begin{bmatrix} \cos t & \sin t \\ -\sin t & \cos t \end{bmatrix} \begin{bmatrix} u \\ v \end{bmatrix}$$

and then compute the first average of the equations via Problems 4 and 5. Analyze the equations. See McGehee and Meyer (1974).

7. Consider a Hamiltonian of two degrees of freedom of the form (10.32), $x \in \mathbb{R}^4$. Let $H_0(x)$ be the Hamiltonian of two harmonic oscillators. Change to action–angle variables $(I_1, I_2, \phi_1, \phi_2)$ and let $H_0 = \omega_1 I_1 + \omega_2 I_2$. Use Theorem 10.4.1 to show that the terms in the normal form are of the form $aI_1^{p/2}I_2^{q/2}\cos(r\phi_1 + s\phi_2)$ or $bI_1^{p/2}I_2^{q/2}\sin(r\phi_1 + s\phi_2)$, a and b constants, if and only if $r\omega_1 + s\omega_2 = 0$, and the terms have the d'Alembert character. See Henrard (1970).

8. Consider a Hamiltonian $H(x)$ with general solution $\phi(t, \xi)$. Observe that the ith component of ϕ is the Lie transform of x_i, i.e., $\phi_i(t, \xi) = \mathcal{L}_H(x_i)(\xi)$, where ε is replaced by t.

a) Show that $\phi_i(t, \xi) = \left[x_i + \{x_i, H\}t + \{\{x_i, H\}, H\}t^2/2 + \cdots\right]_{x=\xi}$.

b) Using Maple, Mathematica, etc., write a simple function to compute the time 1 maps given in (10.69) and (10.71) (Make sure that you compute the time series far enough to pick up all the quadratic and cubic terms in the initial conditions.)

9. Prove Theorem 10.3.3, the uniqueness theorem. (Hint: Show that if the normal form is not unique then there are two different Hamiltonians H and K which are both in normal form and a generating function W carrying one into the other. Show that the terms in the series expansion for W must lie in the kernel of Q_i. Then show that this implies that $W \equiv 0$.)

11. Bifurcations of Periodic Orbits

This chapter and Chapter 12 use the theory of normal forms developed in Chapter 9. They contain an introduction to generic bifurcation theory and its applications. Bifurcation theory has grown into a vast subject with a large literature, so, this chapter can only present the basics of the theory. The primary focus of this chapter is the study of periodic solutions, their existence and evolution. Periodic solutions abound in Hamiltonian systems. In fact, a famous Poincaré conjecture is that periodic solutions are dense in a generic Hamiltonian system, a result that was established in the C^1 case by Pugh and Robinson (1983).

11.1 Bifurcations of Periodic Solutions

Recall that in Section 6.4 the study of periodic solutions of a Hamiltonian system was reduced to the study of a one-parameter family of symplectic maps, the Poincaré map in an integral surface. The integral surface is in a level set of the Hamiltonian, and the parameter is the value of the Hamiltonian on that level set. If the Hamiltonian has n degrees of freedom, then the phase space is $2n$-dimensional, and the section in the integral surface has dimension $2n - 2$. This effects a reduction of dimension by two.

Fixed points of the Poincaré map corresponds to periodic solutions of Hamiltonian systems. The questions answered in this section are: (1) When can a fixed point be continued? (2) What typically happens when you cannot continue a fixed point? (3) Are there other periodic points near a fixed point? However, to keep the presentation simple, the discussion is limited to symplectic maps of two dimensions that depend on one-parameter. This corresponds to a two degree of freedom autonomous system or a one degree of freedom periodic system. In two dimensions, a map is symplectic if and only if it is area-preserving and orientation preserving; henceforth just area-preserving maps will be considered. A warning should be given: the proper generalization of the theory presented below would be to symplectic maps not just volume-preserving maps.

© Springer International Publishing AG 2017 — 279
K.R. Meyer, D.C. Offin, *Introduction to Hamiltonian Dynamical Systems and the N-Body Problem*, Applied Mathematical Sciences 90,
DOI 10.1007/978-3-319-53691-0_11

Even the restriction to area-preserving maps is not enough for a complete classification, because the number of types of bifurcations is manifold. Therefore, only the "generic case" is considered in this section. The word "generic" can be given a precise mathematical meaning in the context of bifurcation theory, but here only the intuitive meaning is given in order to avoid a long digression. Consider the set of all smooth area-preserving mappings depending on some parameter; then a subset of that set is generic if it has two properties: it is open and it is dense. A subset is open if a small smooth perturbation of a mapping in the subset is also in the subset. So the defining properties of elements of the subset are not sensitive to small perturbations, or the elements are "stable" under perturbations. A subset is dense if any element in the set can be approximated by an element of the subset. The set of area-preserving mappings satisfying the properties listed in the propositions given below can be shown to be generic: see Meyer (1970).

11.1.1 Elementary Fixed Points

Let $P : \mathbb{O} \times \mathbb{I} \to \mathbb{O} : (x, \mu) \to P(x, \mu)$ be a smooth function where $\mathbb{I} = (-\mu_0, \mu_0)$, $\mu_0 > 0$, is an interval in $\mathbb{R}$, and $\mathbb{O}$ is an open neighborhood of the origin in $\mathbb{R}^2$. For fixed $\mu \in \mathbb{I}$, let $P_\mu = P(\cdot, \mu) : \mathbb{O} \to \mathbb{O}$ be area-preserving, so, $\det(\partial P(x, \mu)/\partial x) \equiv 1$. Let the origin be a fixed point of P when $\mu = 0$; i.e., $P(0, 0) = 0$. The eigenvalues of $A = \partial P(0, 0)/\partial x$ are called the *multipliers* of the fixed point. In two dimensions, the eigenvalues of the symplectic matrix, A, are (1) both equal to $+1$, or (2) both equal to -1, or (3) real reciprocals, or (4) on the unit circle. If the multipliers are different from $+1$, the fixed point is called *elementary*.

Proposition 11.1.1. *An elementary fixed point can be continued. That is, if $x = 0$ is an elementary fixed point for P when $\mu = 0$, then there exists a $\mu_1 > 0$ and a smooth map $\xi : (-\mu_1, \mu_1) \to \mathbb{O}$ with $P(\xi(\mu), \mu) \equiv \xi(\mu)$. Moreover, the multipliers of the fixed point $\xi(\mu)$ vary continuously with μ, so, if $x = 0$ is elliptic (respectively, hyperbolic) when $\mu = 0$, then so is $\xi(\mu)$ for small μ.*

Proof. The implicit function theorem applies to $G(x, \mu) = P(x, \mu) - x = 0$, because $G(0, 0) = 0$, and $\partial G(0, 0)/\partial x = A - I$ is nonsingular, so, there is a $\xi(\mu)$ such that $G(\xi(\mu), \mu) = P(\xi(\mu), \mu) - \xi(\mu) = 0$. The multipliers of $\xi(\mu)$ are the eigenvalues of $\partial P(\xi(\mu), \mu)/\partial x$, and the eigenvalues of a matrix vary continuously (not always smoothly) with a parameter.

In particular, an elliptic (respectively hyperbolic) fixed point can be continued to an elliptic (respectively hyperbolic) fixed point.

There are several figures in this chapter. These figures show the approximate placement of the fixed points and their type, elliptic or hyperbolic, as parameters are varied. That is all they are meant to convey. They are drawn as if the diffeomorphism was the time-one map of a differential equation. Thus, for example, the drawing of an elliptic point shows concentric circles about the fixed point. Do not assume that the circles are invariant curves for the map. These curves suggest that the mapping approximately rotates the points. Figure 11.1 shows two depictions of an elliptic fixed point. The one on the left shows that the points near the elliptic point move a discrete distance and is a more accurate depiction, whereas the figure on the right indicates invariant curves. The figure on the right is slightly misleading, but is less cluttered and therefore is used in this chapter.

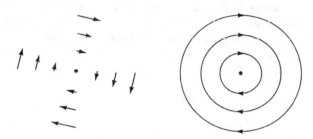

Figure 11.1. Rendering of elliptic fixed points.

11.1.2 Extremal Fixed Points.

Consider the case when the multipliers are equal to $+1$. In this case, the simple implicit function theorem argument fails and for a good reason. Many different things can happen depending on the nonlinear function, so, the simple conclusions of Proposition 11.1.1 may not hold in this case. As an extreme, consider the case when $A = I$ and $P(\mu, x) = x + \mu p(x)$, where $p(x)$ is an arbitrary function. The fixed points of $P(\mu, x)$ for $\mu \neq 0$ are the zeros of $p(x)$; because $p(x)$ is arbitrary, the fixed point set can be quite complicated: in fact, it can be any closed set in $\mathbb{R}^2$. In light of this potential complexity, only the typical or generic situation for a one-parameter family is considered.

Definition. The origin is an *extremal fixed point* for P when $\mu = 0$, if there are symplectic coordinates (u, v) so that $P : (\mu, u, v) \rightarrow (u', v')$, where

$$\begin{bmatrix} u' \\ v' \end{bmatrix} = \begin{bmatrix} 1 & \alpha \\ 0 & 1 \end{bmatrix} \begin{bmatrix} u \\ v \end{bmatrix} + \mu \begin{bmatrix} \gamma \\ \delta \end{bmatrix} + \begin{bmatrix} \cdots \\ \beta u^2 + \cdots \end{bmatrix} + \cdots \tag{11.1}$$

and $\alpha = \pm 1, \beta \neq 0$ and $\delta \neq 0$. First, note that it is assumed that when $\mu = 0$ the linear mapping is already in Jordan normal form, the matrix is not simple $(\alpha = \pm 1)$, and that one nonlinear term is nonzero. Second, because $\delta \neq 0$, the perturbation does not leave the origin as a fixed point. We are considering the unfolding of a shear fixed point. It is not necessary to put the full map into normal form. However, if (11.1) is in the normal form as discussed in Section 10.6, then the assumption that $\beta \neq 0$ means that the first nonlinear term in the normal form appears with a nonzero coefficient.

Proposition 11.1.2. *Let $0 \in \mathbb{O} \subseteq \mathbb{R}^2$ be an extremal fixed point for P when $\mu = 0$. Then there is a smooth curve $\sigma : (-\tau_2, \tau_2) \to \mathbb{I} \times \mathbb{O} : \tau \to (\bar{\mu}(\tau), \xi(\tau))$ of fixed points of P, $P(\bar{\mu}(\tau), \xi(\tau)) = \xi(\tau)$, with $\tau = 0$ giving the extremal fixed point, $\tau(0) = (0, 0)$.*

The extremal point divides the curve of fixed points into two arcs. On one arc the fixed points are all elliptic, and on the other, the fixed points are all hyperbolic. Moreover, the parameter μ achieves a nondegenerate maximum or minimum at the extremal fixed point; so, there are two fixed points when μ has one sign and no fixed points when μ has the other.

Proof. The equations to be solved are

$$0 = u' - u = \alpha v + \mu \gamma + \cdots,$$

$$0 = v' - v = \mu \delta + \beta u^2 + \cdots.$$

Because $\alpha \neq 0$ and $\delta \neq 0$, these equations can be solved for v and μ as a function of u. (Note the difference between this proof and the proof of Proposition 11.1.1: one of the variables solved for in this proof is the parameter.) The solution is of the form $\bar{v}(u) = O(u^2)$ and $\bar{\mu}(u) = (-\beta/\delta)u^2 + O(u^2)$, so, the map is $\sigma : \tau \to (\bar{\mu}(\tau), \tau, \bar{v}(\tau))$. The extreme point is obtained when $\tau = 0$. Note that if $\beta \delta > 0$, then $\bar{\mu}$ obtains a nondegenerate maximum when $\tau = 0$, and if $\beta \delta < 0$ then $\bar{\mu}$ obtains a nondegenerate minimum when $\tau = 0$.

The Jacobian of the map along this solution is

$$\begin{bmatrix} 1 & \alpha \\ 2\beta\tau & 1 \end{bmatrix} + \cdots,$$

and so the multipliers are $1 \pm (2\alpha\beta\tau)^{1/2} + \cdots$. Hence, when $\alpha\beta > 0$, the fixed point is elliptic for $\tau < 0$ and hyperbolic for $\tau > 0$ and vice versa when $\alpha\beta < 0$.

Figure 11.2 shows the curve σ in $\mathbb{I} \times \mathbb{O}$. $\mathbb{I}$ is the horizontal axis and $\mathbb{O}$ is depicted as a one-dimensional space, the vertical axis. In the case considered, the μ achieves a nondegenerate maximum on the curve at the origin.

Consider the case depicted in Figure 11.2. For μ negative there are two fixed points in $\mathbb{O}$, one elliptic and one hyperbolic, see Figure 11.3a. As μ approaches zero through negative values these fixed points come together until they collide and become a degenerate fixed point when $\mu = 0$; see Figure 11.3b. For positive μ there are no fixed points in $\mathbb{O}$; see Figure 11.3c.

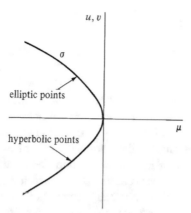

Figure 11.2. The curve σ.

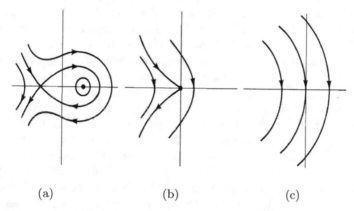

Figure 11.3. Extremal fixed point.

11.1.3 Period Doubling

The solutions given by the implicit function theorem are locally unique, so, there is a neighborhood of an elementary or an extremal fixed point that contains no other fixed points. But, there may be periodic points of higher period near one of these fixed points. There are no periodic points near a hyperbolic or extremal fixed point (see Problem section), but there may be one near an elliptic fixed point.

Let $x = 0$ be an elementary fixed point for P when $\mu = 0$, so, by Proposition 11.1.1, there is a smooth curve $\xi(\mu)$ of fixed points. This fixed point can be shifted to the origin by considering $P'(u, \mu) = P(u + \xi(\mu), \mu) - \xi(\mu)$. Assume that this shift has been done, and revert to the original notation, i.e., assume that $P(0, \mu) \equiv 0$.

Let $P_\mu(x) = P(x, \mu) = Ax + \cdots$; then

$$P_\mu^k(x) = P_\mu \circ P_\mu \circ \cdots \circ P_\mu(x) = A^k x + \cdots.$$

A *k-periodic point* satisfies the equation $P_\mu^k(x) = A^k x + \cdots = x$ which has a unique solution, $x = 0$, unless $A^k - I$ is singular, or one of the eigenvalues of A is a kth root of unity. Thus k-periodic points may exist near a fixed point with a multiplier that is a kth root of unity. In fact, generically they do bifurcate from fixed points whose multipliers are k^{th} roots of unity, in this subsection, the case when $k = 2$ is considered, i.e., when the multipliers are -1.

Let us see what the generic assumptions are for this case. The map $\mathcal{C}$: $\mu \to \partial P(0, \mu)/\partial x$ is a curve in $Sp(2, \mathbb{R})$, the set of all 2×2 real matrices with determinant equal to $+1$. $Sp(2, \mathbb{R})$ is a three-dimensional space because there is one algebraic identity among the four entries of the matrix. Let $\mathcal{M}$ be a subspace of $Sp(2, \mathbb{R})$. If $\mathcal{M}$ is a discrete set of points or a curve, then a small perturbation of $\mathcal{C}$ would miss $\mathcal{M}$, and if $\mathcal{C}$ already misses $\mathcal{M}$, then a small perturbation of $\mathcal{C}$ would still miss $\mathcal{M}$. Thus one open and dense condition (a generic condition) is for $\mathcal{C}$ to miss a discrete set or a curve in $Sp(2, \mathbb{R})$. If $\mathcal{M}$ is a surface in the three-dimensional space $Sp(2, \mathbb{R})$, then the curve $\mathcal{C}$ would in general hit $\mathcal{M}$ in a discrete set of points and cross the surface with nonzero velocity. This is generic when $\mathcal{M}$ is a surface.

The set of matrices $\mathcal{M}_2 = \{A \in Sp(2, \mathbb{R}) :$ trace $A = -2\}$ consists of matrices in $Sp(2, \mathbb{R})$ with eigenvalues equal to -1. It is a surface because the matrices satisfy the additional algebraic identity, trace $A = -2$. The set $\{-I\} \in \mathcal{M}_2$ is a discrete point; thus, generically, the curve $\mathcal{C}$ intersects $\mathcal{M}_2 \backslash \{-I\}$ in a discrete set of points, and at these points, $d(\text{trace}\,\mathcal{C}(\mu))/d\mu \neq 0$. Thus along a curve of elementary fixed points, there are isolated points where the multipliers are -1, and the Jacobian is not simple. At these points, the map can be put into the normal form described in Chapter 10. It is also generic for the first term in the normal form to appear with a nonzero coefficient. This informal discussion leads to the following definition.

Definition. The origin is a *transitional or a flip periodic point* for P at $\mu = 0$ if there are symplectic coordinates (u, v) so that $P : (u, v) \to (u', v')$, where

$$\begin{bmatrix} u' \\ v' \end{bmatrix} = \begin{bmatrix} -1 & \alpha \\ 0 & -1 \end{bmatrix} \begin{bmatrix} u \\ v \end{bmatrix} + \mu \begin{bmatrix} a & b \\ c & d \end{bmatrix} \begin{bmatrix} u \\ v \end{bmatrix} + \begin{bmatrix} \cdots \\ \beta u^3 + \cdots \end{bmatrix} + \cdots,$$

and $\alpha = \pm 1$, $c \neq 0$, $\beta \neq 0$.

There are three conditions in this definition. First, when $\mu = 0$, the multipliers are -1, and the Jacobian matrix is not diagonalizable, $\alpha \neq 0$. Second, inasmuch as

$$\det\left\{\begin{bmatrix} -1 & \alpha \\ 0 & -1 \end{bmatrix} + \mu \begin{bmatrix} a & b \\ c & d \end{bmatrix} + \cdot \right\} = 1 - \mu(a + d + \alpha c) + \cdots = 1, \quad (11.2)$$

$c = -(a+b)/\alpha$, and so the condition $c \neq 0$ implies that the derivative of the trace of the Jacobian is nonzero. Third, $\beta \neq 0$ is the condition that the first term in the normal form when $\mu = 0$ is nonzero. It is not necessary that the map be put into normal form completely, simply eliminate all the quadratic terms, and then assume that $\beta \neq 0$.

Proposition 11.1.3. *Let the origin be a transitional fixed point for P when $\mu = 0$. Let μ be small. If $\alpha c > 0$, then the origin is a hyperbolic fixed point when $\mu > 0$ and the origin is an elliptic fixed point when $\mu < 0$ (vice versa when $\alpha c < 0$).*

If $\beta c > 0$ (respectively, $\beta c < 0$), then there exists a periodic orbit of period 2 for P_μ when $\mu < 0$ (respectively $\mu > 0$), and there does not exist a periodic orbit for $\mu \geq 0$ (respectively, $\mu \leq 0$). As μ tends to zero from the appropriate side, the period-2 orbit tends to the transition fixed point.

For fixed μ, the stability type of the fixed point and the period-2 orbit are opposite. That is, if for fixed μ the origin is elliptic, then the periodic point is hyperbolic and vice versa. (See Figure 11.4.)

Remark. The fixed point is called a transition point because the stability type of the fixed point changes from hyperbolic to elliptic, or vice versa. At the transition point, a new period 2 point appears on one side of $\mu = 0$; this is called *period doubling* in the literature. One says that the period-2 point bifurcates from the transition point.

Proof. By (11.2) the trace of the Jacobian at the origin is $-2 + \mu(a+d) + \cdots = -2 - \mu\alpha c + \cdots$ which implies the first part of the proposition. Compute that the second iterate of the map is $(u, v) \rightarrow (u'', v'')$, where

$$u'' = u - 2\alpha v + \cdots,$$

$$v'' = v - 2\mu c u + \mu(\alpha c - 2d)v - 2\beta u^3 + \cdots.$$

Because $\alpha \neq 0$, the equation $u'' - u = -2\alpha v + \cdots = 0$ can be solved for v as a function of μ and u. Call this solution $\bar{v}(u, \mu)$. The lowest-order terms in $\bar{v}$ are of the form $g\mu u$ and $g'u^3$, where g and g' are constants. Substitute this solution into the equation $v'' - v$ to get

$$v'' - v = -2\mu c u - 2\beta u^3 + \cdots.$$

The origin is always a fixed point, thus u is a common factor, and therefore the equation to solve is

$$(v'' - v)/u = -2\mu c - 2\beta u^2 + \cdots.$$

Because $c \neq 0$, this equation can be solved for μ as a function of u; call this solution $\bar{\mu}(u) = -(\beta/c)u^2 + \cdots$. If $\beta c > 0$, then there are two real solutions, $u_\pm(\mu) = \pm\sqrt{-c\mu/\beta} + \cdots$ for $\mu < 0$, and none for $\mu \geq 0$, and vice versa when $\beta c < 0$. Thus $(u_\pm(\mu), \bar{v}(u_\pm(\mu), \mu))$ are two fixed points of P_μ^2, but because they are not the origin, they are not fixed points of P_μ. Therefore, they are periodic points of period 2. The Jacobian is

$$\frac{\partial(u'', v'')}{\partial(u, v)} = \begin{bmatrix} 1 & -2\alpha \\ 0 & 1 \end{bmatrix} + \begin{bmatrix} \cdots & \cdots \\ -2c\mu - 6\beta u^2 & \cdots \end{bmatrix} + \cdots.$$

The multipliers are $1 \pm +4\sqrt{-\alpha c \mu} + \cdots$ because $u^2 = -c\mu/\beta + \cdots$ along these solutions, and so are hyperbolic if $\alpha c \mu < 0$ and elliptic when $\alpha c \mu > 0$.

There are two basic cases. Case A: the periodic point is elliptic, and case B: the periodic point is hyperbolic. These are depicted in Figure 11.4. In the figure it is assumed that $\alpha c > 0$ and case A is when $\beta c > 0$ and case B is when $\beta c < 0$.

Case A

Case B

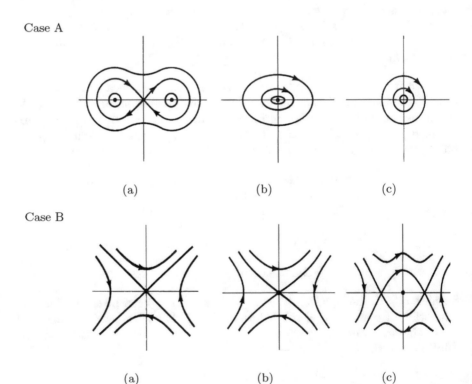

(a)   (b)   (c)

Figure 11.4. Transitional point. Case A: (a) $\mu < 0$; (b) $\mu = 0$; (c) $\mu > 0$. Case B: (a) $\mu < 0$; (b) $\mu = 0$; (c) $\mu > 0$.

11.1.4 k-Bifurcation Points

From the discussion of the last section, periodic points are likely near a fixed point that has multipliers which are k^{th} roots of unity. In the last subsection the generic case of a fixed point with multiplier -1, a square root of unity, was discussed, and in this section, the remaining cases are discussed. Recall the normal forms given in Section 10.6.

Definition. The origin is a k-bifurcation point, $k \geq 3$, for P when $\mu = 0$, if there are symplectic action–angle coordinates (I, ϕ) so that $P : (I, \phi, \mu) \to (I', \phi')$, where

$$I' = I - 2\gamma I^{k/2} \sin(k\phi) + \cdots,$$

$$\phi' = \phi + (2\pi h/k) + \alpha\mu + \beta I + \cdots + \gamma I^{(k-2)/2} \cos(k\phi) + \cdots, \tag{11.3}$$

and

$$\alpha \neq 0, \gamma \neq 0 \text{ when } k = 3,$$
$$\alpha \neq 0, \gamma \neq 0, \beta \pm \gamma \neq 0 \text{ when } k = 4,$$
$$\alpha \neq 0, \beta \neq 0, \gamma \neq 0 \text{ when } k \geq 5.$$

The linearized map is $I' = I$, $\phi' = \phi + (2\pi h/k) + \alpha\mu$. So when $\mu = 0$, the multipliers are $\exp(\pm 2\pi hi/k)$, a kth root of unity. The assumption $\alpha \neq 0$ is the assumption that the multipliers pass through the kth root of unity with nonzero velocity. When $k \geq 5$, the terms with ϕ dependence are higher order, and the map when $\mu = 0$ is of the form $I' = I + \cdots$, $\phi' = \phi + (2\pi h/k) + \beta I + \cdots$. The assumption that $\beta \neq 0$ is the twist assumption, and a map satisfying this assumption is called a *twist map*. Twist maps are discussed in the next section, and in Chapter 12. The assumption that $\gamma \neq 0$ is the assumption that the first angle-dependent term in the normal form appears with a nonzero coefficient. This term is referred to as the *resonance term*, and it is very important to the bifurcation analysis given below. The resonance term is of lower order than the twist term when $k = 3$ and vice versa when $k \geq 5$. When $k = 4$, they are both of the same order. Therefore, the cases $k = 3$ and $k = 4$ are special and will be treated separately.

Proposition 11.1.4. *Let the origin be a 3-bifurcation point for P when $\mu = 0$. Let μ be small. Then there is a hyperbolic periodic orbit of period 3 that exists for both positive and negative values of μ and the periodic point tends to the 3-bifurcation point as $\mu \to 0$ from either side. (See Figure 11.5.)*

Proof. Compute the third iterate of the map P_μ as $P_\mu^3 : (I, \phi) \to (I^3, \phi^3)$, where

$$I^3 = I - 2\gamma I^{3/2} \sin(3\phi) + \cdots,$$

$$\phi^3 = \phi + 2\pi h + 3\alpha\mu + 3\gamma I^{1/2} \cos(3\phi) + \cdots.$$

The origin is always a fixed point, so, I is a common factor in the formula for I^3. Because $\gamma \neq 0$, the equation $(I^3 - I)/(-2\gamma I^{3/2}) = \sin(3\phi) + \cdots$ can be solved for six functions $\phi_j(I, \mu) = j\pi/3 + \cdots, j = 0, 1, \ldots, 5$. For even j we have $\cos 3\phi_j = +1 + \cdots$, and for odd j we have $\cos 3\phi_j = -1 + \cdots$. Substituting these solutions into the ϕ equation gives $(\phi^3 - \phi - 2h\pi)/3 = \alpha\mu \pm \gamma I^{1/2} + \cdots$. The equations with a plus sign have a positive solution for I when $\alpha\gamma\mu$ is negative, and the equations with the minus sign have a positive solution for I when $\alpha\gamma\mu$ is positive. The solutions are of the form $I_j^{1/2} = \mp\alpha\mu/\gamma + \cdots$. Compute the Jacobian along these solutions to be

$$\frac{\partial(I^3, \phi^3)}{\partial(I, \phi)} = \begin{bmatrix} 1 & 0 \\ 0 & 1 \end{bmatrix} + \begin{bmatrix} 0 & \mp 6\gamma I_j^{3/2} \\ (\pm 3\gamma/2)I_j^{-1/2} & 0 \end{bmatrix} + \cdots$$

and so the multipliers are $1 \pm 3|\gamma|I_j^{1/2} + \cdots$, and the periodic points are all hyperbolic.

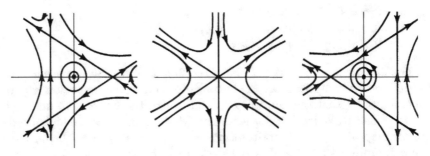

Figure 11.5. A 3-bifurcation point.

Proposition 11.1.5. *Let the origin be a k-bifurcation point, $k \geq 5$, for P when $\mu = 0$. Let μ be small. Then when $\alpha\beta < 0$ (respectively, $\alpha\beta > 0$) there exist an elliptic and also a hyperbolic periodic orbit of period k for $\mu > 0$ (respectively, $\mu < 0$) and no periodic orbit of period k when $\mu < 0$ (respectively, $\mu > 0$). As $\mu \to 0$ from the appropriate side, both the elliptic and hyperbolic orbits tend to the k-bifurcation point. (See Figure 11.6.)*

Remark. These periodic orbits are said to bifurcate from the fixed point when $\mu = 0$. Each orbit consists of k points, and there are exactly two periodic orbits, one elliptic and one hyperbolic.

Proof. Compute the kth iterate of the map P_μ as $P_\mu^k : (I, \phi) \to (I^k, \phi^k)$, where

$$I^k = I - 2\gamma I^{k/2} \sin(k\phi) + \cdots,$$

$$\phi^k = \phi + 2h\pi + \alpha k\mu + \beta kI + \cdots. \tag{11.4}$$

The origin is a fixed point for all μ, thus the first equation is divisible by $I^{k/2}$. By the implicit function theorem, there are $2k$ solutions of $(I^k - I)/(-2\gamma I^{k/2}) = \sin(k\phi) + \cdots = 0$; call them $\phi_j(I, \mu) = j\pi/k + \cdots$. Substitute these solutions into the equation $(\phi^k - \phi - 2h\pi)/k = \alpha\mu + \beta I + \cdots = 0$. For each of the ϕ_j, this second equation has a solution $I_j = -\alpha\mu/\beta + \cdots$ that gives a positive I provided $\alpha\beta\mu < 0$.

The Jacobian at these solutions is

$$\frac{\partial(I^k, \phi^k)}{\partial(I, \phi)} = \begin{bmatrix} 1 & 0 \\ 0 & 1 \end{bmatrix} + \begin{bmatrix} 0 & -2k\gamma I_j^{k/2}\cos(k\phi_j) \\ k\beta & 0 \end{bmatrix} + \cdots,$$

and so the multipliers are $1 \pm k\sqrt{\pm 2\gamma\beta I_j^{k/2}} + \cdots$, where the plus sign inside the square root is taken for even j because $\cos(k\phi_j) = +1 + \cdots$, and the minus sign inside the square root is taken for odd j.

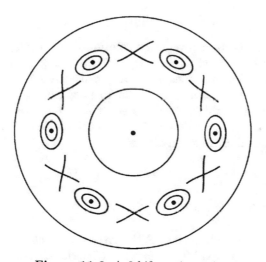

Figure 11.6. A 6-bifurcation point

When $k = 4$ there are two cases depending on the relative size of the twist term and the resonance term. Sometimes it is like $k = 3$ and sometimes it is like $k \geq 5$.

Proposition 11.1.6. *Let the origin be a 4-bifurcation point for P when $\mu = 0$. Let μ be small. Then:*

Case A. If $\beta \pm \gamma$ have different signs, then there is a hyperbolic periodic orbit of period 4 that exists for both positive and negative μ and tends to the 4-bifurcation point as $\mu \to 0$ from either side.

Case B. If $\beta \pm \gamma$ have the same sign, then when $\alpha(\beta \pm \gamma) < 0$ (respectively, $\alpha(\beta \pm \gamma) > 0$), there exists an elliptic and a hyperbolic periodic orbit of period

4 *for* $\mu > 0$ *(respectively,* $\mu < 0$*) and no periodic orbit of period 4 when* $\mu < 0$ *(respectively,* $\mu > 0$*). As* $\mu \to 0$ *from the appropriate side, both the elliptic and hyperbolic orbits tend to the 4-bifurcation point.*

Proof. Compute the fourth iterate of the map P_μ as $P_\mu^4 : (I, \phi) \to (I^4, \phi^4)$, where

$$I^4 = I - 2\gamma I^2 \sin(4\phi) + \cdots,$$

$$\phi^4 = \phi + 2h\pi + 4\alpha\mu + 4\{\beta + \gamma \cos(4\phi)\}I + \cdots.$$

Because the origin is a fixed point for all μ, the first equation is divisible by I^2. By the implicit function theorem there are eight solutions of $(I^4 - I)/(-2\gamma I^2) = \sin(4\phi) + \cdots = 0$; call them $\phi_j(I, \mu) = j\pi/4 + \cdots$. Substitute these solutions into the equation

$$(\phi^4 - \phi - 2h\pi)/4 = \alpha\mu + \{\beta + \gamma \cos(4\phi_j)\}I + \cdots =$$

$$\alpha\mu + \{\beta \pm \gamma\}I + \cdots = 0.$$

For each of the ϕ_j, this equation has a solution $I_j = -\alpha\mu/\{\beta \pm \gamma\} + \cdots$ which gives a positive I provided $\alpha\{\beta \pm \gamma\}\mu < 0$. So if $\beta \pm \gamma$ have different signs then one group of four solutions exists for positive μ and the other group for negative μ; this is Case A and is similar to Proposition 11.1.4. If $\beta \pm \gamma$ are of the same sign, all eight solutions exist for μ on one side on 0; this is Case B and is similar to Proposition 11.1.5. The calculation of the multipliers is similar to the calculations given above.

11.2 Schmidt's Bridges

In Section 9.3, the circular orbits of the Kepler problem were continued into the restricted problem to give two families of periodic solutions for small values of the mass ratio parameter μ. These families are known as the direct and retrograde orbits, depending on whether they rotate in the same or opposite direction as the primaries in the fixed coordinate system. In Section 9.7, some of the elliptic periodic solutions of the 2-body problem were continued into the restricted problem as symmetric periodic orbits.

Schmidt (1972) showed that these elliptic periodic solutions lie on families of symmetric periodic solutions that connect the direct and retrograde orbits. That is, for small μ, there is a smooth family of symmetric periodic solutions of the restricted problem, $\phi(t, \mu, \alpha)$, where α is the parameter of the family such that for $-1 < \alpha < +1$, $\phi(t, 0, \alpha)$ is an elliptic periodic solution, $\phi(t, 0, -1)$ is a direct circular periodic solution, and $\phi(t, 0, +1)$ is a retrograde circular periodic solution of the Kepler problem in rotating coordinates. Of course, this family contains a collision orbit, but there is a natural way to continue a family through a collision. Such a family is called a *bridge* of periodic solutions (connecting the direct and retrograde orbits).

Here just one of Schmidt's bifurcations is presented. Consider the restricted problem for small μ in Poincaré coordinates (see Section (8.9)), so, the Hamiltonian is

$$H = -\frac{1}{2P_1^2} - P_1 + \frac{1}{2}(Q_2^2 + P_2^2) + O(\mu). \tag{11.5}$$

These coordinates are valid in a neighborhood of the direct circular orbits when $\mu = 0$. Recall that Q_2 is an angular coordinate, and when $\mu = 0$, the direct circular orbits are $Q_2 = P_2 = 0$. In Section 9.3 these periodic orbits were continued into the restricted problem for small μ, and these solutions have Q_2, P_2 coordinates that are $O(\mu)$. This result is reproved below.

The condition for an orthogonal crossing of the line of syzygy in these coordinates is

$$Q_1 = m\pi, \qquad Q_2 = 0,$$

where m is an integer.

So let $Q_1(t, p_1, p_2, \mu)$, $Q_2(t, p_1, p_2, \mu)$, $P_1(t, p_1, p_2, \mu)$, $P_2(t, p_1, p_2, \mu)$ be the solution that satisfies $Q_1 = Q_2 = 0$, $P_1 = p_1$, $P_2 = p_2$ when $t = 0$. Then the equations to solve for a symmetric T-periodic solution are

$$Q_1(T/2, p_1, p_2, \mu) = (1/p_1^3 - 1)T/2 - m\pi + O(\mu) = 0,$$
$$Q_2(T/2, p_1, p_2, \mu) = p_2 \sin(T/2) + O(\mu) = 0. \tag{11.6}$$

The direct circular orbits correspond to $m = \pm 1$; take $+1$ for definiteness. When $\mu = 0$ these equations have a solution $p_1^{-3} = j$ (arbitrary), $p_2 = 0$, $T = 2\pi/(j-1)$. Because

$$\frac{\partial(Q_1, Q_2)}{\partial(t, p_2)} = \begin{bmatrix} \frac{1}{2}(j-1) & 0 \\ 0 & \sin(\pi/(j-1)) \end{bmatrix},$$

which is not zero when $j \neq (s+1)/s$, $s = 1, 2, 3, \ldots$, the implicit function theorem implies that these solutions can be continued into the restricted problem for small μ. This is a second proof of the existence of the direct circular orbits.

Assume that the Q_2, P_2 coordinates have been shifted so that the circular orbits are at $Q_2 = P_2 = 0$ for all small μ. This only affects the $O(\mu)$ terms in (11.5). Let k and n be relatively prime integers. The first equation in (11.6) has a solution $T = 2\pi n$, $p_1^{-3} = k/n$, $m = k - n$ when $\mu = 0$, and because $\partial Q_1/\partial t = (1/p_1^{-3} - 1)/2 = (k-n)/2n \neq 0$ it can be solved for $T = T(p_1, p_2, \mu) = 2(k-n)\pi/(1/p_1^{-3} - 1) + O(\mu)$. Substitute this solution into the second equation in (11.6) to get

$$Q_2(T/2, p_1, p_2, \mu) = p_2 \sin\left\{ \frac{(k-n)\pi}{(1/p_1^3 - 1)} \right\} + O(\mu) = 0 \tag{11.7}$$

as the equation to be solved. Because the circular orbit has been shifted to the Q_2, P_2 origin, equation (11.7) is satisfied when $p_2 = 0$, so, p_2 is a factor. Thus to solve (11.7) it is enough to solve

$$\sin\left\{\frac{(k-n)\pi}{(1/p_1^3 - 1)}\right\} + O(\mu) = 0.$$

This equation has a solution, $p_1 = p_1(p_2, \mu) = (n/k)^{1/3} + O(\mu)$, again by the implicit function theorem. This gives rise to a periodic solution for all p_2 that are small including $p_2 = 0$. So this family is parameterized by p_2, $0 \le p_2 \le \delta$ (small), for μ small. The period of the solutions in this family is approximately $2n\pi$ for $p_2 \ne 0$. Where $p_2 = 0$, this periodic solution is the direct circular orbit established before.

11.3 Bifurcations in the Restricted Problem

Many families of periodic solutions of the restricted problem have been studied and numerous bifurcations have been observed. Most of these bifurcations are generic one-parameter bifurcations. Other bifurcations seem to be generic in either the class of symmetric solutions or generic two parameter bifurcations. We claim that these bifurcations can be carried over to the 3-body problem mutatis mutandis using the reduction found in Section 9.6. Because there are a multitude of different bifurcations and they are all generalized in a similar manner we illustrate only one simple case, the 3-bifurcation.

Use the notation of Section 9.6 and let G the Hamiltonian of the restricted problem and H_R the Hamiltonian of the reduced problem three body problem (9.11). We have scaled r, R so that $\nu = 1$.

Let $p(t, h)$ be a smooth family of nondegenerate periodic solutions of the restricted problem parameterized by G, i.e., $G(p(t, h)) = h$, with period $\tau(h)$. When $h = h_0$ let the periodic solution be $p_0(t)$ with period τ_0; so $p_0(t) = p(t, h_0)$ and $\tau_0 = \tau(h_0)$. We say that the τ_0-periodic solution $p_0(t)$ of the restricted problem is a 3-bifurcation orbit if the cross-section map $(\psi, \Psi) \longrightarrow (\psi', \Psi')$ in the surface $G = h$ for this periodic orbit can be put into the normal form

$$\psi' = \psi + (2\pi k/3) + \alpha(h - h_0) + \beta\Psi^{1/2}\cos(3\psi) + \cdots$$

$$\Psi' = \Psi - 2\beta\Psi^{3/2}\sin(3\psi) + \cdots$$

$$T = \tau_0 + \cdots$$

and $k = 1$ or 2, and α and β are nonzero constants. In the above ψ, Ψ are normalized action–angle coordinates in the cross-section and intersect $G = h$, and T is the first return time for the cross-section. The periodic solution

$p(t, h)$ corresponds to the point $\Psi = 0$. The multipliers of the periodic solution $p_0(t)$ are $+1, +1, e^{+2k\pi i/3}, e^{-2k\pi i/3}$ (cube roots of unity) so the periodic solution is a nondegenerate elliptic periodic solution. Thus this family of periodic solutions can be continued into the reduced problem provided τ_0 is not a multiple of 2π by the result of Section 9.6.

The above assumptions imply that the periodic solution $p(t, h)$ of the restricted problem undergoes a bifurcation. In particular, there is a one-parameter family, $p_3(t, h)$, of hyperbolic periodic solution of period $3\tau_0 + \cdots$ whose limit is $p_0(t)$ as $h \to h_0$.

Theorem 11.3.1. *Let $p_0(t)$ be a 3-bifurcation orbit of the restricted problem that is not in resonance with the harmonic oscillator, i.e., assume that $3\tau_0 \neq 2n\pi$, for $n \in \mathbb{Z}$. Let $\tilde{p}(t, h, \varepsilon)$ be the $\tilde{\tau}(h, c)$-periodic solution which is the continuation into the reduced problem of the periodic solution $p(t, h)$ for small c. Thus $\tilde{p}(t, h, \varepsilon) \longrightarrow (p(t, h), 0, 0)$ and $\tilde{\tau}(h, \varepsilon) \longrightarrow \tau(h)$ as $\varepsilon \longrightarrow 0$.*

Then there is a function $\tilde{h}_0(\varepsilon)$ with $\tilde{h}_0(0) = h_0$ such that $\tilde{p}(t, \tilde{h}_0(\varepsilon), \varepsilon)$ has multipliers

$$+1, +1, e^{+2k\pi i/3}, e^{-2k\pi i/3}, e^{+\tau i} + O(\varepsilon), e^{-\tau i} + O(\varepsilon);$$

i.e., exactly one pair of multipliers is cube roots of unity.

Moreover, there is a family of periodic solutions of the reduced problem, $\tilde{p}_3(t, h, \varepsilon)$ with period $3\tilde{\tau}(h, \varepsilon) + \cdots$ such that $\tilde{p}_3(t, h, \varepsilon) \longrightarrow (p_3(t, h), 0, 0)$ as $\varepsilon \longrightarrow 0$ and $\tilde{p}_3(t, h, \varepsilon) \longrightarrow \tilde{p}(t, \tilde{h}_0(\varepsilon), \varepsilon)$ as $h \longrightarrow \tilde{h}_0(\varepsilon)$. The periodic solutions of the family $\tilde{p}_3(t, h, \varepsilon)$ are hyperbolic–elliptic, i.e., they have two multipliers equal to $+1$, two multipliers that are of unit modulus, and two multipliers that are real and not equal to ± 1.

Proof. Because the Hamiltonian of the reduced problem is $H_R = G + \frac{1}{2}(r^2 + R^2) + O(\varepsilon)$ we can compute the cross-section map for this periodic solution in the reduced problem for $\varepsilon = 0$. Use as coordinates ψ, Ψ, r, R in this cross section and let $\eta = h - h_0$. The period map is $(\psi, \Psi, r, R) \longrightarrow (\psi', \Psi', r', R')$, where

$$\psi' = \psi'(\psi, \Psi, r, R, \eta, \varepsilon) = \psi + (2\pi k/3) + \alpha\eta + \beta\Psi^{1/2}\cos(3\psi) + \cdots,$$

$$\Psi' = \Psi'(\psi, \Psi, r, R, \eta, \varepsilon) = \Psi - 2\beta\Psi^{3/2}\sin(3\psi) + \cdots,$$

$$\begin{bmatrix} r' \\ R' \end{bmatrix} = \begin{bmatrix} r'(\psi, \Psi, r, R, \eta, \varepsilon) \\ R'(\psi, \Psi, r, R, \eta, \varepsilon) \end{bmatrix} = B\begin{bmatrix} r \\ R \end{bmatrix} + \cdots,$$

and

$$B = \begin{bmatrix} \cos\tau & \sin\tau \\ -\sin\tau & \cos\tau \end{bmatrix}.$$

The periodic solution of the restricted problem is nondegenerate therefore it can be continued into the reduced problem and so we may transfer the fixed point to the origin, i.e., $\Psi = r = R = 0$ is fixed.

Because $\alpha \neq 0$ we can solve $\psi'(0, 0, 0, 0, \eta, \varepsilon) = 2\pi k/3$ for η as a function of ε to get $\tilde{\eta}(\varepsilon) = h - \tilde{h}_0(\varepsilon)$. This defines the function $\tilde{h}_0$.

Compute the third iterate of the period map to be

$$(\psi, \Psi, r, R) \longrightarrow (\psi^3, \Phi^3, r^3, R^3),$$

where

$$\psi^3 = \psi + 2\pi k + 3\alpha\eta + 3\beta\Psi^{1/2}\cos(3\psi) + \cdots,$$

$$\Psi^3 = \Psi - 2\beta\Psi^{3/2}\sin(3\psi) + \cdots,$$

$$\begin{bmatrix} r^3 \\ R^3 \end{bmatrix} = B^3 \begin{bmatrix} r \\ R \end{bmatrix} + \cdots.$$

Because $3\tau \neq 2k\pi$ the matrix $B^3 - I$ is nonsingular, where I is the 2×2 identity matrix. Thus we can solve the equations $r^3 - r = 0$, $R^3 - R = 0$ and substitute the solutions into the equations for $\psi^3 - \psi = 0$, $\Psi^3 - \Psi = 0$.

The origin is always a fixed point, so, Ψ is a common factor in the formula for Ψ^3. Inasmuch as $\beta \neq 0$, the equation

$$(\Psi^3 - \Psi)/(-2\beta\Psi^{3/2}) = \sin(3\psi) + \cdots$$

can be solved for six functions $\psi_j(\Psi, h) = j\pi/3 + \cdots$, $j = 0, 1, \ldots, 5$. For even j, $\cos 3\psi_j = +1 + \cdots$, and for odd j, $\cos 3\psi_j = -1 + \cdots$. Substituting these solutions into the ψ equation gives

$$(\psi^3 - \psi - 2h\pi)/3 = \alpha\eta \pm \beta\Psi^{1/2} + \cdots.$$

The equations with a plus sign have a positive solution for Ψ when $\alpha\beta\eta$ is negative, and the equations with the negative sign have a positive solution for Ψ when $\alpha\beta\eta$ is positive. The solutions are of the form $\Psi_j^{1/2} = \mp\alpha\eta/\beta$. Compute the Jacobian along these solutions to be

$$\frac{\partial(\Psi^3, \psi^3)}{\partial(\Psi, \psi)} = \begin{bmatrix} 1 & 0 \\ 0 & 1 \end{bmatrix} + \begin{bmatrix} 0 & \mp 6\beta\Psi_j^{3/2} \\ \pm(3\beta/2)\Psi_j^{-1/2} & 0 \end{bmatrix},$$

and so the multipliers are $1 \pm 3|\alpha\eta|$, and the periodic points are all hyperbolic-elliptic.

There are many other types of generic bifurcations, e.g., extremal, period doubling, k-bifurcations with $k > 3$, etc. If such a bifurcation occurs in the restricted problem and the period of the basic periodic orbit is not a multiple of 2π then a similar bifurcation takes place in the reduced three body problem also. The proofs are essentially the same as the proof given above.

11.4 Hamiltonian–Hopf Bifurcation

One of the most interesting bifurcations occurs in the restricted problem at the libration point $\mathcal{L}_4$ as the mass ratio parameter passes through the critical mass ratio of Routh, μ_1. Recall that the linearized equations at $\mathcal{L}_4$ have two pairs of pure imaginary eigenvalues, $\pm\omega_1 i$, $\pm\omega_2 i$ for $0 < \mu < \mu_1$, eigenvalues $\pm i\sqrt{2}/2$ of multiplicity two for $\mu = \mu_1$, and eigenvalues $\pm\alpha\pm\beta i$, $\alpha \neq 0, \beta \neq 0$ for $\mu_1 < \mu \leq 1/2$; see Section 4.5. For $\mu < \mu_1$ and μ near μ_1, Lyapunov's center theorem, 9.2.1, establishes the existence of two families of periodic solutions emanating from the libration point $\mathcal{L}_4$. But for $\mu_1 < \mu \leq 1/2$, the stable manifold theorem, 6.5.1, asserts that there are no periodic solutions near $\mathcal{L}_4$. What happens to these periodic solutions as μ passes through μ_1?

In a interesting paper, Buchanan (1941) proved, up to a small computation, that there are still two families of periodic solutions emanating from the libration point $\mathcal{L}_4$ even when $\mu - \mu_1$. This is particularly interesting, because the linearized equations have only one family. The small computation of a coefficient of a higher-order term was completed by Deprit and Henrard (1969), thus showing that Buchanan's theorem did indeed apply to the restricted problem. Palmore (1969) investigated the question numerically and was led to the conjecture that the two families detach as a unit from the libration point and recede as μ increases from μ_1. Finally, Meyer and Schmidt (1971) proved a general theorem that established Palmore's conjecture using the calculation Deprit and Deprit–Bartholomê (1967). Subsequently, this theorem has been proved again by several authors by essentially the same method. It has become known as the Hamiltonian–Hopf bifurcation theorem.

By the discussion in Section 2.3, the normal form for a quadratic Hamiltonian (linear Hamiltonian system) with eigenvalues $\pm\omega i$, $\omega \neq 0$, with multiplicity two, which is nonsimple, is

$$Q_0 = \omega(\xi_2\eta_1 - \xi_1\eta_2) + (\delta/2)(\xi_1^2 + \xi_2^2), \tag{11.8}$$

where $\delta = \pm 1$ which gives rise to the linear system of equations $\dot{z} = A_0 z$, where

$$z = \begin{bmatrix} \xi_1 \\ \xi_2 \\ \eta_1 \\ \eta_2 \end{bmatrix}, \qquad A_0 = \begin{bmatrix} 0 & \omega & 0 & 0 \\ -\omega & 0 & 0 & 0 \\ -\delta & 0 & 0 & \omega \\ 0 & -\delta & -\omega & 0 \end{bmatrix}. \tag{11.9}$$

Consider a smooth quadratic perturbation of Q_0; i.e., a quadratic Hamiltonian of the form $Q(\nu) = Q_0 + \nu Q_1 + \cdots$, where ν is the perturbation parameter. By the discussion in Sections 10.4 and 10.5, there are four invariants that are important in the theory of normal forms for this problem, namely,

$$\begin{aligned} \Gamma_1 &= \xi_2\eta_1 - \xi_1\eta_2, & \Gamma_2 &= \tfrac{1}{2}(\xi_1^2 + \xi_2^2), \\ \Gamma_3 &= \tfrac{1}{2}(\eta_1^2 + \eta_2^2), & \Gamma_4 &= \xi_1\xi_2 + \eta_1\eta_4. \end{aligned}$$

The higher-order terms in $Q(\nu)$ are in normal form if they are functions of Γ_1 and Γ_3 only. Assume that $Q(\nu)$ is normalized through terms in ν, so that $Q_1 = a\Gamma_1 + b\Gamma_3$ or

$$Q(\nu) = \omega\Gamma_1 + \delta\Gamma_2 + \nu(a\Gamma_1 + b\Gamma_3) + \cdots .$$

Change to complex coordinates by

$$y_1 = \xi_1 + i\xi_2, \qquad y_2 = \xi_1 - i\xi_2,$$

$$y_3 = \eta_1 - i\eta_2, \qquad y_4 = \eta_1 + i\eta_2.$$

This change of coordinates is symplectic with multiplier 2. Note that the reality conditions are $y_1 = \bar{y}_2$ and $y_3 = \bar{y}_4$. We keep the form of $Q(\nu)$ and also make the change in the Γ's, so that,

$$\Gamma_1 = i(y_2 y_4 - y_1 y_3), \quad \Gamma_2 = y_1 y_2, \quad \Gamma_3 = y_3 y_4.$$

The equations of motion are $\dot{w} = (B_0 + \nu B_1 + \cdots)w$, where

$$w = \begin{bmatrix} y_1 \\ y_2 \\ y_3 \\ y_4 \end{bmatrix}, \qquad B_0 = \begin{bmatrix} -\omega i & 0 & 0 & 0 \\ 0 & \omega i & 0 & 0 \\ 0 & -\delta & \omega i & 0 \\ -\delta & 0 & 0 & -\omega i \end{bmatrix},$$

$$B_1 = \begin{bmatrix} -ai & 0 & 0 & b \\ 0 & ai & b & 0 \\ 0 & 0 & ai & 0 \\ 0 & 0 & 0 & -ai \end{bmatrix}.$$

The characteristic polynomial of $Q(\nu)$ is

$$\{\lambda^2 + (\omega + \nu a)^2\}^2 + 2\nu b\delta\{\lambda^2 - (\omega + \nu a)^2\} + \nu^2 b^2 \delta^2,$$

which has roots

$$\lambda = \pm(\omega + \nu a)i \pm \sqrt{-b\delta\nu} + \cdots . \tag{11.10}$$

So the coefficient a controls the way the eigenvalues move in the imaginary direction, and the coefficient b controls the way the eigenvalues split off the imaginary axis. The assumption that $b \neq 0$ means that the eigenvalues move off the imaginary axis when $b\delta\nu < 0$.

Now consider a nonlinear Hamiltonian system depending on the parameter ν which has $Q(\nu)$ as its quadratic part and when $\nu = 0$ has been normalized in accordance with the discussion in Section 10.4 (Sokol'skii normal form) through the fourth-order terms, i.e., consider

$$H(\nu) = \omega\Gamma_1 + \delta\Gamma_2 + \nu(a\Gamma_1 + b\Gamma_3) + \frac{1}{2}(c\Gamma_1^2 + 2d\Gamma_1\Gamma_3 + e\Gamma_3^2) + \cdots, \tag{11.11}$$

where here the ellipsis stands for terms that are at least second-order in ν or fifth-order in the y's. Scale the variables by

$$y_1 \to \epsilon^2 y_1, \qquad y_2 \to \epsilon^2 y_2,$$

$$y_3 \to \epsilon y_3, \qquad y_4 \to \epsilon y_4, \tag{11.12}$$

$$\nu \to \epsilon^2 \nu,$$

which is symplectic with multiplier ϵ^3, so, the Hamiltonian becomes

$$H = \omega\Gamma_1 + \epsilon(\delta\Gamma_2 + \nu b\Gamma_3 + \frac{1}{2}e\Gamma_3^2) + O(\epsilon^2). \tag{11.13}$$

The essential assumption is that all the terms shown actually appear, i.e., $\omega \neq 0$, $\delta = \pm 1$, $b \neq 0$, $e \neq 0$. The equations of motion are

$$\dot{y}_1 = -\omega i y_1 + \epsilon\{\nu b y_4 + e(y_3 y_4)y_4\},$$

$$\dot{y}_2 = \omega i y_2 + \epsilon\{\nu b y_3 + e(y_3 y_4)y_3)\},$$

$$\dot{y}_3 = \omega i y_3 - \epsilon\delta y_2, \tag{11.14}$$

$$\dot{y}_4 = -\omega i y_4 - \epsilon\delta y_1.$$

Note that the $O(\epsilon^2)$ terms have been dropped for the time being. Equations (11.14) are of the form

$$\dot{u} = Cu + \epsilon f(u, \nu), \tag{11.15}$$

where u is a 4-vector, f is analytic in all variables,

$$f(0, \nu) = 0, \quad C = \mathrm{diag}(-\omega i, \omega i, \omega i, -\omega i),$$

$$\exp(CT) = I, \quad f(e^{Ct}u, \nu) = e^{Ct}f(u, \nu),$$

where $T = 2\pi/\omega$. This last property is the characterization of the normal form in the case where the matrix of the linear part is simple; see Theorem 10.4.1. The scaling has achieved this property to first-order in ϵ. For the moment, continue to ignore the $O(\epsilon^2)$ terms. Let τ be a parameter (period correction parameter or detuning); then $u(t) = e^{(1-\epsilon\tau)Ct}v$, v a constant vector, is a solution if and only if

$$D(\tau, v, \nu) = \tau Cv + f(v, \nu) = 0. \tag{11.16}$$

Thus if v satisfies (11.16), then $e^{(1-\epsilon\tau)Ct}v$ is a periodic solution of (11.15) of period $T/(1 - \epsilon\tau) = T(1 + \epsilon\tau + \cdots)$. For Equations (11.14) with the $O(\epsilon^2)$ terms omitted, one calculates

$$D(\tau, v, \nu) = \begin{bmatrix} -i\omega\tau v_1 + \nu b v_4 + er^2 v_4 \\[2mm] i\omega\tau v_2 + \nu b v_3 + er^2 v_3 \\[2mm] i\omega\tau v_3 - \delta v_2 \\[2mm] -i\omega\tau v_4 - \delta v_1 \end{bmatrix} = 0, \qquad (11.17)$$

where $r^2 = v_3 v_4$. Solving for v_1 from the last equation, substituting it into the first equation, and canceling the v_4 yields

$$\omega^2 \tau^2 - \delta e r^2 = \delta b \nu. \qquad (11.18)$$

A solution of (11.18) gives rise to a 3-parameter family of periodic solutions (2-parameter family of periodic orbits) of (11.14) in the following way. Choose v_3 arbitrary, i.e., $v_3 = \alpha_1 + i\alpha_2$, where α_1 and α_2 are parameters. Then $v_4 = \alpha_1 - i\alpha_2$, $r^2 = \alpha_1^2 + \alpha_2^2$. Take τ arbitrary, i.e., $\tau = \alpha_3$. Then solve for v_1 and v_2 by the last two equations in (11.17). Fixing r determines a circle of periodic solutions that corresponds to one periodic orbit; thus, the 2-parameter family of periodic orbits is parameterized by r and τ.

The analysis depends on the sign of the two quantities δe and δb, especially δe. There are two qualitatively different cases: Case A when δe is positive, and Case B when δe is negative.

Case A: $\delta e > 0$; see Figure 11.7.

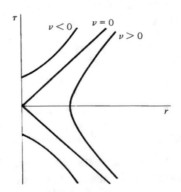

Figure 11.7. Graph of (11.17) when δe is positive.

For definiteness, let δb be positive because the contrary case is obtained by changing the sign of ν. Figure 11.7 is drawn under this convention. For fixed ν, the graph of (11.18) is a hyperbola (two lines through the origin when $\nu = 0$), but only the part where $r \geq 0$ is of interest. The parameter τ is the correction to the period. By the paragraph above, a fixed solution of (11.18), $r \neq 0$, fixes the length of v_3 and so fixes the special coordinates

v_1, v_2, v_3, v_4 up to a circle. Thus a point in the τ, r plane, $r \neq 0$, on the graph of (11.18) corresponds to a periodic orbit of (11.14) with period $T/(1 - \epsilon\tau)$. $r = 0$ corresponds to the libration point at the origin.

By (11.10), when $\nu > 0$ and small, the eigenvalues of the linear part, $B_0 + \nu B_1 + \cdots$, are two distinct pairs of pure imaginary numbers, so, Lyapunov's center theorem implies that there are two families of periodic solutions emanating out of the origin. When $\nu > 0$, the graph of (11.18) is two curves emanating out of the line $r = 0$. This corresponds to two families of periodic solutions of (11.14) emanating out of the origin, and hence, it corresponds to the two Lyapunov families.

When $\nu = 0$, the graph of (11.18) is two lines emanating out of the origin, which again corresponds to two families of periodic solutions emanating out of the origin. In this case, these two families correspond to the two families of Buchanan (1941). When $\nu < 0$, the graph of (11.18) is a single curve that does not pass through the origin and thus corresponds to a single family of period orbits of (11.14) that do not pass through the origin.

Case A summary: The two Lyapunov families emanate from the origin when $\delta b \nu$ is positive. These families persist when $\nu = 0$ as two distinct families of periodic orbits emanating from the origin. As $\delta b \nu$ becomes negative, the two families detach from the origin as a single family and move away from the origin.

Case B: $\delta e < 0$; See Figure 11.8.

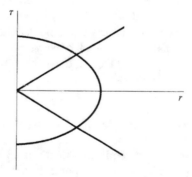

Figure 11.8. Graph of (11.17) when δe is negative.

For definiteness let δb be positive as before. Figure 11.8 is drawn under this convention. For fixed $\nu > 0$, the graph of (11.18) is an ellipse.

By (11.10) when $\nu > 0$ and small, the eigenvalues of the linear part, $B_0 + \nu B_1 + \cdots$, are two distinct pairs of pure imaginary numbers, so, Lyapunov's center theorem implies that there are two families of periodic solutions emanating out of the origin. These families correspond to the upper and lower halves of the ellipse. In this case, the two Lyapunov families are globally connected. As ν tends to zero, this family shrinks to the origin and disappears. For $\nu < 0$ there are no such periodic solutions.

Case B summary: The two Lyapunov families emanate from the origin when $\delta b \nu$ is positive and are globally connected. These families shrink to the origin as $\delta b \nu$ tends to zero through positive values. When $\delta b \nu$ is negative, there are no such periodic solutions.

Now we show that these conclusions remain valid when the $O(\epsilon^2)$ terms in (11.14) are present. If the $O(\epsilon^2)$ terms are included, (11.14) is of the form $\dot{u} = Cu + f(u, \nu) + O(\epsilon^2)$. Let $\phi(t, v, \nu, \epsilon)$ be the general solution of this equation with $\phi(0, v, \nu, \epsilon) = v$. Let $\psi(v, \tau, \nu, \epsilon)$ be this solution after a time $T(1 + \epsilon\tau)$; i.e.,

$$\psi(v, \tau, \nu, \epsilon) = \phi(T(1 + \epsilon\tau), v, \nu, \epsilon) = v + \epsilon D(\tau, v, \nu) + O(\epsilon^2),$$

$$= v + \epsilon\{\tau C v + f(v, \nu)\} + O(\epsilon^2).$$

A periodic solution corresponds to a solution of $\psi(v, \tau, \nu, \epsilon) = v$, and so the equation to be solved is

$$\mathcal{D}(v, \tau, \nu, \epsilon) = (\psi(v, \tau, \nu, \epsilon) - v)/\epsilon = D(\tau, v, \nu) + O(\epsilon) = 0. \tag{11.19}$$

Equation (11.19) is dependent because Equation (11.14) admits H as an integral. Because $H(v + \epsilon D) = H(v)$, it follows by the mean value theorem that $\nabla H(v^*)\mathcal{D} = 0$, where v^* is a point between v and $v + \epsilon D$. Because $\nabla H(v^*) = (-i\omega y_3, \ldots) + \cdots$, if $\mathcal{D}_2 = \mathcal{D}_3 = \mathcal{D}_4 = 0$, then $\mathcal{D}_1 = 0$, except maybe when $y_3 = 0$. Thus only the last three equations in (11.19) need to be solved because the solutions sought must have $y_3 \neq 0$.

From the last two equations in (11.19), one solves, by the implicit function theorem for v_1 and v_2 to get $v_1 = -i\omega\tau\delta v_4 + \cdots$, $v_2 = i\omega\tau\delta v_3 + \cdots$. Substitute these solutions into the second equation to get

$$d(v_3, v_4, \tau, \nu, \epsilon) = (\omega^2\tau^2 - e\delta r^2 - b\delta\nu)(-\delta v_3) + \epsilon g(v_3, v_4, \tau, \epsilon) = 0. \tag{11.20}$$

Because the origin is always an equilibrium point, d and g vanish when v_3 and v_4 are zero. Let $v_3 = re^{i\theta}$, $v_4 = re^{-i\theta}$, and divide (11.20) by $(-\delta r)$ to get

$$d'(r, \theta, \tau, \nu, \epsilon) = (\omega^2\tau^2 - e\delta r^2 - b\delta\nu)e^{i\theta} + \epsilon g'(r, \theta, \tau, \epsilon) = 0.$$

Because $e\delta \neq 0$, this equation can be solved for r^2 to get $r^2 = R(\theta, \tau, \nu, \epsilon) = (\omega^2\tau^2 - b\delta\nu)/e\delta + \cdots$ for all θ, all $\tau, |\tau| < \tau_0$, and all $\epsilon, |\epsilon| < \epsilon_0$. So $r = \sqrt{R}$ yields a real solution when $R > 0$. The analysis of the sign of R leads to the same qualitative conclusions as before.

To get back to the unscaled equations, fix $\epsilon = \epsilon_0/2$ for all times. The scaling is global and so it is invertible globally. Trace back from the solution $r^2 = R(\theta, \tau, \nu, \epsilon_0/2)$ to get $\xi_1, \xi_2, \eta_1, \eta_2$ as functions of θ, τ, ν.

Theorem 11.4.1. *Consider a Hamiltonian of the form* (11.11) *with* $\omega \neq 0$, $\delta = \pm 1$, $b \neq 0$, $e \neq 0$.

Case A: $\delta e > 0$. *The two Lyapunov families emanate from the origin when* $\delta b \nu$ *is small and positive. These families persist when* $\nu = 0$ *as two distinct families of periodic orbits emanating from the origin. As* $\delta b \nu$ *becomes negative, the two families detach from the origin as a single family and recede from the origin.*

Case B: $\delta e < 0$. *The two Lyapunov families emanate from the origin when* $\delta b \nu$ *is small and positive, and the families are globally connected. This global family shrinks to the origin as* $\delta b \nu$ *tends to zero through positive values. When* $\delta b \nu$ *is small and negative, there are no periodic solutions close to the origin.*

One can compute the multipliers approximately to show that in Case A the periodic solutions are elliptic. In Case B, the periodic solutions are initially elliptic as they emanate from the origin but go through extremal bifurcations to become hyperbolic. See the Problems section or Meyer and Schmidt (1971).

Case A occurs in the restricted problem at $\mathcal{L}_4$ as the mass ratio parameter μ passes through the critical mass ratio μ_1. This theorem verifies the numeric experiments of Palmore.

11.5 Problems

1. Let $Q = Q(q, p)$, $P = P(q, p)$ define a smooth area-preserving diffeomorphism of a neighborhood of the origin $q = p = 0$.

   a) Show that $\Omega = (P - p)d(Q + q) - (Q - q)d(P + p)$ is a closed form in q, p (i.e., $d\Omega = 0$), and so by Poincaré's lemma, there is a function $G = G(q, p)$ defined in a neighborhood of the origin such that $dG = \Omega$.

   b) Let the origin be a fixed point whose multipliers are not -1, so, $\xi = Q + q$, $\eta = P + p$ defines new coordinates, and $\partial G / \partial \xi = P - p$, $\partial G / \partial \eta = Q - q$. Thus a fixed point corresponds to critical points of G. Show that if $G = \{\alpha \xi^2 + 2\beta \xi \eta + \gamma \eta^2\}/2$ with $4\Delta = \alpha \gamma - \beta^2 \neq -1$, then

   $$\frac{\partial(Q, P)}{\partial(q, p)} = \frac{1}{1 + 4\Delta} \begin{bmatrix} (1 - \beta^2) - \alpha \gamma & -2\gamma \\ 2\alpha & (1 + \beta^2) - \alpha \gamma \end{bmatrix}.$$

   Thus a maximum or minimum of G corresponds to an elliptic fixed point, and a saddle point corresponds to a hyperbolic fixed point.

   c) Draw the level surfaces of $G = q^2/2 + \epsilon p + p^3/3$ as the parameter varies.

   d) Show that the fixed point is an extremal fixed point if and only if $G = \partial G / \partial q = \partial G / \partial p = \partial^2 G / \partial p^2 = \partial^2 G / \partial q \partial p = 0$ and $\partial^2 G / \partial q^2 \neq 0$, $\partial^2 G / \partial p \partial \epsilon \neq 0$, $\partial^3 G / \partial p^3 \neq 0$. See Meyer (1970).

2. Consider the forced Duffing equation at 1–3 resonance. That is, consider

$$H_* = I + \epsilon\{\gamma I^2 \cos^4 \phi + AI^{1/2} \cos \phi \cos 3t\} + (\epsilon^2/2)\delta I \ cos^2\phi,$$

which is the forced Duffing equation written in action–angle variables. Note that the Hamiltonian is $2\pi/3$-periodic in t. The normalized Hamiltonian is

$$H^* = \epsilon\frac{3}{\gamma/8}I^2$$
$$+\frac{\epsilon^2}{64}\{-2A\gamma I^{3/2} \cos 3(t + \phi) + 17\gamma^2 I^3 + 16\delta I - A^2\} + \cdots.$$

a) Write the Hamiltonian, the normalized Hamiltonian, and the equations of motion in rectangular coordinates.
b) Analyze the normalized systems. Remember to bring the harmonic to the origin.

3. The bifurcation for the forced Duffing equation at 1–2 resonance is not, as predicted in the section, a generic bifurcation, and this is due to the fact that this equation has additional symmetries because the Hamiltonian is even. Consider a Hamiltonian like the forced Duffing equation, but has a cubic term in the Hamiltonian (a quadratic in the equations). That is, consider

$$H_* = I + \epsilon\{\kappa I^{3/2} \cos^3 \phi + AI^{1/2} \cos \phi \cos pt\} + (\epsilon^2/2)\delta I \cos^2 \phi.$$

Note that the Hamiltonian is π-periodic in t. The normalized Hamiltonian is

$$H^* = I + (\epsilon^2/48)\{-12A\kappa I \cos 2(t + \phi) + 45\kappa^2 I^2 + 24\delta I - 4A^2\} + \cdots$$

a) Write the Hamiltonian, the normalized Hamiltonian, and the equations of motion in rectangular coordinates.
b) Analyze the normalized systems. Remember to bring the harmonic to the origin.

4. Using Mathematica, Maple, or an algebraic processor of your choice, write a normalization routine that normalizes Duffing's equation at $q - p$ resonance, i.e., write a program that normalizes

$$H_* = qI + \epsilon\{\delta I \cos^2 \phi + \gamma I^2 \cos^4 \phi + AI^{1/2} \cos \phi \cos pt\}.$$

Analyze the cases $p/q = 1/3, 2/3, 3/4$, etc.

5. Consider the proof in Section 11.4.
a) Show that the Jacobian of D in (11.17) is

$$\frac{\partial D}{\partial v} = \begin{bmatrix} i\omega\tau & 0 & ev^2 & vb + 2er^2 \\ 0 & i\omega\tau & vb + 2er^2 & ev_3^2 \\ 0 & -\delta & i\omega\tau & 0 \\ -\delta & 0 & 0 & -i\omega\tau \end{bmatrix}.$$

b) Show that because of the dependency of the equations, the Jacobian is singular. Also show that the determinant of the minor obtained by deleting the first row and third column is ev_3^2, which is nonzero if $v_3 \neq 0$.

c) Show that the multipliers of the solutions found in Section 11.4 are of the form $1, 1, 1 + \epsilon\mu_1 + \cdots, 1 + \epsilon\mu_2 + \cdots$, where μ_1 and μ_2 are the nonzero eigenvalues of $\partial D/\partial v$.

d) Show that in Case A of Section 11.4, i.e., $\delta e > 0$, that the periodic solutions found are elliptic.

6. Use the notation of Section 11.4.

a) Show that the value of the Hamiltonian (11.13) along a solution of (11.18) is $H = -2\omega^2 \tau \delta r^2 = (2\omega^2/e)\{\delta b \nu \tau - \omega^3 \tau^3\}$.

b) Show that in Case A the periodic solutions can be parameterized by the Hamiltonian.

7. Consider a Hamiltonian of the form

$$H = k\omega I_1 + \omega I_2 + \frac{1}{2}(AI_1^2 + 2BI_1I_2 + CI_2^2) + \cdots,$$

where $I_i = (x_i^2 + y_i^2)/2, \omega > 0$, k is a nonzero integer, and the ellipsis represents terms of degree at least 5 in the xs and ys.

a) Show that Lyapunov's center theorem implies the existence of a family of periodic solutions (the short period family) of approximate period $2\pi/k\omega$ that emanate from the origin when $|k| > 1$.

b) Use the ideas of Section 11.4 to show that there is a family of periodic solutions (the long period family) of approximate period $2\pi/\omega$ that emanates from the origin when $B - kC \neq 0$.

c) Using the normal form calculations for the restricted problem at $\mathcal{L}_4$, show that the long period family exists even when $\mu = \mu_i$, for $i = 4, 5, 6, \ldots$.

12. Stability and KAM Theory

This chapter contains an introduction to the question of the stability and instability of orbits of Hamiltonian systems, and in particular the classical Lyapunov theory and the celebrated KAM theory. This could be the subject of a complete book, so, the reader will find only selected topics presented here. The main example is the five libration points of the restricted problem for various values of the mass ratio parameter. The stability/instability is completely solved for this problem. Of course other examples are touched on.

Consider the differential equation

$$\dot{z} = f(z), \tag{12.1}$$

where f is a smooth function from an open set $O \subset \mathbb{R}^m$ into $\mathbb{R}^m$. Let the equation have an equilibrium point at $\zeta_0 \in O$, so, $f(\zeta_0) = 0$. Let $\phi(t, \zeta)$ be the general solution of (12.1). The equilibrium point ζ_0 is said to be positively (respectively, negatively) stable, if for every $\epsilon > 0$ there is a $\delta > 0$ such that $\|\phi(t, \zeta) - \zeta_0\| < \epsilon$ for all $t \geq 0$ (respectively, $t \leq 0$) whenever $\|\zeta - \zeta_0\| < \delta$. The equilibrium point ζ_0 is said to be stable if it is both positively and negatively stable. In many books "stable" means positively stable, but the above convention is the common one in the theory of Hamiltonian differential equations. The equilibrium ζ_0 is unstable if it is not stable. The adjectives "positively" and "negatively" can be used with "unstable" also. The equilibrium ζ_0 is asymptotically stable, if it is positively stable, and there is an $\eta > 0$ such that $\phi(t, \zeta) \to \zeta_0$ as $t \to +\infty$ for all $\|\zeta - \zeta_0\| < \eta$.

Recall the one result already given on stability in Theorem 1.2.2, which states that a strict local minimum or maximum of a Hamiltonian is a stable equilibrium point. So for a general Newtonian system of the form $H = p^T M p / 2 + U(q)$, a strict local minimum of the potential U is a stable equilibrium point. It has been stated many times that an equilibrium point of U that is not a minimum is unstable. Laloy (1976) showed that for

$$U(q_1, q_2) = \exp(-1/q_1^2) \cos(1/q_1) - \exp(-1/q_2^2)\{\cos(1/q_2) + q_2^2\},$$

© Springer International Publishing AG 2017
K.R. Meyer, D.C. Offin, *Introduction to Hamiltonian Dynamical Systems and the N-Body Problem*, Applied Mathematical Sciences 90, DOI 10.1007/978-3-319-53691-0_12

the origin is a stable equilibrium point, and yet the origin is not a local minimum for U. See Taliaferro (1980) for some positive results along these lines.

Henceforth, let the equilibrium point be at the origin. A standard approach is to linearize the equations, i.e., write (12.1) in the form

$$\dot{z} = Az + g(z),$$

where $A = \partial f(0)/\partial z$ and $g(z) = f(z) - Az$; so, $g(0) = \partial g(0)/\partial z = 0$. The eigenvalues of A are called the exponents (of the equilibrium point). If all the exponents have negative real parts, then a classical theorem of Lyapunov states that the origin is asymptotically stable. By Proposition 2.3.1, the eigenvalues of a Hamiltonian matrix are symmetric with respect to the imaginary axis, so, this theorem never applies to Hamiltonian systems. In fact, because the flow defined by a Hamiltonian system is volume-preserving, an equilibrium point can never be asymptotically stable.

Lyapunov also proved that if one exponent has positive real part then the origin is unstable. Thus for the restricted 3-body problem the Euler collinear libration points, $\mathcal{L}_1$, $\mathcal{L}_2$, $\mathcal{L}_3$, are always unstable, and the Lagrange triangular libration points, $\mathcal{L}_4$ and $\mathcal{L}_5$, are unstable for $\mu_1 < \mu < 1 - \mu_1$ by the results of Section 4.5.

Thus a necessary condition for stability of the origin is that all the eigenvalues be pure imaginary. It is easy to see that this condition is not sufficient in the non-Hamiltonian case. For example, the exponents of

$$\dot{z}_1 = z_2 + z_1(z_1^2 + z_2^2),$$

$$\dot{z}_2 = -z_1 + z_2(z_1^2 + z_2^2)$$

are $\pm i$, and yet the origin is unstable. (In polar coordinates, $\dot{r} = r^3 > 0$.) However, this equation is not Hamiltonian.

In the second 1917 edition of Whittaker's book on dynamics, the equations of motion about the Lagrange point $\mathcal{L}_4$ are linearized, and the assertion is made that the libration point is stable for $0 < \mu < \mu_1$ on the basis of this linear analysis. In the third edition of Whittaker (1937) this assertion was dropped, and an example due to Cherry (1928) was included. A careful look at Cherry's example shows that it is a Hamiltonian system of two degrees of freedom, and the linearized equations are two harmonic oscillators with frequencies in a ratio of $2 : -1$. The Hamiltonian is in the normal form given in Theorem 10.4.1, i.e., in action–angle variables, Cherry's example is

$$H = 2I_1 - I_2 + I_1 I_2^{1/2} \cos(\phi_1 + 2\phi_2). \tag{12.2}$$

Cherry explicitly solves this system, but we show the equilibrium is unstable as a consequence of Chetaev's theorem 12.1.2.

12.1 Lyapunov and Chetaev's Theorems

In this section we present the parts of classical Lyapunov stability theory as it pertains to Hamiltonian systems. Consider the differential equation (12.1).

Return to letting ζ_0 be the equilibrium point. Let $V : O \to \mathbb{R}$ be smooth where O is an open neighborhood of the equilibrium point ζ_0. One says that V is positive definite (with respect to ζ_0) if there is a neighborhood $Q \subset O$ of ζ_0 such that $V(\zeta_0) < V(z)$ for all $z \in O \setminus \{\zeta_0\}$. That is, ζ_0 is a strict local minimum of V. Define $\dot{V} : O \to \mathbb{R}$ by $\dot{V}(z) = \nabla V(z) \cdot f(z)$.

Theorem 12.1.1 (Lyapunov's Stability Theorem). *If there exists a function V that is positive definite with respect to ζ_0 and such that $\dot{V} \leq 0$ in a neighborhood of ζ_0 then the equilibrium ζ_0 is positively stable.*

Proof. Let $\epsilon > 0$ be given. Without loss of generality assume that $\zeta_0 = 0$ and $V(0) = 0$. Because $V(0) = 0$ and 0 is a strict minimum for V, there is an $\eta > 0$ such that $V(z)$ is positive for $0 < \|z\| \leq \eta$. By taking η smaller if necessary we can ensure that $\dot{V}(z) \leq 0$ for $\|z\| \leq \eta$ and that $\eta < \epsilon$ also.

Let $M = \min\{V(z) : \|z\| = \eta\}$. Because $V(0) = 0$ and V is continuous, there is a $\delta > 0$ such that $V(z) < M$ for $\|z\| < \delta$ and $\delta < \eta$. We claim that if $\|\zeta\| < \delta$ then $\|\phi(t, \zeta)\| \leq \eta < \epsilon$ for all $t \geq 0$.

Because $\|\zeta\| < \delta < \eta$ there is a t^* such that $\|\phi(t, \zeta)\| < \eta$ for all $0 \leq t < t^*$ and t^* is the smallest such number. Assume t^* is finite and so $\|\phi(t^*, \zeta)\| = \eta$. Define $v(t) = V(\phi(t, \zeta))$ so $v(0) < M$ and $\dot{v}(t) \leq 0$ for $0 \leq t \leq t^*$ and so $v(t^*) < M$. But $v(t^*) = V(\phi(t^*, \zeta)) \geq M$ which is a contradiction and so t^* is infinite.

Consider the case when (12.1) is Hamiltonian; i.e., of the form

$$\dot{z} = J\nabla H(z), \tag{12.3}$$

where H is a smooth function from $O \subset \mathbb{R}^{2n}$ into $\mathbb{R}$. Again let $z_0 \in O$ be an equilibrium point and let $\phi(t, \zeta)$ be the general solution.

Corollary 12.1.1 (Dirichlet's stability theorem 1.2.2). *If z_o is a strict local minimum or maximum of H, then z_0 is a stable equilibrium for (12.3).*

Proof. Because $\pm H$ is an integral we may assume that H has a minimum. Because $\dot{H} = 0$ the system is positively stable. Reverse time by replacing t by $-t$. In the new time $\dot{H} = 0$, so the system is positively stable in the new time or negatively stable in the original time.

For the moment consider a Hamiltonian system of two degrees of freedom that has an equilibrium at the origin and is such that the linearized equations look like two harmonic oscillators with distinct frequencies ω_1, ω_2, $\omega_i \neq 0$. The quadratic terms of the Hamiltonian can be brought into normal form by a linear symplectic change of variables so that the Hamiltonian is of the form

$$H = \pm\frac{\omega_1}{2}(x_1^2 + y_1^2) \pm \frac{\omega_2}{2}(x_2^2 + y_2^2) + \cdots.$$

If both terms have the same sign then the equilibrium is stable by Dirichlet's Theorem. However, in the restricted problem at Lagrange triangular libration points $\mathcal{L}_4$ and $\mathcal{L}_5$ for $0 < \mu < \mu_1$ the Hamiltonian is of the above form, but the signs are opposite.

Theorem 12.1.2 (Chetaev's theorem). *Let $V : O \to \mathbb{R}$ be a smooth function and Ω an open subset of O with the following properties.*

- $\zeta_0 \in \partial\Omega$.
- $V(z) > 0$ for $z \in \Omega$.
- $V(z) = 0$ for $z \in \partial\Omega$.
- $\dot{V}(z) = V(z) \cdot f(z) > 0$ for $z \in \Omega$.

Then the equilibrium solution ζ_0 of (12.1) is unstable. In particular, there is a neighborhood Q of the equilibrium such that all solutions which start in $Q \cap \Omega$ leave Q in positive time.

Proof. Again we can take $\zeta_0 = 0$. Let $\epsilon > 0$ be so small that the closed ball of radius ϵ about 0 is contained in the domain O, and let $Q = \Omega \cap \{\|z\| < \epsilon\}$. We claim that there are points arbitrarily close to the equilibrium point which move a distance at least ϵ from the equilibrium.

Q has points arbitrarily close to the origin, so for any $\delta > 0$ there is a point $p \in Q$ with $\|p\| < \delta$ and $V(p) > 0$.

Let $v(t) = V(\phi(t,p))$. Either $\phi(t,p)$ remains in Q for all $t \geq 0$ or $\phi(t,p)$ crosses the boundary of Q for the first time at a time $t^* > 0$.

If $\phi(t,p)$ remains in Q, then $v(t)$ is increasing because $\dot{v} > 0$, and so $v(t) \geq v(0) > 0$ for $t \geq 0$. The closure of $\{\phi(t,p) : t \geq 0\}$ is compact and $\dot{v} > 0$ on this set so $\dot{v}(t) \geq \kappa > 0$ for all $t \geq 0$. Thus $v(t) \geq v(0) + \kappa t \to \infty$ as $t \to \infty$. This is a contradiction because $\phi(t,p)$ remains in an ϵ neighborhood of the origin and v is continuous.

If $\phi(t,p)$ crosses the boundary of Q for the first time at a time $t^* > 0$, then $\dot{v}(t) > 0$ for $0 \leq t < t^*$ and $v(t^*) \geq v(0) > 0$. Because the boundary of Q consist of the points q where $V(q) = 0$ or where $\|q\| = \epsilon$, it follows that $\|v(t^*)\| = \epsilon$.

Cherry's counterexample in action–angle coordinates is

$$H = 2I_1 - I_2 + I_1^{1/2}I_2\cos(\phi_1 + 2\phi_2). \tag{12.4}$$

To see that the origin is unstable, consider the Chetaev function

$$W = -I_1^{1/2}I_2\sin(\phi_1 + 2\phi_2),$$

and compute that

$$\dot{W} = 2I_1I_2 + \frac{1}{2}I_2^2.$$

Let Ω be the region where $W > 0$. In Ω, $I_2 \neq 0$, so, $\dot{W} > 0$ in Ω. The region Ω has points arbitrarily close to the origin, so Chetaev's theorem shows that the origin is unstable even though the linearized system is stable.

Theorem 12.1.3 (Lyapunov's instability theorem). *If there is a smooth function $V : O \to \mathbb{R}$ that takes positive values arbitrarily close to ζ_0 and is such that $\dot{V} = V \cdot f$ is positive definite with respect to ζ_0 then the equilibrium ζ_0 is unstable.*

Proof. Let $\Omega = \{z : V(z) > 0\}$ and apply Chetaev's theorem.

As the first application consider a Hamiltonian system of two degrees of freedom with an equilibrium point such that the exponents of this system at the equilibrium point are $\pm \omega i, \pm \lambda$, $\omega \neq 0$, $\lambda \neq 0$; i.e., one pair of pure imaginary exponents and one pair of real exponents. For example, the Hamiltonian of the restricted problem at the Euler collinear libration points $\mathcal{L}_1$, $\mathcal{L}_2$, and $\mathcal{L}_3$ is of this type. We show that the equilibrium point is unstable. Specifically, consider the system

$$H = \frac{\omega}{2}(x_1^2 + y_1^2) + \lambda x_2 y_2 + H^{\dagger}(x, y) \tag{12.5}$$

where $H^{\dagger}$ is real analytic in a neighborhood of the origin in $\mathbb{R}^4$ in its displayed arguments and of at least third degree. Here we have assumed that the equilibrium is at the origin and that the quadratic terms are already in normal form. As we have already seen, Lyapunov's center theorem 9.2.1 implies that the system admits an analytic surface called the Lyapunov center filled with periodic solutions.

Theorem 12.1.4. *The equilibrium at the origin for the system with Hamiltonian (12.5) is unstable. In fact, there is a neighborhood of the origin such that any solution which begins off the Lyapunov center leaves the neighborhood in both positive and negative time. In particular, the small periodic solutions given on the Lyapunov center are unstable.*

Proof. There is no loss of generality by assuming that λ is positive. The equations of motion are

$$\dot{x}_1 = \omega y_1 + \frac{\partial H^{\dagger}}{\partial y_1} \qquad \dot{y}_1 = -\omega x_1 - \frac{\partial H^{\dagger}}{\partial x_1}$$

$$\dot{x}_2 = \lambda x_2 + \frac{\partial H^{\dagger}}{\partial y_2} \qquad \dot{y}_2 = -\lambda y_2 - \frac{\partial H^{\dagger}}{\partial x_2}.$$

We may assume that the Lyapunov center has been transformed to the coordinate plane $x_2 = y_2 = 0$; i.e., $\dot{x}_2 = \dot{y}_2 = 0$ when $x_2 = y_2 = 0$. That means that $H^{\dagger}$ does not have a term of the form $x_2(x_1^n y_1^m)$ or of the form $y_2(x_1^n y_1^m)$.

Consider the Chetaev function $V = \frac{1}{2}(x_2^2 - y_2^2)$ and compute

$$\dot{V} = \lambda(x_2^2 + y_2^2) + x_2\frac{\partial H^\dagger}{\partial y_2} - y_2\frac{\partial H^\dagger}{\partial y_2}$$

$$= \lambda(x_2^2 + y_2^2) + W(x, y).$$

We claim that in a sufficiently small neighborhood Q of the origin $\|W(x,y)\| \leq (\lambda/2)(x_2^2 + y_2^2)$ and so $\dot{V} > 0$ on $Q \setminus \{x_2 = y_2 = 0\}$; i.e., off the Lyapunov center. Let $H^\dagger = H_0^\dagger + H_2^\dagger + H_3^\dagger$ where $H_0^\dagger$ is independent of x_2, y_2, $H_2^\dagger$ is quadratic in x_2, y_2, and $H_3^\dagger$ is at least cubic in x_2, y_2. $H_0^\dagger$ contributes nothing to W; $H_2^\dagger$ contributes to W a function that is quadratic in x_2, y_2 and at least linear in x_1, y_1, and so can be estimated by $O(\{x_1^2 + y_1^2\}^{1/2})O(\{x_2^2 + y_2^2\})$; and $H_3^\dagger$ contributes to W a function that is cubic in x_2, y_2 and so is $O(\{x_2^2 + y_2^2\}^{3/2})$. These estimates prove the claim.

Let $\Omega = \{x_2^2 > y_2^2\} \cap Q$ and apply Chetaev's theorem to conclude that all solutions which start in Ω leave Q in positive time. If you reverse time you will conclude that all solutions which start in $\Omega^- = \{x_2^2 < y_2^2\} \cap Q$ leave Q in negative time.

Proposition 12.1.1. *The Euler collinear libration points $\mathcal{L}_1$, $\mathcal{L}_2$, and $\mathcal{L}_3$ of the restricted 3-body problem are unstable. There is a neighborhood of these points such that there are no invariant sets in this neighborhood other than the periodic solutions on the Lyapunov center manifold.*

As the second application consider a Hamiltonian system of two degrees of freedom with an equilibrium point such that the exponents of this system at the equilibrium point are $\pm\alpha \pm \beta i$, $\alpha \neq 0$, i.e., two exponents with positive real parts and two with negative real parts. For example, the Hamiltonian of the restricted problem at the Lagrange triangular points $\mathcal{L}_4$ and $\mathcal{L}_5$ is of this type when $\mu_1 < \mu < 1 - \mu_1$. We show that the equilibrium point is unstable. Specifically, consider the system

$$H = \alpha(x_1 y_1 + x_2 y_2) + \beta(y_1 x_2 - y_2 x_1) + H^\dagger(x, y), \tag{12.6}$$

where $H^\dagger$ is real analytic in a neighborhood of the origin in $\mathbb{R}^4$ in its displayed arguments and of at least third degree. Here we have assumed that the equilibrium is at the origin and that the quadratic terms are already in normal form.

Theorem 12.1.5. *The equilibrium at the origin for the system with Hamiltonian (12.6) is unstable. In fact, there is a neighborhood of the origin such that any nonzero solution leaves the neighborhood in either positive or negative time.*

Proof. We may assume $\alpha > 0$. The equations of motion are

$$\dot{x}_1 = \alpha x_1 + \beta x_2 + \frac{\partial H^\dagger}{\partial y_1}, \qquad \dot{x}_2 = -\beta x_1 + \alpha x_2 + \frac{\partial H^\dagger}{\partial y_2},$$

$$\dot{y}_1 = -\alpha y_1 + \beta y_2 - \frac{\partial H^\dagger}{\partial x_1}, \qquad \dot{y}_2 = -\beta y_1 - \alpha y_2 + \frac{\partial H^\dagger}{\partial x_2}.$$

Consider the Lyapunov function

$$V = \frac{1}{2}(x_1^2 + x_2^2 - y_1^2 - y_2^2)$$

and compute

$$\dot{V} = \alpha(x_1^2 + x_2^2 + y_1^2 + y_2^2) + W.$$

where W is at least cubic. Clearly V takes on positive values close to the origin and $\dot{V}$ is positive definite, so all solutions in $\{(x,y) : V(x,y) > 0\}$ leave a small neighborhood in positive time. Reversing time shows that all solutions in $\{(x,y) : V(x,y) < 0\}$ leave a small neighborhood in positive time.

Proposition 12.1.2. *The triangular equilibrium points $\mathcal{L}_4$ and $\mathcal{L}_5$ of the restricted 3-body problem are unstable for $\mu_1 < \mu < 1 - \mu_1$. There is a neighborhood of these points such that there are no invariant sets in this neighborhood other than the equilibrium point itself.*

The classical references on stability are Lyapunov (1892) and Chetaev (1934), but a very readable account can be found in LaSalle and Lefschetz (1961). The text by Markeev (1978) contains many of the stability results for the restricted problem given here and below plus a discussion of the elliptic restrict problem and other systems.

12.2 Local Geometry

Much of the rest of this chapter treats the stability of an equilibrium point of a Hamiltonian system of two degrees of freedom where the classical theorems of Dirichlet, Lyapunov, and Chetaev do not apply. The motivating example is $\mathcal{L}_4$ of the restricted three-body problem when $0 < \mu \leq \mu_1$.
    Consider

$$H = \omega_1 I_1 \pm \omega_2 I_2,$$

where ω_1, ω_2 are positive numbers and $I_1, I_2, \theta_1, \theta_2$ are action-angle variables. Here H is the Hamiltonian of two harmonic oscillators with frequencies ω_1, ω_2. In the $+$ case the Hamiltonian is positive definite and Dirichlet's stability theorem 1.2.2 applies. Also in this case the local geometry was discussed in Section 1.4. So consider the $-$ case.

The projection $\mathcal{P} : (I_1, I_2, \theta_1, \theta_2) \to (I_1, I_2)$ maps $\mathbb{R}^4$ onto Q, the closed positive quadrant of $\mathbb{R}^2$. We must be careful to observe the conventions of polar coordinates as discussed in Section 1.4. Above each point in the interior of Q lies a two-dimensional torus, T^2, above each nonzero point on an axis lies a circle, S^1, and above the origin $O \in Q$ sits the origin in $\mathbb{R}^4$.

Look at a neighborhood of the origin in the level set $H = 0$. The origin is a critical point for H, so you cannot expect it to look like a ball in $\mathbb{R}^3$. The level set $H = 0$ is a line through the origin in the quadrant Q, so a typical neighborhood N would be all the points above the line $H = 0$ for $I_1 < b, I_2 < b, b > 0$ a constant. Above a point in the interior of Q lies a torus, T^2, which gets smaller as you approach the origin. Above the origin in Q is the origin in $\mathbb{R}^4$, so the tori shrink down to a point.

Thus a model for a neighborhood of the origin in $H = 0$ is a solid torus with its center circle identified to a point.

The equations of motion are

$$\dot{I}_1 = 0, \quad \dot{\theta}_1 = -\omega_1, \quad \dot{I}_2 = 0, \quad \dot{\theta}_2 = -\omega_2.$$

$\theta_2 = 0$ is a cross section and the period is $T = 2\pi/\omega_2$. Let I_1, θ_1 be coordinates in this cross section and so the section map is

$$I_1 \to I_1, \qquad \theta_1 \to \theta_1 - 2\pi \frac{\omega_1}{\omega_2}.$$

This map takes circles to circles and rotates each by a fixed amount. Another geometric picture of a neighborhood of the origin will be presented in Section 12.6.

12.3 Moser's Invariant Curve Theorem

We return to questions about the stability of equilibrium points later, but now consider the corresponding question for maps. Let

$$F(z) = Az + f(z) \tag{12.7}$$

be a diffeomorphism of a neighborhood of a fixed point at the origin in $\mathbb{R}^m$; so, $f(0) = 0$ and $\partial f(0)/\partial z = 0$. The eigenvalues of A are the multipliers of the fixed point.

The fixed point 0 is said to be stable if for every $\epsilon > 0$ there is a $\delta > 0$ such that $\|F^k(z)\| < \epsilon$ for all $\|z\| < \delta$ and $k \in \mathbb{Z}$.

We reduce several of the stability questions for equilibrium points of a differential equation to the analogous question for fixed points of a diffeomorphism. Let us specialize by letting the fixed point be the origin in $\mathbb{R}^2$ and by letting (12.7) be area-preserving (symplectic). Assume that the origin is an elliptic fixed point, so, A has eigenvalues λ and $\lambda^{-1} = \bar{\lambda}$, $\mid \lambda \mid= 1$. If $\lambda = 1$, -1, $\sqrt[3]{1} = e^{2\pi i/3}$, or $\sqrt[4]{1} = i$ then typically the origin is unstable; see Meyer (1971) and the Problems.

Therefore, let us consider the case when λ is not an mth root of unity for $m = 1, 2, 3, 4$. In this case, the map can be put into normal form up through terms of order three, i.e., there are symplectic action–angle coordinates, I, ϕ, such that in these coordinates, $F : (I, \phi) \to (I', \phi')$, where

$$I' = I + c(I, \phi),$$
$$\phi' = \phi + \omega + \alpha I + d(I, \phi),$$

(12.8)

and $\lambda - \exp(\omega i)$, and c, d arc $O(I^{3/2})$. We do not need the general results because we construct the maps explicitly in the applications given below.

For the moment assume that c and d are zero, so, the map (12.8) takes circles $I = I_0$ into themselves, and if $\alpha \neq 0$, each circle is rotated by a different amount. The circle $I = I_0$ is rotated by an amount $\omega + \alpha I_0$. When $\omega + \alpha I_0 = 2\pi p/q$, where p and q are relatively prime integers, then each point on the circle $I = I_0$ is a periodic point of period q.

If $\omega + \alpha I_0 = 2\pi\delta$, where δ is irrational, then the orbits of a point on the circle $I = I_0$ are dense ($c = d = 0$ still). One of the most celebrated theorems in Hamiltonian mechanics states that many of these circles persist as invariant curves. In fact, there are enough invariant curves encircling the fixed point that they assure the stability of the fixed point. This is the so-called "invariant curve theorem."

Theorem 12.3.1 (The invariant curve theorem). *Consider the mapping* $F : (I, \phi) \to (I', \phi')$ *given by*

$$I' = I + \epsilon^{s+r} c(I, \phi, \epsilon),$$
$$\phi' = \phi + \omega + \epsilon^s h(I) + \epsilon^{s+r} d(I, \phi, \epsilon),$$

(12.9)

where (i) c and d are smooth for $0 \leq a \leq I < b < \infty$, $0 \leq \epsilon \leq \epsilon_0$, and all ϕ, (ii) c and d are 2π-periodic in ϕ, (iii) r and s are integers $s \geq 0, r \geq 1$, (iv) h is smooth for $0 \leq a \leq I < b < \infty$, (v) $dh(I)/dI \neq 0$ for $0 \leq a \leq I < b < \infty$, and (vi) if Γ is any continuous closed curve of the form $\Xi = \{(I, \phi) : I = \Theta(\phi), \Theta : \mathbb{R} \to [a, b]$ continuous and 2π-periodic $\}$, then $\Xi \cap F(\Xi) \neq \emptyset$.

Then for sufficiently small ϵ, there is a continuous F-invariant curve Γ of the form $\Gamma = \{(I, \phi) : I = \Phi(\phi), \Phi : \mathbb{R} \to [a, b]$ continuous and $2\pi - periodic\}$.

Remarks.

1. The origin of this theorem was in the announcements of Kolmogorov who assumed the map was analytic, and the invariant curve was shown to be analytic. In the original paper by Moser (1962), where this theorem was

proved, the degree of smoothness required of c, d, h was very large, C^{333}, and the invariant curve was shown to be continuous. This spread led to a great deal of work to find the least degree of differentiability required of c, d, and h to get the most differentiability for the invariant curve. However, in the interesting examples, c, d, and h are analytic, and the existence of a continuous invariant curve yields the necessary stability.

2. The assumption (v) is the twist assumption discussed above, and the map is a perturbation of a twist map for small ϵ.

3. Assumption (vi) rules out the obvious example where F maps every point radially out or radially in. If F preserves the inner boundary $I = a$ and is area-preserving, then assumption (vi) is satisfied.

4. The theorem can be applied to any subinterval of $[a, b]$, therefore the theorem implies the existence of an infinite number of invariant curves. In fact, the proof shows that the measure of the invariant curves is positive and tends to the measure of the full annulus $a \leq I \leq b$ as $\epsilon \to 0$.

5. The proof of this theorem is quite technical. See Siegel and Moser (1971) for a complete discussion of this theorem and related results.

The following is a slight modification of the invariant curve theorem that is needed later on.

Corollary 12.3.1. *Consider the mapping $F : (I, \phi) \to (I', \phi')$ given by*

$$I' = I + \epsilon c(I, \phi, \epsilon),$$
$$\phi' = \phi + \epsilon h(\phi)I + \epsilon^2 d(I, \phi, \epsilon), \tag{12.10}$$

where (i) c and d are smooth for $0 \leq a \leq I < b < \infty$, $0 \leq \epsilon \leq \epsilon_0$, and all ϕ, (ii) c and d are 2π-periodic in ϕ, (iii) $h(\phi)$ is smooth and 2π-periodic in ϕ, and (iv) if Γ is any continuous closed curve of the form $\Xi = \{(I, \phi) : I = \Theta(\phi), \Theta : \mathbb{R} \to [a, b]$ continuous and 2π-periodic, then $\Xi \cap F(\Xi) \neq \emptyset$.

If $h(\phi)$ is nonzero for all ϕ then for sufficiently small ϵ, there is a continuous F-invariant curve Γ of the form $\Gamma = \{(I, \phi) : I = \Phi(\phi), \Phi : \mathbb{R} \to [a, b]$ continuous and $2\pi - periodic\}$.

Proof. Consider the symplectic change of variables from the action–angle variables I, ϕ to the action–angle variables J, ψ defined by the generating function

$$S(J, \phi) = JM^{-1} \int_0^\phi \frac{d\tau}{h(\tau)}, \qquad M = \int_0^{2\pi} \frac{d\tau}{h(\tau)}.$$

So

$$\psi = \frac{\partial S}{\partial J} = M^{-1} \int_0^\phi \frac{d\tau}{h(\tau)}, \qquad I = \frac{\partial S}{\partial \phi} = \frac{MJ}{h(\phi)},$$

and the map in the new coordinates becomes

$$J' = J + O(\epsilon), \qquad \psi' = \psi + \epsilon MJ + O(\epsilon^2).$$

The theorem applies in the new coordinates.

12.4 Morris' Boundedness Theorem

In 1976 G. R. Morris proved

Theorem 12.4.1. *All solutions of*

$$\ddot{x} + 2x^3 = p(t),$$

are bounded when $p(t)$ is continuous and periodic.

This gave rise to an abundance of generalizations – see Dieckerhoff and Zehnder (1987), Meyer and Schmidt (2016), and the references therein. All the proofs in these papers use Moser's invariant curve theorem to prove the existence of invariant curves near infinity for the period map.

Let the period of p be 1. The Hamiltonian for this equation is

$$H = \frac{1}{2}(y^2 + x^4) - xp(t),$$

and in action-angle variables (K, κ) given in Section 8.5 the Hamiltonian is

$$H = \frac{3}{4}K^{4/3} - \frac{3}{2}K^{1/3}sl\,(\kappa)p(t).$$

The equations of motion are

$$\dot{K} = -\frac{3}{2}K^{1/3}cl\,(\kappa)p(t),$$

$$\dot{\kappa} = -K^{1/3} + \frac{1}{2}K^{-2/3}sl\,(\kappa)p(t).$$

Let $\Lambda = K^{1/3}$ so that the equations become

$$\dot{\Lambda} = -\frac{1}{2}\Lambda^{-1}cl\,(\kappa)p(t),$$

$$\dot{\kappa} = -\Lambda + \frac{1}{2}\Lambda^{-2}sl\,(\kappa)p(t).$$

First note that since sl, cl, and p are all uniformly bounded, these equations are analytic for all κ, t and $\Lambda > 0$.

Integrate from 0 to -1 to get the period map $\mathcal{P} : (\Lambda, \kappa) \to (\Lambda^*, \kappa^*)$ where

$$\Lambda^* = \Lambda + F(\Lambda, \kappa),$$

$$\kappa^* = \kappa + \Lambda + G(\Lambda, \kappa),$$

and

$$F(\Lambda, \kappa) = O(\Lambda^{-1}), \quad G(\Lambda, \kappa) = O(\Lambda^{-2}).$$

An *encircling curve* is a curve, $\mathcal{C}$, of the form $K = \phi(\kappa)$ (or $\Lambda = \phi(\kappa)$) were ϕ is continuous, 4τ-periodic, positive, and near a circle about the origin. The invariant curve theorem requires the image of an encircling curve must intersect itself. The coordinates (K, κ) are symplectic and so the period map is area preserving and thus the image under $\mathcal{P}$ of any encircling curve must intersect itself in the (K, κ) coordinates. The map from K to L is invertible from $K > 0$ to $L > 0$, so the image of an encircling curve in (Λ, κ) coordinates must intersect itself. The curve is invariant if $\mathcal{P}(\mathcal{C}) = \mathcal{C}$.

Let $\mathcal{A}(a)$ be the annulus $\{(\Lambda, \kappa) : 1 < a \leq \Lambda \leq 2a\}$. From the above on $\mathcal{A}$

$$|F| = O(a^{-1}), \quad |G| = O(a^{-2}).$$

Moser's theorem says there is a $\delta > 0$ depending on the given data such that if $|F| < \delta$ and $|G| < \delta$ then there is an invariant encircling curve in $\mathcal{A}(a)$. From the above there is an a^* such that for all $a > a^*$ we have $|F| < \delta$, $|G| < \delta$ on $\mathcal{A}(a)$. So the period map $\mathcal{P}$ has arbitrarily large invariant encircling curves, all solutions are bounded and Morris' Theorem in established.

12.5 Arnold's Stability Theorem

The invariant curve theorem can be used to establish a stability result for equilibrium points as well. In particular, we prove Arnold's stability theorem using Moser's invariant curve theorem.

As discussed above, the only way an equilibrium point can be stable is if the eigenvalues of the linearized equations (the exponents) are pure imaginary. Arnold's theorem addresses the case when exponents are pure imaginary, and the Hamiltonian is not positive definite.

Consider the two degrees of freedom case, and assume that the Hamiltonian has been normalized a bit. Specifically, consider a Hamiltonian H in the symplectic coordinates x_1, x_2, y_1, y_2 of the form

$$H = H_2 + H_4 + \cdots + H_{2N} + H^\dagger, \tag{12.11}$$

where

1. H is real analytic in a neighborhood of the origin in $\mathbb{R}^4$,
2. H_{2k}, $1 \le k \le N$, is a homogeneous polynomial of degree k in I_1, I_2, where $I_i = (x_i^2 + y_i^2)/2$, $i = 1, 2$,
3. $H^\dagger$ has a series expansion that starts with terms at least of degree $2N+1$,
4. $H_2 = \omega_1 I_1 - \omega_2 I_2$, ω_i nonzero constants,
5. $H_4 = \frac{1}{2}(AI_1^2 + 2BI_1I_2 + CI_2^2)$, A, B, C, constants.

There are several implicit assumptions in stating that H is of the above form. Because H is at least quadratic, the origin is an equilibrium point. By (4), H_2 is the Hamiltonian of two harmonic oscillators with frequencies ω_1 and ω_2, so, the linearized equations of motion are two harmonic oscillators. The sign convention is to conform with the sign convention at $\mathcal{L}_4$. It is not necessary to assume that ω_1 and ω_2 are positive, but this is the interesting case when the Hamiltonian is not positive definite. H_{2k}, $1 \le k \le N$, depends only on I_1 and I_2, so, H is assumed to be in Birkhoff normal form (Corollary 10.4.1) through terms of degree $2N$. This usually requires the nonresonance condition $k_1\omega_1 + k_2\omega_2 \ne 0$ for all integers k_1, k_2 with $|k_1| + |k_2| \le 2N$, but it is enough to assume that H is in this normal form.

Theorem 12.5.1 (Arnold's stability theorem). *The origin is stable for the system whose Hamiltonian is (12.11), provided that for some k, $1 \le k \le N$, $D_{2k} = H_{2k}(\omega_2, \omega_1) \ne 0$ or, equivalently, provided that H_2 does not divide H_{2k}. In particular, the equilibrium is stable if*

$$D_4 = \frac{1}{2}\{A\omega_2^2 + 2B\omega_1\omega_2 + C\omega_1^2\} \ne 0. \tag{12.12}$$

Moreover, arbitrarily close to the origin in $\mathbb{R}^4$, there are invariant tori and the flow on these invariant tori is the linear flow with irrational slope.

Proof. Assume that $D_2 = \cdots = D_{2N-2} = 0$ but $D_{2N} \ne 0$; so, there exist homogeneous polynomials F_{2k}, $k = 2, \ldots, N - 1$, of degree $2k$ such that $H_{2k} = H_2 F_{2k-2}$. The Hamiltonian (12.11) is then

$$H = H_2(1 + F_2 + \cdots + F_{2N-4}) + H_{2N} + H^\dagger.$$

Introduce action–angle variables $I_i = (x_i^2 + y_i^2)/2$, $\phi_i = \arctan(y_i/x_i)$, and scale the variables by $I_i = \epsilon^2 J_i$, where ϵ is a small scale variable. This is a symplectic change of coordinates with multiplier ϵ^{-2}, so, the Hamiltonian becomes

$$H = H_2 F + \epsilon^{2N-2} H_{2N} + O(\epsilon^{2N-1}),$$

where

$$F = 1 + \epsilon^2 F_2 + \cdots + \epsilon^{2N-4} F_{2N-4}.$$

Fix a bounded neighborhood of the origin, say $\mid J_i \mid \leq 4$, and call it O so that the remainder term is uniformly $O(\epsilon^{2N+1})$ in O. Restrict your attention to this neighborhood henceforth. Let h be a new parameter that lies in the bounded interval $[-1, 1]$. Because $F = 1 + \cdots$, one has

$$H - \epsilon^{2N-1}h = KF,$$

where

$$K = H_2 + \epsilon^{2N-2}H_{2N} + O(\epsilon^{2N-1}).$$

Because $F = 1 + \cdots$, the function F is positive on O for sufficiently small ϵ so the level set when $H = \epsilon^{2N-1}h$ is the same as the level set when $K = 0$. Let $z = (J_1, J_2, \phi_1, \phi_2)$, and let ∇ be the gradient operator with respect to these variables. The equations of motion are

$$\dot{z} = J\nabla H = (J\nabla K)F + K(J\nabla F).$$

On the level set where $K = 0$, the equations become

$$\dot{z} = J\nabla H = (J\nabla K)F.$$

For small ϵ, F is positive, so, reparameterize the equation by $d\tau = Fdt$, and the equation becomes

$$z' = J\nabla K(z),$$

where $' = d/d\tau$.

In summary, it has been shown that in O for small ϵ, the flow defined by H on the level set $H = \epsilon^{2N-1}h$ is a reparameterization of the flow defined by K on the level set $K = 0$. Thus it suffices to consider the flow defined by K. To that end, the equations of motion defined by K are

$$J_i' = O(\epsilon^{2N-1}),$$

$$\phi_1' = \omega_1 - \epsilon^{2N-2}\frac{\partial H_{2N}}{\partial J_1} + O(\epsilon^{2N-1}), \tag{12.13}$$

$$\phi_2' = +\omega_2 - \epsilon^{2N-2}\frac{\partial H_{2N}}{\partial J_2} + O(\epsilon^{2N-1}).$$

From these equations, the Poincaré map of the section $\phi_2 \equiv 0 \bmod 2\pi$ in the level set $K = 0$ is computed, and then the invariant curve theorem can be applied.

From the last equation in (12.13), the first return time T required for ϕ_2 to increase by 2π is given by

$$T = \frac{2\pi}{\omega_2}\left(1 + \frac{\epsilon^{2N-2}}{\omega_2}\frac{\partial H_{2N}}{\partial J_2}\right) + O(\epsilon^{2N-1}).$$

Integrate the ϕ_1 equation in (12.13) from $\tau = 0$ to $\tau = T$, and let $\phi_1(0) = \phi_0$, $\phi_1(T) = \phi^*$ to get

$$\phi^* = \phi_0 + \left(-\omega_1 - \epsilon^{2N-2} \frac{\partial H}{\partial J_1} \right) T + O(\epsilon^{2N-1})$$

$$= \phi_0 - 2\pi \left(\frac{\omega_1}{\omega_2} \right) - \epsilon^{2N-2} \left(\frac{2\pi}{\omega_2} \right) \left(\omega_2 \frac{\partial H_{2N}}{\partial J_1} + \omega_1 \frac{\partial H_{2N}}{\partial J_2} \right) + O(\epsilon^{2N-1}).$$

$$(12.14)$$

In the above, the partial derivatives are evaluated at (J_1, J_2). From the relation $K = 0$, solve for J_2 to get $J_2 = (\omega_1/\omega_2)J_1 + O(\epsilon^2)$. Substitute this into (12.14) to eliminate J_2, and simplify the expression by using Euler's theorem on homogeneous polynomials to get

$$\phi^* - \phi_0 + \alpha + \epsilon^{2N-2} \beta . J_1^{N-1} + O(\epsilon^{2N-1}), \qquad (12.15)$$

where $\alpha = -2\pi(\omega_1/\omega_2)$ and $\beta = -2\pi(N/\omega_2^{N+1})H_{2N}(\omega_2, \omega_1)$. By assumption, $D_{2N} = H_{2N}(\omega_2, \omega_1) \neq 0$, so, $\beta \neq 0$. Along with (12.15), the equation $J_1 \to J_1 + O(\epsilon^{2N-1})$ defines an area-preserving map of an annular region, say $1/2 \leq J_1 \leq 3$ for small ϵ. By the invariant curve theorem for sufficiently small ϵ, $0 \leq \epsilon \leq \epsilon_0$, there is an invariant curve for this Poincaré map of the form $J_1 = \rho(\phi_1)$, where ρ is continuous, 2π periodic, and $1/2 \leq \rho(\phi_1, \epsilon) \leq 3$ for all ϕ_1. For all ϵ, $0 \leq \epsilon \leq \epsilon_0$, the solutions of (12.13) which start on $K = 0$ with initial condition $J_1 < 1/2$ must have J_1 remaining less than 3 for all τ. Because on $K = 0$ one has that $J_2 = (\omega_1/\omega_2)J_1 + \cdots$, a bound on J_1 implies a bound on J_2. Thus there are constants c and k such that if $J_1(\tau), J_2(\tau)$ satisfy the equations (12.13), start on $K = 0$, and satisfy $| J_i(0) | \leq c$, then $| J_i(\tau) | \leq k$ for all τ and for all $h \in [-1, 1], 0 \leq \epsilon \leq \epsilon_0$.

Going back to the original variables $(I_1, I_2, \phi_1, \phi_2)$, and the original Hamiltonian H, this means that for $0 \leq \epsilon \leq \epsilon_0$, all solutions of the equations defined by the Hamiltonian (12.11) which start on $H = \epsilon^{2N-1}h$ and satisfy $| I_i(0) | \leq \epsilon^2 c$ must satisfy $| I_i(t) | \leq \epsilon^2 k$ for all t and all $h \in [-1, 1], 0 \leq \epsilon \leq \epsilon_0$. Thus the origin is stable. The invariant curves in the section map sweep out an invariant torus under the flow.

Arnold's theorem was originally proved independent of the invariant curve theorem, see Arnold (1963a,b), but the proof given here is taken from Meyer and Schmidt (2016). Actually, in Arnold's original works the stability criterion was $AC - B^2 \neq 0$, which implies a lot of invariant tori, but is not sufficient to prove stability; see the interesting example in Bruno (1987).

The coefficients A, B, and C of Arnold's theorem for the Hamiltonian of the restricted 3-body problem were computed by Deprit and Deprit–Bartholomê (1967) specifically to apply Arnold's theorem. These coefficients were given in Section 10.5. For $0 < \mu < \mu_1, \mu \neq \mu_2, \mu_3$ they found that

$$D_4 = -\frac{36 - 541\omega_1^2\omega_2^2 + 644\omega_1^4\omega_2^4}{8(1 - 4\omega_1^2\omega_2^2)(4 - 25\omega_1^2\omega_2^2)}.$$

D_4 is nonzero except for one value $\mu_c \approx 0.010, 913, 667$ which seems to have no mathematical significance (it is not a resonance value), and has no astronomical significance (it does not correspond to the earth–moon system, etc.)

In Meyer and Schmidt (2016), the normalization was carried to sixth-order using an algebraic processor, and $D_6 = P/Q$ where

$$P = -\frac{3105}{4} + \frac{1338449}{48}\sigma - \frac{48991830}{1728}\sigma^2 + \frac{7787081027}{6912}\sigma^3$$

$$-\frac{2052731645}{1296}\sigma^4 - \frac{1629138643}{324}\sigma^5$$

$$+\frac{1879982900}{81}\sigma^6 + \frac{368284375}{81}\sigma^7,$$

$$Q = \omega_1\omega_2(\omega_1^2 - \omega_2^2)^5(4 - 25\sigma)^3(9 - 100\sigma),$$

$$\sigma = \omega_1^2\omega_2^2,$$

From this expression one can see that $D_6 \neq 0$ when $\mu = \mu_c$ ($D_6 \approx 66.6$). So by Arnold's theorem and these calculations we have the following.

Proposition 12.5.1. *In the restricted 3-body problem the libration points $\mathcal{L}_4$ and $\mathcal{L}_5$ are stable for $0 < \mu < \mu_1, \mu \neq \mu_2, \mu_3$.*

12.6 Singular Reduction

For the restricted problem at $\mathcal{L}_4$ it remains for us to discuss stability in the cases of 1:−1, 2:−1, and 3:−1 resonance, which will be treated in this and the next three sections. Here we introduce a procedure for understanding the geometry of the problem by reducing it to the analysis of a flow on a two-dimensional orbifold which gives rise to nice figures. Refer back to Section 1.4. This method applies to systems that are in normal form. In the next three sections we will give a rigorous proof of the stability/instability for these remaining resonance cases.

There will be four related Hamiltonians: H is for the full system, $\mathbb{H}$ is for the quadratic part of the system, $\bar{H}$ is for the truncated normalizes and $\bar{R}$ the reduced average system. This section deals with the systems defined by $\bar{H}$ on the orbifold defined by $\mathbb{H}$. The full system, H, will be treated in the following sections.

Let

$$\mathbb{H} = \frac{1}{2}[k(x_1^2 + y_1^2) - (x_2^2 + y_2^2)] = kI_1 - I_2$$

where k is a positive integer, $z = (x_1, x_2, y_1, y_2)$ are rectangular variables, and $I_1, I_2, \theta_1, \theta_2$ are action-angle variables. The equations of motion are

$$\dot{I}_1 = 0, \quad \dot{\theta}_1 = -k, \quad \dot{I}_2 = 0, \quad \dot{\theta}_2 = 1.$$

This system has three independent invariants or integrals, namely $I_1, I_2, \theta_1 + k\theta_2$, which is enough since three independent invariants in a four-dimensional system specify an orbit.

A fundamental set of polynomial invariants associated with the $k : -1$ resonance are

$$
\begin{aligned}
a_1 &= I_1 = x_1^2 + y_1^2, \\
a_2 &= I_2 = x_2^2 + y_2^2, \\
a_3 &= I_1^{1/2} I_2^{k/2} \cos(\theta_1 + k\,\theta_2) = \Re[(x_1 + \mathbf{i}\,y_1)(x_2 + \mathbf{i}\,y_2)^k], \\
a_4 &= I_1^{1/2} I_2^{k/2} \sin(\theta_1 + k\,\theta_2) = \Im[(x_1 + \mathbf{i}\,y_1)(x_2 + \mathbf{i}\,y_2)^k],
\end{aligned}
$$

subject to the constraint

$$a_3^2 + a_4^2 = a_1\, a_2^k, \tag{12.16}$$

which follows from the trigonometric identity $\cos^2 \phi + \sin^2 \phi = 1$ and the restrictions $a_1 \geq 0$, $a_2 \geq 0$.

The Poisson brackets associated with the invariants are given in Table 12.1. Note that all Poisson brackets are polynomial, since k is a positive integer.

For the flow defined by $\mathbb{H}$ consider the set $\mathbb{N} = \{z \in \mathbb{R}^4 : \mathbb{H}(z) = h\}$ which is a smooth invariant submanifold of dimension 3 except possibly at $z = 0$. The orbit space $\mathbb{O}$ for $\mathbb{H}$ is the quotient space obtained from $\mathbb{N}$ by identifying orbits to a point. Let $\Pi : \mathbb{N} \to \mathbb{O}$ be the projection. Thus if $p \in \mathbb{O}$ then $\Pi^{-1}(p) \in \mathbb{N}$ is a circle (a periodic solution of the system defined by $\mathbb{H}$) or maybe just the origin. This quotient spaces is nice since all solutions are periodic and so $\mathbb{N}$ is foliated by circles. This orbit space is a symplectic manifold except at the origin.

An orbit of the system is uniquely specified by the four invariants subject to the constraint (12.16) and so the orbit space $\mathbb{O}$ is determined by the

Table 12.1. Poisson brackets among the invariants a_1, a_2, a_3, a_4. The a_i's of the first column must be put in the left-hand side of the bracket, whereas the a_i's of the top row are placed on the right-hand side of the brackets

$\{\,,\}$	a_1	a_2	a_3	a_4
a_1	0	0	$-2\,a_4$	$2\,a_3$
a_2	0	0	$-2k\,a_4$	$2k\,a_3$
a_3	$2\,a_4$	$2k\,a_4$	0	$a_2^{k-1}(k^2\,a_1 + a_2)$
a_4	$-2\,a_3$	$-2k\,a_3$	$-a_2^{k-1}(k^2\,a_1 + a_2)$	0

constraint and the integral $\mathbb{H} = k a_1 - a_2 = h$. Solve the integral for a_2 and substitute into the constraint to get the orbit space equation

$$a_3^2 + a_4^2 = a_1(k\, a_1 - h)^k, \tag{12.17}$$

which defines a surface in the (a_1, a_3, a_4)-space and is a representation of the orbit space $\mathbb{O}$. Note that the surface is a surface of revolution, so let ρ, ψ be polar coordinates in the (a_3, a_4)-plane so that the equation becomes

$$\rho^2 = a_1(k\, a_1 - h)^k.$$

This surface of revolution is unbounded and it is smooth when the right-hand side is positive. The region where the surface is smooth is called the plateau. As always $a_1 \geq 0$ but $a_2 \geq 0$ implies $a_1 \geq h/k$. See the figures below.

When $h < 0$, the right-hand side of the orbit space equation is zero only at $a_1 = 0$ and nearby $\rho \sim c\, a_1^{1/2}$ with c a positive constant. Thus the surface $\mathbb{O}$ is smooth at $a_1 = 0$.

When $h = 0$ the right-hand side is zero at $a_1 = 0$ and nearby $\rho \sim c\, a_1^{(k+1)/2}$ with $c > 0$ a constant. Thus the surface is cone-like when $k = 1$ and is cusp-like when $k > 1$.

When $h > 0$ the right-hand side is zero at $a_1 = h/k$ and nearby $\rho \sim c\,(k\, a_1 - h)^{k/2}$ where $c > 0$ is a constant. Thus the surface is smooth at $a_1 = h/k$ when $k = 1$, is cone-like when $k = 2$ and is cusp-like when $k > 2$.

Now let $\bar{H} = \mathbb{H} + \cdots$ be a polynomial Hamiltonian system of degree $k+1$ that is in normal form with quadratic part $\mathbb{H}$. (The $k = 1$ case is a little different so will be treated later.) Since $\bar{H}$ is in normal form $\{\bar{H}, \mathbb{H}\} = 0$ or $\mathbb{H}$ is an integral for the system defined by $\bar{H}$ and also it can be written in terms of a_1, a_2, a_3, a_4.

The invariants a_1 and a_2 are the action variables I_1 and I_2 and are of degree 2 in z. The invariants a_3 and a_4 depend on the angle $\theta_1 + k\theta_2$ and are of degree $k + 1$ in z. As we have seen the terms that contain angles are of prime importance in determining the existence and nature of some of the periodic solutions.

We consider the generic case where the angle term appears at the lowest degree, that is, at degree $k + 1$. A linear combination of a_3 and a_4 can be combined into one by a shift of θ_1, i.e.,

$$\alpha a_3 + \beta a_4 = I_1^{1/2} I_2^{k/2} [\alpha \cos(\theta_1 + k\theta_2) + \beta \sin(\theta_1 + k\theta_2)]$$
$$= G I_1^{1/2} I_2^{k/2} \cos(\theta_1 + k\theta_2 - \tilde{\theta}),$$

where $G = \sqrt{\alpha^2 + \beta^2}$ and $\tan\tilde{\theta} = \beta/\alpha$. Shift θ_1 by $\theta_1 \to \theta_1 + \tilde{\theta}$.

Scale by $z \to \varepsilon z$ which is symplectic with multiplier ε^{-2}. This scaling indicates we are working near the equilibrium when ε is small. Thus

$$\bar{H} = \mathbb{H} + \sum_{j=2}^{l} \varepsilon^{2j-2} \hat{H}_k^j(I_1, I_2) + \varepsilon^{k-1} G I_1^{1/2} I_2^{k/2} \cos(\theta_1 + k\theta_2),$$

with $2l \leq k+1$ and $\hat{H}_k^j$ is a polynomial in I_1, I_2 of degree j. Here we have separated out the single angle term $a_3 = I_1^{1/2} I_2^{k/2} \cos(\theta_1 + k\theta_2)$.

Since $\bar{H}$ is in normal form $\mathbb{H}$ is an integral, so hold it fixed by setting

$$h = \mathbb{H} = kI_1 - I_2 = ka_1 - a_2.$$

Solve for $a_2 = ka_1 - h$, $I_2 = kI_1 - h$ and so

$$\bar{H} = h + \sum_{j=2}^{l} \varepsilon^{2j-2} \tilde{H}_k^j(a_1, h) + \varepsilon^{k-1} G a_3.$$

Pass to the averaged system on the orbit space by dropping the constant term and when $k > 2$ dividing by ε^2 (time scaling), so the reduced averaged system is

$$\bar{R} - \sum_{j=2}^{l} \varepsilon^{2j-4} \tilde{H}_k^j(a_1, h) + \varepsilon^{k-3} G a_3. \tag{12.18}$$

When $k = 2$ the averaged system on the orbit space is obtained by dropping the constant term and dividing by ε (time scaling), so the reduced averaged system is

$$\bar{R} = G a_3. \tag{12.19}$$

Using the table of Poisson brackets given above we can obtain the reduced averaged flow on the orbit space by using $\dot{a}_i = \{a_i, \bar{H}\}, i = 1, 3, 4$. A critical point $p \in \mathbb{O}$ of this flow corresponds to a periodic solution $P = \Pi^{-1}(p) \in \mathbb{N}$ of the averaged system or to the origin. Likewise an orbit of the reduced averaged system which tends to the critical point p corresponds to a surface in $\mathbb{N}$ filled with orbits tending to the periodic solution P.

Case $k = 2$: The averaged system in $2 : -1$ resonance is

$$\bar{H} = 2a_1 - a_2 + \varepsilon G a_3 = 2I_1 - I_2 + \varepsilon G I_1^{1/2} I_2 \cos(\theta_1 + 2\theta_2),$$

and the reduced averaged Hamiltonian is $\bar{R} = G a_3$ and since we assume $G \neq 0$ by scaling time again $G = 1$, so $\bar{R} = a_3$.

The geometry of the problem is obtained by considering the intersection of two surfaces in $\mathbb{R}^3$. The first surface, the orbit space $\mathbb{O}$, is given by the orbit space equation

$$a_3^2 + a_4^2 = a_1(2a_1 - h)^2, \qquad a_1 \geq 0, \quad a_1 \geq \frac{1}{2}h,$$

for various values of h and the second surface is the reduced averaged Hamiltonian

$$\bar{R} = a_3 = \bar{h},$$

for various values of $\bar{h}$. Near the origin in a-space the orbit space is a paraboloid of revolution when $h < 0$ and hence smooth, it is rotated cusp when $h = 0$, and it is a cone when $h > 0$ — see Figure 12.1.

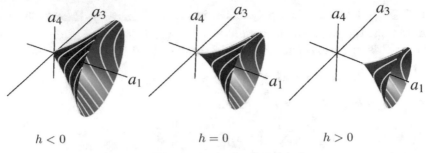

$$h < 0 \qquad\qquad h = 0 \qquad\qquad h > 0$$

Figure 12.1. $\mathbb{O}$ when $k = 2$

Of course $\bar{R} = a_3$ =constant is just a plane parallel to the a_1, a_4 coordinate plane. The flow on the orbit spaces is defined by the equations of motion

$$\dot{a}_1 = \{a_1, \bar{R}\} = -2a_4, \ \dot{a}_3 = \{a_3, \bar{R}\} = 0, \ \dot{a}_4 = \{a_4, \bar{R}\} = -a_2(4a_1 + a_2).$$

Recall that on the orbit space $a_2 = 2a_1 - h$. For an equilibrium point one must have $a_4 = 0$ and $a_2(4a_1 + a_2) = 0$ and since not both a_1 and a_2 can be zero or negative the conditions for an equilibrium are $a_4 = a_2 = 0$. But on the orbit space $a_2 = 0$ only when $a_1 = h/2$ and that only occurs when $h \geq 0$.

Look at the flow lines in Figure 12.1. The flow lines lie in $\bar{R} = a_3 =$ constant and a_4 is decreasing.

When $h = 0$ the origin in a-space corresponds to the origin in $\mathbb{R}^4$. There is an orbit on $\mathbb{O}$ tending to the origin as $t \to +\infty$ and there is an orbit on $\mathbb{O}$ tending to the origin as $t \to -\infty$. These represent a surface of solutions that spiral to the origin as $t \to \pm\infty$. Thus the equilibrium point is unstable.

When $h > 0$ there is an equilibrium on $\mathbb{O}$ at $a_1 = h/2$, $a_3 = a_4 = 0$. This gives rise to a periodic solution of period $T \sim \pi$ for each $h \geq 0$. These solutions are the short periodic family given by Lyapunov's center theorem. Note that here too there is an orbit on $\mathbb{O}$ tending to the equilibrium as $t \to \pm\infty$. Thus the solutions in the short period family are unstable.

When $h < 0$ there are no equilibria and so all solutions recede far away as $t \to \pm\infty$. Thus, there is no other period family when $h < 0$.

Case $k = 3$: The averaged $3 : -1$ system is

$$\hat{H} = 3I_1 - I_2 + \frac{\varepsilon^2}{2}(AI_1^2 + 2BI_1I_2 + CI_2^2) + \varepsilon^2 G I_1^{1/2} I_2^{3/2} \cos(\theta_1 + 3\theta_2)$$

where A, B, C, G are constants. Introduce the constants $D = \frac{1}{2}(A + 6B + 9C)$ and $R = B + 3C$.

In KAM theory D is called the twist coefficient. Note that in the present $3 : -1$ example the twist coefficient D and the angle coefficient G are both defined at the same order of ε, i.e., at order ε^2. In the previous $2 : -1$ example the angle coefficient is of lower order than the twist coefficient. So we are now looking at the case when the twist and the angle are competing.

Passing to the reduced averaged system we get

$$\hat{R} = Da_1^2 - Rha_1 + Ga_3$$

which is defined on the orbit space, $\mathbb{O}$,

$$a_3^2 + a_4^2 = a_1(3a_1 - h)^3, \quad a_1 \geq 0, \; 3a_1 \geq h.$$

Again the geometry of the problem is depicted by the intersection in $\mathbb{R}^3$ of the surface of the orbit space for various values of h and the surfaces of the reduced averaged Hamiltonian $\hat{R} = \bar{h}$ for various values of $\bar{h}$. Near the origin in a-space the orbit space is smooth when $h < 0$, it is a rotated parabola when $h = 0$ and it is a rotated cusp when $h > 0$. The surface $\mathbb{H} = \bar{h}$ =constant is just a translation of a parabola — translated in the a_4 direction.

The associated vector field of $\bar{R}$ (the reduced system)

$$
\begin{aligned}
\dot{a}_1 &= \{a_1, \bar{R}\} = -2Ga_4, \\
\dot{a}_3 &= \{a_3, \bar{R}\} = 2a_4(2Da_1 - Rh), \\
\dot{a}_4 &= \{a_4, \bar{R}\} = -2a_3(2Da_1 - Rh) - Ga_2^2(9a_1 + a_2) \\
&= -2a_3(2Da_1 - Rh) - G(3a_1 - h)^2(12a_1 - h),
\end{aligned}
$$

gives the precise flow on the orbit space. First find a critical point of these equations on the orbit space. Clearly $a_4 = 0$ (we consider $G \neq 0$) and so we must find a solution of the equations

$$
\begin{aligned}
a_3^2 &= a_1(3a_1 - h)^3, \\
2a_3(2Da_1 - Rh) &+ G(3a_1 - h)^2(12a_1 - h) = 0.
\end{aligned}
$$

Solve the second equation for a_3, square it, then substitute in a_3^2 from the first equation, cancel some terms and expand to get

$$16(D^2 - 27G^2)a_1^3 + 8h(27G^2 - 2DR)a_1^2 + h^2(4R^2 - 27G^2)a_1 + G^2h^3 = 0.$$

This is a cubic polynomial in a_1 with parameters D, G, R and h, but we can reduce by one the number of parameters by defining $\alpha = D/G$, $\beta = R/G$ and dividing the cubic polynomial by G^2.

We seek roots $a_1 \geq \min\{0, h/3\}$. The number of roots changes when the resultant of the polynomial with its derivative with respect to a_1 vanishes, i.e., when

$$1024(\alpha^2 - 27)(\alpha - 6\beta)^2(729 + 108\alpha^2 + 648\beta^2 - 48\beta^4 + 8\alpha\beta(-81 + 4\beta^2))h^6 = 0.$$

Also the number of critical points of the equations on the orbit space changes when one root of the cubic polynomial meets the peak at $a_1 = h/3$ when $h \geq 0$, i.e., when after replacing a_1 by $h/3$ in the cubic we get:

$$\frac{4}{27}(2\alpha - 3\beta)^2 h^3 = 0.$$

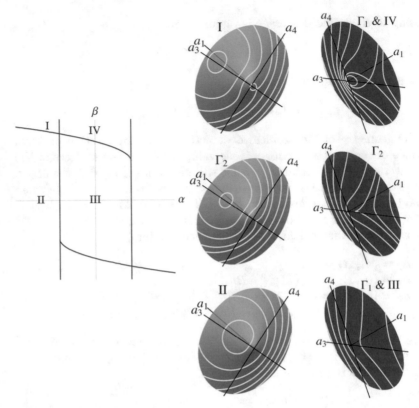

Figure 12.2. Bifurcation diagram and flows for the $3:-1$ resonance when $h < 0$

We will distinguish the cases $h < 0$, $h = 0$, and $h > 0$ and analyze the equilibria and bifurcations as functions of α and β.

The case $h = 0$ is simple because the above vanishes and the resultant becomes

$$16a_1^3(\alpha^2 - 27) = 0.$$

When $h \neq 0$ we have the two bifurcation diagrams appearing in Figures 12.2 and 12.3. The equations for the bifurcation lines become:

Γ_1: $\alpha^2 - 27 = 0$ (red lines);
Γ_2: $729 + 108\alpha^2 + 648\beta^2 - 48\beta^4 + 8\alpha\beta(-81 + 4\beta^2) = 0$ (blue curves);
Γ_3: $2\alpha - 3\beta = 0$ (green curve).

The diagrams are symmetric with respect to the origin. The blue curves correspond to a saddle-center or extremal bifurcation of critical points, thus an extremal bifurcation of periodic orbits. On the red lines, the leading term of the cubic vanishes, so only two zeros are possible. The green curve is a bifurcation of the peak.

Case $h < 0$, Figure 12.2: The surface $\mathbb{O}$ is smooth, so all the equilibria are in the plateau and correspond to 2π-periodic orbits. In Region I the

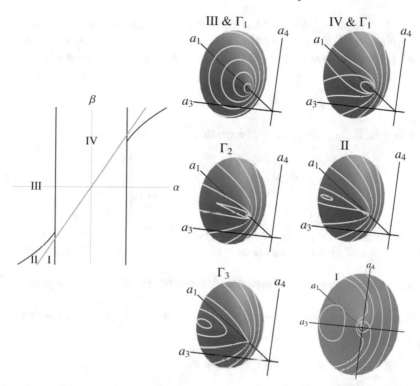

Figure 12.3. Bifurcation diagram and flows for the $3 : -1$ resonance when $h > 0$

cubic polynomial has 3 different positive roots, so we have 3 equilibria on the surface: two centers and one saddle. On the blue curve one of the centers and the saddle collide giving rise to an extremal critical point that disappears in Region II, where only the other center survives. In Region IV, only one center and the saddle are present. On the blue curve, this center and this saddle collide in another extremal critical point thus there are no equilibria in Region III.

Case $h > 0$, Figure 12.3: Here the peak $a_1 = h/3$ is always an equilibrium. We are interested in roots of the cubic which are bigger than $h/3$. In Region I the cubic has 2 roots that are different from $h/3$. So, there is 3 equilibria in total. The peak is a center, there is one saddle close to the peak and another center is relatively far from the other two points. On the green curve the saddle collides with the peak and the situation of Region I is recovered in Region II. On the blue curve the saddle and the center (that is not the peak) collide and then they disappear in Region III, where only the peak stays as a center up to the red line. After crossing the red line, in Region IV, a saddle appears and the peak continues to be a center.

Case $k = 1$: We placed the $1 : -1$ at the end because the preliminary work is slightly different. We start with a Hamiltonian matrix A that has eigenvalues $\pm i$ with multiplicity two and A is not diagonalizable. The standard normal form for the Hamiltonian is

$$\mathbf{H} = x_2 y_1 - x_1 y_2 + \frac{\delta}{2}(x_1^2 + x_2^2)$$

where $\delta = \pm 1$. The linear system of equations is $\dot{z} = Az$, where

$$A = \begin{bmatrix} 0 & 1 & 0 & 0 \\ -1 & 0 & 0 & 0 \\ -\delta & 0 & 0 & 1 \\ 0 & -\delta & -1 & 0 \end{bmatrix}, \qquad z = \begin{bmatrix} x_1 \\ x_2 \\ y_1 \\ y_2 \end{bmatrix}.$$

The characteristic polynomial of A is $p(\lambda) = (\lambda^2 + 1)^2$ with repeated eigenvalues $\pm i$. But not all solutions are 2π periodic, since there are secular terms like $t \sin t$, $t \cos t$.

The four invariants usually associated with this Hamiltonian are just

$$b_1 = x_2 y_1 - x_1 y_2, \quad b_2 = \frac{1}{2}(x_1^2 + x_2^2), \quad b_3 = \frac{1}{2}(y_1^2 + y_2^2), \quad b_4 = x_1 y_1 + x_2 y_2,$$

with the constraint

$$b_1^2 + b_4^2 = 4 b_2 b_3.$$

The nonzero Poisson brackets are

$$\{b_2, b_3\} = -\{b_3, b_2\} = b_4, \quad \{b_2, b_4\} = -\{b_4, b_2\} = 2b_2,$$
$$\{b_4, b_3\} = -\{b_3, b_4\} = 2b_3.$$

Consider the nonlinear Hamiltonian system which has $\mathbb{H} = b_1 + \delta b_2$ as its quadratic part and has been normalized to Sokol'skii normal form through the fourth–order terms, i.e., let

$$\bar{H} = b_1 + \delta b_2 + (\alpha b_1^2 + 2\beta b_1 b_3 + \gamma b_3^2).$$

Use the scaling

$$x_1 \to \varepsilon^2 x_1, \qquad x_2 \to \varepsilon^2 x_2,$$
$$y_1 \to \varepsilon y_1, \qquad y_2 \to \varepsilon y_2,$$

which is symplectic with multiplier ε^{-3}, so, the Hamiltonian becomes

$$\bar{H} = b_1 + \varepsilon(\delta b_2 + \gamma b_3^2).$$

Now with $\varepsilon = 0$ all solutions are periodic with least period 2π, so the orbit space is an orbifold manifold as before. Let

$$\bar{H} = b_1 + \varepsilon(\delta b_2 + \gamma b_3^2)$$

play the role of the averaged system.

The orbit space, $\mathbb{O}$, is specified by the invariants subject to the constraint and

$$\bar{R}_{\varepsilon=0} = b_1 = h.$$

Thus the equation of the orbit space is

$$h^2 + b_4^2 = 4 b_2 b_3$$

which is the equation of a two sheeted hyperboloid in b_2, b_3, b_4 space when $h \neq 0$, but we only look at the sheet where $b_2 > 0$, $b_3 > 0$. This one sheet represents both $h > 0$ and $h < 0$.

The reduced averaged Hamiltonian is

$$\bar{R} = \delta b_2 + \gamma b_3^2 = \bar{h}.$$

The Hamiltonian $\bar{R}$ has critical point on the orbit space if $b_4 = 0$ and $\gamma b_3 = \delta b_2$ and so a critical point exists if $\gamma h^2 = \delta b_2^2$. Hence there is a critical point of $\bar{R}$ on $\mathbb{O}$ if and only if δ and γ have the same sign.

The equations of motion are

$$\dot{b}_2 = \{b_2, \bar{R}\} = 2\gamma b_3 b_4, \quad \dot{b}_3 = \{b_3, \bar{R}\} = 0, \quad \dot{b}_4 = \{b_4, \bar{R}\} = -2\delta b_2 + 4\gamma b_3^2.$$

Since b_3 is a positive constant, say $b_3 = p$, and then the equations in b_2, b_4 is a harmonic oscillator when $\delta\gamma > 0$.

Thus, *in the case of $1 : -1$ resonance there are two families of nearly 2π elliptic periodic solution emanated from the origin when $\delta\gamma > 0$. One family exists for $\mathcal{H} > 0$ and one for $\mathcal{H} < 0$. There are no near by 2π periodic solutions when $\delta\gamma < 0$* (Figure 12.4).

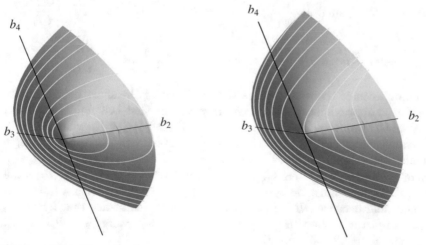

Figure 12.4. Flows in the $1 : -1$ resonance. On the left: sign(δ)=sign(γ). On the right: sign(δ)$\neq$sign(γ).

12.7 2:-1 Resonance

In this section we consider a system when the linear system is in $2 : -1$ resonance, i.e., when the linearized system has exponents $\pm i\omega_1$ and $\pm i\omega_2$ with $\omega_1 = 2\omega_2$. Let $\omega = \omega_2$. By the discussion in Section 10.5 the normal form for the Hamiltonian is a function of I_1, I_2 and the single angle $\phi_1 + 2\phi_2$. Assume the system has been normalized through terms of degree three, i.e., assume the Hamiltonian is of the form

$$H = 2\omega I_1 - \omega I_2 + \delta I_1^{1/2} I_2 \cos \psi + H^\dagger, \qquad (12.20)$$

where $\psi = \phi_1 + 2\phi_2$, $H^\dagger(I_1, I_2, \phi_1, \phi_2) = O((I_1 + I_2)^2)$. Notice this Hamiltonian is just a perturbation of Cherry's example. Lyapunov's center theorem assures the existence of one family of periodic solutions emanating from the origin, the short period family with period approximately π/ω.

Theorem 12.7.1. *If in the presence of $2 : -1$ resonance, the Hamiltonian system is in the normal form (12.20) with $\delta \neq 0$ then the equilibrium is unstable. In fact, there is a neighborhood O of the equilibrium such that any solution starting in O and not on the Lyapunov center leaves O in either positive or negative time. In particular, the small periodic solutions of the short period family are unstable.*

   Remark. If $\delta = 0$ then the Hamiltonian can be put into normal form to the next order and the stability of the equilibrium may be decidable on the bases of Arnold's theorem, Theorem 12.5.1.

Proof. The equations of motion are

$$\dot{I}_1 = -\delta I_1^{1/2} I_2 \sin \psi + \frac{\partial H^\dagger}{\partial \phi_1}, \qquad \dot{\phi}_1 = -2\omega - \frac{\delta}{2} I_1^{-1/2} I_2 \cos \psi - \frac{\partial H^\dagger}{\partial I_1},$$

$$\dot{I}_2 = -2\delta I_1^{1/2} I_2 \sin \psi + \frac{\partial H^\dagger}{\partial \phi_2}, \qquad \dot{\phi}_2 = \omega - \delta I_1^{1/2} \cos \psi - \frac{\partial H^\dagger}{\partial I_2}.$$

Lyapunov's center theorem ensures the existence of the short period family with period approximately $\pi/2\omega$. We may assume that this family has been transformed to the plane where $I_2 = 0$. So $\partial H^\dagger/\partial \phi_2 = 0$ when $I_2 = 0$. The Hamiltonian (12.20) is a real analytic system written in action–angle variables thus the terms in $H^\dagger$ must have the d'Alembert character, i.e., a term of the form $I_1^{\alpha/2} I_2^{\beta/2} \cos k(\phi_1 + 2\phi_2)$ must have $\beta \geq 2k$ and $\beta \equiv 2k \bmod 2$ so in particular β must be even. Thus I_2 does not appear with a fractional exponent and because $\partial H^\dagger/\partial \phi_2 = 0$ when $I_2 = 0$ this means that $\partial H^\dagger/\partial \phi_2$ contains a factor I_2. Let $\partial H^\dagger/\partial \phi_2 = I_2 U_1(I_1, I, 2, \psi)$ where $U_1 = O(I_1 + I_2)$.
   Consider the Chetaev function

$$V = -\delta I_1^{1/2} I_2 \sin \psi$$

and compute

$$\dot{V} = \delta^2 \left\{ \frac{1}{2} I_2^2 + 2 I_1 I_2 \right\} + W,$$

where

$$W = -\delta \left\{ \frac{1}{2} I_1^{-1/2} I_2 \sin \psi \frac{\partial H^\dagger}{\partial \psi_1} + I_1^{1/2} \sin \psi \frac{\partial H^\dagger}{\partial \psi_2} \right.$$

$$\left. - I_1^{1/2} I_2 \cos \psi \frac{\partial H^\dagger}{\partial I_1} - 2 I_1^{1/2} I_2 \cos \psi \frac{\partial H^\dagger}{\partial I_2} \right\}$$

Because $\partial H^\dagger / \partial \phi_2 = I_2 U_1$, $W = I_2 U_2$ where $U_2 = O((I_1 + I_2)^{3/2}$ and

$$\dot{V} = \delta^2 I_2 (\frac{1}{2} I_2 + 2 I_1 + U_2).$$

Thus there is a neighborhood O where $\dot{V} > 0$ when $I_2 \neq 0$. Apply Chetaev's theorem with $\Omega = O \cap \{V > 0\}$ to conclude that all solutions which start in Ω leave O in positive time. By reversing time we can conclude that all solutions which start in $\Omega' = O \cap \{V < 0\}$ leave O in negative time.

When

$$\mu - \mu_2 = \frac{1}{2} - \frac{1}{30} \sqrt{\frac{611}{3}} \approx 0.0242939$$

the exponents of the Lagrange equilateral triangle libration point $\mathcal{L}_4$ of the restricted 3-body problem are $\pm 2\sqrt{5}i/5$, $\pm\sqrt{5}i/5$ and so the ratio of the frequencies ω_1/ω_2 is 2. Expanding the Hamiltonian about $\mathcal{L}_4$ when $\mu = \mu_2$ in a Taylor series through cubic terms gives

$$H = \frac{1}{14} \left\{ 5x_1^2 - 2\sqrt{611}x_1 x_2 - 25x_2^2 - 40x_1 y_2 + 40x_2 y_1 + 20y_1^2 + 20y_2^2 \right\}$$

$$\frac{1}{240\sqrt{3}} \left\{ -7\sqrt{611}x_1^3 + 135x_1^2 x_2 + 33\sqrt{611}x_1 x_2^2 + 135x_2^3 \right\} + \cdots.$$

Using Mathematica we can put this Hamiltonian into the normal form (12.20) with

$$\omega = \frac{\sqrt{5}}{5} \approx 0.447213, \qquad \delta = \frac{11\sqrt{11}}{18\sqrt[4]{5}} \approx 1.35542,$$

and so we have the following.

Proposition 12.7.1. *The libration point $\mathcal{L}_4$ of the restricted 3-body problem is unstable when $\mu = \mu_2$.*

12.8 3:-1 Resonance

In this section we consider a system when the linear system is in $3 : -1$ resonance, i.e., $\omega_1 = 3\omega_2$. Let $\omega = \omega_2$. By the discussion in Section 10.5 the normal form for the Hamiltonian is a function of I_1, I_2 and the single angle $\phi_1 + 3\phi_2$. Assume the system has been normalized through terms of degree four, i.e., assume the Hamiltonian is of the form

$$H = 3\omega I_1 - \omega I_2 + \delta I_1^{1/2} I_2^{3/2} \cos\psi + \frac{1}{2}\{AI_1^2 + 2BI_1 I_2 + CI_2^2\} + H^\dagger, \quad (12.21)$$

where $\psi = \phi_1 + 3\phi_2$, $H^\dagger = O((I_1 + I_2)^{5/2})$. Let

$$D = A + 6B + 9C, \quad (12.22)$$

and recall from Arnold's theorem the important quantity $D_4 = \frac{1}{2}D\omega^2$.

Theorem 12.8.1. *If in the presence of $3 : -1$ resonance, the Hamiltonian system is in the normal form (12.21) and if $6\sqrt{3}|\delta| > |D|$ then the equilibrium is unstable, whereas, if $6\sqrt{3}|\delta| < |D|$ then the equilibrium is stable.*

Proof. Introduce the small parameter ϵ by scaling the variables $I_i \to \epsilon I_i$, $i = 1, 2$ which is symplectic with multiplier ϵ^{-1}, the Hamiltonian becomes

$$H = 3\omega I_1 - \omega I_2 + \epsilon\{\delta I_1^{1/2} I_2^{3/2} \cos\psi + \frac{1}{2}(AI_1^2 + 2BI_1 I_2 + CI_2^2)\} + O(\epsilon^2),$$

and the equations of motion are

$$\dot{I}_1 = -\epsilon \delta I_1^{1/2} I_2^{3/2} \sin\psi + O(\epsilon^2),$$

$$\dot{I}_2 = -3\epsilon \delta I_1^{1/2} I_2^{3/2} \sin\psi + O(\epsilon^2),$$

$$\dot{\phi}_1 = -3\omega - \epsilon\{\frac{1}{2}\delta I_1^{-1/2} I_2^{3/2} \cos\psi + (AI_1 + BI_2)\} + O(\epsilon^2),$$

$$\dot{\phi}_2 = \omega - \epsilon\{\frac{3}{2}\delta I_1^{1/2} I_2^{1/2} \cos\psi + (BI_1 + CI_2)\} + O(\epsilon^2).$$

Instability. Consider the Chetaev function

$$V = -\delta I_1^{1/2} I_2^{3/2} \sin\psi$$

and compute

$$\dot{V} = \epsilon\left\{\delta^2\left(\frac{1}{2}I_2^3 + \frac{9}{2}I_1 I_2^2\right)\right.$$

$$\left. - \delta I_1^{1/2} I_2^{3/2}(AI_1 + BI_2 + 3BI_1 + 3CI_2)\cos\psi\right\} + O(\epsilon^2).$$

Consider the flow in the $H = 0$ surface. Solve $H = 0$ for I_2 as a function of I_1, ϕ_1, ϕ_2 to find $I_2 = 3I_1 + O(\epsilon)$. On the $H = 0$ surface we find

$$V = -3\sqrt{3}\delta I_1^2 \sin \psi + O(\epsilon)$$

and

$$\dot{V} = \epsilon\{\delta^2 I_1^3 (54 - \delta^{-1} 3^{3/2}(A + 6B + 9C)\cos\psi)\} + O(\epsilon^2).$$

If $54 > |\delta^{-1} 3^{3/2} D|$ or $6\sqrt{3}|\delta| > |D|$ the function $\dot{V}$ is positive definite in the level set $H = 0$. Because V takes positive and negative values close to the origin in the level set $H = 0$, Chetaev's theorem implies that the equilibrium is unstable.

Stability. Now we compute the cross section map in the level set $H = \epsilon^2 h$ where $-1 \le h \le 1$ and the section is defined by $\phi_2 \equiv 0 \bmod 2\pi$. We use (I_1, ϕ_1) as coordinates in this cross-section. From the equation $H = \epsilon^2 h$ we can solve for I_2 to find that $I_2 = 3I_1 + O(\epsilon)$. Integrating the equation for ϕ_2 we find that the return time T is

$$T = \frac{2\pi}{\omega - \epsilon\{3\sqrt{3}/2\delta \cos\psi + (B + 3C)\}I_1} + \cdots$$

$$= \frac{2\pi}{\omega}\left\{1 + \frac{\epsilon}{\omega}\left(\frac{3\sqrt{3}}{2}\delta \cos\psi + (B + 3C)\right)I_1\right\} + \cdots.$$

Integrating the ϕ_1 equation from $t = 0$ to $t = T$ gives the cross-section map of the form $P : (I_1, \phi_1) \to (I_1', \phi_1')$, where

$$I_1' = I_1 + O(\epsilon),$$

$$\phi_1' = \phi_1 + \frac{2\pi\epsilon}{\omega}\{(A + 6B + 9C)I_1 - 6\sqrt{3}\delta \cos 3\phi_1\} + O(\epsilon^2). \tag{12.23}$$

By hypothesis the coefficient of I_1 in (12.23) is nonzero and so Corollary 12.3.1 implies the existence of invariant curves for the section map. The stability of the equilibrium follows now by the same argument as found in the proof of Arnold's stability theorem 12.5.1.

When

$$\mu = \mu_3 = \frac{1}{2} - \frac{\sqrt{213}}{30} \approx 0.0135160$$

the exponents of the Lagrange equilateral triangle libration point $\mathcal{L}_4$ of the restricted 3-body problem are $\pm 3\sqrt{10}i/10, \pm\sqrt{10}i/10$ and so the ratio of the frequencies ω_1/ω_2 is 3.

Using Mathematica we can put this Hamiltonian into the normal form (12.21) with

$$\omega = \frac{\sqrt{10}}{10} \approx 0.316228, \qquad \delta = \frac{3\sqrt{14277}}{80} \approx 4.48074$$

$$A = \frac{309}{1120}, \qquad B = -\frac{1219}{560}, \qquad C = \frac{79}{560}.$$

From this we compute

$$6\sqrt{3}|\delta| \approx 46.5652 > |D| \approx 8.34107,$$

and so we have the following.

Proposition 12.8.1. *The libration point $\mathcal{L}_4$ of the restricted 3-body problem is unstable when $\mu = \mu_3$.*

That the Lagrange point $\mathcal{L}_4$ is unstable when $\mu = \mu_2, \mu_3$ was established in Markeev (1966) and Alfriend (1970, 1971). Hagel (1996) analytically and numerically studied the stability of $\mathcal{L}_4$ in the restricted problem not only at μ_2 but near μ_2 also.

12.9 1:-1 Resonance

A one-parameter problem, such as the restricted 3-body problem, generically can have an isolated equilibriums point with exponents with multiplicity two, but in this case the matrix of the linearized system is not diagonalizable. Thus the equilibrium at $\mathcal{L}_4$ when $\mu = \mu_1$ is typical of an equilibrium in a one-parameter family. An equilibrium with exponents with higher multiplicity or an equilibrium such that the linearized system is diagonalizable is degenerate in a one-parameter family.

Consider a system in the case when the exponents of the equilibrium are $\pm i\omega$ with multiplicity two and the linearized system is not diagonalizable. The normal form for the quadratic part of such a Hamiltonian was given as

$$H_2 = \omega(x_2 y_1 - x_1 y_2) + \frac{\delta}{2}(x_1^2 + x_2^2), \qquad (12.24)$$

where $\omega \neq 0$ and $\delta = \pm 1$. The linearized equations are

$$\begin{bmatrix} \dot{x}_1 \\ \dot{x}_2 \\ \dot{y}_1 \\ \dot{y}_2 \end{bmatrix} = \begin{bmatrix} 0 & \omega & 0 & 0 \\ -\omega & 0 & 0 & 0 \\ -\delta & 0 & 0 & \omega \\ 0 & -\delta & -\omega & 0 \end{bmatrix} \begin{bmatrix} x_1 \\ x_2 \\ y_1 \\ y_2 \end{bmatrix}.$$

Recall that the normal form in this case depends on the four invariants

$$\Gamma_1 = x_2 y_1 - x_1 y_2, \qquad \Gamma_2 = \tfrac{1}{2}(x_1^2 + x_2^2),$$
$$\Gamma_3 = \tfrac{1}{2}(y_1^2 + y_2^2), \qquad \Gamma_4 = x_1 y_1 + x_2 y_2,$$

and that $\{\Gamma_1, \Gamma_2\} = 0$ and $\{\Gamma_1, \Gamma_3\} = 0$. The system is in Sokol'skii normal form if the higher-order terms depend on the two quantities Γ_1 and Γ_3 only; that is, the Hamiltonian is of the form

$$H = \omega\Gamma_1 + \delta\Gamma_2 + \sum_{k=2}^{\infty} H_{2k}(\Gamma_1, \Gamma_3), \qquad (12.25)$$

where here H_{2k} is a polynomial of degree k in two variables.

Consider a system which is in Sokol'skii's normal form up to order four, i.e., consider the system

$$H = \omega\Gamma_1 + \delta\Gamma_2$$

$$+ \{A\Gamma_3^2 + B\Gamma_1\Gamma_3 + C\Gamma_1^2\} \qquad (12.26)$$

$$+ H^\dagger(x_1, x_2, y_1, y_2)$$

where A, B, and C are constants and $H^\dagger$ is at least fifth order in its displayed arguments.

Theorem 12.9.1. *If in the presence of 1:-1 resonance the system is reduced to the form (12.26) with $\delta A < 0$ then the equilibrium is unstable, but if $\delta A > 0$ then the equilibrium is stable.*

Proof (Unstable). Introduce a small parameter ϵ by the scaling

$$x_1 \to \epsilon^2 x_1, \qquad x_2 \to \epsilon^2 x_2,$$

$$\qquad (12.27)$$

$$y_1 \to \epsilon y_1, \qquad y_2 \to \epsilon y_2,$$

which is symplectic with multiplier ϵ^{-3} so the Hamiltonian (12.25) is

$$H = \omega(x_2 y_1 - x_1 y_2) + \epsilon\left\{\frac{\delta}{2}(x_1^2 + x_2^2) + A(y_1^2 + y_2^2)^2\right\} + O(\epsilon^2). \qquad (12.28)$$

The equations of motion are

$$\dot{x}_1 = \omega x_2 + \epsilon 4A(y_1^2 + y_2^2)y_1 + \cdots,$$

$$\dot{x}_2 = -\omega x_1 + \epsilon 4A(y_1^2 + y_2^2)y_2 + \cdots,$$

$$\dot{y}_1 = \omega y_2 - \epsilon\delta x_1 + \cdots,$$

$$\dot{y}_2 = -\omega y_1 - \epsilon\delta x_2 + \cdots.$$

Consider the Lyapunov function

$$V = \delta\Gamma_4 = \delta(x_1 y_1 + x_2 y_2),$$

and compute

$$\dot{V} = \epsilon\{-\delta^2(x_1^2 + x_2^2) + 4\delta A(y_1^2 + y_2^2)^2\} + O(\epsilon^2).$$

So V takes on positive and negative values and $\dot{V}$ is negative on $Q' = \{0 < x_1^2 + x_2^2 + y_1^2 + y_2^2 < 1\}$ and for some $\epsilon = \epsilon_0 > 0$. Thus by Lyapunov's instability theorem 12.1.3 all solutions in $\{V > 0\} \cap Q'$ leave Q' in positive time. By reversing time we see that all solutions $\{V < 0\} \cap Q'$ leave Q' in negative time.

In the original unscaled variables all solutions that start in

$$Q = \{0 < \epsilon_0^{-2}(x_1^2 + x_2^2) + \epsilon_0^{-1}(y_1^2 + y_2^2) < 1\}$$

leave Q in either positive or negative time.

Proof (Stable). Our problem is a small perturbation of the Hamiltonian Γ_1. Make the linear symplectic change of coordinates

$$\begin{aligned}
x_1 &= \tfrac{1}{\sqrt{2}}(u_1 - v_2), & y_1 &= \tfrac{1}{\sqrt{2}}(u_2 + v_1), \\
x_2 &= \tfrac{1}{\sqrt{2}}(u_2 - v_1), & y_2 &= \tfrac{1}{\sqrt{2}}(u_1 + v_2),
\end{aligned} \tag{12.29}$$

so

$$\Gamma_1(u, v) = \frac{1}{2}(-u_1^2 - v_1^2 + u_2^2 + v_2^2), \tag{12.30}$$

thus manifesting that the Hamiltonian is equivalent to two harmonic oscillators in $1 : -1$ resonance whose geometry was discussed in Sections 12.2 and 12.6.

Let us compute the total section map to the flow. We start by scaling variables as in (12.27) obtaining the Hamiltonian (12.28) and make the change (12.29) to get

$$H = \frac{\omega}{2}(-u_1^2 - v_1^2 + u_2^2 + v_2^2)$$

$$+ \frac{\varepsilon}{32}\left(8\delta\big((u_2 - v_1)^2 + (u_1 - v_2)^2\big) + A\big((u_2 + v_1)^2 + (u_1 + v_2)^2\big)^2\right) \tag{12.31}$$

$$+ O(\varepsilon^2).$$

Then introduce the standard action–angle coordinates:

$$\begin{aligned}
u_1 &= (2\tilde{I}_1)^{1/2}\cos\tilde{\theta}_1, & v_1 &= (2\tilde{I}_1)^{1/2}\sin\tilde{\theta}_1, \\
u_2 &= (2\tilde{I}_2)^{1/2}\cos\tilde{\theta}_2, & v_2 &= (2\tilde{I}_2)^{1/2}\sin\tilde{\theta}_2,
\end{aligned}$$

to get:

$$\begin{aligned}
H = {}& \omega(\tilde{I}_2 - \tilde{I}_1) \\
& + \frac{\varepsilon}{8}\Big(4\delta(\tilde{I}_1 + \tilde{I}_2) + A(\tilde{I}_1^2 + 4\tilde{I}_1\tilde{I}_2 + \tilde{I}_2^2) - 2A\tilde{I}_1\tilde{I}_2\cos(2(\tilde{\theta}_1 + \tilde{\theta}_2)) \\
& + 4\tilde{I}_1^{1/2}\tilde{I}_2^{1/2}\big(-2\delta + A(\tilde{I}_1 + \tilde{I}_2)\big)\sin(\tilde{\theta}_1 + \tilde{\theta}_2)\Big) + O(\varepsilon^2).
\end{aligned} \tag{12.32}$$

Next make the symplectic change:

$$\tilde{I}_1 = \tfrac{1}{2}(I_1 + I_2), \qquad \tilde{I}_2 = \tfrac{1}{2}(-I_1 + I_2),$$
$$\tilde{\theta}_1 = \theta_1 + \theta_2, \qquad \tilde{\theta}_2 = -\theta_1 + \theta_2,$$

arriving at the Hamiltonian:

$$H = -\omega I_1 + \frac{\varepsilon}{16} H_1 + O(\varepsilon^2), \tag{12.33}$$

where

$$H_1 = -AI_1^2 + 8\delta I_2 + 3AI_2^2 + A(I_1^2 - I_2^2)\cos 4\theta_2 \\ +4(AI_2 - 2\delta)(I_2^2 - I_1^2)^{1/2}\sin 2\theta_2. \tag{12.34}$$

Note that $I_1 = -\Gamma_1$ and $I_2 \geq 0$.

The equations of motion are:

$$\dot{I}_1 = O(\varepsilon^2), \qquad\qquad \dot{I}_2 = \frac{\varepsilon}{16}\frac{\partial H_1}{\partial \theta_2} + O(\varepsilon^2),$$
$$\dot{\theta}_1 = \omega - \frac{\varepsilon}{16}\frac{\partial H_1}{\partial I_1} + O(\varepsilon^2), \qquad \dot{\theta}_2 = -\frac{\varepsilon}{16}\frac{\partial H_1}{\partial I_2} + O(\varepsilon^2). \tag{12.35}$$

Next compute the section map in the energy level $H = 0$. In that level $I_1 = O(\varepsilon)$. Take the cross section to be $\theta_1 = 0$ and use I_2, θ_2 as coordinates in that section. For ethic reasons we drop the subscript 2 in the following. The first return time is $T = 2\pi/\omega + O(\varepsilon)$ and so the section map is $\mathcal{P} : (I, \theta) \to (I^*, \theta^*)$ with

$$I^* = I + \nu\frac{\partial L}{\partial \theta} + O(\nu^2),$$
$$\theta^* = \theta - \nu\frac{\partial L}{\partial I} + O(\nu^2), \tag{12.36}$$

where $\nu = \varepsilon\pi/(8\omega)$ and

$$L = 8\delta I(1 - \sin 2\theta) + AI^2(3 - \cos 4\theta + 4\sin 2\theta). \tag{12.37}$$

We will treat L like a generating function for $\mathcal{P}$. Note that L is π–periodic in θ. Treating ν as small is equivalent to treating ε as small.

The map $\mathcal{P}$ is a symplectic diffeomorphism (symplectomorphism) of a special form. We have written this diffeomorphism in this form to illustrate its connection to the following Hamiltonian differential equations

$$\frac{dI}{d\nu} = \frac{\partial L}{\partial \theta}, \qquad \frac{d\theta}{d\nu} = -\frac{\partial L}{\partial I}. \tag{12.38}$$

For fixed ν the solutions of these equations is of the form (12.36) with possible different $O(\nu^2)$ terms. Since in general it is easier to change variables in a differential equation than in a diffeomorphism we shall exploit this form of the equations.

The point $I = 0$ is a fixed point for the diffeomorphism $\mathcal{P}$ in (12.36) and an equilibrium point for the equations (12.38). Rewriting (12.37) we have

$$L = 16\delta I \sin^2(\theta - \pi/4) + 8AI^2 \sin^4(\theta + \pi/4). \qquad (12.39)$$

Now we introduce coordinates q and p such that

$$16\delta I \sin^2(\theta - \pi/4) = \alpha q^2, \qquad 8AI^2 \sin^4(\theta + \pi/4) = \beta p^4,$$

where $\alpha = 8\delta$ and $\beta = A/4$. That is, we define the symplectic change:

$$I = \frac{1}{2}(q^2 + p^2), \qquad \theta = \tan^{-1}\left(\frac{p+q}{p-q}\right),$$

whose inverse is:

$$q = \pm\sqrt{I}\left(\cos\theta - \sin\theta\right), \qquad p = \pm\sqrt{I}\left(\cos\theta + \sin\theta\right).$$

Applying this change to L gives

$$L = 8\delta q^2 + 2Ap^4. \qquad (12.40)$$

The terms inside $O(\nu^2)$ of (12.36) are analytic functions in q and p; indeed they are polynomials in the two coordinates.

Scale q and p as follows:

$$q \to \frac{q}{16\delta\sqrt{\delta A}}, \qquad p \to \frac{p}{2\sqrt{\delta A}},$$

which is a symplectic change with multiplier $32\delta^2 A$ so that

$$L = \frac{1}{2}(q^2 + p^4). \qquad (12.41)$$

This reminds us of the lemniscate coordinates discussed in Section 8.5 and used previously in Section 12.4. Change to (K, κ) coordinates by

$$q = -K^{2/3}cl\left(\kappa\right), \qquad p = K^{1/3}sl\left(\kappa\right),$$

which is symplectic with multiplier $3/2$. So

$$L = \frac{3}{4}K^{4/3}.$$

In these coordinates the map (12.36) becomes $\mathcal{P} : (K, \kappa) \to (K^*, \kappa^*)$ where

$$K^* = K + \nu\frac{\partial L}{\partial \kappa} + O(\nu^2) = K + O(\nu^2)$$

$$\kappa^* = \kappa - \nu\frac{\partial L}{\partial K} + O(\nu^2) = \kappa - \nu K^{1/3} + O(\nu^2).$$

On the annulus $\mathcal{A} = \{1 < K < 2\}$ the map $\mathcal{P}$ is smooth and is a twist map. So Moser's invariant curve theorem implies that there are invariant curves in $\mathcal{A}$ for ν small.

When

$$\mu = \mu_1 = \frac{1}{2}(1 - \sqrt{69}/9) \approx 0.0385209$$

the exponents of the libration point $\mathcal{L}_4$ of the restricted 3-body problem are two pair of pure imaginary numbers. Schmidt (1990) put the Hamiltonian of the restricted 3-body problem at $\mathcal{L}_4$ into the normal form (12.25) with

$$\omega = \frac{\sqrt{2}}{2}, \qquad \delta = 1, \qquad A = \frac{59}{864}.$$

The value for A agrees with the independent calculations in Niedzielska (1994) and Goździewski and Maciejewski (1998). It differs from the numeric value given in Markeev (1978). These quantities in a different coordinate system were also computed by Deprit and Henrard (1968). By these considerations and calculations we have the following.

Proposition 12.9.1. *The libration point $\mathcal{L}_4$ of the restricted 3-body problem is stable when $\mu = \mu_1$.*

The problem of showing stability in the case of $1 : -1$ resonance has been treated by several authors, but some of the arguments are in question. See the paper by Lerman and Markova (2009) for a critical review of the literature and a detailed exhausting proof of this stability result. The simpler proof given here is from Meyer et al. (2012).

12.10 Stability of Fixed Points

The study of the stability of a periodic solution of a Hamiltonian system of two degrees of freedom can be reduced to the study of the Poincaré map in an energy level (i.e., level surface of the Hamiltonian). We summarize some results and refer the reader to the Problems or Meyer (1971) or Cabral and Meyer (1999) for the details. The proofs for the results given below are similar to the proofs given above.

We consider diffeomorphisms of the form

$$F : N \subset \mathbb{R}^2 \to \mathbb{R}^2 : z \to f(z), \tag{12.42}$$

where N is a neighborhood of the origin in $\mathbb{R}^2$, and F is a smooth function such that

$$F(0) = 0, \qquad \det \frac{\partial F}{\partial z}(z) \equiv 1.$$

The origin is a fixed point for the diffeomorphism because $F(0) = 0$, and it is orientation-preserving and area-preserving because $\det \partial F/\partial z \equiv 1$. This map should be considered as the Poincaré map associated with a periodic solution of a two degree of freedom Hamiltonian system.

The fixed point 0 is stable if for every $\epsilon > 0$ there is a $\delta > 0$ such that $|F^k(z)| \leq \epsilon$ for all $k \in \mathbb{Z}$ whenever $|z| \leq \delta$. The fixed point is unstable if it is not stable.

The linearization of this map about the origin is $z \to Az$ where A is the 2×2 matrix $(\partial f / \partial x)(0)$. The eigenvalues λ, λ^{-1} of A are called the multipliers of the fixed point. There are basically four cases: (i) hyperbolic fixed point with multipliers real and $\lambda \neq \pm 1$, (ii) elliptic fixed point with multipliers complex conjugates and $\lambda \neq \pm 1$, (iii) shear fixed point with $\lambda = +1$ and A is not diagonalizable, and (iv) flip fixed point with $\lambda = -1$ and A is not diagonalizable.

Proposition 12.10.1. *A hyperbolic fixed point is unstable.*

In the hyperbolic case one need only to look at the linearization; in the other case one must look at higher-order terms. In the elliptic case we can change to action–angle coordinates (I, ϕ) so that the map $F : (I, \phi) \to (I', \phi')$ is in normal form up to some order. In the elliptic case the multipliers are complex numbers of the form $\lambda^{\pm 1} = \exp \pm \omega i \neq \pm 1$.

Proposition 12.10.2. *If $\lambda^{\pm 1} = \exp \pm 2\pi i/3$ (the multipliers are cube roots of unity) the normal form begins*

$$I' = I + 2\alpha I^{3/2} \sin(3\phi) + \cdots, \phi' = \phi \pm (\pi/3) + \alpha I^{1/2} \cos(3\phi) + \cdots.$$

If $\alpha \neq 0$ the fixed point is instable.

If $\lambda^{\pm 1} = \pm i$ (the multipliers are fourth roots of unity) the normal form begins

$$I' = I + 2\alpha I^2 \sin(4\phi) + \cdots, \phi' = \phi \pm \pi/2 + \{\alpha \cos(4\phi) + \beta\} I + \cdots.$$

If $\alpha > \beta$ the fixed point is unstable, but if $\alpha < \beta$ the fixed point is stable.

If λ is not a cube or fourth root of unity then the normal form begins

$$I' = I + \cdots, \phi' = \phi \pm \omega + \beta I + \cdots.$$

If $\beta \neq 0$ then the fixed point is stable.

Proposition 12.10.3. *For a shear fixed point the multipliers are both $+1$ and A is not diagonalizable. The first few terms of the normal form $F :$ $(u, v) \to (u', v')$ are*

$$u' = u \pm v - \cdots, \quad v' = v - \beta u^2 + \cdots.$$

If $\beta \neq 0$ then the fixed point is unstable.

Proposition 12.10.4. *For a flip fixed point the multipliers are both -1 and A is not diagonalizable. The first few terms of the normal form $F : (u, v) \to (u', v')$ are*

$$u' = -u - v + \cdots, \quad v' = -v + \beta u^3 + \cdots.$$

If $\beta > 0$ the fixed point is stable and if $\beta < 0$ the fixed point is unstable.

12.11 Applications to the Restricted Problem

In Chapter 9, a small parameter was introduced into the restricted problem in three ways. First the small parameter was the mass ratio parameter μ, second the small parameter was a distance to a primary, and third the small parameter was the reciprocal of the distance to the primaries.

In all three cases an application of the invariant curve theorem can be made. Only the first and third are given here, inasmuch as the computations are easy in these cases.

12.11.1 Invariant Curves for Small Mass

The Hamiltonian of the restricted problem (4.1) for small μ is

$$H = \frac{\|y\|^2}{2} - x^T K y - \frac{1}{\|x\|} + O(\mu).$$

For $\mu = 0$ this is the Hamiltonian of the Kepler problem in rotating coordinates. Be careful that the $O(\mu)$ term has a singularity at the primaries. When $\mu = 0$ and Delaunay coordinates are used, this Hamiltonian becomes

$$H = -\frac{1}{2L^3} - G$$

and the equations of motion become

$$\dot{\ell} = 1/L^3, \qquad \dot{L} = 0,$$

$$\dot{g} = -1, \qquad \dot{G} = 0.$$

The variable g, the argument of the perihelion, is an angular variable. $\dot{g} = -1$ implies that g is steadily decreasing from 0 to -2π and so $g \equiv 0 \bmod 2\pi$ defines a cross-section. The first return time is 2π. Let ℓ, L be coordinates in the intersection of the cross-section $g \equiv 0$ and the level set $H = $ constant. The Poincaré map in these coordinates is

$$\ell' = \ell + 2\pi/L^3, \qquad L' = L.$$

Thus when $\mu = 0$ the Poincaré map in the level set is a twist map. By the invariant curve theorem some of these invariant curves persist for small μ.

12.11.2 The Stability of Comet Orbits

Consider the Hamiltonian of the restricted problem scaled as was done in Section 9.5 in the discussion of comet orbits, i.e., the Hamiltonian 9.7. In Poincaré variables it is

$$H = -P_1 + \frac{1}{2}(Q_2^2 + P_2^2) - \epsilon^3 \frac{1}{2P_1^2} + O(\epsilon^5),$$

where Q_1 is an angle defined modulo 2π, P_1 is a radial variable, and Q_1, P_1 are rectangular variables. For typographical reasons drop, but don't forget, the $O(\epsilon^5)$ terms. The equations of motion are

$$\dot{Q}_1 = -1 + \epsilon^3/P_1^3, \qquad \dot{P}_1 = 0,$$

$$\dot{Q}_2 = P_2, \qquad\qquad \dot{P}_2 = -Q_2.$$

The circular solutions are $Q_2 = P_2 = 0 + O(\epsilon^5)$ in these coordinates. Translate the coordinates so that the circular orbits are exactly $Q_2 = P_2 = 0$; this does not affect the displayed terms in the equations. The solutions of the above equations are

$$Q_1(t) = Q_{10} + t(-1 + \epsilon^3/P_1^3), \qquad P_1(t) = P_{10},$$

$$Q_2(t) = Q_{20}\cos t + P_{20}\sin t, \qquad P_2(t) = -Q_{20}\sin t + P_{20}\cos t.$$

Work near $P_1 = 1, Q_2 = P_2 = 0$ for ϵ small. The time for Q_1 to increase by 2π is

$$T = 2\pi/\mid -1 + \epsilon^3/P_1^3 \mid = 2\pi(1 + \epsilon^3 P_1^{-3}) + O(\epsilon^6).$$

Thus

$$Q' = Q_2(T) = Q\cos 2\pi(1 + \epsilon^3 P_1^{-3}) + P\sin 2\pi(1 + \epsilon^3 P_1^{-3})$$
$$= Q + \nu P P_1^{-3} + O(\nu^2),$$

$$P' = P_2(T) = -Q\sin 2\pi(1 + \epsilon^3 P_1^{-3}) + P\cos 2\pi(1 + \epsilon^3 P_1^{-3})$$
$$= -\nu Q P_1^{-3} + P + O(\nu^2),$$

where $Q = Q_{20}, P = P_{20}$, and $\nu = 2\pi\epsilon^3$. Let $H = 1$, and solve for P_1 to get

$$P_1 = -1 + \frac{1}{2}(Q^2 + P^2) + O(\nu),$$

and hence

$$P_1^{-3} = -1 - \frac{3}{2}(Q^2 + P^2) + O(\nu),$$

Substitute this back to get

$$Q' = Q + \nu P(-1 - \tfrac{3}{2}(Q^2 + P^2)) + O(\nu^2)$$

$$P' = P - \nu Q(-1 - \tfrac{3}{2}(Q^2 + P^2)) + O(\nu^2).$$

This is the section map in the energy surface $H = 1$. Change to action–angle variables, $I = (Q^2 + P^2)/2$, $\phi = \tan^{-1}(P/Q)$, to get

$$I' = I + O(\nu^2), \qquad \phi' = \phi + \nu(-1 - 3I) + O(\nu^2).$$

This is a twist map. Thus, the continuation of the circular orbits into the restricted problem is stable.

12.12 Problems

1. Let F be a diffeomorphism defined in a neighborhood O of the origin in $\mathbb{R}^m$, and let the origin be a fixed point for F. Let V be a smooth real-valued function defined on O, and define $\Delta V(x) = V(F(x)) - V(x)$.
   a) Prove that if the origin is a minimum for V and $\Delta V(x) \leq 0$ on O, then the origin is a stable fixed point.
   b) Prove that if the origin is a minimum for V and $\Delta V(x) < 0$ on $O \backslash \{0\}$, then the origin is an asymptotically stable fixed point.
   c) State and prove the analog of Chetaev's theorem.
   d) State and prove the analog of Lyapunov's instability theorem.

2. Let $F(x) = Ax$ and $V(x) = x^T S x$, where A and S are $n \times n$ matrices, and S is symmetric.
   a) Show that $\Delta V(x) = x^T R x$, where $R = A^T S A - S$.
   b) Let $\mathcal{S}$ be the linear space on all $m \times m$ symmetric matrices and $\mathcal{L} = \mathcal{L}_A : \mathcal{S} \to \mathcal{S}$ be the linear map $\mathcal{L}(S) = A^T S A - S$. Show that $\mathcal{L}$ is invertible if and only if $\lambda_i \lambda_j \neq 1$ for all $i, j = 1, \ldots, m$, where $\lambda_1, \ldots, \lambda_m$ are the eigenvalues of A.
   (Hint: First prove the result when $A = \mathrm{diag}(\lambda_1, \ldots, \lambda_m)$. Then prove the result when $A = D + \epsilon N$, where D is simple (diagonalizable), and N is nilpotent, $N^m = 0$, $SN = NS$, and ϵ is small. Use the Jordan canonical form theorem to show that A can be assumed to be $A = D + \epsilon N$.)
   c) Let A have all eigenvalues with absolute value less than 1. Show that $S = \sum_0^\infty (A^T)^i R A^i$ converges for any fixed R. Show S is symmetric if R is symmetric. Show S is positive definite if R is positive definite. Show that $\mathcal{L}(S) = -R$, so, $\mathcal{L}^{-1}$ has a specific formula when all the eigenvalues of A have absolute value less than 1.

3. Let $F(x) = Ax + f(x)$, where $f(0) = \partial f(0)/\partial x = 0$.
   a) Show that if all the eigenvalues of A have absolute value less than 1, then the origin is asymptotically stable. (Hint: Use Problems 1 and 2.)
   b) Show that if A has one eigenvalue with absolute value greater than 1 then the origin is a positively unstable fixed point.

4. Let $r = 1, s = 0$, and $h(I) = \beta I$, $\beta \neq 0$ in formulas of the invariant curve theorem.
   a) Compute F^q, the qth iterate of F, to be of the form $(I, \phi) \to (I", \phi")$ where
   $$I" = I + O(\epsilon), \qquad \phi" = \phi + q\omega + q\beta I + O(\epsilon).$$
   b) Let $2\pi p/q$ be any number between $\omega + \beta a$ and $\omega + \beta b$, so $2\pi p/q = \omega + \beta I_0$ where $a < I_0 < b$. Show that there is a smooth curve $\Gamma_\epsilon = \{(I, \phi) : I = \Phi(\phi, \epsilon) = I_0 + \cdots\}$ such that F^q moves points on Γ only in the radial direction, i.e., $\Phi(\phi)$ satisfies $\phi" - \phi - 2\pi p = 0$. (Hint: Use the implicit function theorem.)

c) Show that because F^q is area-preserving, $\Gamma \cap F^q(\Gamma)$ is nonempty, and the points of this intersection are fixed points of F^q or q-periodic points of F.

5. Consider the forced Duffing's equation with Hamiltonian

$$H = \frac{1}{2}(q^2 + p^2) + \frac{\gamma}{4}q^4 + \gamma^2 \cos \omega t,$$

where ω is a constant and $\gamma \neq 0$ is considered as a small parameter. This Hamiltonian is periodic with period $2\pi/\omega$ for small ϵ. If $\omega \neq 1, 2, 3, 4$, the system has a small (order γ^2) $2\pi/\omega$ periodic solution called the harmonic. The calculations in Section 10.3 show the period map was shown to be

$$I' = I + O(\gamma),$$

$$\phi' = \phi - 2\pi/\omega - (3\pi\gamma/2\omega)I + O(\gamma^2),$$

where the fixed point corresponding to the harmonic has been moved to the origin. Show that the harmonic is stable.

6. Using Poincaré elements show that the continuation of the circular orbits established in Section 8.1 (Poincaré orbits) are of twist type and hence stable.

7. Consider the various types of fixed points discussed in Section 11.1 and prove the propositions in 12.10. That is:
   a) Show that extremal points are unstable.
   b) Show that 3-bifurcation points are unstable.
   c) Show that k-bifurcation points are stable if $k \geq 5$.
   d) Transitional and 4-bifurcation points can be stable or unstable depending on the case. Figure out which case is unstable. (The stability conditions are a little harder.) See Meyer (1971) or Cabral and Meyer (1999).

13. Variational Techniques

Recently, an interesting example of a highly symmetric T-periodic solution of the Newtonian planar 3-body problem with equal masses was rediscovered by Chenciner and Montgomery (2000). This particular family of solutions was investigated earlier by Moore (1993) using numerical techniques, however the methods of Chenciner and Montgomery were analytical, using global variational methods applied to families of curves which encode the 12th order symmetry group $D_3 \times \mathbb{Z}_2$. As it turned out, the method provided a breakthrough in analyzing global families of periodic orbits of the N-body problem. Chenciner–Montgomery's orbit is referred to as the figure eight as the configuration component of the orbit is realized by the three masses chasing one another around a fixed figure eight curve in the plane with the masses distinguished on the eight by a $T/3$ phase shift. Such solutions are now referred to as choreographies after C. Simó coined this expression to describe the amazing dance that these particles maintain under their mutual gravitational attraction. Apparently the first such choreography discovered was by Lagrange almost 240 years ago when he discovered the equilateral triangular solutions of the 3-body problem. The uniformly rotating Lagrangian solutions are a choreography on a circle. Many other choreographies with arbitrary number of equal masses have been identified by numerical minimization for symmetric and nonsymmetric curves; Simó (2002). The figure eight is distinguished by its stability, as well as having a purely analytic argument for its existence. A continuous family may be chosen so that either energy or period may be used to parameterize the family.

The figure eight solutions are not close to any integrable cases of the 3-body problem. Based on numerical evidence by Simó (2002), Chenciner and Montgomery (2000) have announced orbital (elliptic) stability of this orbit and Roberts (2007) has given a convincing numerical proof of elliptic linear stability based on a geometric argument that factors the monodromy matrix into the sixth power of the "first hit map." The first hit map arises by considering the symmetry reduced monodromy matrix acting on an isoenergetic cross-section modulo a discrete symmetry group.

K.R. Meyer, D.C. Offin, *Introduction to Hamiltonian Dynamical Systems and the N-Body Problem*, Applied Mathematical Sciences 90, DOI 10.1007/978-3-319-53691-0_13

The analytic variational argument constructs one-twelfth of the orbit by a reduced variational principle on a zero angular momentum level set, with the remaining segments of the orbit constructed from reflections and rotations. One of the main analytical obstacles to overcome in this argument is to show that the orbit segment in a fundamental domain has no collision singularities. The variational technique applied to the Newtonian action functional fairly easily yields a minimizing curve, however the real possibility arises that the minimizing curve includes collision as was understood already by Poincaré in his research on the 3-body problem.

In Section 7.7 we discussed reduction by symmetries and the reduced space that carries a symplectic structure. In the setting of the figure eight, we have occasion to study this construction more closely. The zero momentum level set can be realized as a cotangent bundle by projection along an $SO(2)$ symmetry acting diagonally on position and momenta of the phase space resulting in the reduced space $T^*(M/SO(2))$. This space gives the setting of the variational problem. More recent methods for constructing the eight include a topological shooting method of Moeckel (2007) based on the Conley index. One tantalizing feature of this family is that perturbation analysis cannot be extended from known solutions to include this family. Moreover the stability of the orbit cannot be deduced from continuation arguments except through numerical results; see for example Simó (2002) and Roberts (2007).

Because this groundbreaking work, other families of symmetric periodic solutions have been found using a combination of variational methods Chenciner and Venturelli (2000), Chenciner and Montgomery (2000), Ferrario and Terracini (2004), Cabral and Offin (2008), Chen (2001) which avoid collision singularities. Moreover there have now been hundreds of choreographies found by numerical techniques; see Simó (2002). The question of collision for solutions defined by variational methods in the action functional with Newtonian potential is an important component of the analysis, that has been resolved to large extent by a result of Marchal (2002) that was used by Ferrario and Terracini (2004) to investigate more thoroughly the interplay between the symmetry group and the variational method for collision-free solutions.

Our goal in this chapter is to explain the variational construction of this type of symmetric orbit, and to show that the variational argument can be used to find other periodic orbits, having the same braid structure as the orbit of Chenciner–Montgomery, but which due to the symmetry invariance may be seen in fact to be hyperbolic (in the nondegenerate setting). This last work brings in the ideas mentioned earlier in Section 5.6 on the Maslov index of sections of Lagrange planes along a closed phase curve.

13.1 The N-Body and the Kepler Problem Revisited

We recall the basic ingredients of the N-body problem; see Chapter 3. We pay particular attention to the variational aspects of the periodic solutions of the N-body problem. The configuration $\mathbf{r} = (r_1, \ldots, r_N)$ describes spatial positions of N masses $m_1, \ldots, m_N$ where $r_i \in \mathbb{R}^3$. Interaction between the masses is determined by the Newtonian potential function

$$V(\mathbf{r}) = -\sum_{i<j} \frac{m_i m_j}{\|r_i - r_j\|}$$

on the set of noncollision configurations (where $r_i \neq r_j, \ i \neq j$). The Hamiltonian for the N-body problem is the sum of kinetic plus potential

$$H(r_1, \ldots, r_N, p_{r_1}, \ldots, p_{r_N}) = \sum_i^N \frac{1}{2m_i} \|p_{r_i}\|^2 + V(\mathbf{r}).$$

We study the Hamiltonian vector field X_H in coordinates

$$\dot{r}_i = \frac{1}{m_i} p_{r_i} \qquad \dot{p}_{r_i} = \frac{\partial U(\mathbf{r})}{\partial r_i}, \quad i = 1, \ldots, N \tag{13.1}$$

where the function $U(q) = -V(q)$ is called the force function or self-potential. The vector field X_H given by Equations (13.1) generates a noncomplete flow ϕ_t on the phase space T^*M, some solutions will have collision singularities in finite time.

Let us recall some details from Section 7.7 on symmetries and reduction. The Galilean group of translations in $\mathbb{R}^3$, as well as the group of simultaneous rotation of all masses about a fixed point may be lifted to the phase space as a symplectic action that leaves H invariant. By Noether's theorem 3.4.1, the integrals corresponding to the group of translations of the motion are the total linear momentum,

$$l = p_1 + p_2 + p_3.$$

whereas those of the rotation group yield the angular momentum vector $\mathbf{J} = \sum^N r_i \times p_{r_i}$. Because H is translation invariant, we recognize that this means that any periodic solutions for Equations (13.1) will occur in continuous families parameterized by the elements of the group. For the variational theory, it is convenient to remove the translation degeneracy by fixing the total linear momentum to be zero, and identifying the configurations that differ by a translation. This fixes the center of mass at the origin of $\mathbb{R}^3$, and we are led to consider the configuration manifold modulo translations

$$M = \left\{ \mathbf{r} = (r_1, \ldots, r_N) \mid \sum m_i r_i = 0, r_i \neq r_j, \ i \neq j \right\}. \tag{13.2}$$

The angular momentum level set $\mathbf{J}^{-1}(c)$ is invariant under the flow, and the reduced space $\mathbf{J}^{-1}(c)/G_c$, where $G_c = SO(2)$ is the subgroup of $SO(3)$

which fixes $\mathbf{J}^{-1}(c)$, is a symplectic space equipped with the flow of the reduced Hamiltonian vector field. For the planar 3-body problem which is discussed more fully in the next section, this reduced space is six dimensional. After fixing a value h of the reduced energy, we are situated on a five-dimensional manifold $\Xi(h,c)$, which carries the essential dynamics of the planar 3-body problem.

We recall the fact that the planar N-body problem always admits uniformly rotating solutions that generalize the circular rotational solutions of the Kepler equation. These solutions are called relative equilibria. We recall the construction of these orbits from Chapter 3. We let $I(q)$ denote the moment of inertia of the configuration

$$I(q) = \sum_{i}^{N} m_i \langle r_i, r_i \rangle. \tag{13.3}$$

Recall (Section 3.5) the configuration q_0 is a central configuration if q_0 is a critical point of $U(q)$ restricted to the level set $I^{-1}(\frac{1}{2})$. Solutions to the N-body equations (13.1) of the form

$$q(t) = a(t)q_0, \quad a(t) \in \mathbb{C}, \quad q_0 \in M \tag{13.4}$$

for fixed $q_0 = (r_1, \ldots, r_N) \in M$, require that

$$\ddot{a} = -\frac{a}{|a|^3} \frac{U(q_0)}{I(q_0)} = -\frac{a}{|a|^3} \frac{U(q_0)}{\mathbf{r}^2},$$

where $|q_0|^2 = I(q_0) = \mathbf{r}^2$. This differential equation for $a(t)$ can be scaled as follows (using $\mathbf{r}$ as scaling parameter).

$$\mathbf{r}\ddot{a} = -\frac{\mathbf{r}a}{(\mathbf{r}|a|)^3}\mathbf{r}U(q_0) = -\frac{\mathbf{r}a}{(\mathbf{r}|a|)^3}U\left(\frac{q_0}{\|q_0\|}\right).$$

Therefore $a(t)$ must satisfy the Kepler equation

$$\ddot{a}(t) = -\frac{a}{|a|^3}\tilde{U}(q_0), \quad \tilde{U}(q) = U\left(\frac{q}{\|q\|}\right). \tag{13.5}$$

All central configurations q_0 admit homothetic solutions $q(t) = a(t)q_0$, $a(t) \in \mathbb{R}$ satisfies (13.5). Such homothetic solutions end in total collapse. Ejection orbits are the time reversal of collision orbits. Coplanar central configurations admit in addition homographic solutions where each of the N masses executes a similar Keplerian ellipse of eccentricity e, $0 \le e \le 1$. When $e = 1$ the homographic solutions degenerate to a homothetic solution that includes total collapse, together with a symmetric segment corresponding to ejection. When $e = 0$, the relative equilibrium solutions are recovered consisting of uniform circular motion for each of the masses about the common center of mass.

Turning to the variational properties of homographic solutions (13.4), we define the action for absolutely continuous T-periodic curves $q(t)$ in the collision less configuration manifold M,

$$\mathcal{A}_T(q) = \int_0^T \sum \frac{1}{2m_i} \|p_{r_i}\|^2 + U(\mathbf{r})dt, \quad q \in \Lambda_T. \tag{13.6}$$

where the function space of such curves with square integrable derivatives is denoted by Λ_T. We use the period T as a parameter in this functional, inasmuch as we wish to compare solutions with different periods.

Because the work of Sundman (1913) on the 3-body problem, it is known that the action $\mathcal{A}_T$ stays finite along collision orbits of Equation (13.1). For this reason, special care must be taken when considering the variational problem. Due to the exclusion of collision points in M, the space Λ_T is not complete. If the action stays finite along a sequence of curves tending to collision, then such a limiting curve on the boundary of the functional space Λ_T could provide the minimizing loop. This is exactly the situation discovered by Gordon (1970) for the Kepler problem.

Gordon found that for fixed period T, the actual minimizing solutions are arranged in families of ellipses, parameterized by eccentricity e, including the degenerate ellipse when eccentricity $e = 1$.

All of these elliptical paths have the same action. The collision ejection orbit appears as a limit case on the boundary of the family of curves, and in fact realizes the minimizing action! He also found that the minimum value of the action functional (realized by elliptical families having period T) could be computed as

$$\mathcal{A}_T^K(q) = \int_0^T \frac{1}{2}\|\dot{q}\|^2 + \frac{\mu}{\|q\|}dt = 3(2\pi)^{1/3}(\mu)^{2/3}T^{1/3}, \tag{13.7}$$

where we have added the superscript K to denote the Keplerian action integral. The homographic and homothetic solutions $q(t) = a(t)q_0$ (see Equation (13.5)) share the property in which the action $\mathcal{A}_T(q)$ along such a trajectory q, can be determined as uncoupled Keplerian orbits. Substituting the homothetic curves into the action functional (13.6) we find that

$$\mathcal{A}_T(a(t)q_0) = \mathbf{r}^2 \int_0^T \frac{1}{2}\dot{a}^2 + \frac{1}{|a(t)|}\frac{U(q_0)}{\mathbf{r}^2}dt = 3(2\pi)^{1/3}\left(\tilde{U}(q_0)\right)^{2/3}T^{1/3},$$

which we recognize as a scaled version of the Keplerian action (13.7).

13.2 Symmetry Reduction for Planar 3-Body Problem

In this section we explain the unique geometry of the variational problem whose solutions are the figure eight choreographies. The planar 3-body problem has 6 well-known integrals that reduce the dimension of the phase space from 12 to 6. In addition the quotient by the rotational group $SO(2)$ reduces the dimension by one more. We discuss the topology and geometry of the reduced configuration space, which is crucial for the variational construction of the figure eight.

The 3-body problem describes the dynamical system which consists of three masses interacting through their mutual gravitational attraction. The planar 3-body problem restricts the three masses to the coordinate plane $z = 0$ in $\mathbb{R}^3$. Let $q_i \in \mathbb{R}^2$ denote the position of the ith body whose mass is m_i, and whose velocity is v_i. Interpreting this as a complex variable, we set $q = (q_1, q_2, q_3) \in \mathbb{C}^3$ which describes the configuration of the system and $p = (m_1 v_1, m_2 v_2, m_3 v_3) \in \mathbb{C}^3$ the momentum vector. The configuration of the system should be thought of as an oriented triangle in the plane, with the three masses at the vertices. The dynamical system is governed by Hamilton's equations (13.1) with $N = 3$. The force function is

$$U(q) = \frac{m_1 m_2}{\|r_1 - r_2\|} + \frac{m_1 m_3}{\|r_1 - r_3\|} + \frac{m_2 m_3}{\|r_2 - r_3\|}. \tag{13.8}$$

Fixing the center of mass at the origin restricts the configuration and the momentum to 4-dimensional subspaces of $\mathbb{C}^3$ respectively. The action of the rotation group $SO(2)$ on phase space

$$R \cdot (q, p) = (Rq_1, Rq_2, Rq_3, Rp_1, Rp_2, Rp_3), \quad R = e^{i\theta} \tag{13.9}$$

gives a further reduction on the level set of the angular momentum vector $\mathbf{J}^{-1}(c)$.

The Hamiltonian itself is invariant under the flow, and setting the value $H = h$, we reduce the dimension by one more. Finally, we quotient the resulting energy-momentum level set by the action of $SO(2)$. This reduces the dimension by one as well, leaving the five-dimensional quotiented energy–momentum set $\Xi(h, \omega)$, Moeckel (1988). In the quotiented manifold $\Xi(h, \omega)$, we identify all phase points that differ by a simultaneous rotation of configuration and momentum. The reduced configuration space is obtained by projecting out the momenta. We think of a point in the reduced configuration space as consisting of equivalence classes of oriented triangles in the plane, up to translation and rotation. We introduce a simple geometrical model of the reduced configuration space that facilitates the understanding of the variational problem.

In addition to the $SO(2)$ invariant kinetic and potential energies, the polar moment of inertia (13.3) can be used for understanding the reduced configuration space. The function $\sqrt{I} = r$ measures the size of the oriented triangle formed by the three masses. Notice that the system undergoes triple collision when $r = 0$.

The configuration space consists of all possible configurations

$$M = \{(r, q) \mid r \geq 0, I(q) = r^2, \sum m_i r_i = 0\}. \tag{13.10}$$

Inasmuch as we are interested in collision-free configurations only, we set

$$\tilde{M} = M/\{(r,q)|q_i = q_j, \text{ for } i \neq j\}. \tag{13.11}$$

The reduced configuration space is defined by identifying configurations that differ by a fixed rotation. This gives us the quotient space $M/SO(2)$. Before the removal of the collisions, the space of configurations (13.10) is homeomorphic to $\mathbb{R}^+ \times S^2$. This can be seen from the fact that the three-sphere, $I(q) = r^2$, fibers as a circle bundle over the two-sphere, using the Hopf fibration (Section 1.4). The shape sphere is defined as

$$\mathcal{S} = \left\{ q \mid I(q) = 1, \sum m_i r_i = 0 \right\} / S^1. \tag{13.12}$$

The double collisions correspond to rays emanating from the origin, and triple collision corresponds to the origin $q = 0$, because we have restricted the center of mass at the origin (Equation (13.10)). Each ray representing double collision configurations pierces the shape sphere in only one point. We should think of points of $\mathcal{S}$ describing equivalence classes of oriented triangles, normalized so that $I(q) = 1$.

The most important features of the shape sphere (13.12), as described in Moeckel (1988), are three longitudinal circles on the sphere, which meet at the north and south poles, and the meridian circle $\mathcal{E}$ of the equator. The longitudinal circles correspond to the three branches of oriented isosceles configurations, with the ith mass on the symmetry line. We denote these circles by M_i, $i = 1, 2, 3$.

At the north pole, the longitudinal circles M_i come together at the Lagrangian central configuration which is an equilateral triangle. There are two of these equilateral configurations, one at each pole, corresponding to a choice of orientation of the three masses. Each of the longitudinal circles M_i meet the equator in another central configuration, the Eulerian collinear configuration E_i, and also in a double collision point. The equator $\mathcal{E}$ describes the totality of collinear configurations. On the equator there are three double collision points, separated by three symmetric degenerate isosceles configurations E_i, which were mentioned above. For the 3-body problem, all of the central configurations, that is, solutions of the dynamical equations that correspond to a rigid rotation of the initial configuration, were discovered by Euler (collinear) and Lagrange (equilateral). These solutions exist for all choices of the masses m_i.

When the force function U is restricted to the shape sphere, the main features that were described above can be given a more analytical interpretation. The function $U|_{I=1}$ takes its absolute minimum at the two Lagrange central configurations. The Eulerian collinear points E_i are also critical points for this function, but are now saddle points. Finally, the force function takes its supremum $+\infty$ at the three double collision points on the equator.

The Hill's regions $C(h, \omega)$ of the reduced configuration space are the projected images of the energy–momentum sets $\Xi(h, \omega)$ onto the reduced configuration space $M/SO(2)$. The topological structure of these sets and their bifurcations give enormous insight into the kinds of orbits that can be imagined, and sometimes proven to exist mathematically. Without any restriction on energy, we mention here the fact that the zero momentum set $\omega^{-1}(0)$ projects onto the entire reduced configuration space. That is, if we restrict the total momentum to be zero, then any possible reduced configuration can be realized. This is not true for nonzero momentum in general by Moeckel (1988).

Finally, we describe the reflection symmetry of the shape sphere $\mathcal{S}$, which is important. This is the reflection symmetry about the equator which can be realized as follows. If we identify each q_i as a complex variable, then

$$M = \left\{ q = (q_1, q_2, q_3) \in \mathbb{C}^3 \mid \sum m_i q_i = 0 \right\}. \tag{13.13}$$

The reflection across the first coordinate axis on M is

$$\sigma(q) = \bar{q} = (\bar{q}_1, \bar{q}_2, \bar{q}_3), \tag{13.14}$$

and the fixed point set of σ is

$$\mathbf{Fix}\ \sigma = \{ q = (q_1, q_2, q_3) \mid q_i = \bar{q}_i \}. \tag{13.15}$$

When the reflection is dropped to the reduced configuration space $M/SO(2)$, the resulting map is called the reflection across the line of masses (also called the syzygy axis); see Meyer (1981b) and the fixed point set (13.15) projects onto the equatorial plane $\mathcal{E}$ of collinear configurations. The reduced symmetry σ restricted to the shape sphere $\mathcal{S}$ fixes the equator $\mathcal{E}$, and reflects the northern hemisphere onto the southern hemisphere. Without causing confusion, we refer henceforth to the reduced symmetry with the same notation as the reflection σ.

In the inertial frame, that is, the fixed coordinate system before quotienting the group $SO(2)$, we can visualize the reduced symmetry σ by first rotating the configuration into a position so that σ is represented by reflection across the first coordinate axis. After describing a general principle of reduction on the zero momentum set, we need to describe how this $\mathbb{Z}_2$ symmetry generated by σ on $M/SO(2)$ can be extended into the phase space by describing how it should act on velocity.

13.3 Reduced Lagrangian Systems

The dynamics of the problem were described in the last section by appealing to the Hamiltonian structure of the problem. In contrast, the variational structure is contained on the Lagrangian side. The Lagrangian for the 3-body problem is

$$L(q, v) = \frac{1}{2} \sum_{i=1}^{3} m_i v_i^2 + U(q),$$

where the force function U is the same as given in equation (13.8). The Hamiltonian equations (13.1) are equivalent to the Euler–Lagrange equations (1.30).

As explained in Chapter 1, the Euler–Lagrange equations (1.30) are also the equations for critical points of the action functional,

$$A(q) = \int_0^T L(q, \dot{q}) dt, \quad q \in H^1[0, T], \tag{13.16}$$

where the Sobolev space $H^1[0, T]$ consists of absolutely continuous parameterized arcs with values in the configuration manifold M and with $L^2[0, T]$ derivatives $\dot{q}$. This space is the completion of the space of C^2 curves that we considered in Chapter 1, and is necessary for the existence theory we use here. The boundary conditions are suppressed for the moment.

The variational principle can be reduced modulo translations as well as the reduction by rotations explained in the previous section. It turns out that a variational principle using only reduced configurations modulo translations and rotations is equivalent to restricting the momentum value at 0. To explain this equivalence, we make a slight detour and discuss a slightly more general version based on a Lie group G of symmetries acting on the configuration manifold.

Let us consider more generally, a C^3 mechanical Lagrangian system on the tangent bundle of a Riemannian manifold M,

$$L : TM \longrightarrow \mathbb{R}$$

$$L(q, v) = \frac{1}{2} K_q(v) + U(q),$$

where $K_q(v) = \|v\|_q^2$ is twice the kinetic energy, and $U : M \longrightarrow \mathbb{R}$ is the force function. We denote the Riemannian metric on $T_q M$ associated with the kinetic energy by $K_q(v, w)$ for tangent vectors $v, w \in T_q M$. We also assume that there is a Lie group G acting freely and properly on M, which is denoted $g(q) = g \cdot q$, for $g \in G$. In this case we have the principal bundle

$$\pi : M \longrightarrow M/G$$

The Lie algebra of G is denoted $\mathfrak{G}$, and the pairing between $\mathfrak{G}$ and $\mathfrak{G}^*$ is denoted $< \cdot, \cdot >$. Moreover, it is assumed that the group action lifts to the tangent bundle of M, by isometrics on TM,

$$g \cdot (q, v) = (g \cdot q, Tg \cdot v).$$

Finally, we assume that the Lagrangian L is G-invariant

$$g^*L = L, \quad g \in G.$$

In this setting the equivariant momentum map, which generalizes the angular momentum $\mathbf{J}(\mathbf{q}, \mathbf{p})$, is given by

$$\mathbf{J} : TM \longrightarrow \mathfrak{G}^*, \tag{13.17}$$

$$\langle \mathbf{J}(q, v), \xi \rangle = K_q(v, X_\xi), \quad \xi \in \mathfrak{G} \tag{13.18}$$

where $X_\xi(q)$ denotes the infinitesimal generator of the one-parameter subgroup action on M, associated with the Lie algebra element $\xi \in \mathfrak{G}$. Noether's theorem states that Equation (13.18) is a conservation law for the system.

We use the velocity decomposition in Saari (1988), but in this more general setting. For fixed momentum value $\mathbf{J}(q, v) = \mu$, the velocity decomposition is given by

$$v_q = hor_q v + ver_q v, \tag{13.19}$$

where,

$$ver_q v = X_\xi, \quad \text{and} \quad hor_q = v - ver_q v, \tag{13.20}$$

and $\xi = \xi_q \in \mathcal{G}$ is the unique Lie algebra element such that

$$\mathbf{J}(q, X_\xi) = \mu. \tag{13.21}$$

The uniqueness of the element $\xi \in G$ given by Equation (13.21) is given in the case of the N-body problem in Saari (1988), and the general case considered here may be found in Marsden (1992), Arnold (1990). or Section 7.7.

From Equations (13.18) through (13.21), it follows that the space of horizontal vectors

$$Hor_q = \{(q, v) | \mathbf{J}(q, v) = 0\} \tag{13.22}$$

and the space of vertical vectors

$$Ver_q = ker(T_q \pi) \tag{13.23}$$

are orthogonal complementary subspaces with respect to the Riemannian metric K_q. It is also clear, due to the equivariance of $\mathbf{J}$, that the horizontal vectors are invariant under the G-action on TM. The mechanical connection of Marsden (1992) is the principal connection on the bundle $\pi : M \longrightarrow M/G$,

$$\alpha : TM \longrightarrow \mathcal{G}, \quad (q, v) \mapsto \xi, \quad X_\xi = ver_q v. \tag{13.24}$$

In the setting of the planar N-body problem, the vertical component of the velocity is that velocity which corresponds to an instantaneous rigid rotation, and which has the angular momentum value μ.

For completeness, we summarize the argument in Arnold (1990) which gives the description of the reduced space $\mathbf{J}^{-1}(0)/G$.

Theorem 13.3.1. *The zero momentum set* $\mathbf{J}^{-1}(0)$, *modulo the group orbits, has a natural identification as the tangent bundle of the reduced configuration space,*

$$\mathbf{J}^{-1}(0)/G = T(M/G). \tag{13.25}$$

Proof. It suffices to notice that when $\mu = 0$, the vertical component of velocity is zero, $ver_q v = 0$. Thus the points of $\mathbf{J}^{-1}(0)$ naturally project onto points in $T(M/G)$ with the projection in the velocity component along the subspace (13.23) orthogonal to the horizontal space (13.22).

With the velocity decomposition given by Equations (13.19) through (13.21), it is now a simple matter to see how the Lagrangian drops on the momentum level set $\mathbf{J}^{-1}(0)$. First of all, the kinetic energy decomposes naturally, because (13.22) and (13.23) are orthogonal subspaces,

$$K_q(v) = K_q^{red}(v) + K_q^{rot}(v), \tag{13.26}$$

where

$$K_q^{red}(v) = K_q(hor_q v), \quad K_q^{rot}(v) = K_q(ver_q v). \tag{13.27}$$

This allows us to define the reduced Lagrangian on the reduced space $\mathbf{J}^{-1}(0)/G$,

$$L^{red} : T(M/G) \longrightarrow \mathbb{R}, \quad L^{red}(q, v) = K_q^{red}(\tilde{v}) + U(\tilde{q}), \tag{13.28}$$

where $(\tilde{q}, \tilde{v})$ is an arbitrary element of $\mathbf{J}^{-1}(0)$ that projects to $(q, v) \in T(M/G)$. It is important when considering the properties of extremals to consider the projected metric on $T(M/SO(2))$, which is called the reduced metric,

$$K_q^{red}(v, w) = K_q(\tilde{v}, \tilde{w}), \quad (v, w) \in T(M/SO(2)), \quad (\tilde{v}, \tilde{w}) \in Hor_q. \tag{13.29}$$

We have the reduced variational principle, based on the space $H^1[0, T]$ of parameterized curves in the reduced configuration space $M/SO(2)$.

$$A^{red}(x) = \int_0^T L^{red}(x, \dot{x}), \quad x \in H_T^1(M/SO(2)). \tag{13.30}$$

This action functional has the familiar property that critical points (with respect to certain boundary conditions that are suppressed here) correspond to solutions of the reduced Euler–Lagrange equations. Moreover, the action functional (13.30) is equivariant with respect to the $\mathbb{Z}^2$ symmetry generated by σ (Equation (13.14)), as well as the dihedral symmetry which we consider in the next section.

13.4 Discrete Symmetry with Equal Masses

The figure eight periodic orbit discovered by Chenciner and Montgomery (2000) has the discrete symmetry group $\mathbb{Z}_2 \times \mathcal{D}^3$. This symmetry drops to the reduced space $\mathbf{J}^{-1}(0)/SO(2)$. We have already described the $\mathbb{Z}_2$ symmetry on $M/SO(2)$ generated by reflection across the syzygy axis, σ in Equation (13.14). The elements of $\mathcal{D}^3$, on the other hand, are generated by interchanging two of the equal masses. We explain how the symmetry σ can be extended to a time-reversing symmetry on the reduced space $T(M/SO(2))$. For a general treatment of time reversing symmetries, see Meyer (1981b). Secondly we show that if $\lambda \in \mathcal{D}^3$, then the product $\sigma\lambda$ generates a time-reversing symmetry on reduced phase space. Moreover the $\mathbb{Z}_2$ subgroup generated by such a product has a fixed point set corresponding to the normal bundle of one of the meridian circles on the shape sphere.

We extend the reflection symmetry σ by isometries on the reduced space $T(M/SO(2))$,

$$\Sigma : T(M/SO(2)) \longrightarrow T(M/SO(2)), \quad (q, v) \mapsto (\sigma(q), -d\sigma(v)). \quad (13.31)$$

This symmetry leaves the reduced Lagrangian invariant, and reverses the symplectic form on $T(M/SO(2))$. By standard arguments is possible to see that the Hamiltonian flow ϕ_t is time reversible with respect to Σ

$$\Sigma\phi_t = \phi_{-t}\Sigma \quad (13.32)$$

and that the fixed point set is

$$\mathbf{Fix}(\Sigma) = \{(q, v) \mid q \in \mathcal{E}, (q, v) \perp \mathcal{E}\}, \quad (13.33)$$

where $\mathcal{E}$ denotes the equatorial circle of the shape sphere (13.12).

Next, we consider the $\mathcal{D}^3$ symmetry, generated by interchanging two of the masses. Using the complex notation established in Equation (13.13), we define the reflection symmetry on the reduced configuration space $M/SO(2)$,

$$\lambda_{1,2} : M/SO(2) \longrightarrow M/SO(2), \quad q = (q_1, q_2, q_3) \mapsto (q_2, q_1, q_3) \in \mathbb{C}^3, \quad (13.34)$$

which effects an interchange between masses m_1 and m_2. When the masses are equal this symmetry leaves the force function U (13.8) invariant, and also extends to the reduced space by isometrics,

$$\Lambda_{1,2} : T(M/SO(2)) \longrightarrow T(M/SO(2)), \quad (q, v) \mapsto (\lambda_{1,2}(q), d\lambda_{1,2}(v)). \quad (13.35)$$

The symmetry $\Lambda_{1,2}$ on the reduced space $T(M/SO(2))$ leaves the reduced Lagrangian L^{red} invariant, which implies that the symmetry takes orbits to orbits. Clearly, there are three generators of this type for the 3-body problem with equal masses. These three group elements form the generators of the dihedral group of order three $\mathcal{D}^3$.

The effect of the interchange symmetry $\lambda_{1,2}$ on the shape sphere $\mathcal{S}$, Equation (13.12), can be described as follows. The meridian circle M_3 intersects the equator in the Eulerian configuration E_3 (the mass m_3 in the middle) and the double collision point between masses m_1 and m_2. The reflection $\lambda_{1,2}$ rotates the shape sphere through angle π, about the axis through these two points, holding these two points fixed, so that the northern hemisphere is rotated onto the southern hemisphere.

Now we want to consider the group element $\sigma\lambda$. In the statement of the theorem below, the condition $(q,v) \perp M_3$ means that v, the principal part of the tangent vector (q,v), is orthogonal in the reduced metric (13.29) to $T_q M_3$.

Theorem 13.4.1. *The reflection* $\Sigma \cdot \Lambda_{1,2}$ *is a time-reversing symmetry on* $T(M/SO(2))$, *which leaves the Lagrangian* L^{red} *invariant and such that*

$$\mathbf{Fix}(\Sigma \cdot \Lambda_{1,2}) = \{(q,v) | (q,v) \perp M_3\} \, .$$

Proof. Whereas both symmetries σ and $\lambda_{1,2}$ were lifted to the tangent space by isometrics, only Σ is a time-reversing symmetry on $T(M/SO(2))$ (it reverses the canonical symplectic form) whereas $\Lambda_{1,2}$ is a symplectic symmetry (it fixes the symplectic form). The product therefore, is a time-reversing, antisymplectic symmetry. The fixed point set of $\Sigma\Lambda_{1,2}$ can be found by direct calculation. Recall our earlier observation, that the meridian circle M_3 is invariant under the symmetry $\lambda_{1,2}$. This implies that M_3 is fixed by the symmetry $\sigma\lambda_{1,2}$, because σ takes M_3^+ to M_3^-. Finally, it is not difficult to see using (13.31),(13.33) that any tangent vector $(q,v) \in T_q M_3$ is reversed by $\Sigma\Lambda_{1,2}$, and any normal vector $(q,v) \perp M_3$ is fixed by this map.

13.5 The Variational Principle

The periodic orbit of Chenciner and Montgomery is described using the reduced action functional (13.30), making use of the reduced symmetries discussed in the last section.

We first consider certain boundary conditions for the variational problem. Recall that by E_1 we denote the ray of configurations emanating from the origin, which consist of all Eulerian collinear configurations with the mass m_1 in the middle. The two-dimensional manifold M_3^+ denotes the portion of the meridian plane M_3 that projects into the northern hemisphere of the shape sphere $\mathcal{S}$ (Equation (13.12)), between rays corresponding to double collision (of masses m_1 and m_2) and the Eulerian configuration E_3.

The variational problem that is introduced by Chenciner–Montgomery is

$$A_{red}(\alpha) = \min_{E_{1,2}} A_{red}(x), \quad E_{1,2} = \left\{ x \in H^1[0,T] \,|\, x(0) \in E_1, x(T) \in M_3^+ \right\}.$$

$$(13.36)$$

The variational problem (13.36) enjoys the advantage of a well-understood existence theory, which is easily applied in this setting. Although our main goal is that of describing the properties of the solution, it is convenient to include the discussion on existence, because this argument is used again below.

Summary of existence proof. The existence of a solution to the variational problem (13.36) can be deduced using standard arguments (originally due to Tonelli) based on the fact that the reduced action is bounded below, coercive on the function space $E_{1,2}$, and is weakly lower-semicontinuous in the velocity. A minimizing sequence has bounded L^2 derivatives, which converge weakly to an L^2 function. The minimizing sequence can thereby be shown to be equi-Lipschitzian, and hence has a uniformly convergent subsequence. The limit curve provides a minimizing solution, using the property of weak lower-semicontinuity.

Thus the existence of a solution $\alpha(t)$ of the reduced Euler–Lagrange equations on $T(M/SO(2))$ is assured which joins in an optimal way the manifolds E_1 and M_3. Using the transversality property (see Section 1.9 where these conditions are discussed for the case of periodic boundary conditions) at the endpoints of such an arc, we deduce that

$$\alpha(0) \in E_1, \quad (\alpha(T), \dot{\alpha}(T)) \perp M^3. \tag{13.37}$$

The more difficult problem to overcome is the avoidance of collision singularities; that is, we want to ensure that $\alpha : [0, T] \longrightarrow \tilde{M}/SO(2)$, where $\tilde{M}$ is defined as the set of configurations that exclude collisions (13.11). The difficulty with the variational construction of $\alpha(t)$ arises because, as explained above for the Keplerian action, the collision orbits have finite action and could thus compete as the limiting case of a minimizing sequence. The basic argument introduced by Chenciner–Montgomery explains how the variational principle excludes such collision trajectories. This argument is used again later in a different setting. Thus we can summarize this argument here.

The noncollision of minimizing curves is based on comparing the action with a much simpler problem, that of the 2-body problem. To prepare for this comparison, the collision action A_2 of the two-body problem is introduced

$A_2(T) =$
$inf\{$ action of Keplerian collision orbit in time T between two masses$\}$.

This action is computed by using evaluations of action integrals (13.7) for the collision trajectories in the 2-body problem; see Gordon (1970).

Now the comparison with the values of the reduced action functional can be described following the argument from Chenciner and Montgomery (2000). The key observation is that the reduced action can be viewed as parameter-dependent (on the three masses m_i) and the reduced action along any arc is lowered by setting one of the masses to zero. Because $A(m_1, m_2, m_3; x) > A(m_1, m_2, 0; x)$, it follows that if $x(t)$ solves the variational problem (13.36), and $x(t)$ has a collision singularity, then $A(x) > A_2$. Using careful numerical length estimates, it is proven in Chenciner and Montgomery (2000) that

$$\min_{E_{1,2}} A_{red}(x) < A_2, \qquad (13.38)$$

A different argument in Chen (2001) using analytical methods gives the same inequality as (13.38).

Now suppose that α which satisfies (13.36) suffers a collision in the time interval $[0, T]$. Then $A^{red}(\alpha) \geq A_2$ which contradicts Equation (13.38). Thus the solution of the variational problem (13.36) has no collision singularities.

At this juncture the extremal arc $\alpha(t)$, having been shown to be collision-free, may be considered on its maximal interval of existence. In particular the dynamical features of $\alpha(t)$ and the symmetry structure that encodes them now come to the fore.

Using the full symmetry group $\mathbb{Z}_2 \times \mathcal{D}^3$, the arc α can be extended to give a periodic solution. The first extension uses property (13.37) together with Theorem (13.4.1), which extends the arc from E_1 to E_2.

Lemma 13.5.1. *The arc $\alpha_1(t)$ has a symmetric extension $\alpha_2(t)$ to the interval $[0, 2T]$, relative to the meridian circle M_3, so that*

$$\alpha_2(0) \in E_1, \quad (\alpha_2(T), \dot{\alpha}_2(T)) \perp M^3, \quad \alpha(2T) \in E_2.$$

Proof. We recall that the fixed point set of $\sigma\lambda_{1,2}$ is the meridian circle M_3, and $\sigma\lambda_{1,2}$ takes E_1 to E_2. It follows from (13.37) and the extension of $\sigma\lambda_{1,2}$ to TM, that

$$(\alpha_1(T), \dot{\alpha}_1(T)) \in \mathrm{FIX}(\Sigma\Lambda_{1,2}).$$

Therefore the symmetric extension $\alpha_2(t)$ on the interval $[0, 2T]$ is a solution of the (reduced) Euler–Lagrange equations, and satisfies the boundary conditions specified.

The remainder of the figure eight orbit can now be constructed by applying the interchange symmetries of $\mathcal{D}^3$. The extension between E_2 and E_3 is obtained by reparameterizing the symmetric arc $\lambda_{1,3}\alpha_2(t)$, $0 \leq t \leq 2T$. Recall that the endpoints of $\alpha_2(t)$ are E_1, E_2 respectively, and that the symmetry $\lambda_{1,3}$ fixes E_2, and exchanges E_1 and E_3. Thus we define $\alpha_3(t) = \lambda_{1,3}\alpha_2(2T - t)$, and $\alpha_4(t) = \lambda_{2,3}\alpha_2(2T - t)$. This last arc joins $E_3 = \alpha_4(0)$ to $E_1 = \alpha_4(2T)$. If we add the resulting arcs together (using the obvious parameterization) $\sigma\alpha_1(t) + \alpha_2(t) + \alpha_3(t) + \alpha_4(t)$ we find an extremal that intersects M_3 orthogonally at its endpoints. Thus this combined arc can be continued by the time-reversing symmetry $\sigma\lambda_{1,2}$ to give a closed orbit having minimal period $12T$.

13.6 Isosceles 3-Body Problem

In this section we discuss some global results on existence and stability for symmetric periodic solutions of the isosceles 3-body problem, see Cabral and Offin (2008). The isosceles 3-body problem can be described as the special motions of the 3-body problem whose triangular configurations always describe an isosceles triangle Wintner (1944). It is known that this can only occur if two of the masses are the same, and the third mass lies on the symmetry axis described by the binary pair. The symmetry axis can be fixed or rotating. We consider below the case where the symmetry axis is fixed.

We assume that $m_1 = m_2 = m$. The constraints for the isosceles problem can be formulated as

$$\sum m_i r_i = 0, \quad < r_1 - r_2, \mathbf{e}_3 > = 0, \quad < (r_1 + r_2), \mathbf{e}_i > = 0, \quad i = 1, 2, \tag{13.39}$$

where $\mathbf{e}_1, \mathbf{e}_2, \mathbf{e}_3$ denote the standard orthogonal unit vectors of $\mathbb{R}^3$. We consider the three-dimensional collisionless configuration manifold M_{iso}, modulo translations

$$M_{\text{iso}} = \{\mathbf{q} = (r_1, r_2, r_3) \mid r_i \neq r_j, \quad \mathbf{q} \text{ satisfies}(13.39)\}, \tag{13.40}$$

and restrict the potential V to the manifold M_{iso}, $\tilde{V} = V|_{M_{\text{iso}}}$. When all three masses lie in the horizontal plane, the third mass must be at the origin and the three masses are collinear. The set of collinear configurations is two-dimensional, and is denoted by $\mathcal{S}$.

There is a $\mathbb{Z}_2$ symmetry $\sigma : M_{\text{iso}} \to M_{\text{iso}}$, $\mathbf{r} \to \sigma\mathbf{r}$ across the plane of collinear configurations that leaves the potential $\tilde{V}$ invariant. An elementary argument shows that orbits which cross this plane orthogonally are symmetric with respect to σ. We can lift σ to T^*M_{iso} as a symplectic symmetry of H, namely $\mathbf{R}(q, p) = (\sigma q, \sigma p)$.

We use cylindrical coordinates (r, θ, z) on the manifold M_{iso}, where z denotes the vertical height of the mass m_3 above the horizontal plane, and (r, θ) denotes the horizontal position of mass m_1, relative to the axis of symmetry. The corresponding momenta in the fiber $T_q^* M_{\text{iso}}$ are denoted (p_r, p_θ, p_z). In cylindrical coordinates, the Hamiltonian is

$$H = \frac{p_r^2}{4m} + \frac{p_\theta^2}{4mr^2} + \frac{p_z^2}{2m_3(\frac{m_3}{2m} + 1)} - \frac{m^2}{2r} - \frac{2mm_3}{\sqrt{r^2 + z^2(1 + \frac{m_3}{2m})^2}}. \tag{13.41}$$

With these coordinates, the symmetry σ takes $(r, \theta, z) \to (r, \theta, -z)$. The plane of collinear configurations is now identified with the fixed-point plane Fix $\sigma/\{z = 0, r = 0\}$.

M_{iso} has an $SO(2)$ action that rotates the binary pair around the fixed symmetry axis and leaves $\tilde{V}$ invariant, namely $e^{i\theta}\mathbf{q} = (e^{i\theta}r_1, e^{i\theta}r_2, r_3)$. Lifting this as a symplectic diagonal action to T^*M_{iso} we find that H is equivariant, which gives the angular momentum of the system as a conservation law.

The reduced space $\mathbf{J}^{-1}(c)/SO(2) \approx T^*(M_{\mathrm{iso}}/SO(2))$ comes equipped with a symplectic structure together with the flow of the reduced Hamiltonian vector field obtained by projection along the $SO(2)$ orbits; see Meyer (1973) and Marsden (1992). Setting $p_\theta = c$ and substituting in Equation (13.41) gives $H_c = H(r, 0, z, p_r, c, p_z)$ and the reduced Hamiltonian vector field X_{H_c} on $T^*(M_{\mathrm{iso}}/SO(2))$ is

$$\dot{r} = \frac{\partial H_c}{\partial p_r}, \quad \dot{z} = \frac{\partial H_c}{\partial p_z}, \quad \dot{p}_r = -\frac{\partial H_c}{\partial r}, \quad \dot{p}_z = -\frac{\partial H_c}{\partial z}.$$

13.7 A Variational Problem for Symmetric Orbits

We now turn to an analytical description of periodic orbits, using a symmetric variational principle. In general when we consider a family of periodic orbits parameterized by the period T, it is not possible to specify the functional relation with the energy, nor for that matter angular momentum.

Consider the fixed time variational problem

$$\mathcal{A}_T^{\mathrm{iso}}(x) = \inf_{\Lambda_{\mathrm{iso}}} \mathcal{A}_T(q),$$

$$\mathcal{A}_T^{\mathrm{iso}}(q) = \int_0^T \sum \frac{1}{2m_i} \|p_{r_i}\|^2 + U(\mathbf{r}) \ dt, \tag{13.42}$$

$$\Lambda_{\mathrm{iso}} = \left\{ q \in H^1([0, T], M_{\mathrm{iso}}) \mid q(T) = \sigma e^{i2\pi/3} q(0) \right\}.$$

The function space Λ_{iso} contains certain paths that execute rotations and oscillations about the fixed point plane $\mathcal{S}$. It is not difficult to see that $\mathcal{A}_T^{\mathrm{iso}}$ is coercive on Λ_{iso}, using the boundary conditions given. Indeed if $q \in \Lambda_{\mathrm{iso}}$ tends to ∞ in M_{iso}, then the length l of q will also tend to ∞ due to the angular separation of the endpoints $q(0), q(T)$. An application of Holder's inequality then implies that the average kinetic energy of q will tend to ∞ as well. This coercivity of the functional $\mathcal{A}_T^{\mathrm{iso}}$ on the function space Λ_{iso} together with the fact that $\mathcal{A}_T^{\mathrm{iso}}$ is bounded below implies that the solution of the variational problem (13.42) exists by virtue of Tonelli's theorem.

We also show that the solution of (13.42) is collision-free, and can be extended to a periodic integral curve of X_H, provided that the masses satisfy the inequality

$$\frac{m^2 + 4mm_3}{3\sqrt{2}} < (m^2 + 2mm_3) \sqrt{\frac{m + 2m_3}{2m + m_3}}. \tag{13.43}$$

Theorem 13.7.1. *A minimizing solution of (13.42) is collision free on the interval $0 \le t \le T$, for all choices of the masses m, m_3.*

Proof. The argument rests on comparing the collision–ejection homothetic paths that also satisfy the boundary conditions of Λ_{iso}, for 3-body central configurations. First of all, notice that collinear collision with a symmetric congruent ejection path rotated by $2\pi/3$ will belong to Λ_{iso}, because $\sigma = id$ on $\mathcal{S}$. Moreover, symmetric homothetic equilateral paths also belong to Λ_{iso}, provided that we ensure that the congruent ejection path is rotated by $e^{i2\pi/3}$ so as to satisfy the conditions of Λ_{iso}.

Denote the collinear collision–ejection curve in Λ_{iso} by $q_1(t)$, which consists of the homothetic collinear collision-ejection orbit in $\mathcal{S}$, with collision at $T_1 = T/2$, and so that $q_1(T) = e^{i2\pi/3} q_1(0)$. Because this path gives the same action as the action of an individual collision–ejection orbit in time T, we can compare this with the action of $\frac{1}{3}$ the uniformly rotating collinear relative equilibrium having period $3T$. Using the concavity of (13.7) in the period T it can easily be seen that the path $q_1(t)$, $0 \le t \le T$ has action that is strictly bigger than the corresponding action of the path which consists of $\frac{1}{3}$ collinear relative equilibrium. Thus, the collinear collision–ejection path in Λ_{iso} is not globally minimizing. Now we let $q_1(t)$ denote the collinear relative equilibrium.

We wish to compare this action with the action of a symmetric homothetic equilateral path. In this case we let $q_0(t)$ denote a homothetic path for the equilateral configuration q_0, with collision at time $T/2$, together with a congruent symmetric segment consisting of the path $\sigma e^{i2\pi/3} q_0(t + T/2)$.

We now proceed to compare the action for the two types of motion of the 3-body problem described above, which are based on the two central configurations consisting of the equilateral triangle, denoted q_0, and that of the collinear relative equilibrium, denoted q_1. This remark is used now to make the comparison between the two types of motion in M_{iso}. Our first important observation is that the symmetric homothetic collision-ejection path $q_0(t)$ described above, has the same action as a periodic homothetic collision–ejection path, with the same period T. In turn, the periodic homothetic collision–ejection path has the same action as the uniformly rotating equilateral configurations having the same period T, $q(t) = e^{i2\pi t/T} q_0$.

Next we compute the force function $U(q)$, and the moment of inertia (13.3) for the two types of configurations, equilateral and collinear. We denote the common mutual distance for the two configurations corresponding to period T rotation by l_0 (equilateral), and l_1 (collinear). A direct computation yields

$$U(q_0) = \frac{m^2 + 2mm_3}{l_0}, \quad I(q_0) = \frac{m^2 + 2mm_3}{2m + m_3} l_0^2 \qquad (13.44)$$

whereas for collinear configurations,

$$U(q_1) = \frac{m^2 + 4mm_3}{2l_1}, \quad I(q_1) = 2ml_1^2. \qquad (13.45)$$

We can compute the action functional on each of the two uniformly rotating configurations, equilateral and collinear. Using the expression (13.7) for the action, we have

$$\mathcal{A}_T^{\text{iso}}(q_0) = 3(2\pi)^{1/3} \left(\tilde{U}(q_0) \right)^{2/3} T^{1/3}$$

$$\frac{1}{3}\mathcal{A}_{3T}^{\text{iso}}(q_1) = (2\pi)^{1/3} \left(\tilde{U}(q_1) \right)^{2/3} (3T)^{1/3}.$$

We argue that $q_0(t)$ does not fulfill the requirements for a global minimizer of the action in Λ_{iso} whenever the inequality $\tilde{U}(q_1) < 3\tilde{U}(q_0)$ is met. This can be tested using the expressions (13.44) and (13.45) to obtain

$$\tilde{U}_0 = U(q_0)\mathbf{r_0} = \sqrt{\frac{(m^2 + 2mm_3)^3}{2m + m_3}},$$

$$\tilde{U}_1 = U(q_1)\mathbf{r_1} = \frac{(m^2 + 4mm_3)\sqrt{m}}{\sqrt{2}}.$$

The inequality $\tilde{U}(q_1) < 3\tilde{U}(q_0)$ is equivalent to (13.43). To analyze this inequality further, notice that the expression appearing on the right of the inequality (13.43) satisfies

$$\sqrt{\frac{m + 2m_3}{2m + m_3}} > \frac{1}{\sqrt{2}}.$$

It is now a simple exercise to deduce that (13.43) holds for all choices of the masses.

Now, we make the simple argument which shows that the solution of the isosceles variational problem outlined here cannot have any collision singularities. The minimizing curve $x(t) \in \Lambda_{\text{iso}}$, if it does contain collision singularities, must consist of arcs of the N-body equations (13.1) which abut on collision. But each of these arcs beginning or ending with collision would have to have zero angular momentum by the results of Sundman (1913). This implies that the symmetry condition stated in the definition of Λ_{iso} could not be fulfilled, unless $x(t)$ were a symmetric homothetic collision–ejection orbit. However, both the collinear and the equilateral homothetic orbits in Λ_{iso} are not globally minimizing, provided that inequality (13.43) is fulfilled.

We now address the question of whether the solution of (13.42) gives a new family of periodic solutions, and in particular whether the relative equilibrium solutions might provide a solution. We employ a technique similar to that used in Chenciner and Venturelli (2000). As was discussed in the section on geometry of reduction above, the shape sphere $\{I = \frac{1}{2}\}$ describes the similarity classes of configurations up to rotation and dilation. If $q(t)$ denotes such a collinear relative equilibria solution, it is possible to construct a periodic vector field $\zeta(t)$, tangent to Λ_{iso} at $q(t)$, and which points in the tangent direction of the shape sphere, transverse to the equatorial plane of collinear configurations. The fact that the collinear central configurations are saddle points, only minimizing $U(q) = -\tilde{V}$ over the collinear configurations in the sphere $\{I = \frac{1}{2}\}$, implies that $d^2 \mathcal{A}_T^{\text{iso}}(q) \cdot \zeta < 0$. Hence $q(t)$ cannot be an absolute minimizer for the isosceles action $\mathcal{A}_T^{\text{iso}}$.

Using the equivariance of the symmetries with respect to the flow of (13.1) we can see the following.

Theorem 13.7.2. *The solution $q(t)$ to the variational problem (13.42) may be extended so as to satisfy the relation $q(t + T) = \sigma e^{i2\pi/3} q(t)$. The corresponding momentum $p(t)$ satisfies the same symmetry $p(t+T) = \sigma e^{i2\pi/3} p(t)$. Together, the pair $(q(t), p(t))$ may be extended to a $6T$ periodic orbit of the Hamiltonian vector field X_H for isosceles Hamiltonian (13.41) which undergoes two full rotations and six oscillations in each period, and which is not the collinear relative equilibrium in $\mathcal{S}$.*

Proof. Critical points of $\mathcal{A}_T^{\text{iso}}$ on Λ_{iso} must satisfy the transversality condition

$$\delta \mathcal{A}_T^{\text{iso}}(q) \cdot \xi = \langle \xi, p \rangle|_0^T = 0,$$

where $\xi(t)$ is a variation vector field along $q(t)$ satisfying $\xi(T) = \sigma e^{i2\pi/3} \xi(0)$. Therefore

$$\langle \sigma e^{-i2\pi/3} p(T) - p(0), \xi(0) \rangle = 0,$$

which implies that $p(T) = \sigma e^{i2\pi/3} p(0)$ because $\xi(0)$ is arbitrary.

Now $\sigma e^{i2\pi/3}$ generates a symplectic subgroup of order 6 on $T^* M_{\text{iso}}$, which fixes the Hamiltonian H. Therefore, $\sigma e^{i2\pi/3}(q(t), p(t))$ is also an integral curve of the Hamiltonian vector field X_H. Let $(x(t), y(t)) = (q(t+T), p(t+T))$ denote the time shifted integral curve of X_H. Then $(x(0), p(0)) = (q(T), p(T)) = \sigma e^{i2\pi/3}(q(0), p(0))$. By uniqueness of the initial condition, we conclude that $(x(t), y(t)) = \sigma e^{i2\pi/3}(q(t), p(t))$ as stated in the theorem.

Iterating the symmetry $\sigma e^{i2\pi/3}$ shows

$$(q(2T), p(2T)) = e^{i4\pi/3}(q(0), p(0)),$$

and

$$(q(T), p(T)) = \sigma e^{i2\pi/3}(q(0), p(0))$$
$$(q(2T), p(2T)) = \sigma e^{i2\pi/3}(q(T), p(T))$$
$$(q(3T), p(3T)) = \sigma e^{i2\pi/3}(q(2T), p(2T))$$
$$= \sigma e^{i6\pi/3}(q(0), p(0)),$$

therefore $q(6T), p(6T) = e^{i12\pi/3}(q(0), p(0))$ which shows as well as periodicity, that the orbit undergoes two full rotations and six oscillations before closing.

For given integers (M, N) we can study the more general variational problem

$$\mathcal{A}_T^{\text{iso}}(x) = \inf_{\Lambda_{(M,N)}} \mathcal{A}_T(q), \quad \mathcal{A}_T^{\text{iso}}(q) = \int_0^T \sum \frac{1}{2m_i}\|p_{r_i}\|^2 + U(\mathbf{r}) \ dt, \quad (13.46)$$

$$\Lambda_{(M,N)} = \left\{ q \in H^1([0,T], M_{\text{iso}}) \mid q(T) = \sigma e^{i2M\pi/N} q(0) \right\}.$$

The function space $\Lambda_{(M,N)}$ contains certain paths that execute M rotations and N oscillations about the fixed point plane $\mathcal{S}$ before closing. Using similar techniques to those above, it is shown in Cabral and Offin (2008) that the following generalization of the families of periodic orbits occur.

Theorem 13.7.3. *The solution $q(t)$ to the variational problem (13.46) is collision-free on the interval $[0,T]$ provided that the inequality $M\tilde{U}_1 < N\tilde{U}_0$. This occurs in the equal mass case, provided that $M < \frac{3\sqrt{2}}{5}N$, and in the case when $m_3 = 0$ when $M < N$. The solution $q(t)$ may be extended so as to satisfy the condition $q(t + T) = \sigma e^{i2M\pi/N} q(t)$, and together with $p(t)$ gives a NT-periodic integral curve of (13.1) in the case where N is even, and an 2NT-periodic integral curve in the case where N is odd.*

13.8 Instability of the Orbits and the Maslov Index

In this section we discuss the application of the Maslov index of the periodic orbits discussed above, to consider the question of stability.

Theorem 13.8.1. *The σ-symmetric periodic orbit that extends the solution $q(t)$ to the variational problem (13.46) is unstable, and hyperbolic on the reduced energy–momentum surface $H^{-1}(h)$ whenever (q,p) is nondegenerate in the reduced energy surface.*

The proof uses the second variation of the action, and symplectic properties of the reduced space $\mathbf{J}^{-1}(c)/SO(2)$ where $\mathbf{J}(q,p) = c$. The Maslov index of *invariant Lagrangian curves* is an essential ingredient. More complete details on the Maslov index in this context are given in Offin (2000) and that of symplectic reduction in Marsden (1992).

The functional and its differentials evaluated along a critical curve $q(t)$ in the direction $\xi \in T_{q(t)}\Lambda_{(M,N)}$ are

$$\mathcal{A}_T^{\text{iso}}(q) = \int_0^T \sum \frac{1}{2m_i}\|p_{r_i}\|^2 + U(\mathbf{r})dt,$$

$$\delta\mathcal{A}_T^{\text{iso}}(q) \cdot \xi = \sum_i \langle p_{r_i}, \xi_i \rangle|_0^T + \int_0^T \sum_i \langle -\frac{d}{dt}p_{r_i} + \frac{\partial U}{\partial q_i}, \xi_i \rangle dt,$$

$$\delta^2\mathcal{A}_T^{\text{iso}}(q)(\xi,\xi) = \sum_i \langle \eta_i, \xi_i \rangle|_0^T + \int_0^T \sum_{i,j} \langle -\frac{d}{dt}\eta_i + \frac{\partial^2 U}{\partial q_i \partial q_j}\xi_j, \xi_i \rangle dt.$$

We defined the Jacobi field along $q(t)$ as a variation of the configuration $\xi(t)\partial/\partial q$ which together with the variation in momenta $\eta(t)\partial/\partial p$, satisfies the equations

$$\frac{d\xi_i}{dt} = m_i\eta_i, \quad \frac{d\eta_i}{dt} = \sum_j \frac{\partial^2 U}{\partial q_i\partial q_j}\xi_j. \tag{13.47}$$

Such Jacobi fields are used to study stability properties of $(q(t), p(t))$. We are particularly interested in the Jacobi fields $\xi(t)$ that satisfy the boundary relation $\xi(T) = \sigma e^{i2M\pi/N}\xi(0)$, because these are natural with respect to the variational problem (13.46). We show below that an important subset of them will correspond to the variations within $\Lambda_{(M,N)}$ belonging to the tangent of this space at $q(t)$, and moreover may be used to decide the stability of $(q(t), p(t))$.

Due to symmetry invariance of the flow of the Hamiltonian vector field X_H, it is possible to see that the second variation $\delta^2\mathcal{A}_T^{\text{iso}}(q)$ will always have degeneracies (zero eigenvalues) in the direction of the constant Jacobi field $\xi(t)\partial/\partial q = r^{-1}\partial/\partial\theta$. Such degeneracies can be removed by considering variations in the reduced configuration space $M_{\text{iso}}/SO(2)$. Moreover, by conservation of energy and angular momentum, further degeneracies of the second variation are given by variations $\xi(t)\partial/\partial q$ so that $\zeta(t) = \xi(t)\partial/\partial q + \eta(t)\partial/\partial p$ is transverse to the energy momentum surface of the periodic orbit $(q(t), p(t))$. In other words, we can only expect nondegenerate effects in the second variation if we choose variations $\zeta(t) = \xi(t)\partial/\partial q + \eta(t)\partial/\partial p$ which modulo their rigid rotations by $e^{i\theta}$ are tangent to this energy–momentum surface. To effect this kind of reduction of Jacobi fields, it is most useful to return to our discussion of the symmetry reduced space from Section 13.2.

The reduced space is defined to be the set of equivalence classes of configurations and momenta on the c-level set of the angular momentum up to rigid rotation; that is, $P_c = \mathbf{J}^{-1}(c)/SO(2)$. Using the cylindrical coordinates of the symmetric mass m_1 introduced earlier, it is easily seen that P_c is a symplectic space which is symplectomorphic to $T^*(M_{\text{iso}}/SO(2))$. The reduced symplectic form on P_c is the canonical one in these coordinates. Let H_c denote the reduced Hamiltonian on the reduced space P_c. In reduced cylindrical coordinates, this can be computed by simply substituting $\theta = 0$ and $p_\theta = c$ in the original Hamiltonian for the isosceles problem (as we mentioned earlier in Section 13.2 on the description of the isosceles problem)

$$H_c(r, z, p_r, p_z) = H(r, 0, z, P_r, c, p_z).$$

Now we consider the reduced energy–momentum space $H_c^{-1}(h)$. The directions tangent to this manifold thus become the natural place to look for positive directions of the second variation.

We therefore consider an essential direction for the second variation, those Jacobi fields in $T_{q(t)}\Lambda_{(M,N)}$ which together with the conjugate variations $\eta(t)$ will project along the rigid rotations to variations that are everywhere tangent to the energy surface $H^{-1}(h)$ in the reduced space $\mathbf{J}^{-1}(c)/SO(2)$. This means that if

$$\overline{\xi}(t)\frac{\partial}{\partial q} = \xi_r(t)\frac{\partial}{\partial r} + \frac{\xi_\theta(t)}{r}\frac{\partial}{\partial \theta} + \xi_z(t)\frac{\partial}{\partial z}$$

is our Jacobi field in cylindrical coordinates, where $r(t)$ is the radial component of the configuration of $q(t)$, then the symmetry reduced Jacobi field is just

$$\xi(t) = \xi_r(t)\frac{\partial}{\partial r} + \xi_z(t)\frac{\partial}{\partial z}.$$

Notice in addition that $\xi_\theta(t) = 1$ for all Jacobi variations. This procedure is therefore obviously reversible, if

$$\xi(t) = \xi_r(t)\frac{\partial}{\partial r} + \xi_z(t)\frac{\partial}{\partial z}$$

is a reduced Jacobi variation then

$$\overline{\xi}(t) = \xi_r(t)\frac{\partial}{\partial r} + \frac{1}{r(t)}\frac{\partial}{\partial \theta} + \xi_z(t)\frac{\partial}{\partial z}$$

is a solution to (13.47).

We need to consider the projection of the reduced space into the reduced configuration space

$$\pi : \mathbf{J}^{-1}(c)/SO(2) \longrightarrow M_{\text{iso}}/SO(2),$$

and denote the vertical space of the projection at $x = (q,p)$ by $V|_{(x,p)} = \ker d_x\pi$. Recall that a subspace λ of tangent variations to $\mathbf{J}^{-1}(c)/SO(2)$ is called Lagrangian if $\dim\lambda = 2$, and $\omega|_\lambda = 0$. We consider the invariant Lagrangian subspaces of Jacobi fields and conjugate variations $\zeta(t) = \xi(t)\partial/\partial q + \eta(t)\partial/\partial p$ which are tangent everywhere to $H^{-1}(h)$ within $\mathbf{J}^{-1}(c)/SO(2)$, and for which $\overline{\xi}(t), \overline{\eta}(t)$ satisfy (13.47). Because this is a three-dimensional manifold, we find that every two-dimensional invariant Lagrangian curve that is tangent to $H_c^{-1}(h)$ includes the flow direction X_{H_c} and one transverse direction field,

$$\lambda_t = \text{span}\,\langle\zeta(t), X_{H_c}(z(t))\rangle, \quad dH_c(\zeta(t)) = 0.$$

A focal point of the Lagrangian plane λ_0 is the value $t = t_0$, where $d_x\pi : \lambda_{t_0} \to M_{\text{iso}}/SO(2)$ is not surjective. These Lagrangian singularities correspond to the vanishing of the determinant

$$D(t_0) = \det\begin{bmatrix} \xi_r(t_0) & p_r(t_0) \\ \xi_z(t_0) & p_z(t_0) \end{bmatrix} = 0, \tag{13.48}$$

where (ξ_r, ξ_z) denotes the reduced configuration component of a reduced variational vector field along $(q(t), p(t))$. Now we study the invariant Lagrangian curve λ_t^* of reduced energy–momentum tangent variations

$$\lambda_0^* = \{\zeta(0) = \xi(0)\partial/\partial q + \eta(0)\partial/\partial p \mid \xi(T) = \sigma\xi(0), \ dH_c(\xi(0), \eta(0)) = 0\}.$$
$$(13.49)$$

Evidently, the subspace λ_0^* is not empty, because $X_{H_c}(q(0), p(0)) \in \lambda_0^*$.

Lemma 13.8.1. *If the periodic integral curve $(q(t), p(t))$ is nondegenerate on the reduced energy-momentum manifold $H^{-1}(h)$ within $\mathbf{J}^{-1}(c)/SO(2)$, then λ_0^* is Lagrangian.*

Proof. We need to show that $\dim \lambda_0^* = 2$, and that $\omega|_{\lambda_0^*} = 0$. The last condition follows immediately from the first and from the fact that variations within λ_0^* are tangent to $H^{-1}(h)$. We prove the first condition on the dimension of λ_0^*. The key to this is to make the following observations on the mapping $P - \sigma$,

$$(P - \sigma)T_x H_c^{-1}(h) \subset T_x H_c^{-1}(h), \quad x = (q(0), p(0))$$
$$\ker (P - \sigma) = \langle X_{H_c}\rangle(x)\rangle$$
$$(P - \sigma)\lambda_0^* \subset \ker d_{Px}\pi.$$

Both the Poincaré mapping and the symmetry σ lifted to the cotangent bundle leave the energy surface $H^{-1}(h)$ invariant. The first observation then follows by projecting from $T^*(M_{iso})$ onto the reduced energy–momentum manifold. The second property follows exactly from the condition on nondegeneracy of the periodic orbit $(q(t), p(t))$, because vectors in the kernel will give rise to periodic solutions of the linearized equations that are tangent to P_c. For $\zeta \in \lambda_0^*$, the last condition can be seen from the computation

$$d_{Px}\pi(P - \sigma)\zeta = (\xi(T) - \sigma\xi(0))\partial/\partial q = 0.$$

From the first two conditions we see that $P - \sigma$ is an isomorphism when restricted to $T_x H^{-1}(h)/\langle X_{H_c}\rangle(x)\rangle$, and this can be used to define the transverse variation $\zeta(0) \in \lambda_0^*$ modulo $\langle X_{H_c}\rangle(x)\rangle$ which, by virtue of the third condition, must be the preimage under $P - \sigma$ of a vertical vector in $T_x H^{-1}(h)$.

Next we state the second-order necessary conditions in terms of reduced energy–momentum variations, in order that $q(t)$ is a minimizing solution of the variational problem (13.46). We recall that the symplectic form and the symplectic symmetry σ drop to $\mathbf{J}^{-1}(c)/SO(2)$, and we denote these without confusion, respectively, by ω and σ. Similarly, we let $\mathcal{P}$ denote the symplectic map that is the relative Poincaré map for the reduced Jacobi fields $(\xi(t), \eta(t)) \mapsto (\xi(t + T), \eta(t + T))$.

Proposition 13.8.1. *If the curve $q(t)$ is a collision-free solution of the variational problem* (13.46), *and when projected by π is nondegenerate as a periodic integral curve of X_{H_c} then λ_0^* has no focal points in the interval $[0, T]$, and*

$$\omega(\lambda_0^*, \sigma \lambda_T^*) = \omega(\lambda_0^*, \sigma \mathcal{P} \lambda_0^*) > 0.$$

Proof. The fact that λ_0^* has no focal points on $[0, T]$ is classical. From the expression (13.47) for the second variation we may deduce that

$$
\begin{aligned}
\delta^2 \mathcal{A}_T^{\text{iso}}(q)(\xi, \xi) &= \Sigma_i \langle \eta_i, \xi_i \rangle |_0^T \\
&= \langle \sigma e^{-i2M\pi/N} \eta_i(T) - \eta_i(0), \xi(0) \rangle \\
&= \omega(\lambda_0^*, \sigma e^{-i2M\pi/N} \lambda_T^*) \\
&> 0.
\end{aligned}
$$

The last follows because $\sigma e^{-i2M\pi/N} \mathcal{P}(\xi(0), \eta(0)) = (\xi(0), \sigma e^{-i2M\pi/N} \eta(T))$. Moreover, because we are working in the reduced energy–momentum space, we can drop the action of the rotation $e^{-i2M\pi/N}$ and the final inequality reads $\omega(\lambda_0^*, \sigma \mathcal{P} \lambda_0^*) > 0$.

Now we consider the reduced Lagrangian planes $\lambda_0^*, \sigma \lambda_T^* = \sigma \mathcal{P} \lambda_0^*$.

Lemma 13.8.2. *The Lagrange planes $(\sigma \mathcal{P})^n \lambda_0^*$ have no focal points in the interval $0 \leq t \leq T$.*

Proof This argument proceeds by showing that the successive iterates $(\sigma \mathcal{P})^n \lambda_0^*$ have a particular geometry in the reduced space of Lagrangian planes. This geometry then allows a simple comparison between the focal points of λ_0^* and that of $\mathcal{P}^n \lambda_0^*$.

Recall that the Lagrange planes $(\sigma \mathcal{P})^n \lambda_t^*$ are all generated by a single transverse variational vector field $(\xi(t), \eta(t))$, so that at $t = 0$ we need to consider only the initial conditions $(\xi(0), \eta(0))$. We observe from Proposition (13.8.1) that $\omega(\lambda_0^*, \sigma \mathcal{P} \lambda_0^*) > 0$ that is $\omega((\xi(0), \eta(0)), \sigma \mathcal{P}(\xi(0), \eta(0))) > 0$. Recall that the reduced symplectic form ω restricted to the tangent of the level set $H^{-1}(h)$ is nothing more than the signed area form in the reduced plane of transverse vector fields $(\xi(t), \eta(t))$ for which $dH_c(\xi(t), \eta(t)) = 0$. The fact that $\omega > 0$ on the pair $\lambda_0^*, \sigma \mathcal{P} \lambda_0^*$ implies an orientation of these subspaces in the plane. In particular, $\sigma \mathcal{P} \lambda_0^*$ is obtained from λ_0^* by rotating counterclockwise, by an angle less than π. Moreover, $\sigma \mathcal{P} \lambda_0^*$ must lie between λ_0^* and the positive vertical $V|_{(x,p)}$, due to the fact that both Lagrange planes have the same horizontal component $\xi(0)$. Now, as t changes over the interval $[0, T]$, the vertical Lagrange plane $V|_{(x,p)}$ rotates initially clockwise, and the Lagrange plane λ_0^* cannot move through the vertical $V|_{(x,p)}$ due to the fact that λ_0^* is focal point free in this interval. The comparison between $\lambda_0^*, \sigma \mathcal{P} \lambda_0^*$, mentioned above, amounts to the statement that the first focal point of $\sigma \mathcal{P} \lambda_0^*$ must come after the first focal point of λ_0^*. Due to Proposition (13.8.1), we infer that $\sigma \mathcal{P} \lambda_0^*$ is focal point free on the interval $[0, T]$.

The argument given between $\lambda_0^*, \sigma \mathcal{P} \lambda_0^*$ can be repeated for $\sigma \mathcal{P} \lambda_0^*, (\sigma \mathcal{P})^2 \lambda_0^*$ because $\omega(\sigma \mathcal{P} \lambda_0^*, (\sigma \mathcal{P})^2 \lambda_0^*) > 0$, by application of the symplectic mapping $\sigma \mathcal{P}$. Moreover, as we have just shown above, $\sigma \mathcal{P} \lambda_0^*$ is focal point free on $[0, T]$ so that comparing with $(\sigma \mathcal{P})^2 \lambda_0^*$ and using the orientation supplied by the symplectic form ω indicates that $(\sigma \mathcal{P})^2 \lambda_0^*$ is focal point free on the interval $[0, T]$ as well.

This argument is applied successively to each of the iterates $(\sigma \mathcal{P})^n \lambda_0^*$.

Lemma 13.8.3. *The reduced Lagrange plane λ_0^* of transverse variations is focal point free on the interval $0 \leq t < \infty$.*

Proof The argument proceeds by showing that λ_0^* is focal point free on each of the intervals $[0, T], [T, 2T], \ldots$. This property holds for the first interval $[0, T]$, due to the second order conditions in Proposition (13.8.1). The next interval and succeeding ones can be explained because $(\sigma \mathcal{P})^n \lambda_0^*$, is focal point free on $[0, T]$. In particular, $\sigma \mathcal{P} \lambda_0^*$ is focal point free on $[0, T]$ implies that $\mathcal{P} \lambda_0^*$ is focal point free on $[0, T]$, because $\sigma V|_{(x,p)} = V|_{(x,p)}$. Therefore λ_0^* is focal point free on $[0, 2T]$. The iterates $(\sigma \mathcal{P})^{n+1} \lambda_0^*$ are also focal point free on $[0, T]$, together with the fact that $\sigma \mathcal{P} = \mathcal{P} \sigma$ implies similarly that λ_0^* is focal point free on $[0, (n+1)T]$. This concludes the proof.

Because there is no rotation of the Lagrange planes λ_t^* (Lemma 13.8.3), we can ask what obstruction there is to prevent this. The answer given in the next theorem is that the Poincaré map must have real invariant subspaces.

Theorem 13.8.2. *Under the assumptions of Proposition 13.8.1, there are (real) invariant subspaces for the reduced Poincaré map $\mathcal{P}^2$. These subspaces are transverse when $\delta^2 \mathcal{A}_T(q)$ is nondegenerate when restricted to the subspace in the reduced space of tangential variations $T_{q(t)} \Lambda_{(M,N)}$ which are also tangent to the reduced energy surface $H_c^{-1}(h)$.*

Proof. The proof proceeds by examining the iterates $(\sigma \mathcal{P})^n \lambda_0^*$ of the subspace λ_0^* of tangential Jacobi variations in the reduced energy–momentum space $H_c^{-1}(h)$. By Lemma 13.8.2 and the fact that $\omega((\sigma \mathcal{P})^n \lambda_0^*, (\sigma \mathcal{P})^{n+1} \lambda_0^*) > 0$, the iterates $(\sigma \mathcal{P})^n \lambda_0$ must have a limit subspace $\beta = \lim_{n \to \infty} (\sigma \mathcal{P})^n \lambda_0^*$. The subspace β is thereby Lagrangian, and invariant for the symplectic map $\sigma \mathcal{P}$. Therefore this implies that inasmuch as $\sigma \mathcal{P} = \mathcal{P} \sigma$,

$$\sigma \mathcal{P} \beta = \beta$$
$$\mathcal{P} \beta = \sigma \beta$$
$$\mathcal{P}^2 \beta = \mathcal{P} \sigma \beta$$
$$= \sigma \mathcal{P} \beta$$
$$= \beta.$$

Because λ_0^* is focal point free on the interval $0 \le t < \infty$, and $V|_{(x,p)}$ can have no focal points before λ_0^*, it follows that $V|_{(x,p)}$ can have no focal points in $0 < t < \infty$. However more is true, because the forward iterates of $V|_{(x,p)}$ cannot cross any of the subspaces $(\sigma \mathcal{P})^n \lambda_0^*$ it is not difficult to see that the subspace β can be also represented as the forward limit of the iterates $\mathcal{P}^n$ of the vertical space, $\beta = \lim_{n \to \infty} \mathcal{P}^n V|_{(x,p)}$. It follows that β represents the reduced transverse directions of the stable manifold of (q, p).

Using Lemma 13.8.3 it follows that backward iterates of the vertical space $V|_{(x,p)}$ under $\mathcal{P}$ cannot cross the subspace λ_0^*. Therefore the unstable manifold α may be represented as the limit in backward time, $\alpha = \lim_{n \to \infty} \mathcal{P}^{-n} V|_{(x,p)}$.

Finally, to show transversality of the subspaces β, α we can use the fact that in the case where (q, p) is nondegenerate, $\omega(\lambda_0^*, \sigma \mathcal{P} \lambda_0^*) > 0$, by virtue of Proposition 13.8.1. This implies that in the case of nondegeneracy $\omega(\alpha, \beta) > 0$, which implies transversality.

13.9 Remarks

In this chapter we have chosen two simple yet interesting and complex examples from the global study of periodic solutions of the 3-body problem. In the first example, the symmetry group of the orbit contains a dihedral component $D_3 \times \mathbb{Z}_2$. In the second example, the symmetry group $\mathbb{Z}_6$ is far simpler and consequently we can say a great deal about the stability type of the orbit families. The Maslov theory is an ideal topic to apply in this example. This technique for studying global stability of periodic families was first applied to the case of $\mathbb{Z}_2$ symmetry generated by a time reversing symmetry in Offin (2000). The analytic stability analysis of the figure eight at this time remains a mystery, yet it seems tantalizingly close to resolution. Other orbits that are noteworthy and which fall into the category of those determined by symmetric variational principles are interesting objects of current research. These include the "crazy eights" or figure eights with less symmetry, Ferrario and Terracini (2004), and the "hip-hop" family of equal mass $2N$-body problem as well as some of the fascinating examples described in Simó (2002) and Ferrario and Terracini (2004). The hip-hop orbits, discovered initially by Chenciner and Venturelli (2000), have spatio-temporal symmetry group $\mathbb{Z}_2$, named by Chenciner as the Italian symmetry. This family has also been analyzed using the Maslov theory for stability. They fall into the category of a cyclic symmetry group without time reversal, similar to the isosceles example we studied above. The stability type of the hip-hop family is identical to that of the isosceles families, hyperbolic on its energy–momentum surface, when nondegenerate. A recent result of Buono and Offin (2008) treats this case of cyclic symmetry group in general, and again the families of periodic orbits in this category are all hyperbolic whenever they are nondegenerate on their energy–momentum surface. A forthcoming paper by Buono, Meyer, and Offin analyzes the dihedral group symmetry of the crazy eights. As a final comment we mention that other methods have been developed Dell'Antonio (1994), for analyzing stability of periodic orbits in Hamiltonian and Lagrangian systems that are purely convex in the phase variables.

References

Abraham, R. and Marsden, J. 1978: *Foundations of Mechanics*, Benjamin-Cummings, London.

Alfriend, J.

  1970: The stability of the triangular Lagrangian points for commensurability of order 2, *Celest. Mech.* 1, 351–59.

  1971: Stability of and motion about L_4 at three-to-one commensurability, *Celest. Mech.*, 4, 60–77.

Arenstorf, R. F. 1968: New periodic solutions of the plane three-body problem corresponding to elliptic motion in the lunar theory, *J. Diff. Eqs.*, 4, 202–256.

Arnold, V. I.

  1963a: Proof of A. N. Kolmogorov's theorem on the preservation of quasiperiodic motions under small perturbations of the Hamiltonian, *Russian Math. Surveys*, 18(5), 9–36.

  1963b: Small divisor problems in classical and celestial mechanics, *Russian Mathematical Surveys*, 18(6), 85–192.

  1978: *Mathematical Methods of Classical Mechanics*, Springer-Verlag, New York.

  1985: The Sturm theorems and symplectic geometry, *Functional Anal. Appl.*, 19(4), 251–259

  1990: *Dynamical Systems IV*, Encyclopedia of Mathematics, 4, Springer-Verlag, New York.

Barrar, R. B. 1965: Existence of periodic orbits of the second kind in the restricted problem of three–bodies, *Astron. J.*, 70(1), 3–4.

Bialy, M. 1991: On the number of caustics for invariant tori of Hamiltonian systems with two degrees of freedom, *Ergodic Theory Dyn. Sys.*, 11(2), 273–278.

Birkhoff, G. D.

  1927: *Dynamical Systems*, Coloq. 9, Amer. Math. Soc., Providence.

Bondarchuk, V. S. 1984: Morse index and deformations of Hamiltonian systems, *Ukrain. Math. Zhur.*, 36, 338–343.

© Springer International Publishing AG 2017
K.R. Meyer, D.C. Offin, *Introduction to Hamiltonian Dynamical Systems and the N-Body Problem*, Applied Mathematical Sciences 90,
DOI 10.1007/978-3-319-53691-0

Bruno, A. D. 1987: Stability of Hamiltonian systems, *Mathematical Notes*, 40(3), 726–730.

Buchanan, D. 1941: Trojan satellites–limiting case, *Trans. of the Royal Soc. of Canada*, 35, 9–25.

Buono, L. and Offin D. C. 2008: Instability of periodic solutions with spatio–temporal symmetries in Hamiltonian systems, preprint.

Cabral, H. and Meyer, K. R. 1999: Stability of equilibria and fixed points of conservative systems, *Nonlinearity*, 12, 1999, 1351–1362.

Cabral, H. and Offin D. C. 2008: Hyperbolicity for symmetric periodic solutions of the isosceles three body problem, preprint.

Chen, Kuo–Chang 2001: On Chenciner–Montgomery's orbit in the three-body problem, *Discrete Contin. Dyn. Sys.*, 7(1), 85–90.

Chenciner, A. and Montgomery, R. 2000: On a remarkable periodic orbit of the three body problem in the case of equal masses, *Ann. Math.*, 152, 881–901.

Chenciner, A. and Venturelli, A. 2000: Minima de l'intégrale d'action du problème Newtonien de 4 corps de masses égales dans $\mathbb{R}^3$: orbites 'Hip–Hop', *Celest. Mech. Dyn. Astr.*, 77, 139–152.

Cherry, T. M. 1928: On periodic solutions of Hamiltonian systems of differential equations, *Phil. Trans. Roy. Soc. A*, 227, 137–221.

Chetaev, N. G. 1934: Un théorème sur l'instabilité, *Dokl. Akad. Nauk SSSR*, 2, 529–534.

Chevally, C. 1946: *Theory of Lie Groups*, Princeton University Press, Princeton, NJ.

Chicone, C. 1999: *Ordinary Differential Equations with Applications*, Texts in Applied Mathematics 34, Springer, New York.

Conley, C. and Zehnder, E. 1984: Morse–type index theory for flows and periodic solutions for Hamiltonian systems, *Comm. Pure Appl. Math.*, 37, 207–253.

Contreras, G., Gaumbado, J.-M., Itturaga, R., and Paternain, G. 2003: The asymptotic Maslov index and its applications, *Erg Th. Dyn. Sys.*, 23, 1415–1443.

Crowell, R. and Fox, R. 1963: *Introduction to Knot Theory*, Ginn, Boston.

Cushman, R., Deprit, A., and Mosak, R. 1983: Normal forms and representation theory, *J. Math. Phys.*, 24(8), 2102–2116.

Dell'Antonio, G. F. 1994: Variational calculus and stability of periodic solutions of a class of Hamiltonian systems, *Rev. Math. Phys.*, 6, 1187–1232

Deprit, A. 1969: Canonical transformation depending on a small parameter, *Celest. Mech. 72*, 173–79.

Deprit, A. and Deprit–Bartholomê, A. 1967: Stability of the Lagrange points, *Astron. J.* 72, 173–79.

Deprit, A. and Henrard, J.
    1968: A manifold of periodic solutions, *Adv. Astron. Astrophy.* 6, 6–124.

1969: Canonical transformations depending on a small parameter. *Celest. Mech.*, 1, 12–30.

Dieckerhoff, R. and Zehnder, E. 1987: Boundedness of solutions via the twist-theorem, *Annali Della Scu. Norrm. Super. di Piza*, 14, 79–95.

Dirichlet, G. L. 1846: Über die stabilitát des gleichgewichts, *J. Reine Angew. Math.*, 32, 85–88.

Duistermaat, J. J. 1976: On the Morse index in variational calculus, *Adv. Math.*, 21, 173–195.

Elphick, C., Tirapegui, E., Brachet, M., Coullet, P., and Iooss, G. 1987: A simple global characterization for normal forms of singular vector fields, *Physica D*, 29, 96–127.

Ferrario D. and Terracini S. 2004: On the existence of collisionless equivariant minimizers for the classical n–body problem, *Inv. Math.* 155, 305–362.

Flanders, H. 1963: *Differential Forms with Applications to Physical Sciences*, Academic Press, New York.

Gordon, W. B. 1970: A minimizing property of Keplerian orbits, *Amer. J. Math.*, 99, 961–971.

Goździewski, K. and Maciejewski, A. 1998: Nonlinear stability of the Lagrange libration points in the Chermnykh problem, *Celest. Mech.*, 70(1), 41–58.

Hadjidemetriou, J. D. 1975: The continuation of periodic orbits from the restricted to the general three–body problem, *Celest. Mech.*, 12, 155 174.

Hagel, J. 1996: Analytical investigation of non–linear stability of the Lagrangian point L_4 around the commensurability 1:2, *Celest, Mech. Dynam. Astr.*, 63(2), 205–225.

Hale, J. K. 1972: *Ordinary Differential Equations*, John Wiley, New York.

Hartman, P. 1964: *Ordinary Differential Equations*, John Wiley, New York.

Henrard, J. 1970: On a perturbation theory using Lie transforms, *Celest. Mech.*, 3, 107–120.

Hénon, M. 1997: *Generating families in the restricted three-body problem*, Springer Verlag, LNP 52.

Hestenes, M. R. 1966: *Calculus of Variations and Optimal Control*, John Wiley, New York.

Hubbard, J. and West, B. 1990: *Differential Equations: A Dynamical Systems Approach*, Springer-Verlag, New York.

Kamel, A. 1970: Perturbation method in the theory of nonlinear oscillations, *Celest. Mech.*, 3, 90–99.

Kummer, M.

1976: On resonant non linearly coupled oscillators with two equal frequencies, *Comm. Math. Phy.* 48, 53–79.

1978: On resonant classical Hamiltonians with two equal frequencies, *Comm. Math. Phy.* 58, 85–112.

Laloy, M. 1976: On equilibrium instability for conservative and partially dissipative mechanical systems, *Int. J. Non–Linear Mech.*, 2, 295–301.

LaSalle, J. P. and Lefschetz, S. 1961: *Stability by Liapunov's Direct Method with Applications*, Academic Press, New York.

Laub, A. and Meyer, K. R. 1974: Canonical forms for symplectic and Hamiltonian matrices, *Celest. Mech.* 9, 213–238.

Lerman, L. and Markova, A. 2009: On stability at the Hamiltonian Hopf bifurcation, *Regul. Chaotic Dyn.*, 14, 148–162.

Liu, J. C. 1985: The uniqueness of normal forms via Lie transforms and its applications to Hamiltonian systems, *Celest. Mech.*, 36(1), 89–104.

Long, Y. 2002: *Index Theory for Symplectic Paths with Applications*, Birhäuser Verlag, Basel.

Lyapunov, A. 1892: Probleme général de la stabilité du mouvement, *Ann. of Math. Studies 17*, Princeton University Press, Princeton, NJ. (Reproduction in 1947 of the French translation.)

Marchal, C. 2002: How the method of minimization of action avoids singularities, *Celest. Mech. Dyn. Astron.*, 83, 325–353

Markeev, A. P.
> 1966: On the stability of the triangular libration points in the circular bounded three body problem, *Appl. Math. Mech.* 33, 105–110.
>
> 1978: *Libration Points in Celest. Mech. and Space Dynamics* (in Russian), Nauka, Moscow.

Markus, L. and Meyer, K. 1974: Generic Hamiltonian systems are neither integrable nor ergodic, *Mem. Am. Math. Soc.*, 144.

Marsden, J. 1992: *Lectures on Mechanics*, LMS Lecture Note Series 174, Cambridge University Press, Cambridge, UK.

Marsden, J. and Weinstein, A. 1974: Reduction of symplectic manifolds with symmetry, *Rep. Mathematical Phys.*, 5(1), 121–130.

Mathieu, E. 1874: Mémorire sur les equations différentieles canoniques de la mácanique, *J. de Math. Pures et Appl.*, 39, 265–306.

Mawhin, J. and Willem, M. 1984: Multiple solutions of the periodic boundary value problem for some forced pendulum-type equations, *J. Diff. Eqs.*, 52, 264–287.

McCord, C., Meyer, K. R. and Wang, Q. 1998: Integral manifolds of the spatial three body problem, *Mem. Amer. Math. Soc.*, 628, 1–91.

McGehee, R. and Meyer, K. R. 1974: Homoclinic points of area preserving diffeomorphisms, *Amer. J. Math.*, 96(3), 409–21.

Meyer, K. R.
> 1970: Generic bifurcation of periodic points, *Trans. Amer. Math. Soc.*, 149, 95–107.
>
> 1971: Generic stability properties of periodic points, *Trans. Amer. Math. Soc.*, 154, 273–77.
>
> 1973: Symmetries and integrals in mechanics., *Dynamical Systems* (Ed. M. Peixoto), Academic Press, New York, 259–72.
>
> 1981a: Periodic orbits near infinity in the restricted N–body problem, *Celest. Mech.*, 23, 69–81.

1981b: Hamiltonian systems with a discrete symmetry, *J. Diff. Eqs.*, 41(2), 228–38.

1984b: Normal forms for the general equilibrium, *Funkcialaj Ekvacioj*, 27(2), 261–71.

1999: *Periodic Solutions of the N–Body Problem*, Lecture Notes in Mathematics 1719, Springer, New York.

Meyer, K. R. and Hall, G. R. 1991: *Introduction to Hamiltonian Dynamical Systems and the N–Body Problem*, Springer–Verlag, New York.

Meyer, K. R., Palacián, J. and Yanguas, P. 2012: Stability of a Hamiltonian System in a Limiting Case, *Regular and Chaotic Dynamics*, 17(1), 24–35.

Meyer, K., Palacian J. and Yanguas, P. 2015: The elusive Liapunov periodic solutions, *Qual. Th. Dyn. Sys.*, 14, 381–401.

Meyer, K. R. and Schmidt, D. S.

1971: Periodic orbits near L_4 for mass ratios near the critical mass ratio of Routh, *Celest. Mech.*, 4, 99–109.

1982: The determination of derivatives in Brown's lunar theory, *Celest. Mech.* 28, 201–07.

1986: The stability of the Lagrange triangular point and a theorem of Arnold, *J. Diff. Eqs.*, 62(2), 222–36.

2016: Bounding solutions of a forced oscillator, submitted.

Moeckel, R.

1988: Some qualitative features of the three body problem, *Hamiltonian Dynamical Systems*, Contemp. Math. 181, Amer. Math. Soc., 1–22.

2007: Shooting for the eight – a topological existence proof for a figure-eight orbit of the three-body problem, preprint.

Moore, C. 1993: Braids in classical gravity, *Phys. Rev. Lett.*, 70, 3675–3679.

Morris, G. R. 1976: A class of boundedness in Littlewood's problem on oscillatory differential equations, *Bull. Austral. Math. Soc.*, 14, 71–93.

Morse, M. 1973: *Variational Analysis, Critical Extremals and Sturmian Extensions*, Wiley–Interscience, New York.

Moser, J. K.

1956: The analytic invariants of an area-preserving mapping near a hyperbolic fixed point, *Comm. Pure Appl. Math.*, 9, 673–692.

1962: On invariant curves of area-preserving mappings of an annulus, *Nachr. Akad. Wiss. Gottingen Math. Phys.*, Kl. 2, 1–20.

1970: Regularization of Kepler's problem and the averaging method on a manifold, *Comm. Pure Appl. Math.*, 23(4), 609–636.

Moser, J. K. and Zehnder, E. J. 2005: *Notes on Dynamical Systems*, American Mathematical Society, Providence, RI.

Moulton, F. R. 1920: *Periodic Orbits*, Carnegie Institute of Washington, Washington DC.

Niedzielska, Z. 1994: Nonlinear stability of the libration points in the photogravitational restricted three–body problem, *Celest. Mech.*, 58, 203–213.

Noether, E. 1918: Invariante Variationsprobleme, *Nachr. D. König. Gesellsch. D. Wiss. Zu Göttingen*, Math-phys. Klasse, 235–257.

Offin, D. C.
    1990: Subharmonic oscillations for forced pendulum type equations, *Diff. Int. Eqs.*, 3, 615–629.
    2000: Hyperbolic minimizing geodesics, *Trans. Amer Math. Soc.*, 352(7), 3323–3338.
    2001: Variational structure of the zones of stability, *Diff. and Int. Eqns.*, 14, 1111–1127.

Palis, J. and de Melo, W. 1980: *Geometric Theory of Dynamical Systems*, Springer–Verlag, New York.

Palmore, J. I. 1969: *Bridges and Natural Centers in the Restricted Three Body Problem*, University of Minnesota Report.

Poincaré, H.
    1885: Sur les courbes definies par les equations differentielles, *J. Math. Pures Appl.*, 4, 167–244.
    1899: *Les methods nouvelles de la mechanique celeste*, Gauthier–Villar, Paris.

Pollard, H. 1966: *Mathematical Introduction to Celestial Mechanics*. Prentice–Hall, New Jersey.

Pugh, C. and Robinson, C. 1983: The C^1 closing lemma, including Hamiltonians, *Erg. Th. and Dyn. Sys.*, 3, 261–313.

Roberts, G. E. 2007: Linear Stability analysis of the figure eight orbit in the three body problem, *Erg. Theory Dyn. Sys.*, 27, 6, 1947–1963.

Robinson, C. 1999: *Dynamical Systems, Stability, Symbolic Dynamics and Chaos* (2nd ed.), CRC Press, Boca Raton, FL.

Rolfsen, D. 1976: *Knots and Links*, Publish or Perish Press, Berkeley, CA.

Rüssman, H. 1959: Uber die Existenz einer Normalform inhaltstreuer elliptischer Transformationen, *Math. Ann.*, 167, 55–72.

Saari, D.
    1971: Expanding gravatitational systems, *Trans. Amer. Math. Soc.*, 156, 219–240.
    1988: Symmetry in n–particle systems, *Contemporary Math.*, 81, 23–42.
    2005: *Collisions, Rings and Other Newtonian N–Body Problems*, American Mathematical Society, Providence, RI.

Schmidt, D. S.
    1972: Families of periodic orbits in the restricted problem of three bodies connecting families of direct and retrograde orbits, *SIAM J. Appl. Math.*, 22(1), 27–37.
    1990: Transformation to versal normal form, *Computer Aided Proofs in Analysis* (Ed. K. R. Meyer and D. S. Schmidt), IMA Series 28, Springer–Verlag, New York.

Sibuya, Y. 1960: Note on real matrices and linear dynamical systems with periodic coefficients, *J. Math. Anal. Appl.* 1, 363–72.

Siegel, C. L. and Moser, J. K. 1971: *Lectures on Celestial Mechanics*, Springer-Verlag, Berlin.

Simó, C. 2002: Dynamical properties of the figure eight solution of the three body problem. *Celestial Mechanics*, Contemp. Math. 292, American Mathematical Society, Providence, RI, 209–228.

1978: Proof of the stability of Lagrangian solutions for a critical mass ration, *Sov. Astron. Lett.*, 4(2), 79–81.

Spivak, M. 1965: *Calculus on Manifolds*, W. A. Benjamin, New York.

Strömgren, E. 1935: Connaissance actuelle des orbites dans le problème des trois corps, *Bull. Astron.*, 9, 87–130.

Sundman, K. F. 1913: Mémoire sur le problème des trois corps, *Acta Math.*, 36, 105–179.

Szebehely, V. 1967: *Theory of Orbits*, Academic Press, New York.

Taliaferro, S. 1980: Stability of two dimensional analytic potential, *J. Diff. Eqs.* 35, 248–265.

Weyl, H. 1948: *Classical Groups*, Princeton University Press, Princeton, NJ.

Whittaker, E. T. 1937: *A Treatise on the Analytical Dynamics of Particles and Rigid Bodies,* Cambridge University Press, Cambridge, UK.

Whittaker, E. T. and Watson, G. N. 1927: *A Course of Modern Analysis,* (4th ed), Cambridge University Press, Cambridge, UK.

Williamson, J.

    1936: On the algebraic problem concerning the normal forms of linear dynamical systems, *Amer. J. Math.*, 58, 141–63.

    1937: On the normal forms of linear canonical transformations in dynamics, *Amer. J. Math.*, 59, 599–617.

    1939: The exponential representation of canonical matrices, *Amer. J. Math.*, 61, 897–911.

Wintner, A. 1944: *The Analytic Foundations of Celestial Mechanics*, Princeton University Press, Princeton, NJ.

Yakubovich, V. A. and Starzhinskii, V. M. 1975: *Linear Differential Equations with Periodic Coefficients, 1 and 2*, John Wiley, New York.

Index

$\mathcal{L}_4$, 92–99, 228, 266–267, 295–301
2-body problem, 62–66, 200
3-body problem, 200, 233–236, 345–372
– angular momentum, 346–352
– linear momentum, 347

Abraham, Ralph, 169
action–angle variables, 6, 200–204
algebra
– Lie, 26, 186
amended potential, 25, 85
anomaly
– mean, 215, 217
– true, 66, 207–220
argument of perigee, 207
Arnold's stability theorem, 316–320

Barrar, Richard, 237
bifurcation
– Hamiltonian–Hopf, 301
– periodic solutions, 279–290

Cabral, Hildeberto, 339, 344, 346–365
canonical coordinates, 53
center of mass, 63, 68, 235
central configuration(s), 73–78,
    348–364
central force, 3, 190
characteristic polynomial, 42, 91, 93
Chenciner, Alain, 345–371
Cherry's example, 308–309
Chetaev's Theorem, 308, 343
Chicone, Carmen, 143
closed form, 181
collapse, 79–80
complexification, 43
conjugate variable, 2

Conley, Charlie, 137
conservative, 2
coordinate(s),
– action–angle, 6, 200–204
– canonical, 6
– complex, 211–214
– Delaunay, 214–215, 217–219, 236–238
– ignorable, 25, 154, 199
– Jacobi, 63, 69, 197–200, 235
– lemniscate, 205
– Poincaré, 216, 291
– polar, 5, 206–208
– pulsating, 219–223
– rotating, 78, 229, 78–237
– spherical, 24, 25, 208–211
– symplectic, 6, 25, 52, 41–57
Coriolis forces, 78
cotangent bundle, 177
covector, 38, 170, 174–182
cross section, 10–14, 157

d'Alembert character, 201
Darboux's Theorem, 185
degrees of freedom, 2
Delaunay elements, 214–215, 217–219
Deprit, André, 66, 245–266, 295
det A=+1, 35, 116, 132, 174
determinant, 32, 35, 42, 170–174
differential forms, 177–182, 195–197
Dirichlet's theorem, 4, 307
Duffing's equation, 17, 201, 238, 344

eigenvalues, 113, 41–116
Einstein convention, 175–177
elliptic
– equilibrium, 146
– fixed point, 148

elliptic restricted problem, 101–102,
    221–223
equilibrium(a), 4
– elementary, 226
– Euler collinear, 91–92, 228–229, 310,
    351–357
– hyperbolic, 93, 162
– Lagrange equilateral, 92–99,
    228–229, 311, 320, 330, 331, 334, 339
– N-body, 73–80
– relative, 25, 75, 348–364
– restricted problem, 87–90
– stable/instable, 4, 93, 307–311,
    316–339
Euler, L., 78
Euler–Lagrange's equation, 18–25,
    353–359
exact form, 181
exponents, 52, 155, 226–228
exterior algebra, 169–174

fixed point(s),
– elementary, 280
– elliptic, 243, 272
– extremal, 281
– flip, 244, 275–276
– hyperbolic, 243, 271–272
– k-bifurcation, 287–290
– normal form, 267–276
– period doubling, 283–286
– shear, 244, 274–275, 282
– stability, 312–315, 339–340, 343–344
Floquet–Lyapunov theorem, 50–52, 267

generating function(s), 196–197
geodesic, 27
gravitational constant, 62
group
– Lie, 185–187
– orthogonal, 60
– special linear, 60
– special orthogonal, 60
– symplectic, 31, 60, 129–133
– unitary, 130

Hadjidemetriou, John, 234
Hamiltonian
– linear, 29
– matrix, 30–31
– operator, 40
Hamiltonian–Hopf, 301
harmonic oscillator, 4–9, 17
Henrard, Jacques, 245, 295

Hill's
– lunar problem, 58, 99–102
– orbits, 230–232
– regions, 89
holonomic constraint, 184
homoclinic, 165
Hopf fibration, 9
hyperbolic, 146, 148

integrable, 190–191
integral(s), 2, 25, 67, 199, 206, 209,
    213, 214, 224, 347–352
invariant
– curve theorem, 312–315
– plane, 65
– set, 2
involution, 34, 154

J, 2, 29, 38
Jacobi, 73
– constant, 85
– coordinates, 63, 69, 197–200, 235
– field(s), 139–140
– identity, 3, 26, 178

KAM theorem, 312–320
Kepler problem, 3, 80, 206–207,
    229–237, 347–349
Kepler's law(s), 65, 66, 80
kinetic energy, 19, 25, 62
knots, 9
Kronecker delta, 38

Lagrange-Jacobi Formula, 79–80
Lagrangian
– complement, 40
– Grassmannian, 133–141
– manifold, 185
– set of functions, 34
– set of solutions, 34
– splitting, 40
– subspace, 40
lambda lemma, 165
Lang, Serge, 47
lemniscate functions, 15–16, 205,
    336–339
Levi-Civita Regularization, 212–214
Lie
– algebra, 26, 60, 178
– bracket, 178
– product, 31
Lie group, 185
– action, 187

– free, 187
– proper, 187
linearize/linearization, 90–119, 228,
    137–238, 240–268
logarithm, 50, 108–113
long period family, 228
Lyapunov
– center theorem, 227–229
– stability theorems, 307–311, 343–344

Markus, Larry, 129
Marsden, Jerry, 73, 169
Maslov index, 133–141, 365–371
mass ratio, 93
Mathieu transformation, 197
matrix
    fundamental, 32
– Hamiltonian, 30
– infinitesimally symplectic, 30
– normal form, 94–99, 117–129
– skew symmetric, 37
– symplectic, 31
McCord, Chris, 69
Moeckel, Rick, 346–352
moment map, 188
momentum
    angular, 3, 25, 26, 64, 214, 217,
    346–372
– linear, 63, 67, 347
monodromy, 50, 55, 156, 345
Montgomery, Richard, 345–371
Morris, Granger, 315
Moser, Jurgen, 228
Moultom, F. R., 78
multilinear, 169
multipliers, 50, 156, 189, 226, 229, 233,
    293, 294

N-body problem, 61–81
– angular momentum, 68
– central configuration(s), 73–78
– equilibrium, 73–74
– Hamiltonian, 62
– integrals, 67
– linear momentum, 67
Newton's
– headache, 61
– laws, 61–62, 73
Newton's laws, 1
Newtonian system, 18
Noether's Theorem, 69
normal form(s)
– fixed points, 242–244

– summary, 239–242

orbifold, 11, 320
orbit(s)
– circular, 231–233, 290
– comet, 232
– direct, 230, 290
– figure eight, 345–372
– Hill's, 230–232
– Poincaré, 229–230
– retrograde, 230, 290
– space, 187
orbits, 2

Palacián, Jesús, 229
Palmore, Julian, 295
parametric stability, 103–108
pendulum
– equation, 17, 26
– spherical, 24–26
perihelion,perigee, 207
period map, 150
phase space, 2
Poincaré
– conjecture, 279
– continuation method, 225
– elements, 216
– lemma, 181
– map, 150, 157, 160, 279
– orbits, 229–230
Poisson bracket, 3–4, 26, 34, 56–57, 118
Poisson series, 201
polar decomposition, 35–36, 129
Pollard, Harry, 66
potential energy, 1, 18, 19, 24, 25
Pugh, Charles, 279

Ratiu, Tudor, 169
reality conditions, 45, 212
reduced space, 25, 189, 346–349,
    355–370
reduction, 25
relative equilibria, 78, 221
remainder function, 33, 54, 248–251
resonance
– 1 : −1, 334–339
– 2 : −1, 330–331
– 3 : −1, 332–334
restricted 3-body problem, 83–86,
    233–236, 266–267, 290–301, 310–311,
    319–339, 341–342
reversible, 86
Robinson, Clark, 143, 279
rotating coordinates, 78

Saari, Don, 75, 354
Schmidt, Dieter, 90, 230, 290
self-potential, 62
short period family, 228
singular reduction, 320
Sokol'skii normal form, 242, 296,
    334–339
spectra, 41
spectral decomposition, 113–116
stability/instability
– equilibrium(a), 4, 305–339
– fixed point(s), 339–340
– orbit(s), 365, 371
stable manifold, 161–166
Sundaman's theorem, 79, 363
symmetry(ies), 86–87
symplectic
– action, 188
– basis, 37, 174
– coordinates, 52–57
– form, 37, 174
– group, 31, 60, 129–133
– infinitesimally, 30
– linear space, 36–41
– manifold, 169, 183
– matrix, 31–32

– operator, 40
– scaling, 57–58, 101
– structure, 183
– subspace, 39
– symmetry, 69
– transformations, 52–58
– with multiplier, 31
symplectically similar, 41
symplectomorphic, 38
symplectomorphism, 148
syzygy, 85, 168, 352

tangent bundle, 177
topological equivalent, 145, 148
transversal, 165
transversality conditions, 20
twist map, 287

variational equation(s), 55–56
variational vector field, 20, 368–369

Wang, Qiudong (Don), 69, 154
wedge product, 171–174
Weinstein, Alan, 73

Yanguas, Patricia, 229

Printed in the United States
By Bookmasters